Lecture Notes in Computer Science 726

Edited by G. Goos and J. Hartmanis

Advisory Board: W. Brauer   D. Gries   J. Stoer

Thomas Lengauer   (Ed.)

Algorithms – ESA '93

First Annual European Symposium
Bad Honnef, Germany
September 30 - October 2, 1993
Proceedings

Springer-Verlag

Berlin Heidelberg New York
London Paris Tokyo
Hong Kong Barcelona
Budapest

Series Editors

Gerhard Goos
Universität Karlsruhe
Postfach 69 80
Vincenz-Priessnitz-Straße 1
D-76131 Karlsruhe, Germany

Juris Hartmanis
Cornell University
Department of Computer Science
4130 Upson Hall
Ithaca, NY 14853, USA

Volume Editor

Thomas Lengauer
Gesellschaft für Mathematik und Datenverarbeitung mbH
Institut für methodische Grundlagen (I1)
Schloß Birlinghoven, D-53732 Sankt Augustin, Germany

CR Subject Classification (1991): F.2, G, I.3.5

ISBN 3-540-57273-2 Springer-Verlag Berlin Heidelberg New York
ISBN 0-387-57273-2 Springer-Verlag New York Berlin Heidelberg

Typesetting: Camera-ready by author
Printing and binding: Druckhaus Beltz, Hemsbach/Bergstr.
45/3140-543210 - Printed on acid-free paper

Preface

The papers in this volume were presented at the First Annual European Symposium on Algorithms (ESA'93), held September 30–October 2, 1993, in Bad Honnef/Bonn, Germany. The symposium is intended to be an annual series of international conferences, held in early fall, that cover the field of algorithms. Within the scope of the symposium lies all research on algorithms, theoretical as well as applied, that is carried out in the fields of computer science and discrete applied mathematics. The symposium aims to cater to both of these research communities and to intensify the exchange between them.

The program committee met April 30, 1993, and selected the 35 contributed papers in this volume from 101 abstracts submitted in response to the call for papers. The selection was based on originality, quality, and relevance to the study of algorithms. It is anticipated that most of the submissions will appear in a more polished and complete form in scientific journals. The conference program also included invited lectures by Michael Paterson (Coventry): Evolution of an Algorithm, Alexander Schrijver (Amsterdam): Complexity of Disjoint Paths Problems in Planar Graphs, and Michael S. Waterman (Los Angeles): Sequence Comparison and Statistical Significance in Molecular Biology.

We wish to thank all members of the program committee, all those who submitted abstracts for consideration, our referees and colleagues who helped in the evaluations of the abstracts, and the many individuals who contributed to the success of the conference.

We would like to acknowledge the help of the following sponsoring institutions and corporations: Gesellschaft für Informatik (GI), Rheinische Friedrich-Wilhelms-Universität Bonn, Gesellschaft für Mathematik und Datenverarbeitung (GMD), Association for Computing Machinery (ACM) Special Interest Group for Algorithms and Computation Theory, and European Association for Theoretical Computer Science (EATCS).

Finally we would like to thank the following people for their extraordinary engagement in organizing the conference: Christine Harms, Brigitte Hönig, Luzia Sassen-Heßeler, and Egon Wanke.

Bonn, September 1993   Thomas Lengauer

Program Committee

Additional Referees

Table of Contents

The Influence of Lookahead in Competitive Paging Algorithms

(Extended Abstract)

Susanne Albers[*]

Max-Planck-Institut für Informatik, 66123 Saarbrücken, Germany

Abstract. We introduce a new model of lookahead for on-line paging algorithms and study several algorithms using this model. A paging algorithm is *on-line with strong lookahead l* if it sees the present request and a sequence of future requests that contains l pairwise distinct pages. These pages also differ from the page requested by the present request. We show that strong lookahead has practical as well as theoretical importance and significantly improves the competitive factors of on-line paging algorithms. This is the first model of lookahead having such properties. In addition to lower bounds we present a number of deterministic and randomized on-line paging algorithms with strong lookahead which are optimal or nearly optimal.

1 Introduction

In recent years the competitive analysis of on-line algorithms has received much attention [ST85, KMRS88, MMS88, BBKTW90, G91]. Among on-line problems, the *paging problem* is of fundamental interest. Consider a two-level memory system which has a fast memory that can store k pages and a slow memory that can manage, basically, an unbounded number of pages. A sequence of requests to pages in the memory system must be served by a paging algorithm. A request is served if the corresponding page is in fast memory. If the requested page is not stored in fast memory, a *page fault* occurs. Then a page must be evicted from fast memory so that the requested page can be loaded into the vacated location. A paging algorithm specifies which page to evict on a fault. The cost incurred by a paging algorithm equals the number of page faults. A paging algorithm is *on-line* if it determines which page to evict on a fault without knowledge of future requests.

We analyze the performance of on-line paging algorithms using *competitive analysis* [ST85, KMRS88]. In a competitive analysis, the cost incurred by an on-line algorithm is compared to the cost incurred by an *optimal off-line* algorithm. An optimal off-line algorithm knows the entire request sequence in advance and can serve it with minimum cost. Let $C_A(\sigma)$ and $C_{OPT}(\sigma)$ be the cost of the on-line algorithm A and the optimal off-line algorithm OPT on request sequence σ. Then the algorithm A is c-competitive, if there exists a constant a such that

$$C_A(\sigma) \leq c \cdot C_{OPT}(\sigma) + a$$

for all request sequences σ. The *competitive factor* of A is the infimum of all c such that A is c-competitive. If A is a randomized algorithm, then $C_A(\sigma)$ is the

---

[*] *This work was done while the author was a student at the Graduiertenkolleg Informatik, Universität des Saarlandes, and was supported by a graduate fellowship of the Deutsche Forschungsgemeinschaft.*

expected cost incurred by A on request sequence σ. In this paper we evaluate the performance of randomized on-line algorithms only against the *oblivious adversary* (see [BBKTW90] for details). Belady [B66] has exhibited an optimal off-line paging algorithm which is also called the MIN algorithm. On a fault, MIN evicts the page whose next request occurs farthest in the future.

The paging problem (without lookahead) has been studied intensively. Sleator and Tarjan [ST85] have demonstrated that the well-known replacement algorithms LRU (Least Recently Used) and FIFO (First-In First-Out) are k-competitive. They have also proved that no on-line paging algorithm can be better than k-competitive; hence LRU and FIFO achieve the best competitive factor. Fiat *et al.* [FKLSY91] have shown that no randomized on-line paging algorithm can be better than $H(k)$-competitive against an oblivious adversary. Here $H(k) = \sum_{i=1}^{k} 1/i$ denotes the kth harmonic number. They have also given a simple replacement algorithm, called the MARKING algorithm, which is $2H(k)$-competitive. McGeoch and Sleator [MS91] have proposed a more complicated randomized paging algorithm which achieves a competitive factor of $H(k)$.

In this paper we study the problem of *lookahead* in on-line paging algorithms. An important question is, what improvement can be achieved in terms of competitiveness, if an on-line algorithm knows not only the present request to be served, but also some future requests. This issue is fundamental from the practical as well as the theoretical point of view. In paging systems some requests usually wait in line to be processed by a paging algorithm. One reason is that requests do not necessarily arrive one after the other, but rather in blocks of possibly variable size. Furthermore, if several processes run on a computer, it is likely that some of them incur page faults which then wait for service. Many memory systems are also equipped with prefetching mechanisms, i.e. on a request not only the currently accessed page but also some related pages which are expected to be asked next are demanded to be in fast memory. Thus each request generates a number of additional requests. In fact, some paging algorithms used in practice make use of lookahead [S77]. In the theoretical context a natural question is: What is it worth to know the future?

Previous research on lookahead in on-line algorithms has mostly addressed dynamic location problems and on-line graph problems [CGS89, I90, KT91, HS92]; only very little is known in the area of paging with lookahead. Consider the intuitive model of lookahead, which we call *weak lookahead*. Let $l \geq 1$ be an integer. We say that an on-line paging algorithm has a *weak lookahead of size l*, if it sees the present request to be served and the next l future requests. It is well known that this model cannot improve the competitive factors of on-line paging algorithms. If an on-line paging algorithm has a weak lookahead of size l, then an adversary that constructs a request sequence can simply replicate each request l times in order to make the lookahead useless. The only other result known on paging with lookahead has been developed by Young [Y91]. According to Young, a paging algorithm is *on-line with a resource-bounded lookahead of size l* if it sees the present request and the maximal sequence of future requests for which it will incur l faults. Young presents deterministic and randomized on-line paging algorithms with resource-bounded lookahead l which are $\max\{2k/l, 2\}$-competitive and $2(\ln(k/l) + 1)$-competitive, respectively. However, the model of resource-bounded lookahead is unrealistic in practice.

We now introduce a new model of lookahead which has practical as well as theoretical importance. As we shall see, this model can significantly improve the competitive factors of on-line paging algorithms. Let $\sigma = \sigma(1), \sigma(2), \ldots, \sigma(m)$ be a request sequence of length m. $\sigma(t)$ denotes the request at time t. For a given set S, $card(S)$ denotes the cardinality of S. Let $l \geq 1$ be an integer.

Strong lookahead of size l: The on-line algorithm sees the present request and a sequence of future requests. This sequence contains l pairwise distinct pages which also differ from the page requested by the present request. More precisely, when serving request $\sigma(t)$, the algorithm knows requests $\sigma(t+1), \sigma(t+2), \ldots, \sigma(t')$, where $t' = \min\{s > t | card(\{\sigma(t), \sigma(t+1), \ldots, \sigma(s)\}) = l + 1\}$. The requests $\sigma(s)$, with $s \geq t' + 1$, are not seen by the on-line algorithm at time t.

Strong lookahead is motivated by the observation that in request sequences generated by real programs, subsequences of consecutive requests generally contain a number of distinct pages. Furthermore, strong lookahead is of interest in the theoretical context when we ask how significant it is to know part of the future. An adversary may replicate requests in the lookahead, but nevertheless it has to reveal some really significant information on future requests.

In the following, we always assume that an on-line algorithm has a strong lookahead of fixed size $l \geq 1$. If a request sequence $\sigma = \sigma(1), \sigma(2), \ldots, \sigma(m)$ is given, then for all $t \geq 1$ we define a value $\lambda(t)$. If $card(\{\sigma(t), \sigma(t+1), \ldots, \sigma(m)\}) < l+1$ then let $\lambda(t) = m$; otherwise let $\lambda(t) = \min\{t' > t | card(\{\sigma(t), \sigma(t+1), \ldots, \sigma(t')\}) = l + 1\}$. The *lookahead $L(t)$ at time t* is defined as

$$L(t) = \{\sigma(s) | s = t, t+1, \ldots, \lambda(t)\}.$$

We say that *a page x is in the lookahead at time t* if $x \in L(t)$.

The remainder of this paper is an in-depth study of paging with strong lookahead. Strong lookahead is the first model of lookahead that is significant from a practical and theoretical standpoint and also reduces the competitive factors of on-line paging algorithms. In Section 2 we consider deterministic on-line algorithms and present a variant of the algorithm LRU that, given a strong lookahead of size l, where $l \leq k-2$, achieves a competitive factor of $(k - l)$. We also show that no deterministic on-line paging algorithm with strong lookahead l, $l \leq k - 2$, can be better than $(k - l)$-competitive. Thus our proposed algorithm is optimal. Furthermore, we give another variant of the algorithm LRU with strong lookahead l, $l \leq k-2$, which is $(k-l+1)$-competitive. Interestingly, this algorithm does not exploit full lookahead but rather serves the request sequence in a series of blocks. Section 3 addresses randomized on-line paging algorithms with strong lookahead. We prove that a modification of the MARKING algorithm with strong lookahead l, $l \leq k - 2$, is $2H(k - l)$-competitive. This competitiveness is within a factor of 2 of optimal. In particular, we show that no randomized on-line paging algorithm with strong lookahead l, $l \leq k - 2$, can be better than $H(k - l)$-competitive. Furthermore we present an extremely simple randomized on-line paging algorithm with strong lookahead l, $l \leq k - 2$, which is $(k - l + 1)$-competitive.

2 Deterministic paging with strong lookahead

Unless otherwise stated, we assume in the following that all our paging algorithms are *lazy* algorithms, i.e. they only evict a page on a fault.

Let $k \geq 3$. We consider the important case that an on-line paging algorithm has a strong lookahead of size $l \leq k - 2$. The on-line paging algorithms we present are extensions of the algorithm LRU to our model of strong lookahead.

Algorithm LRU(l): On a fault execute the following steps. Among the pages in fast memory which are not contained in the present lookahead determine the page whose last request occurred least recently. Evict this page and load the requested page.

Theorem 1. *Let $l \leq k - 2$. The algorithm LRU(l) with strong lookahead l is $(k - l)$-competitive.*

Now we prove this theorem. Let $\sigma = \sigma(1), \sigma(2), \ldots, \sigma(m)$ be a request sequence of length m. We assume without loss of generality that LRU(l) and OPT start with an empty fast memory and that on the first k faults, both LRU(l) and OPT load the requested page into the fast memory. Furthermore we assume that σ contains at least $l + 1$ distinct pages. The following proof consists of three main parts. First, we introduce the potential function we use to analyze LRU(l). In the second part, we partition the request sequence σ into a series of phases and then, in the third part, we bound LRU(l)'s amortized cost using that partition.

1. The potential function

We introduce some basic notations. For $t = 1, 2, \ldots, \lambda(1) - 1$, let $\mu(t) = 1$ and for $t = \lambda(1), \lambda(1) + 1, \ldots, m$, let

$$\mu(t) = \max\{t' < t \mid card(\{\sigma(t'), \sigma(t'+1), \ldots, \sigma(t)\}) = l + 1\}.$$

Define

$$M(t) = \{\sigma(s) \mid s = \mu(t), \mu(t) + 1, \ldots, t\}.$$

For a given time t, the set $M(t)$ contains the last $l + 1$ requested pages.

For $t = 1, 2, \ldots, m$, let $S_{LRU(l)}(t)$ be the set of pages contained in LRU(l)'s fast memory after request t, and let $S_{OPT}(t)$ be the set of pages contained in OPT's fast memory after request t. $S_{LRU(l)}(0)$ and $S_{OPT}(0)$ denote the sets of pages which are initially in fast memory, i.e. $S_{LRU(l)}(0) = S_{OPT}(0) = \emptyset$. For the analysis of the algorithm we assign weights to all pages. These weights are updated after each request. Let $w(x, t)$ denote the weight of page x after request t, $1 \leq t \leq m$. The weights are set as follows. If $x \notin S_{LRU(l)}(t)$ or $x \in L(t)$, then

$$w(x, t) = 0.$$

Let $j = card(S_{LRU(l)}(t) \setminus L(t))$. Assign integer weights from the range $[1, j]$ to the pages in $S_{LRU(l)}(t) \setminus L(t)$ such that any two pages $x, y \in S_{LRU(l)}(t) \setminus L(t)$ satisfy

$$w(x, t) < w(y, t)$$

iff the last request to x occurred earlier than the last request to y. For $t = 1, 2, \ldots, m$, let

$$S(t) = S_{LRU(l)}(t) \setminus \{M(t) \cup L(t) \cup S_{OPT}(t)\}.$$

We now define the potential function:

$$\Phi(t) = \sum_{x \in S(t)} w(x, t).$$

Intuitively, $S(t)$ contains those pages which cause LRU(l) to have a higher cost than OPT. Instead of the pages $x \in S(t)$, OPT can store pages in its fast memory which are not contained in $S_{LRU(l)}(t)$ but are requested in the future. The weight $w(x,t)$ of a page $x \in S(t)$ equals the number of faults that LRU(l) must incur before it can evict x.

2. The partitioning of the request sequence

We will partition the request sequence σ into phases, numbered from 0 to p for some p, such that phase 0 contains at most $l+1$ distinct pages and phase i, $i = 1, 2, \ldots, p$, has the following two properties. Let t_i^b and t_i^e denote the beginning and the end of phase i, respectively.

Property 1: Phase i contains exactly $l+1$ distinct pages, i.e.
$$card(\{\sigma(t_i^b), \sigma(t_i^b + 1), \ldots, \sigma(t_i^e)\}) = l + 1.$$
Property 2: For all $x \in S_{LRU(l)}(t_{i-1}^e) \setminus \{L(t_i^b) \cup S_{OPT}(t_{i-1}^e)\}$,
$$w(x, t_i^e) \le k - l - 2.$$

In the following, we describe how to decompose σ. We partition the request sequence starting at the end of σ. Suppose that we have already constructed phases $P(i+1), P(i+2), \ldots, P(p)$. We show how to generate phase $P(i)$. Let $t_i^e = t_{i+1}^b - 1$. (We let $t_p^e = m$ at the beginning of the decomposition.) Now set $t = \mu(t_i^e)$ and compute $S_{LRU(l)}(t - 1) \setminus L(t)$. If $S_{LRU(l)}(t - 1) \setminus L(t) \ne \emptyset$, then let y be the most recently requested page in $S_{LRU(l)}(t-1) \setminus L(t)$. We consider two cases. If $S_{LRU(l)}(t - 1) \setminus L(t) = \emptyset$ or if $S_{LRU(l)}(t - 1) \setminus L(t) \ne \emptyset$ and $y \in S_{OPT}(t - 1)$, then let $t_i^b = t$ and call the i-th phase $P(i) = \sigma(t_i^b), \sigma(t_i^b + 1), \ldots, \sigma(t_i^e)$ a type 1 phase. Otherwise (if $S_{LRU(l)}(t - 1) \setminus L(t) \ne \emptyset$ and $y \notin S_{OPT}(t - 1)$) let t', $t' < t$, be the time when OPT evicted page y most recently. Let $t_i^b = t'$ and call the i-th phase $P(i) = \sigma(t_i^b), \sigma(t_i^b + 1), \ldots, \sigma(t_i^e)$ a type 2 phase.

Lemma 2. *The partition generated above satisfies the following conditions.*
a) Phase $P(0)$ contains at most $l + 1$ distinct pages.
b) Every phase $P(i)$, $1 \le i \le p$, has Property 1 and Property 2.

Proof. First we prove part a). We show that $P(0)$ is a type 1 phase. This immediately implies that $P(0)$ contains at most $l + 1$ vertices. If $P(0)$ was a type 2 phase, then OPT would evict a page on the first request $\sigma(1)$. However, this is impossible because initially the fast memories are empty and on the first k faults both LRU(l) and OPT load the requested page into the fast memory.

Now we prove part b) of the lemma. Consider an arbitrary phase $P(i)$, $1 \le i \le p$. Let $t = \mu(t_i^e)$. If $S_{LRU(l)}(t - 1) \setminus L(t) \ne \emptyset$, then let y be the most recently requested page in $S_{LRU(l)}(t - 1) \setminus L(t)$ and let t'', $t'' < t$, be the time when y was requested most recently. If $P(i)$ is a type 2 phase, then let t', $t' \le t - 1$, be the time when OPT evicted y most recently. (Since $y \notin S_{OPT}(t - 1)$, we have $t'' < t' \le t - 1$.)

We show that $P(i)$ contains exactly $l + 1$ pages. For a type 1 phase there is nothing to show. Suppose $P(i)$ is a type 2 phase. Then $t_i^b = t'$. Let $s \in [t', t - 1]$ be arbitrary and let x be the page requested at time s. We need to show $x \in L(t)$. So assume $x \notin L(t)$. Then by the definition of y, $x \notin S_{LRU(l)}(t - 1)$, i.e. x was evicted by LRU(l) at some time $s' \in [s + 1, t - 1]$. Since y was not evicted by LRU(l) at time s' and y's most recent request was at time $t'' < s$, we must have $y \in L(s') \subseteq \{\sigma(s'), \ldots, \sigma(t - 1), \sigma(t), \ldots, \sigma(t_i^e)\} = \{\sigma(s'), \ldots, \sigma(t - 1)\} \cup L(t)$. But $y \notin \{\sigma(s'), \ldots, \sigma(t - 1)\}$ and $y \notin L(t)$, by the definition of t'' and y. Thus $x \notin L(t)$ is impossible. We conclude that $P(i)$ contains exactly $l + 1$ distinct pages.

It remains to prove that $P(i)$ has Property 2. Consider an arbitrary page $x \in S_{LRU(l)}(t_{i-1}^e) \setminus \{L(t_i^b) \cup S_{OPT}(t_{i-1}^e)\}$. If $w(x, t_i^e) = 0$, then the property clearly holds. Therefore assume $w(x, t_i^e) \geq 1$. By Property 1, $L(t_i^b)$ contains all pages which are requested in $P(i)$. Since $w(x, t_i^e) \geq 1$, we have $x \in S_{LRU(l)}(t_i^e) \setminus L(t_i^e)$ and hence $x \notin L(t_i^b) \cup L(t_i^e) \supseteq L(s)$ for all $s \in [t_i^b, t_i^e]$. Thus, x was a candidate for eviction by LRU(l) throughout $P(i)$, but was not evicted. This implies immediately that all pages requested in $P(i)$, i.e. all pages in $L(t_i^b)$, also belong to $S_{LRU(l)}(t_i^e)$. Using a very similar analysis we can show that $y \neq x$ and $y \in S_{LRU(l)}(t_i^e)$. Hence we have identified $l + 2$ pages in $S_{LRU(l)}(t_i^e)$ which, at time t_i^e, were requested later than x. At time t_i^e, each of these pages has a weight of 0 or a weight which is greater than that of x. Thus, $w(x, t_i^e) \leq k - l - 2$. $\square$

3. Bounding LRU(l)'s amortized cost

Using the partition of σ generated above, we will evaluate LRU(l)'s amortized cost on σ. First we will bound the increase in potential $\sum_{t=1}^m \Phi(t) - \Phi(t-1)$. Then we will estimate LRU(l)'s actual cost in each phase of σ. For $t = 1, 2, \ldots, m$, let

$$N(t) = S(t) \setminus S(t-1).$$

We set $M(0) = L(0) = \emptyset$ and $S(0) = S_{LRU(l)}(0) \setminus \{M(0) \cup L(0) \cup S_{OPT}(0)\}$, which is used in the definition of $N(1) = S(1) \setminus S(0)$.

We present two lemmas which are crucial in analyzing the change in potential $\Phi(t) - \Phi(t-1)$, $1 \leq t \leq m$. Note that

$$\Phi(t) - \Phi(t-1) = \sum_{x \in S(t)} w(x, t) - \sum_{x \in S(t-1)} w(x, t-1)$$

$$= \sum_{x \in N(t)} w(x, t) + \sum_{x \in S(t-1) \cap S(t)} (w(x, t) - w(x, t-1)) - \sum_{x \in S(t-1) \setminus S(t)} w(x, t-1).$$

Lemma 3. Let $1 \leq t \leq m$. If $x \in N(t)$, then $w(x, t) \leq k - l - 1$.

Proof. By the definition of $N(t)$, we have $x \in S_{LRU(l)}(t) \setminus \{M(t) \cup L(t) \cup S_{OPT}(t)\}$. Since $x \notin M(t)$, page x is not requested in the interval $[\mu(t), t]$ and hence $x \in S_{LRU(l)}(\mu(t) - 1)$. We have $x \notin M(t) \cup L(t)$ which implies $x \notin L(s)$ for all s with $\mu(t) \leq s \leq t$. Thus, x has been a candidate for eviction by LRU(l) throughout the interval $[\mu(t), t]$, but was not evicted. It follows that all pages in $M(t)$ must be in $S_{LRU(l)}(t)$. Note that $M(t)$ contains $l + 1$ pages because OPT does not evict a page before the $(k+1)$-st fault. At time t, all pages in $M(t)$ have a weight of 0 or a weight which is greater than $w(x, t)$. Thus $w(x, t) \leq k - l - 1$. $\square$

Lemma 4. Let $1 \leq t \leq m$ and $x \in S(t-1) \cap S(t)$. Then x's weight satisfies $w(x, t-1) \geq w(x, t)$. In particular, if LRU(l) incurs a fault at time t, then $w(x, t-1) > w(x, t)$.

Proof. Note that by the definition of $S(t-1)$ and $S(t)$, we have $x \in S_{LRU(l)}(t-1) \setminus L(t-1)$ and $x \in S_{LRU(l)}(t) \setminus L(t)$. Hence $w(x, t-1) \geq 1$ and $w(x, t) \geq 1$. The inequality $w(x, t-1) \geq w(x, t)$, follows from the following two statements whose proofs we omit.

1) Let y, $y \neq x$, be a page which satisfies $w(y, t-1) = 0$ and $w(y, t) > 0$. Then $w(x, t) < w(y, t)$.
2) Let y, $y \neq x$, be a page which satisfies $w(y, t-1) > 0$ and $w(x, t-1) < w(y, t-1)$. Then $w(y, t) = 0$ or $w(x, t) < w(y, t)$.

Now suppose that LRU(l) incurs a fault at time t. Then, at time t, LRU(l) evicts a page z, $z \neq x$, whose last request occurred earlier than x's last request. Hence $1 \leq w(z, t-1) < w(x, t-1)$. Since the statements 1) and 2) hold, x's weight must decrease after z is evicted, i.e. $w(x, t-1) > w(x, t)$. $\square$

Lemma 3 implies that at any time t, $1 \leq t \leq m$, a page $x \in N(t)$ can cause an increase in potential of at most $k - l - 1$. Thus, for every t, $1 \leq t \leq m$, we have

$$\Phi(t) - \Phi(t-1) = (k - l - 1)card(N(t)) - W(t), \tag{1}$$

where $W(t) = W^1(t) + W^2(t) + W^3(t)$ and

$$W^1(t) = \sum_{x \in N(t)} (k - l - 1 - w(x, t))$$

$$W^2(t) = \sum_{x \in S(t-1) \cap S(t)} (w(x, t-1) - w(x, t))$$

$$W^3(t) = \sum_{x \in S(t-1) \setminus S(t)} w(x, t-1).$$

For all $t = 1, 2, \ldots, m$, we have

$$W^1(t) \geq 0, \ W^2(t) \geq 0, \ W^3(t) \geq 0. \tag{2}$$

Clearly, $W^1(t) \geq 0$ and $W^3(t) \geq 0$. The inequality $W^2(t) \geq 0$ follows from Lemma 4.

Next we estimate $\sum_{t=1}^{m} card(N(t))$ and derive a bound on $\sum_{t=1}^{m} \Phi(t) - \Phi(t-1)$. To each element $x \in N(t)$ we assign the most recent eviction of x by OPT. More formally, let

$$X = \{(x, t) \in (\bigcup_{t=1}^{m} N(t)) \times [1, m] | x \in N(t)\}.$$

We define a function $f : X \longrightarrow [1, m]$. For $(x, t) \in X$ we define

$$f(x, t) = \max\{s \leq t | \text{OPT evicts page } x \text{ at time } s\}.$$

Note that f is well-defined. The next lemma presents two properties of the function f. Part b) will be useful when bounding LRU(l)'s actual cost in each phase of σ. The proof of the lemma is omitted.

Lemma 5. *a) The function f is injective.*
b) Let $(x, t) \in X$ and $f(x, t) = t'$. Let $t \in [t_i^b, t_i^e]$, $0 \leq i \leq p$. If $i = 0$, then $t' \in [t_0^b, t_0^e]$. If $i \geq 1$, then $t' \in [t_{i-1}^b, t_i^e]$.

Let T_{OPT} be the set of all $t \in [1, m]$ such that OPT evicts a page at time t. Note that $C_{OPT}(\sigma) = card(T_{OPT})$. Let $T_{OPT}^1 = \{f(x, t) | (x, t) \in X\}$. By Lemma 5, f is injective and hence

$$\sum_{t=1}^{m} card(N(t)) = card(X) = card(T_{OPT}^1).$$

Thus, by equation (1), we obtain

$$\sum_{t=1}^{m} \Phi(t) - \Phi(t-1) = (k - l - 1)card(T_{OPT}^1) - \sum_{t=1}^{m} W(t). \tag{3}$$

Now we bound LRU(l)'s actual cost in each phase of σ. For $i = 0, 1, \ldots, p$, let $C_{LRU(l)}(i)$ be the actual cost LRU(l) incurs in serving phase $P(i)$, and let $C_{OPT}(i)$ be the cost OPT incurs in serving $P(i)$. Furthermore, let

$$T^2_{OPT} = T_{OPT} \setminus T^1_{OPT}$$

and, for $i = 0, 1, \ldots, p$, let

$$T^2_{OPT}(i) = \{t \in T^2_{OPT} \mid t^b_i \le t \le t^e_i\}.$$

Lemma 6. *a)* $C_{LRU(l)}(0) = C_{OPT}(0)$

b) For $i = 1, 2, \ldots, p$, $\quad C_{LRU(l)}(i) \le C_{OPT}(i) + card(T^2_{OPT}(i-1)) + \sum_{t=t^b_i}^{t^e_i} W(t)$.

Proof. Part a) follows from the fact that phase $P(0)$ contains at most $l+1 < k$ distinct pages and that on the first k faults, both LRU(l) and OPT load the requested page into the fast memory.

In the proof of part b), we consider a fixed $i \in [1, p]$. If $C_{LRU(l)}(i) = 0$, then the inequality clearly holds because, according to line (2), $W(t) \ge 0$ for all $t \in [t^b_i, t^e_i]$. So suppose $C_{LRU(l)}(i) \ge 1$. Let $\tilde{C}(i) = card(S_{LRU(l)}(t^e_{i-1}) \setminus \{L(t^b_i) \cup S_{OPT}(t^e_{i-1})\})$. An easy exercise shows that $C_{LRU(l)}(i) \le C_{OPT}(i) + \tilde{C}(i)$. In the following we sketch how to prove

$$\tilde{C}(i) \le card(T^2_{OPT}(i-1)) + \sum_{t=t^b_i}^{t^e_i} W(t). \tag{4}$$

$C_{LRU(l)}(i) \le C_{OPT}(i) + \tilde{C}(i)$ and inequality (4) imply part b).

We introduce some notations. Let $t \in [t^b_i, t^e_i]$. For $x \in N(t)$, let $W^1(x, t) = k - l - 1 - w(x, t)$. For $x \in S(t-1) \cap S(t)$, let $W^2(x, t) = w(x, t-1) - w(x, t)$ and for $x \in S(t-1) \setminus S(t)$, let $W^3(x, t) = w(x, t-1)$. Note that

$$W^1(t) = \sum_{x \in N(t)} W^1(x, t) \qquad W^2(t) = \sum_{x \in S(t-1) \cap S(t)} W^2(x, t)$$

$$W^3(t) = \sum_{x \in S(t-1) \setminus S(t)} W^3(x, t).$$

For any $x \in N(t)$ ($x \in S(t-1) \cap S(t)$, $x \in S(t-1) \setminus S(t)$) we have

$$W^1(x, t) \ge 0 \quad (W^2(x, t) \ge 0, \ W^3(x, t) \ge 1). \tag{5}$$

The inequality $W^1(x, t) \ge 0$ follows from Lemma 3. Lemma 4 implies $W^2(x, t) \ge 0$. If $x \in S(t-1) \setminus S(t)$, then $x \in S_{LRU(l)}(t-1) \setminus L(t-1)$ and hence $1 \le w(x, t-1) = W^3(x, t)$.

We sketch the main idea of the proof of inequality (4). We show that for each page $x \in S_{LRU(l)}(t^e_{i-1}) \setminus \{L(t^b_i) \cup S_{OPT}(t^e_{i-1})\}$ one of the following two statements holds.

1) There exists a $t' \in T^2_{OPT}(i-1)$ such that OPT evicts page x at time t'.
2) There exists a time $t' \in [t^b_i, t^e_i]$ and a $j \in \{1, 2, 3\}$ such that $W^j(x, t') \ge 1$.

These statements, together with line (5), imply the correctness of inequality (4).

Consider a page $x \in S_{LRU(l)}(t^e_{i-1}) \setminus \{L(t^b_i) \cup S_{OPT}(t^e_{i-1})\}$. We distinguish between two main cases.

Case 1: For $t = t^e_{i-1}, t^b_i, t^b_i + 1, \ldots, t^e_i$, $x \notin S(t)$

We can prove that statement 1) holds. Since $x \in S_{LRU(l)}(t^e_{i-1}) \setminus \{L(t^b_i) \cup S_{OPT}(t^e_{i-1})\}$, we have $x \notin S_{OPT}(t^e_{i-1})$. Let $t' = \max\{s \leq t^e_{i-1} | OPT$ evicts page x at time $s\}$. Using Property 1 and part b) of Lemma 5, we are able to show that $t' \geq t^b_{i-1}$ and $t' \notin T^1_{OPT}$. (A detailed proof is omitted here.) Thus $t' \in T^2_{OPT}(i-1)$.

Case 2: There exists a t, $t^e_{i-1} \leq t \leq t^e_i$, such that $x \in S(t)$

In this case we show that the above statement 2) holds. Let $t_{\min}$ be the smallest $t \in [t^e_{i-1}, t^e_i]$ such that $x \in S(t)$.

Case 2.1: $t_{\min} = t^e_{i-1}$

Let t'' be the time when $LRU(l)$ incurs the first fault during phase $P(i)$. We consider $w(x, t'')$. If $w(x, t'') = 0$, then $x \notin S(t'')$. Hence there must exist a t', $t^b_i \leq t' \leq t''$, such that $x \in S(t'-1) \setminus S(t')$. Thus $W^3(x, t') \geq 1$. If $w(x, t'') \geq 1$, then we can prove that $x \in S(t''-1) \cap S(t'')$. Now Lemma 4 implies $W^2(x, t'') \geq 1$.

Case 2.2: $t_{\min} > t^e_{i-1}$

If $w(x, t_{\min}) < k - l - 1$, then $W^1(x, t_{\min}) \geq 1$. Suppose $w(x, t_{\min}) = k - l - 1$. By Property 2, $w(x, t^e_i) \leq k - l - 2$. Now a simple argument shows that there must exist a time $t' \in [t_{\min} + 1, t^e_i]$ such that $W^2(x, t') \geq 1$ or $W^3(x, t') \geq 1$.

The proof of Lemma 6 is complete.□

Now it is easy to finish the proof of Theorem 1. We estimate $LRU(l)$'s amortized cost. Applying equation (3) and Lemma 6 we can show

$$C_{LRU(l)}(\sigma) + \Phi(m) - \Phi(0) = \sum_{i=0}^{p} C_{LRU(l)}(i) + \sum_{t=1}^{m} \Phi(t) - \Phi(t-1) \leq$$

$$\sum_{i=0}^{p} C_{OPT}(i) + \sum_{i=0}^{p-1} card(T^2_{OPT}(i)) + \sum_{t=t^b_1}^{m} W(t) - \sum_{t=1}^{m} W(t) + (k-l-1)card(T^1_{OPT}).$$

Line (2) implies that $W(t) \geq 0$ for all $t \in [t^b_0, t^e_0]$. Hence

$$C_{LRU(l)}(\sigma) + \Phi(m) - \Phi(0) \leq C_{OPT}(\sigma) + card(T^2_{OPT}) + (k-l-1)card(T^1_{OPT})$$
$$\leq (k-l)C_{OPT}(\sigma).$$

The proof of Theorem 1 is complete.

Next we present another on-line algorithm with strong lookahead. This algorithm does not use full lookahead but rather serves the request sequence in a series of blocks.

Algorithm LRU(l)-blocked: Serve the request sequence in a series of blocks $B(1), B(2), \ldots$, where $B(1) = \sigma(1), \sigma(2), \ldots, \sigma(\lambda(1))$ and $B(i) = \sigma(t^e_{i-1}+1), \sigma(t^e_{i-1}+2), \ldots, \sigma(\lambda(t^e_{i-1}+1))$ for $i \geq 2$. Here t^e_{i-1} denotes the end of block $B(i-1)$. If there occurs a fault while $B(i)$ is processed, then the following rule applies. Among the pages in fast memory which are not contained in $B(i)$ determine the page whose last request occurred least recently. Evict that page.

$LRU(l)$-blocked has the advantage that it updates its information on future requests only once during each block. Thus it can respond to requests faster that $LRU(l)$. Furthermore, $LRU(l)$-blocked takes into account that in practice requests often arrive in blocks. Interestingly, this simpler algorithm is only slightly weaker than $LRU(l)$. Using a very similar analysis as in the proof of Theorem 1, we are able to show

Theorem 7. *Let $l \leq k - 2$. The algorithm LRU(l)-blocked with strong lookahead l is $(k - l + 1)$-competitive.*

The following theorem shows that $LRU(l)$ and $LRU(l)$-blocked are optimal and nearly optimal, respectively.

Theorem 8. *Let A be a deterministic on-line paging algorithm with strong lookahead l, where $l \leq k - 2$. If A is c-competitive, then $c \geq (k - l)$.*

Proof. Let $S = \{x_1, x_2, \ldots, x_{k+1}\}$ be a set of $k + 1$ pages. We assume without loss of generality that $A's$ and OPT's fast memories initially contain $x_1, x_2, \ldots, x_k$. Let $SL = \{x_1, x_2, \ldots, x_l\}$. We construct a request sequence σ consisting of a series of phases. Each phase contains $l + 1$ requests to $l + 1$ distinct pages. The first phase $P(1)$ consists of requests to the pages in SL, followed by a request to the page x_{k+1} which is not in fast memory, i.e. $P(1) = x_1, x_2, \ldots, x_l, x_{k+1}$. Each of the following phases $P(i)$, $i \geq 2$, has the form $P(i) = x_1, x_2, \ldots, x_l, y_i$, where $y_i \in S \setminus SL$ is chosen as follows. Let $z_i \in S$ be the page which is not in A's fast memory after the last request of phase $i - 1$. If $z_i \in S \setminus SL$, then set $y_i = z_i$. Otherwise, if $z_i \in SL$, y_i is an arbitrary page in $S \setminus SL$. The algorithm A incurs a cost of 1 in each phase. During $k - l$ successive phases, OPT's cost is at most 1. □

So far, we have assumed $k \geq 3$ and $l \leq k - 2$, which, of course, is the interesting case. Note that if $l = k - 1$ and the total number of different pages in the memory system equals $k+1$, then LRU(l) achieves a competitive factor of 1 because it behaves like Belady's optimal paging algorithm MIN [B66].

3 Randomized paging with strong lookahead

Suppose a randomized paging algorithm has a strong lookahead of size l. Again, we assume $k \geq 3$ and $l \leq k - 2$. The first algorithm we propose is a slight modification of the MARKING algorithm due to Fiat *et al.* [FKLMSY91]. The MARKING algorithm proceeds in a series of phases. During each phase a set of marked pages is maintained. At the beginning of each phase all pages are unmarked. Whenever a page is requested, that page is marked. On a fault, a page is chosen uniformly at random from among the unmarked pages in fast memory, and that page is evicted. A phase ends immediately before a fault, when there are k marked pages in fast memory.

The modified algorithm with strong lookahead l uses lookahead once during each phase.

Algorithm MARKING(l): At the beginning of each phase execute an initial step: Determine the set S of pages which are in the present lookahead but not in fast memory. Choose $card(S)$ pages uniformly at random from among the pages in fast memory which are not contained in the current lookahead. Evict these pages and load the pages in S. After this initial step proceed with the MARKING algorithm.

Theorem 9. *Let $l \leq k - 2$. The algorithm MARKING(l) with strong lookahead l is $2H(k - l)$-competitive.*

Proof. The idea of the proof is the same as the idea of the original proof of the MARKING algorithm [FKLMSY91]. We assume without loss of generality that MARKING(l)'s and OPT's fast memories initially contain the same k pages. During each phase we compare the cost incurred by MARKING(l) to the cost incurred by the optimal algorithm OPT. Consider an arbitrary phase. We use the same terminology as Fiat *et al.* A page is called *stale* if it is unmarked but was marked in the previous phase, and *clean* if it is neither stale nor marked.

Let c be the number of clean pages and s be the number of stale pages requested in the phase. Note that $c + s = k$. Fiat *et al.* prove that OPT has an amortized cost of at least $c/2$ during the phase.

We evaluate MARKING(l)'s cost during the phase. Serving c requests to clean pages obviously costs c. It remains to bound the expected cost for serving the stale pages. Let s_1 be the number of stale pages contained in the lookahead at the beginning of the phase and let $s_2 = s - s_1$. Then $s_1 + c \geq l + 1$ because every page in the lookahead is either clean or counted in s_1. Thus $s_2 = s - s_1 \leq k - c - (l + 1 - c) = k - l - 1$. Note that serving the first s_1 stale requests does not incur any cost and that we just have to evaluate MARKING(l)'s cost on the following s_2 requests to stale pages. We are able to show that this expected cost is bounded by

$$\frac{c}{k - s_1} + \frac{c}{k - s_1 - 1} + \ldots + \frac{c}{k - s_1 - s_2 + 1} = \frac{c}{k - s_1} + \frac{c}{k - s_1 - 1} + \ldots + \frac{c}{k - s + 1}.$$

The above sum consists of $s_2 \leq k - l - 1$ terms and $\frac{c}{1}$ is missing. Hence the sum is bounded by $c(H(k - l) - 1)$, and we conclude that MARKING(l)'s cost during the phase is bounded from above by $cH(k - l)$. $\square$

This following theorem implies that MARKING(l) is nearly optimal.

Theorem 10. *Let $l \leq k - 2$ and let A be a randomized on-line paging algorithm with strong lookahead l. If A is c-competitive, then $c \geq H(k - l)$.*

Proof. The proof is similar to Raghavan's proof that no randomized on-line paging algorithm without lookahead can be better than $H(k)$-competitive [R89]. So we just sketch the difference. Let $S = \{x_1, x_2, \ldots, x_{k+1}\}$ be a set of $k + 1$ pages and let $SL = \{x_1, x_2, \ldots, x_l\}$. We construct a request sequence which consists of a series of phases. The first phase is of the form $P(1) = x_1, x_2, \ldots, x_l, y_1$, where y_1 is chosen uniformly at random from all pages in $S \setminus SL$. The following phases $P(i)$ equal $P(i) = x_1, x_2, \ldots, x_l, y_i$, where y_i is chosen uniformly at random from $S \setminus \{SL \cup \{y_{i-1}\}\}$. It is possible to partition σ into rounds such that during each round OPT incurs a cost of 1 and any deterministic on-line algorithm with strong lookahead l incurs an expected cost of at least $H(k - l)$. Applying Yao's minimax principle [Y77], we obtain the theorem. $\square$

We conclude this section by presenting another randomized algorithm, called RANDOM(l)-blocked. As the name suggests this algorithm is a variant of the algorithm RANDOM due to Raghavan and Snir [RS89]. On a fault RANDOM chooses a page uniformly at random from among the pages in fast memory and evicts that page. In terms of competitiveness RANDOM(l)-blocked represents no improvement upon the previously presented algorithms with strong lookahead. However, RANDOM(l)-blocked, as the original algorithm RANDOM, is very simple and uses no information on previous requests.

Algorithm RANDOM(l)-blocked: Serve the request sequence σ in a series of blocks. These blocks have the same structure as those in the algorithm LRU(l)-blocked. At the beginning of block $B(i)$ determine the set S_i of pages in $B(i)$ which are not in fast memory. Choose $card(S_i)$ pages uniformly at random from among the pages in fast memory which are not contained in $B(i)$. Evict these pages and load the pages in S_i. Then serve the requests in $B(i)$.

Theorem 11. *Let $l \leq k - 2$. The algorithm RANDOM(l)-blocked with strong lookahead l is $(k - l + 1)$-competitive.*

The proof of the theorem is omitted.

Acknowledgment

The author thanks Kurt Mehlhorn and Ronald Rasch for many useful comments.

References

[B66] L.A. Belady. A study of replacement algorithms for virtual storage computers. *IBM Systems Journal*, 5:78-101, 1966.

[BBKTW90] S. Ben-David, A. Borodin, R.M. Karp, G. Tardos and A. Wigderson. On the power of randomization in on-line algorithms. In *Proc. 22nd Annual ACM Symposium on Theory of Computing*, pages 379-386, 1990. To appear in *Algorithmica*.

[CGS89] F.K. Chung, R. Graham and M.E. Saks. A dynamic location problem for graphs. *Combinatorica*, 9(2):111-131, 1989.

[FKLMSY91] A. Fiat, R.M. Karp, M. Luby, L.A. McGeoch, D.D. Sleator and N.E. Young. Competitive paging algorithms. *Journal of Algorithms*, 12:685-699, 1991.

[G91] E. Grove. The harmonic online k-server algorithm is competitive. In *Proc. 23nd Annual ACM Symposium on Theory of Computing*, pages 260-266, 1991.

[HS92] M.M. Halldórsson and M. Szegedy. Lower bounds for on-line graph coloring. In *Proc. 3rd Annual ACM-SIAM Symposium on Discrete Algorithms*, pages 211-216, 1992.

[I90] S. Irani. Coloring inductive graphs on-line. In *Proc. 31st Annual IEEE Symposium on Foundations of Computer Science*, pages 470-479, 1990.

[KT91] M.-Y. Kao and S.R. Tate. Online matching with blocked input. *Information Processing Letters*, 38:113-116, May 1991.

[KMRS88] A.R. Karlin, M.S. Manasse, L. Rudolph and D.D. Sleator. Competitive snoopy caching. *Algorithmica*, 3(1):79-119, 1988.

[MMS88] M.S. Manasse, L.A. McGeoch and D.D. Sleator. Competitive algorithms for on-line problems. In *Proc. 20th Annual ACM Symposium on Theory of Computing*, pages 322-333, 1988.

[MS91] L.A. McGeoch and D.D. Sleator. A strongly competitive randomized paging algorithm. *Algorithmica*, 6:816-825, 1991.

[R89] P. Raghavan. Lecture notes on randomized algorithms. IBM Research Report No. RC 15340 (# 68237), Yorktown Heights, 1989.

[RS89] P. Raghavan and M. Snir. Memory versus randomization in on-line algorithms. In *Proc. 16th International Colloquium on Automata, Languages and Programming*, Springer Lecture Notes in Computer Science, Vol. 372, pages 687-703, 1989.

[ST85] D.D. Sleator and R.E. Tarjan. Amortized efficiency of list update and paging rules. *Communication of the ACM*, 28:202-208, 1985.

[S77] J.R. Spirn. *Program Behavior: Models and Measurements*. Elsevier, New York, 1977.

[Y77] A.C.-C. Yao. Probabilistic computations: Towards a unified measure of complexity. In *Proc. 17th Annual IEEE Symposium on Foundations of Computer Science*, pages 222-227, 1977.

[Y91] N. Young. *Competitive Paging and Dual-Guided On-Line Weighted Caching and Matching Algorithms*. Ph.D. thesis, Princeton University, 1991. Available as Computer Science Department Technical Report CS-TR-348-91.

An Optimal Algorithm for Shortest Paths on Weighted Interval and Circular-Arc Graphs, with Applications*

Mikhail J. Atallah[†]   Danny Z. Chen[‡]   D. T. Lee[§]

Abstract

We give the first linear-time algorithm for computing single-source shortest paths in a weighted interval or circular-arc graph, when we are given the model of that graph, i.e., the actual weighted intervals or circular-arcs *and* the sorted list of the interval endpoints. Our algorithm solves this problem optimally in $O(n)$ time, where n is the number of intervals or circular-arcs in a graph. An immediate consequence of our result is an $O(qn + n \log n)$ time algorithm for the minimum-weight circle-cover problem, where q is the minimum number of arcs crossing any point on the circle; the $n \log n$ term in this time complexity is from a preprocessing sorting step when the sorted list of endpoints is not given as part of the input. The previously best time bounds were $O(n \log n)$ for this shortest paths problem, and $O(qn \log n)$ for the minimum-weight circle-cover problem. Thus we improve the bounds of both problems. More importantly, the techniques we give hold the promise of achieving similar $\log n$-factor improvements in other problems on such graphs.

1 Introduction

Given a weighted set S of n intervals on a line, a *path* from interval $I \in S$ to interval $J \in S$ is a sequence $\sigma = (J_1, J_2, \ldots, J_k)$ of intervals in S such that $J_1 = I$, $J_k = J$, and J_i and J_{i+1} overlap for every $i \in \{1, \ldots, k-1\}$. The *length* of σ is the sum of the weights of its intervals, and σ is a *shortest path* from I to J if it has the smallest length among all possible I-to-J paths in S. The single-source shortest paths problem is that of computing a shortest path from a given "source" interval to all the other intervals. Our algorithm solves this shortest paths problem on interval and circular-arc graphs optimally in $O(n)$ time, when we are given the model of such a graph, i.e., the actual weighted intervals or circular-arcs *and* the sorted list of the interval endpoints. A node of an interval (resp., circular-arc) graph corresponds to an interval (resp., circular-arc) and an edge is between two nodes in the graph iff the two intervals (resp., circular-arcs) corresponding to these nodes intersect each other. Note that an interval or circular-arc graph with n nodes can have $O(n^2)$ edges. Our algorithm achieves the optimal $O(n)$ time bound by exploiting several geometric properties of this problem and by making use of the special UNION-FIND structure of [6].

---

*This research was supported in part by the Leonardo Fibonacci Institute in Trento, Italy.

[†]Department of Computer Sciences, Purdue University, West Lafayette, IN 47907. mja@cs.purdue.edu. Research supported in part by the Air Force Office of Scientific Research under Contract AFOSR-90-0107 and by the National Science Foundation under Grant CCR-9202807.

[‡]Department of Computer Science and Engineering, University of Notre Dame, Notre Dame, IN 46556. chen@cse.nd.edu.

[§]Department of Electrical Engineering and Computer Science, Northwestern University, Evanston, IL 60208. dtlee@eecs.nwu.edu. Research supported in part by the National Science Foundation under Grant CCR-8901815.

One of the main applications of this shortest paths problem is to the minimum-weight circle-cover problem [10, 3, 2, 9], whose definition we briefly review: Given a set of weighted circular-arcs on a circle, choose a minimum-weight subset of the circular-arcs whose union covers the circle. It is known [3] that the minimum-weight circle-cover problem can be solved by solving q instances of the previously mentioned single-source shortest paths problem, where q is the minimum number of arcs crossing any point on the circle[1] (in [3], a minimum-weight circle-cover is found in $O(qn^2)$ time). It is the circle-cover problem that has the main practical applications, and the study of this shortest-paths problem has mainly been for the purpose of solving the circle-cover problem. However, interval graphs and circular-arc graphs do arise in VLSI design, scheduling, biology, traffic control, and other application areas [4, 7, 8], so that our shortest paths result may be useful in other optimization problems. More importantly, our approach holds the promise of shaving a $\log n$ factor from the time complexity of other problems on such graphs.

Note that, by using our single-source shortest paths algorithm, the *all-pair* shortest paths problem on weighted interval and circular-arc graphs can be solved in $O(n^2)$ time, which is optimal. The previously best time bound for the all-pair shortest paths problem on weighted interval graphs was $O(n^2 \log n)$ (by using [9]). An $O(n^2)$ time and space algorithm for the *unweighted* case of the all-pair shortest paths problem was given in [11], and these bounds have been improved recently by Chen and Lee [5].

We henceforth assume that the intervals are given sorted by their left endpoints, and also sorted by their right endpoints. This is not a limiting assumption in the case of the main application of the shortest paths problem, which is the minimum-weight circle-cover problem. In the latter problem, an $O(n \log n)$ preprocessing sorting step is cheap compared to the previously best bound for solving that problem, which was $O(qn \log n)$ [9] (by using q times the subroutine for solving the shortest paths problem, at a cost of $O(n \log n)$ time each). Using our shortest paths algorithm, the minimum-weight circle-cover problem is solved in $O(qn + n \log n)$ time, where the $n \log n$ term is from the preprocessing sorting step when the sorted list of endpoints is not given as part of the input. Therefore, in order to establish the bound we claim for the minimum-weight circle-cover problem, it suffices to give a linear-time algorithm for the shortest paths problem on interval graphs. The linear-time solution to the shortest paths problem on circular-arc graphs makes use of the solution to the shortest paths problem on interval graphs. Therefore, we mainly focus on the problem of solving, in linear time, the shortest paths problem on interval graphs.

We also henceforth assume, without loss of generality (WLOG), that we are computing the shortest paths from the source interval to only those intervals whose right endpoints are to the right of the right endpoint of the source; the same algorithm that solves this case can, of course, be used to solve the case for the shortest paths to intervals whose left endpoints are to the left of the left endpoint of the source. Clearly we need not worry about paths to intervals whose right endpoints are covered by the source since the problem is trivial for those intervals – the length of the shortest path is simply the sum of the weight of the source plus the weight of the destination.

We consider the shortest paths problem on interval (resp., circular-arc) graphs in which

---

[1]q can be found in $O(n \log n)$ time. See, *e.g.* [12]

the weights of the intervals (resp., circular-arcs) are nonnegative. The minimum-weight circle-cover problem [3], however, does allow circular-arcs to have negative weights. Bertossi [3] has already given a reduction of any minimum-weight circle-cover problem with both negative and nonnegative weights to one with only nonnegative weights (to which the algorithm for computing shortest paths in interval graphs with nonnegative weights is applicable). Therefore it suffices to solve the shortest paths problem on interval graphs for the case of nonnegative weights. Bertossi's reduction introduces zero-weight intervals, so it is important to be able to handle problems with zero-weight intervals.

We only show how to compute the lengths of shortest paths. Our algorithm can be easily modified to handle the computation for actual shortest paths and shortest path trees, in $O(n)$ time and $O(n)$ space.

In the next section, we introduce some terminology needed in the rest of the paper. Sections 3 and 4 consider the special case of the shortest paths problem on interval graphs with only positive weights. In particular, Section 3 presents a preliminary suboptimal algorithm which illustrates our main idea and observations, and Section 4 shows how to implement various computation steps of the preliminary algorithm so that it runs optimally in linear time. Section 5 gives a linear-time reduction that reduces the nonnegative weight case to the positive weight case, and it shows how to use the solution to the shortest paths problem on interval graphs to obtain the solution to that on circular-arc graphs.

2   Terminology

In this section, we introduce some additional terminology.

We say that an interval I *contains* another interval J iff $I \cap J = J$. We say that I *overlaps* with J iff their intersection is not empty, and that I *properly overlaps* with J iff they overlap but neither one contains the other.

An interval I is typically defined by its two endpoints, i.e., $I = [a, b]$ where $a \leq b$ and a (resp., b) is called the *left* (resp., *right*) endpoint of I. A point x is *to the left* (resp., *right*) of interval $I = [a, b]$ iff $x < a$ (resp., $b < x$).

We assume that the input set S consists of intervals $I_1, \ldots, I_n$, where $I_i = [a_i, b_i]$, $b_1 \leq b_2 \leq \cdots \leq b_n$, and that the weight of each interval I_i is $w_i \geq 0$. To avoid unnecessarily cluttering the exposition, we assume that the intervals have distinct endpoints, that is, $i \neq j$ implies $a_i \neq a_j$, $b_i \neq b_j$, $a_i \neq b_j$, and $b_i \neq a_j$ (the algorithm for nondistinct endpoints is a trivial modification of the one we give).

Definition 1 *We use S_i to denote the subset of S that consists of intervals $I_1, I_2, \ldots, I_i$. We assume, WLOG, that the union of all the I_i's in S covers the portion of the line from a_1 to b_n. We also assume, WLOG, that the source interval is I_1.*

Observe that for a set S^* of intervals, the union of all the intervals in S^* may form more than one connected component. If for two intervals I' and I'' in S^*, I' and I'' respectively belong to two different connected components of the union of the intervals in S^*, then there is no path between I' and I'' that uses only the intervals in S^*.

Figure 1: For $i = 1, 2, \ldots, 10$, $w_i = 15, 12, 13, 17, 17, 19, 21, 13, 15, 18$, respectively.

3 A Preliminary Algorithm

This section gives a preliminary, $O(n \log \log n)$ time (hence suboptimal) algorithm for the special case of the shortest paths problem on intervals with positive weights. This should be viewed as a "warm-up" for the next section, which will give an efficient implementation of some of the steps of this preliminary algorithm, resulting in the claimed linear-time bound. In Section 5, we point out how the algorithm for positive-weight intervals can also be used to solve problems with nonnegative-weight intervals.

We begin by introducing definitions that lead to the concept of an *inactive* interval in a subset S_i, then proving lemmas about it that are the foundation of the preliminary algorithm.

Definition 2 *An* extension *of S_i is a set S' that consists of S_i and one or more intervals (not necessarily in S) whose right endpoints are larger than b_i. (There are, of course, infinitely many choices for such an S'.)*

Definition 3 *An interval I_k in S_i ($k \le i$) is inactive in S_i iff for every extension S' of S_i, the following holds: Every $J \in S' - S_i$ for which there is an I_1-to-J path in S' has no shortest I_1-to-J path in S' that uses I_k. An interval of S_i which is not inactive in S_i is said to be active in S_i.*

Intuitively, I_k is inactive in S_i if the other intervals in S_i are such that, as far as any interval J with right endpoint larger than b_i is concerned, I_k is "useless" for computing a shortest I_1-to-J path (in particular, this is true for $J \in \{I_{i+1}, \ldots, I_n\}$). In Figure 1, I_2 is inactive in S_4, I_3 is active in S_4, I_5 is inactive in S_5, I_9 is inactive in S_{10}, and I_{10} is active in S_{10}.

Observe that an interval I_k that is active in S_i, $k \le i$, may be inactive for an S_j with $j > i$, but is certainly active for any S_j with $k \le j \le i$. On the other hand, an interval I_k which is inactive for S_i, $k \le i$, is also inactive for every S_j with $j > i$.

Note that I_i is active in S_i iff there is an I_1-to-I_i path in S_i (i.e., if $\cup_{1 \le k \le i} I_k$ covers the portion of the line from a_1 to b_i).

Lemma 1 *The union of all the active intervals in S_i covers a contiguous portion of the line from a_1 to some b_j, where b_j is the rightmost endpoint of any active interval in S_i.*

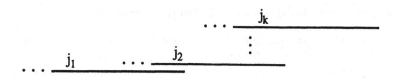

Figure 2: Illustrating Corollary 1: $label_i(j_1) \leq label_i(j_2) \leq \cdots \leq label_i(j_k)$.

Proof. An immediate consequence of the fact that if I_k, $k \leq i$, is active in S_i, then there is an I_1-to-I_k path in S_i. This is because if there is an I_1-to-I_k path in S_i, then there is a shortest I_1-to-I_k path in S_i, implying that every constituent interval of such a shortest I_1-to-I_k path is active in S_i. □

Definition 4 *Let $label_j(i)$, $j \geq i$, denote the length of a shortest I_1-to-I_i path in S that does not use any I_k for which $k > j$. By convention, if $j < i$, then $label_j(i) = +\infty$.*

Observe that for all i, $label_1(i) \geq label_2(i) \geq \cdots \geq label_n(i)$. For an $I_k \in S_i$, if there is no I_1-to-I_k path in S_i, then obviously $label_i(j) = +\infty$, for every $j = k, k+1, \ldots, i$. In Figure 1, $label_9(7) = +\infty$, but $label_{10}(7) = 71$.

Our algorithm is based on the following lemmas.

Lemma 2 *If $i > k$ and $label_i(i) < label_i(k)$, then I_k is inactive in S_i.*

Proof. Since $label_i(i) < label_i(k)$, $label_i(i)$ is not $+\infty$. Hence there is an I_1-to-I_i path in S_i, and there is an I_1-to-I_k path in S_i. Because $label_i(i) < label_i(k)$, it follows that there is a shortest I_1-to-I_i path in S_i that does not use I_k: The union of the intervals on that I_1-to-I_i path contains I_k (because $i > k$), and hence I_k is "useless" for any $J \in S' - S_i$ where S' is an extension of S_i. □

The following are immediate consequences of Lemma 2.

Corollary 1 *Let $I_{j_1}, I_{j_2}, \ldots, I_{j_k}$ be the active intervals in S_i, $j_1 < j_2 < \cdots < j_k \leq i$. Then $label_i(j_1) \leq label_i(j_2) \leq \cdots \leq label_i(j_k)$.*

Figure 2 illustrates Corollary 1. Note that the right endpoints of the active intervals $I_{j_1}, I_{j_2}, \ldots, I_{j_k}$ in S_i are in the same sorted order as that of their labels $label_i(j_1)$, $label_i(j_2)$, $\ldots$, $label_i(j_k)$. Their left endpoints, however, are not necessarily in such a sorted order (in Figure 2, the left endpoints of the intervals are omitted, indicated by marks "...").

Corollary 2 *If I_i contains I_k (hence $i > k$) and $label_i(k) > label_i(i)$, then I_k is inactive in S_i.*

18

Lemma 3 *If $i > k$ and $label_i(i) < label_{i-1}(k)$, then I_k is inactive in S_i.*

Proof. That $label_i(i) < label_{i-1}(k)$ implies that $label_i(i)$ is not $+\infty$. Hence there is an I_1-to-I_i path in S_i, and there is an I_1-to-I_k path in S_i. There are two cases to consider. (i) The shortest I_1-to-I_k path in S_i does not need to use I_i. Then $label_{i-1}(k) = label_i(k)$, and hence $label_i(i) < label_i(k)$. By Lemma 2, I_k is inactive in S_i. (ii) The shortest I_1-to-I_k path in S_i does use I_i. Then $label_i(k) \geq label_i(i) + w_k > label_i(i)$ (since $w_k > 0$). Again by Lemma 2, I_k is inactive in S_i. □

Lemma 4 *If interval I_k, $k > 1$, does not contain any b_j ($j < k$) such that I_j is active in S_{k-1}, then I_k is inactive in S_i for every $i \geq k$.*

Proof. It suffices to prove that I_k is inactive in S_k. Suppose I_k is active in S_k. Then by Lemma 1, the union of all the active intervals in S_k covers the contiguous portion of the line from a_1 to b_k (note that b_k is the rightmost endpoint of any interval in S_k). This implies that I_k contains the right endpoint of at least one active interval in S_k other than I_k. But all the intervals in S_{k-1} that I_k intersects are inactive in S_{k-1}, and hence they remain inactive in S_k, contradicting to that I_k intersects some active intervals in S_k other than I_k. □

We first give an overview of the algorithm. The algorithm scans the intervals in the order $I_1, I_2, \ldots, I_n$ (i.e., the scan is based on the increasing order of the sorted right endpoints of the intervals in S). When the scan reaches I_i, the following must hold before the scan can proceed to I_{i+1}:

(1) All the active intervals in S_i are stored in a tree T.

(2) All the inactive intervals in S_i have been marked as such (possibly at an earlier stage, when the scan was at some $I_{i'}$ with $i' < i$).

(3) If I_k ($k \leq i$) is active in S_i, then the correct $label_i(k)$ is known.

If we can maintain the above invariants, then clearly when the scan terminates at I_n, we already know the desired $label_n(i)$'s for all I_i's which are active in S_n. A postprocessing step will then compute, in linear time, the correct $label_n(i)$'s of the inactive I_i's in S_n (more on this later).

The details of the preliminary algorithm follow next. In this algorithm, the *right* endpoints of the active intervals are maintained in the leaves of the tree structure T, one endpoint per leaf, in sorted order.

1. Initialize T to contain I_1.

2. For $i = 2, 3, \ldots, n$, do the following. Perform a search in T for a_i. This gives the smallest b_j in T that is $> a_i$. If no such b_j exists, then (by Lemma 4) mark I_i as being inactive and proceed to $i+1$. So suppose such a b_j exists. Set $label_i(i) = label_{i-1}(j) + w_i$, and note that this implies that I_j remains active in S_i and has the same label as in S_{i-1}, i.e., $label_i(j) = label_{i-1}(j)$. Next, insert I_i in T (of course b_i is then in the rightmost leaf of T). Then repeatedly check the leaf for I_k which is immediately to the left of the leaf for I_i in T, to see whether I_k is inactive in S_i (by Lemma 3, i.e.,

check whether $label_{i-1}(k) < label_i(i))$, and, if I_k is inactive, then mark it as such, delete it from T, and repeat with the leaf made adjacent to I_i by the deletion of I_k. Note that more than one leaf of T may be deleted in this fashion, but that the deletion process stops short of deleting I_j itself, because it is I_j that gave I_i its current label (i.e., $label_i(i) = label_{i-1}(j) + w_i \geq label_{i-1}(j)$). Of course any I_ℓ whose leaf in T is *not* deleted is in fact active in S_i and already has the correct value of $label_i(\ell)$: It is simply the same as $label_{i-1}(\ell)$ and we need not explicitly update it (the fact that this updating is implicit is important, as we cannot afford to go through all the leaves of T at the iteration for each i).

When Step 2 terminates (at $i = n$), we have the values of the $label_n(\ell)$'s for all the active I_ℓ in S_n. The next step obtains the values of the $label_n(\ell)$'s for the other intervals (those that are inactive in S_n).

3. For every inactive I_i in S_n, find the smallest right endpoint $b_j > a_i$ such that I_j is active in S_n, and set $label_n(i) = label_n(j) + w_i$. Note that by Lemma 1, such an I_j exists and it intersects I_i. This step can be easily implemented by a right-to-left scan of the sorted list of all the endpoints.

The correctness of this algorithm easily follows from the definitions, lemmas, and corollaries preceding it. Note that although a particular iteration in Step 2 may result in many deletions from T, overall there are less than n such deletions. The time complexity of this algorithm is $O(n \log n)$ if we implement T as a 2-3 tree [1], but $O(n \log \log n)$ if we use the data structure of Van Emde Boas [14] (the latter would require normalizing all the $2n$ sorted endpoints so that they are integers between 1 and $2n$). The next section gives an $O(n)$ time implementation of the above algorithm. Note that the main bottleneck is Step 2, since the scan needed for Step 3 obviously takes linear time.

4   A Linear Time Implementation

As observed earlier, the main bottleneck is Step 2 of the preliminary algorithm given in the previous section. We shall implement essentially the same algorithm, but without using the tree T. Instead, we use a UNION-FIND structure [6] where the elements of the sets are integers in $\{1, \ldots, n\}$, with integer i corresponding to interval I_i. Initially, each element i is in a singleton set also named i, that is, initially set i is $\{i\}$. (We often call a set whose name is integer i as set i, with the understanding that set i may contain other elements than i.) During the execution of Step 2, we maintain the following invariants (assume we are at index i in Step 2):

(1) To each currently active interval I_j corresponds a set named j. If $I_{i_1}, I_{i_2}, \ldots, I_{i_k}$ are the active intervals in S_i, $i_1 < i_2 < \cdots < i_k$, then for every $i_j \in \{i_1, i_2, \ldots, i_{k-1}\}$, the indices of the inactive intervals $\{I_\ell \mid i_j < \ell < i_{j+1}\}$ are all in the set whose name is i_{j+1}. Set i_{j+1} consists of the indices of the above-mentioned inactive intervals, and also of the index i_{j+1} of the active interval $I_{i_{j+1}}$. Note that since I_1 is always active, $i_1 = 1$ in the above discussion, and the set whose name is 1 is a singleton (recall that a preprocessing

step has eliminated intervals whose right endpoints are contained in interval I_1). The next invariant is about intervals that are inactive and do not overlap with any active interval.

(2) Let $Loose(S_i)$ denote the subset of the inactive intervals in S_i that do not overlap with any active interval in S_i. In Figure 1, the active intervals in S_9 are I_1, I_3, I_4, and $Loose(S_9)$ consists of intervals $I_5, I_6, \ldots, I_9$. Observe that, based on Lemma 1, every interval in $Loose(S_i)$ is to the right of the union of the active intervals in S_i; furthermore, $Loose(S_i)$ is nonempty iff $I_i \in Loose(S_i)$. If $Loose(S_i)$ is not empty, then let $CC_1, CC_2, \ldots, CC_t$ be the connected components of $Loose(S_i)$: There is a set named j_l for every such CC_l, where I_{j_l} is the *rightmost* interval in CC_l (I_{j_l} is the interval in CC_l having the largest right endpoint); we say that such an inactive I_{j_l} is *special inactive*. The (say) μ elements in set j_l correspond to the μ intervals in CC_l; more specifically, they are the contiguous subset of indices $\{j_l - \mu + 1, j_l - \mu + 2, \ldots, j_l - 1, j_l\}$. Note that $j_l - \mu$ is the set named j_{l-1} if $1 < l \leq t$, and that $j_t = i$.

In Figure 1, for $i = 9$, $CC_1 = \{I_5, I_6, I_7\}$, $CC_2 = \{I_8, I_9\}$, and the special inactive intervals are I_7 and I_9.

(3) An auxiliary stack contains the active intervals $I_{i_1}, I_{i_2}, \ldots, I_{i_k}$ mentioned in item (1) above, with I_{i_k} at the top of the stack. We call it the *active* stack.

In Figure 1, for $i = 9$, the active stack contains I_1, I_3, I_4 (with I_4 at the top of the stack).

(4) Another auxiliary stack contains the special inactive intervals $I_{j_1}, I_{j_2}, \ldots, I_{j_t}$ mentioned in item (2) above, with I_{j_t} at the top of the stack. We call it the *special inactive* stack.

In Figure 1, for $i = 9$, the special inactive stack contains I_7, I_9 (with I_9 at the top of the stack).

A crucial point is how to implement, in Step 2, the search for b_j using a_i as the key for the search. This is closely tied to the way that the above invariants (1)–(4) are maintained. It makes use of some preprocessing information that is described next.

Definition 5 *For every I_i, let $Succ(I_i)$ be the smallest index ℓ such that $a_i < b_\ell$, i.e., $b_\ell = Min\{b_r \mid I_r \in S, a_i < b_r\}$.*

In Figure 1, $Succ(I_5) = 5$, $Succ(I_9) = 8$, and $Succ(I_{10}) = 4$.

Note that $\ell \leq i$, and that $\ell = i$ occurs when I_i does not contain any b_r other than b_i. Also, observe that the definition of the $Succ$ function is static (it does not depend on which intervals are active). The $Succ$ function can easily be precomputed in linear time by scanning right-to-left the sorted list of all the $2n$ interval endpoints.

The significance of the $Succ$ function is that, in Step 2, instead of searching for b_j using a_i as the key for the search, we simply do a FIND($Succ(I_i)$): Let j be the set name returned by this FIND operation. We distinguish 3 cases.

1. If $j = i$, then surely I_i does not overlap with any interval in S_{i-1} and it is inactive in S_i (by Lemma 4). We simply mark I_i as being special inactive, push I_i on the special inactive stack, and move the scan of Step 2 to index $i + 1$.

   In Figure 1, this happens for $i = 2$, $i = 5$, and $i = 8$.

2. If $j < i$ and I_j is active in S_{i-1}, we set $label_i(i) = label_{i-1}(j) + w_i$. Then do the following updates on the two stacks:

   (a) We pop *all* the special inactive intervals I_{i_l} from their stack and, for each such I_{i_l}, we do UNION(i_l, i), which results in the disappearance of set i_l and the merging of its elements with set i; set i retains its old name.

   In Figure 1, for $i = 10$, this results in the disappearance of sets 7 and 9, and the merging of their contents with set 10.

   (b) We repeatedly check whether the top of the active stack, I_{i_k}, is going to become inactive in S_i because of I_i (that is, because $label_i(i) < label_{i-1}(i_k)$). If the outcome of the test is that I_{i_k} becomes inactive, then we do UNION(i_k, i), pop I_{i_k} from the active stack, and continue with $I_{i_{k-1}}$, etc. If the outcome of the test is that I_{i_k} is active in S_i, then we keep it on the active stack, push I_i on the active stack, and move the scan of Step 2 to index $i + 1$.

   In Figure 2, if I_i is active in S_i, $j = j_1$, and $label_i(i) < label_{i-1}(j_2)$, then the sets $j_2, j_3, \ldots, j_k$ disappear and their contents get merged with set i.

3. If $j < i$ and I_j is special inactive in S_{i-1}, then I_i does not overlap with any active interval in S_{i-1} and it is inactive in S_i (by Lemma 4). But, I_i does overlap with one or more inactive intervals in S_{i-1}, including the special inactive interval I_j; more precisely, I_i overlaps with some connected components of $Loose(S_{i-1})$ whose rightmost intervals are contiguously stored in the stack of special inactive intervals. Let these connected components with which I_i overlaps be called, in left to right order, $C_1, C_2, \ldots, C_h$. The rightmost interval of C_1 is I_j. Let $I_{r_2}, I_{r_3}, \ldots, I_{r_h}$ be the rightmost intervals of (respectively) $C_2, C_3, \ldots, C_h$ (of course $I_{r_h} = I_{i-1}$). Observe that the top h intervals in the special inactive stack are $I_j, I_{r_2}, \ldots, I_{r_h}$, with I_{r_h} ($= I_{i-1}$) on top. Because of I_i, all of these h intervals will become inactive in S_i (whereas they were special inactive in S_{i-1}). Their h sets (corresponding to $C_1, C_2, \ldots, C_h$) must be merged into a new, single set having I_i as its rightmost interval. I_i is special inactive in S_i. This is achieved by:

   (a) Popping $I_{r_h}, \ldots, I_{r_2}, I_j$ from the special inactive stack,

   (b) performing UNION(r_h, i), UNION(r_{h-1}, i), $\ldots$, UNION(r_2, i), UNION(j, i), and

   (c) pushing I_i on the special inactive stack.

Observe that the total number of the UNION and FIND operations performed by our algorithm is $O(n)$. It is well-known that a sequence of m UNION and FIND operations on n elements can be performed in $O(m\alpha(m + n, n) + n)$ time [13], where $\alpha(m + n, n)$ is the (very slow-growing) functional inverse of Ackermann's function. Therefore, our algorithm runs within the same time bound. However, it is possible to achieve $O(n)$ time performance for our algorithm, by the following observations.

In our algorithm, every UNION operation involves two set names that are *adjacent* in the sorted order of the currently existing set names. That is, if L is the sorted list of the set names (initially L consists of all the integers from 1 to n), then a UNION operation always involves two adjacent elements of L. Thus the underlying UNION-FIND structure we use satisfies the requirements of the *static tree set union* in [6], in order to result in linear-time performance: It is the *linked list* $LL = (1, 2, \ldots, n)$, where the element in LL that follows element ℓ is $next(\ell) = \ell + 1$, for every $\ell = 1, 2, \ldots, n - 1$ (the requirement in [6] is that the structure be a static tree). Note that the *next* function is static throughout our algorithm. The UNION operation in our algorithm is always of the form $unite(next(\ell), \ell)$, as defined in [6], that is, it concatenates two disjoint but *consecutive* sublists of LL into one contiguous sublist of LL. On this kind of structures, a sequence of m UNION and FIND operations on n elements can be performed in $O(m + n)$ time [6]. Therefore, the time complexity of our algorithm is $O(n)$.

5   Further Extensions

This section sketches how the shortest paths algorithm of the previous sections can be used to solve problems where intervals can have zero weight, and how it can be used to solve the version of the problem where we have circular-arcs rather than intervals on a line.

5.1   Zero-Weight Intervals

The astute reader will have observed that the definitions and the shortest paths algorithm of the previous sections can be modified to handle zero-weight intervals as well. However, doing so would unnecessarily clutter the exposition. Instead, we show in what follows that the shortest paths problem in which some intervals have zero weight can be reduced in linear time to one in which all the weights are positive. Not only does this simplify the exposition, but the reduction used is of independent interest.

Let $P1$ be the version of the problem that has zero-weight intervals, and let Z be the nonempty subset of S that contains all the zero-weight intervals of S. First, observe that in order to solve $P1$, it suffices to solve the problem $P2$ obtained from $P1$ by replacing every connected component CC of Z by a new zero-weight interval that is the union of the zero-weight intervals in CC (because the label of $I \in Z$ in $P1$ is the same as the label of $J = \cup_{I \in CC} I$ in $P2$). Hence it suffices to show how to solve $P2$. In what follows assume that we have already created, in $O(n)$ time, $P2$ from $P1$.

We next show how to obtain, from $P2$, a problem $P3$ such that (i) every interval in $P3$ has a positive weight (and therefore $P3$ can be solved by the algorithm of the previous sections), and (ii) the solution to $P3$ can be used to obtain a solution to $P2$.

Recall that, by the definition of $P2$, two zero-weight intervals in it cannot overlap. $P3$ is obtained from $P2$ by doing the following for each zero-weight interval $J = [a, b]$: "cut out" the portion of the problem in between a and b, that is, first erase, for every interval I of $P2$, the portion of I in between a and b, and then "pull" a and b together so they coincide in $P3$.

This means that in $P3$, J has disappeared, and so has every interval J' that was contained in J. An interval J'' in $P2$ that contained J, or that properly overlapped with J, gets shrunk by the disappearance of its portion that used to overlap with J. For example, if we imagine that the situation in Figure 1 describes problem $P2$, and that J is (say) interval I_4 in Figure 1 (so I_4 has zero weight), then "cutting" I_4 results in the disappearance of I_2 and I_3 and the "bringing together" of I_1 and I_{10} so that, in the new situation, the right endpoint of I_1 coincides with the left endpoint of I_{10}.

Implementation Note: The above-described cutting-out process of the zero-weight intervals can be implemented in linear time by using a linked list to do the cutting and pasting. In particular, if in $P2$ an interval I of positive weight contains many zero-weight intervals $J_1, \ldots, J_k$, the cutting-out of these zero-weight intervals does *not* affect the representation we use for I (although in a geometric sense I is "shorter" afterwards, as far as the linked list representation is concerned, it is unchanged). This is an important point, since it implies that only the endpoints contained in a J_k are affected by the cutting-out of that J_k, and such an endpoint gets updated only once because it is not contained in any other zero-weight interval of $P2$ (recall that the zero-weight intervals of $P2$ are pairwise non-overlapping).

By definition, $P3$ has no zero-weight intervals. So suppose $P3$ has been solved by using the algorithm we gave in the earlier sections. The solution to $P3$ yields a solution to $P2$ in the following way.

- If an interval I is in $P3$ (i.e., I has not been cut out when $P3$ was obtained from $P2$), then its label in $P2$ is exactly the same as its label in $P3$.

- Let $J = [a, b]$ be a zero-weight interval which was cut out from $P2$ when $P3$ was created. (In $P3$, a and b coincide, so in what follows when we refer to "a in $P3$" we are also referring to b in $P3$.) For each such $J = [a, b]$, compute in $P3$ the smallest label of any interval of $P3$ that contains a: This is the label of J in $P2$. This computation can be done for all such J's by one linear-time scan of the endpoints of the active intervals for $P3$.

- Suppose I is a positive-weight interval of $P2$ that was cut out when $P3$ was created, because it was contained in a zero-weight interval J of $P2$. Then the label of I in $P2$ is equal to: (weight of I) + (label of J in $P2$).

5.2 Circular-Arcs

The version of the shortest paths problem where we have circular-arcs on a circle C instead of intervals on a straight line can be solved by two applications of the shortest paths algorithm for intervals: Suppose $I_1 = [a, b]$ is the "source" circular-arc, where a and b are now positions on circle C. (We use the convention of writing a circular-arc as a pair of positions on the circle such that, when going from the first position to the second position along the arc, we travel in the clockwise direction.)

It is not hard to see that the following linear-time procedure solves the shortest paths problem on circular-arc graphs.

- Create a problem on a straight line by "opening" circle C at a. That is, create an n-interval problem by starting at a and traveling clockwise along C, putting the intervals encountered during this trip on a straight line, until the trip is back at a. Intervals that contain a are not included twice in the straight-line problem: Only their first appearance on the clockwise trip is used, and they are "truncated" at a (so that on the line, they appear to begin at a, just like the source I_1). Then solve the straight-line problem so created, by using the algorithm for the interval case. The computation of this step gives each circular-arc a label.

- Repeat the above step with a playing the role of b, and "counterclockwise" playing the role of "clockwise".

- The correct label for a circular-arc is the smaller of the two labels, computed above, for the intervals corresponding to that arc.

Acknowledgement. The authors would like to thank Dr. Jan-Ming Ho for the discussions on this problem.

References

[1] A. V. Aho, J. E. Hopcroft, and J. D. Ullman. *The Design and Analysis of Computer Algorithms*, Addison-Wesley, Reading, Massachusetts, 1974.

[2] M. J. Atallah and D. Z. Chen. "An optimal parallel algorithm for the minimum circle-cover problem," *Information Processing Letters*, 32 (1989), pp. 159–165.

[3] A. A. Bertossi. "Parallel circle-cover algorithms," *Information Processing Letters*, 27 (1988), pp. 133–139.

[4] K. S. Booth and G. S. Lukeher. "Testing for the consecutive ones property, interval graphs, and graph planarity using PQ-tree algorithms," *Journal of Computer and System Sciences*, 13 (1976), pp. 335–379.

[5] D. Z. Chen and D. T. Lee. "Solving the all-pair shortest path problem on interval and circular-arc graphs," Technical Report No. 93-3, Department of Computer Science and Engineering, University of Notre Dame, May 1993.

[6] H. N. Gabow and R. E. Tarjan. "A linear-time algorithm for a special case of disjoint set union," *Journal of Computer and System Sciences*, 30 (1985), pp. 209–221.

[7] M. C. Golumbic. *Algorithmic Graph Theory and Perfect Graphs*, Academic Press, New York, 1980.

[8] U. I. Gupta, D.T. Lee, and J. Y.-T. Leung. "Efficient algorithms for interval graphs and circular-arc graphs," *Networks*, Vol. 12 (1982), pp. 459–467.

[9] O. H. Ibarra, H. Wang, and Q. Zheng. "Minimum cover and single source shortest path problems for weighted interval graphs and circular-arc graphs," *Proc. of 30th Annual Allerton Conf. on Commun., Contr., and Comput.*, 1992, Univ. of Illinois, Urbana, pp. 575–584.

[10] C. C. Lee and D. T. Lee. "On a circle-cover minimization problem," *Information Processing Letters*, 18 (1984), pp. 109–115.

[11] R. Ravi, M.V. Marathe, and C.P. Rangan. "An optimal algorithm to solve the all-pair shortest path problem on interval graphs," *Networks*, Vol. 22 (1992), pp. 21–35.

[12] M. Sarrafzadeh and D. T. Lee, "Restricted track assignment with applications," *Int'l Journal Computational Geometry & Applications*, to appear.

[13] R. E. Tarjan. "A class of algorithms which require nonlinear time to maintain disjoint sets," set union algorithm," *Journal of Computer and System Sciences*, 18 (2) (1979), pp. 110–127.

[14] P. Van Emde Boas. "Preserving order in a forest in less than logarithmic time and linear space," *Information Processing Letters*, 6 (3) (1977), pp. 80–82.

Efficient Self Simulation Algorithms
for Reconfigurable Arrays

Yosi Ben-Asher[1]   Dan Gordon[1]   Assaf Schuster[2]

[1] Dept. of Math. and Comp. Sci., Univ. of Haifa, Haifa 31905, Israel.
[2] Dept. of Comp. Sci., Technion, Haifa 32000, Israel.

Abstract. Perhaps the most basic question concerning a model for parallel computation is the *self simulation problem*: given an algorithm which is designed for a large machine, can it be executed efficiently on a smaller one? In this work we give several positive answers to the self simulation problem on dynamically reconfigurable meshes. We show that the simulation of a reconfiguring mesh by a smaller one can be carried optimally, by using standard methods, on meshes such that buses are established along rows or along columns. A novel technique is shown to achieve asymptotically optimal self simulation on models which allow buses to switch column and row edges, provided that a bus is a "linear" path of connected edges. Finally, for models in which a bus is any sub-graph of the underlying mesh efficient simulations are presented, paying by an extra factor which is polylogarithmic in the size of the simulated mesh. Although the self simulation algorithms are complex and require extensive bookkeeping operations, the required space is asymptotically optimal.

1   Introduction

The basic idea of a reconfigurable network is to enable flexible connection patterns, by allowing nodes to connect and disconnect their adjacent edges in various patterns. This yields a variety of possible topologies for the network, and enables the program to exploit this topological variety in order to speed up the computation.

Informally, a reconfigurable network operates as follows. Essentially, the edges of the network are viewed as building blocks for larger *bus* components. The network dynamically reconfigures itself at each time step, where an allowable configuration is a partition of the network into a set of edge-disjoint buses. A crucial point is that the reconfiguration process is carried out *locally* at each processor (or *switch*) of the network. That is, at the beginning of each step during the execution of a program, each switch in the network fixes its *local configuration* by partitioning its collection of edges into some allowable combination of subsets. Two adjacent edges that are grouped by a switch into the same partition are viewed as if they were (hardware) connected.

There are several reconfiguring models that are considered in the published literature, depending on their switch capabilities. In this work we focus on two dimensional arrays (or, meshes) operating in three of the more popular models:

Horizontal-Vertical Reconfigurable Mesh (HV-RN model): the switches may change the configuration of the network so that buses of different lengths are formed horizontally along rows and vertically along columns. Thus, a single bus cannot "change directions" by using both horizontal and vertical bus components (mesh edges) [10, 8, 16]. A VLSI chip called YUPPIE (Yorktown Ultra Parallel Polymorphic Image Engine) has been implemented to demonstrate the feasibility of this reconfiguration style [11].

Linear Reconfigurable Mesh (LRN model): a bus may consist of any connected path of edges, not only vertical or only horizontal. In this model, however, only "linear" buses are composed, so that a bus component is attached to at most one other bus component at each end. Many results present efficient algorithms on the linear reconfigurable mesh. Some of these algorithms achieve constant running time (even when this is not possible using the popular PRAM model), and some match known $Area \times Time^2$ lower bounds. These results include arithmetic operations [5, 15], sorting and selection [2, 7, 6, 14, 4], and others [13, 2].

General Reconfigurable Mesh (RN model): a configuration of buses is any partition of the network into edge-disjoint subgraphs, so buses are not necessarily linear. Efficient algorithms were presented on the general reconfigurable mesh, including for example a constant time transitive closure algorithm [17]. A version of this model, namely the CAAPP (Content Addressable Array Parallel Processor), consisting of a 2-D array of 512×512 bit-serial processors, was implemented [18].

In [3] the expanding volume of reconfigurable results and architectures was given a theoretical treatment. For example, it was shown that the set of problems computable in constant time on a polynomial size mesh in the linear model is exactly the set of problems computable by a logspace Turing machine. The corresponding set in the general model contains exactly all the problems that are computable by a logspace Turing machine having a symmetric logspace oracle. These results partially explain the existence of a fairly simple connected components algorithm on the general reconfigurable mesh [17], while no such equivalent is known at the linear reconfiguring side.

1.1 This work - Self Simulations

The question we are interested in this work is whether reconfigurable models (in particular two dimensional reconfigurable arrays) can form the basis for the design of massively parallel computers. Perhaps the most basic aspect of this question is the efficiency and ease of algorithms design. Usually, for a particular problem, the solution is given by an algorithm which is suitable for input of size n, where the number of computing processors may be a function of n. It is assumed by the algorithm designer that as many processors as required by his algorithm are simultaneously available for his program. This assumption frees the programmer from the need to know the exact size of the machine he is working on. The assignment of logical processors tasks to the available physical ones is automatically determined by the compiler. In fact, the compiler writes a self simulation program of a large machine having many processors by a smaller machine with less processors. Hence the ability of efficiently achieving the logical to physical mapping is an extremely important property of a model for parallel computation. In its absence, it is not likely that the model will be chosen for a direct implementation on existing architectures.

Despite the large number of efficient algorithms that are known for reconfigurable arrays, none of the models was previously shown to support optimal self simulations. In this work we give several positive answers to this problem. We present asymptotically optimal and almost optimal self simulation results of large reconfigurable-mesh machines by smaller ones. We have the following (informally stated) results. **(1)** Using standard simulation techniques the mesh in the HV-RN model exhibits optimal self simulations (Section 3). **(2)** Although using the same method fails in the LRN model a different algorithm is shown to achieve asymptotically optimal self simulation for that model, too (Section 4). **(3)** A third algorithm presents self simulations in the RN model, paying by an extra slowdown which is polylogarithmic in the size of the simulated mesh (Section 5).

The self simulation algorithms are very complex and require lots of book-keeping operations. We show that the required space for these is asymptotically optimal, too. Yet, to avoid painful reading, we do not cope with constants minimization. Therefor the algorithms may seem wasteful and non-optimized at first glance. In addition, although given for the mesh, the simulation results may be applied to arbitrary rectangles as well.

2 Reconfiguring Models of Computation. Preliminaries.

A reconfigurable network is a network of processors operating synchronously. The processors residing at the nodes of the network perform the same program, taking local decisions and calculations according to the input and locally stored data. Input and output locations are specified by the problem to be solved, so that initially, each input item is available at a single node of the network, and eventually, each output item is stored by one. A single node of the network may consist of a computing unit, a memory unit and a switch with reconnection capability. In the sequel, we use the notions of *switch*, *processor* and a *network node* in an interchangeable manner.

A single time step of a reconfigurable network computation is composed of the following substeps. **Substep 1:** The network selects a *configuration H* of the buses, and reconfigures itself to H. This is done by local decisions taken at each switch. **Substep 2:** One or more of the processors connected by a bus transmit a message on the bus. These processors are called the *speakers* of the bus. **Substep 3:** Several of the processors connected by the bus attempt to read the message transmitted on the bus by the speaker(s). These processors are called the *readers* of the bus. **Substep 4:** Some local computation is taken by every processor.

At each time step, a bus may take one of the following three states. *Idle*: no processor transmits, *Speak*: there is one or more speakers, all sending the same message, *Error*: there is more than one speaker, and two or more messages are different. An *Error* state is detectable by all processors connected by the corresponding bus, but the messages are assumed to be destroyed.

2.1 Switch Operations

The general reconfigurable network model, as presented above, does not specify the exact operation of the switches. In the main part of this paper we consider three basic variants:

General RN (RN model): The switch may partition its collection of edges into any combination of subsets, where all edges in a subset are connected as building blocks for the same bus. Thus the possible network configurations are any partition into edge-disjoint connected subgraphs.

Linear RN (LRN model): The switch may partition its collection of edges into any combination of connected pairs and singletons. Hence buses are of the form of a path (or a cycle) and the global configuration is a partition of the network into paths, or a set of edge-disjoint linear buses.

Horizontal-Vertical RN (HV-RN model): Buses are formed either along rows (horizontally) or along columns (vertically), but may not contain building blocks from both dimensions.

For clarity we omit the description of other switching variants that are considered in the literature. Nevertheless our methods may be applied to some of these models as well.

In this extended abstract we omit the formal definitions of *simulation* and *slowdown*, see [1]. Essentially, these terms coincide with their intuitive counterparts.

2.2 Folding the Mesh

The *reconfigurable mesh* is the underlying network topology which is the most popular in the literature, and which is also the topology considered in the main part of this work. The $n \times n$ reconfigurable mesh, called the *n*-mesh, or the mesh of size n, is composed of an array of n columns and n rows of processors, with edges connecting each processor to its four neighbors (or fewer, for borderline processors). These connections are called $Up, Down, Left$ and $Right$. A configuration is given by splitting these into allowable subsets according to the switch capability. We refer to the processor at the ith row and the jth column as $[i-1, j-1]$. For convenience, we envision the mesh as embedded in the plane so that row 0 is at the bottom, and column 0 is to the left. We note that by paying a quadratic blow-up of the number of processors, any network may be simulated by a 2-dimensional mesh with no slowdown [3, 2].

The following function is sometimes used for mapping large meshes into smaller ones. $FOLD(m) = m \bmod p$ if m div p is even, otherwise $FOLD(m) = p - 1 - (m \bmod p)$. We use $[r, c] \longrightarrow [FOLD(r), FOLD(c)]$ for mapping a processor $[r, c]$ of a large mesh into the *p*-mesh. This has the same effect as that of folding a large page of paper several times into a square of size $p \times p$. A point on the p-sized square "simulates" all points of the folded page that are stabbed when pushing a pin at this point.

3   Self Simulations - HV-RN model

In this section we show how self simulation is performed with optimal asymptotic slowdown on meshes that operate in the HV-RN model. The algorithm is based on the simulation of an n/p-submesh of the n-mesh by a single node of the p-mesh. We assume that n/p is an integral integer.

Lemma 1. *In the HV-RN model, a single processor can simulate the l-mesh with slowdown $4l^2$.*

Theorem 2. *In the HV-RN model the simulation of the n-mesh by a p-mesh can be completed with slowdown $5(n/p)^2 + O(n/p)$.*

We omit the proofs in this extended abstract. The method that is used involves a contraction of a full, connected sub-mesh of the larger mesh, so that it is simulated by a single processor of the smaller mesh. In other words, this is a **contraction mapping** of simulated to simulating processors, which remains fixed throughout the simulation. This is actually the general approach of simulating a processor array with a smaller one, e.g. in the fixed connection network model of computation [9, pp. 234]. The following **All-or-nothing assumption** is also followed in the proof of Theorem 2: any bus in the simulated mesh which crosses a sub-mesh border is simulated in a single step by a bus in the simulating mesh, which is solely dedicated for that purpose (Although the "structure" of the corresponding buses may be different). It can be shown that the combined approach of the contraction mapping and the all-or-nothing assumption fails to achieve efficient self simulations in the LRN model (see [1]). This motivates the technique that is developed in the following section.

4   Self Simulations - LRN model

The main result of this section is the following theorem.

Theorem 3. *The simulation of the n-mesh by the m-mesh in the LRN model is completed with slowdown $\Theta((n/m)^2)$. The simulation algorithm uses $\Theta((n/m)^2)$ extra space at each processor of the m-mesh.*

The algorithm uses a variant of a connected components algorithm for graphs having only linear and non-cyclic components, which may be of interest in its own right. A few preliminary remarks are in order.

1. We first note that a processor writing on a linear bus may do so in two different modes: (a) the bus is connected inside the processor, and the processor simply writes on the bus; and (b) the bus is not connected inside the processor, and the processor writes a message on each end of the bus (in theory, these messages may be different).

2. Our notion of a linear bus allows cycles. Whenever we have a configuration of buses and speakers on the buses, we say that *condition* NBC *holds* if there is no speaker on a cycle. Note that a speaker in mode (b) above cannot be on a cycle, because the bus is not connected inside the processor. Our next observation is that any configuration of linear buses and speakers can be simulated in two steps in a straightforward way by a configuration satisfying NBC as follows:

   Step 1: A processor that broadcasts in mode (a) does not connect the buses inside, and simply broadcasts its message in mode (b) on the two bus ends; it also listens to the two ends. Step 2: If the above processor did not detect an error at either bus end, it connects the bus ends and listens without broadcasting. If it did detect an error, it broadcasts an error message on both bus ends.

3. The leader election problem for a bus is the problem of all the processors agreeing on one of them being a leader. When the bus is linear and not a cycle, this problem can be solved in $O(1)$ time by simulating an n-mesh with a $2n$-mesh as follows: Every processor of the n-mesh is simulated by a 2×2 square of processors on the $2n$-mesh, and every bus is simulated by a double bus on the $2n$-mesh. This double bus can be viewed as having 2 directions (see [1]). The leader is elected by having each endpoint of the bus transmit its *id* towards the other end. All processors on the bus listen, and the processor with the smallest *id* is chosen by all as the leader.

4.1 LCC - linear-connected components

Definition 4. A graph $G = (V, E)$ is called *linear* if the degree of every vertex is ≤ 2, and G is acyclic.

The LCC problem is defined in the following lemma, stating our main result for this subsection.

Lemma 5. *Let $G = (V, E)$ be a linear graph, $\mid V \mid = n$. Assume that the adjacency matrix $M = (m_{i,j})$ of G is stored on the $2n$-mesh, where $m_{i,j}$ is stored on processor $[2i, 2j]$. Then the connected components of G can be found in constant time, and the output is stored such that for every $0 \leq i < n$, $[2i, 0]$ holds some j such that vertices i and j are connected, and two connected vertices hold the same value.*

Proof: We consider an n-mesh in which every processor is the image of the contraction mapping of a 2-submesh of the $2n$-mesh. We call the processors of the given n-mesh v-processors. A bus in the n-mesh can be represented by two paths in the $2n$-mesh, so that a leader may be elected as described above. The rest of the proof is described in terms of the n-mesh, so that $[i, j]$ denotes the v-processor in row i, column j.

We now describe the algorithm solving the LCC problem:

1. Every $[i,j]$ that holds a 1 determines which of the following cases hold: **(a)** It is a unique 1 in its row; **(b)** it is a unique 1 in its column; **(c)** if is not unique in its row, the direction (left or right) of the other 1 in the same row; **(d)** if it is not unique in its column, the direction (up or down) of the other 1 in the same column.

   Note that there are at most two 1's in a row or column, because the degree of every vertex is ≤ 2.

2. If $m_{i,j} = 0$, $[i,j]$ connects the buses $\{L - R, U - D\}$. Else: **(a)** If $m_{i,j}$ is the unique 1 in its row, $[i,j]$ does not connect to anything in its row; **(b)** similar to (a), but for the column; **(c)** if there is another 1 in the same row, $[i,j]$ connects in the direction of that 1. **(d)** similar to (c), but for the column.

Fig. 2.b shows the 1's of a 9×9 adjacency matrix and the corresponding bus configuration. Define G_{LCC} to be the graph formed according to 2 above, i.e., $G_{LCC} = (V_{LCC}, E_{LCC})$, where $V_{LCC} = \{[i,j] \mid m_{i,j} = 1\}$, and

$$E_{LCC} = \left\{ \{[i,j],[k,l]\} \,\middle|\, \begin{array}{l} [i,j],[k,l] \text{ are connected by a} \\ \text{horizontal or a vertical bus segment.} \end{array} \right\}.$$

Denote by G' the *dual* graph of G. Namely, $G' = (E, \{\{e_1, e_2\} | e_1 \cap e_2 \in V\})$.

Claim 6. G_{LCC} *contains exactly two isomorphic copies of* G'.

Proof: Consider a connected component of G', which is a path of maximal length $e_1, e_2, \ldots, e_k$, and $e_i \cap e_{i+1} \in V$. For some $v_i, v_j \in V$, $e_1 = (v_i, v_j)$. Assume w.l.o.g. that $i < j$. The isomorphic copy of $e_1, \ldots, e_k$ is obtained as follows:

$$e_1 \rightarrow [i,j] \quad ; \quad e_2 \rightarrow \begin{cases} [i,k] \text{ if } e_2 = (i,k) \\ [k,j] \text{ if } e_2 = (j,k) \end{cases}$$

The maps of e_1, e_2 are connected in G_{LCC}, either on row i or column j. This construction continues for $e_3, \ldots, e_k$. The second copy of the connected component $e_1, \ldots, e_k$ is obtained by mapping e_1 to $[j,i]$ (when $i < j$), and proceeding in the same manner as above. Note that the two copies of the connected components are *reflections* of each other about the main diagonal of the mesh. ∎

Back to the LCC algorithm: every connected component in G_{LCC} now chooses a unique label as follows: A v-processor $[i,j]$ knows that it is the end of a linear component when it holds a unique 1 in its row or column. Each end $[i,j]$ transmits $\min(i,j)$ to the other end (using the double path mentioned in the preliminaries). Now all v-processors on the linear component choose the minimum number that was transmitted as the component's label. Note that the reflection component about the diagonal will have the same label.

To transmit the information to column 0, we do the following: Every v-processor holding a 1 transmits i (the component label) to the left, provided there is no edge of G_{LCC} to its left - see ⟵ in fig. 2.b. This transition is done by letting every v-processor holding a 0 connect $\{L - R\}$, while v-processors which do have a 1 disconnect their edges ($\emptyset$) and transmit i to the left. Note that between them, the two isomorphic copies of a connected component transmit all necessary information to column 0. The label transmitted to every processor in column 0 is shown in fig. 2.b in parenthesis to its left.

This completes the proof of Lemma 5. ▮

The LCC' problem: Consider now a graph G such that every vertex degree is ≤ 2, but we now allow cycles. The LCC' problem is defined the same as the LCC problem, except that all processors that are on *any* cycle are considered as belonging to the same component. The component label for such processors will be some special value CYCLE. A simple modification of the LCC will solve the LCC' problem: At the stage where each end $[i, j]$ transmits $\min(i, j)$ to the other end, v-processors that are on a cycle will not detect anything on the bus; from this they will conclude that they are on a cycle, and assign themselves the component value CYCLE. For convenience, we henceforth use LCC to refer to LCC'.

4.2 The LRN Simulation Algorithm

If $m \leq 4$, we simulate the n-mesh with one processor. Else, let $p = m/4$. We shall use a p-mesh to traverse the n-mesh. As we have seen, the m-mesh solves the LCC problem on $2p$ vertices in $O(1)$ time, given that the inputs reside in every alternate processor.

We define a mapping of the p-mesh into the m-mesh so that its image can simulate the p-mesh with no slowdown. **Mapping:** Processor $[i, j]$ of the p-mesh is mapped to processor $[4i, 4j]$ of the m-mesh. **Simulation:** Processors of the m-mesh which are not the image of the p-mesh fix their configuration as $\{L - R, U - D\}$. It is straightforward to see that in this way the m-mesh can simulate the p-mesh and can also solve the LCC problem on $2p$ vertices. In the rest of the algorithm description, we use the term p-mesh to refer to the image of the p-mesh in the given m-mesh.

Algorithm LRN SIMULATION (sketch):
The basic idea is to traverse the n-mesh with the p-mesh in snake-like order. At every window position, the following occurs: New bus segments are encountered, old bus segments enter the window, and some old bus segments join up with others (see fig. 1). In addition, some processors may write on a bus. Every new bus segment that is encountered is given some unique *id*, and when bus segments join, the combined segment is given a single *id* - this is where the LCC is used. At the end of the forward traversal of the window, we have all the separate buses, each identified by a unique *id*, and we also have all the necessary broadcast information for each bus. The window is then moved over the n-mesh in the opposite order. At every position, the bus segments in the window are set up, and the broadcast information for each segment is broadcast from one of its endpoints. Note that every bus which is contained entirely inside some window can be handled in a simple manner, so we assume from now on that we are only dealing with buses that cross window boundaries.

According to our previous comments, it suffices to consider the case when condition NBC holds. Note that there may still be cycles, but there will be no speakers on a cycle. Also, any bus cycle eventually results in a cycle in the LCC

Fig. 1. buses A, B and C encountered and joined inside a window.

graph, and the processors on such a cycle will become aware of it by getting the value CYCLE, as explained in the preliminary comment. During the backsweep, such bus cycles will all have the *id* CYCLE, but the broadcast message will be null.

To begin with, we assume that all the bus connections in the n-mesh are stored in the processors of the p-mesh in the folding mapping, as described in Section 2.2. Thus, in moving from one window to the next in the snake-like order, the 2 rows or columns on either side of the boundary are simulated by the same row or column of the window. Information about speakers is also stored in this manner.

In general, every window position is bordered by up to 4 windows, of which some are "old" window positions, and some are "future" window positions. Furthermore, one old window is the immediate predecessor of the current position, and one future window is the immediate successor (called the "next" window). Since the buses are linear, each bus segment may have 2 ends leaving a window to a future window position. The border processor at which a bus end leaves a window retains all relevant information about the bus, including the status of the other end of the bus and the identity of the processor "in charge" of the other end. Whenever 2 or more bus segments merge in a window (due to the LCC operation), they become a single bus segment, with only 2 ends. The processors "in charge" of these ends are informed about the *id* of the single bus segment.

As long as one end of a bus continues from the current window to the next, it retains all relevant information about the bus (including the status of the other end). However, when a bus end does not continue into the next window (e.g., if the bus segment terminates in the window), that same bus may still be encountered in a future window. In this case, a special mechanism (called the *column stack*) is used to convey all necessary information to that bus in the future window position. The precise technical details of all this are too lengthy to be included in this extended abstract - see [1].

Theorem 7. *The algorithm in the proof of Theorem 3 can be modified to handle any order of traversal of the n-mesh by the m-mesh.*

5 Self Simulations - RN model

For the RN model we can show the following result which may not be asymptotically optimal.

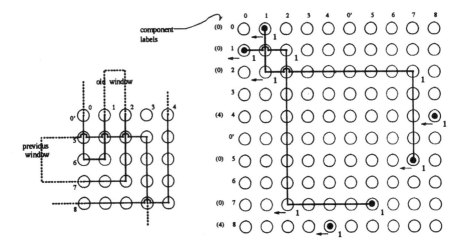

Fig. 2. Example for the LCC process applied to a bus configuration:
a. (left) bus segments in a 5 × 5 window   b. (right) the corresponding LCC graph

Theorem 8. *The simulation of the n-mesh by the p-mesh in the RN model is completed with slowdown $O((n/p)^2 \log n \log(n/p))$. The simulation algorithm uses $O((n/p)^2)$ extra space for temporary storage at each processor of the p-mesh.*

The general method is different from the one used in the LRN model simulation from Section 4. We keep using *windows* for the simulation of *p*-submeshes of the *n*-mesh. Similarly, we keep the method of assigning *id*'s (representatives, labels) to the bus segments that are discovered, from the set of *id*'s of *window boundary* processors. The major difference comes from the way that the book-keeping is handled. The algorithm here is based on iterations, where the basic iteration step is a connected components algorithm, presented in Subsection 5.1. The *i*'th iteration ($i \geq 0$) collects information for bus segments that are contained in *windows* of size $2^i p \times 2^i p$ of the *n*-mesh (A window of size $m \times m$ in the *n* mesh is always one of the $(n/m)^2$ *m*-submeshes that are created by folding the *n*-mesh into an *m*-mesh). The algorithm makes use of LRN simulations as subroutines, relying on the result of Section 4. Thus, for consistency, the folding mapping is also used here for RN simulations of *n*-meshes by a *p*-mesh.

5.1 Connected Components

The algorithm uses the following connected component results, see [1] for detailed proofs.

Lemma 9. *[13, Thm. 4.4] Given the adjacency matrix of an undirected graph with n vertices distributed so that element $[i, j]$ of the matrix is stored in processor $[i, j]$ of the LRN n-mesh, the connected components of the graph can be determined in $O(\log n)$ steps.*

Lemma 10. *Let $G = (V, E)$ be an undirected graph having $\mid V \mid = n$ nodes and $\mid E \mid \leq n/2$ edges. Given the edges of G arbitrarily distributed at the processors of the leftmost column of the LRN n-mesh, such that there is at most one edge at each processor, the connected components of G can be determined in $O(\log n)$ steps.*

The results from Section 4, Lemma 9 and Lemma 10 imply

Corollary 11. *1. Given the adjacency matrix of an undirected graph with n vertices distributed so that element $[i, j]$ of the matrix is stored in processor $[FOLD(i), FOLD(j)]$ of the LRN reconfigurable p-mesh, the connected components of the graph can be determined in $O((n/p)^2 \log n)$ steps.*
2. Let $G = (V, E)$ be an undirected graph having $\mid V \mid = n$ nodes and $\mid E \mid \leq n/2$ edges. Given the edges of G arbitrarily distributed at the processors of the leftmost column of the LRN reconfigurable p-mesh, such that there are at most n/p edges at each processor, the connected components of the graph can be determined in $O((n/p)^2 \log n)$ steps.

5.2 The Algorithm

The self simulation algorithm consists of three phases, sketched informally below.

Algorithm RN SIMULATION (**sketch**):

Phase (1) This phase consists of $\log(n/p)$ iterations, gathering configuration information and constructing a spanning forest for the set of buses. During iteration 0, each window is simulated by the given p-mesh and a *representative* for each bus segment is elected. During iteration $i > 0$ all windows of size $2^i p \times 2^i p$ are considered, deducing information for each of them out of the information of its four composing sub-windows. This is achieved by computing the connected components of a graph representing the bus configuration.

Phase (2) In this phase the data gathered throughout Phase (1) is used to associate a message (or an error indication) with each processor which is a bus representative in some window (of size $p \times p$).

Phase (3) The p-mesh is moved through all windows (of size $p \times p$). Each window is simulated for a single step, in which the representative of each of its buses transmits the appropriate message.

Our goal in the detailed description of the phases and the data management is to show that phase (1) terminates in $O((n/p)^2 \log n \log(n/p))$ steps, phase (2) terminates in $O((n/p)^2 \log(n/p))$ steps, and phase (3) takes only $O((n/p)^2)$ steps. We also consider the space requirements: if we define a single unit of space by the number of bits $(+sizeof(id))$ required by the algorithm at each processor of the original n-mesh, then the simulation requires $O((n/p)^2)$ space units at each processor of the simulating p-mesh, which is asymptotically optimal.

The technical details of how all this is carried cannot be given here for lack of space, see [1].

Acknowledgements: David Peleg made the initial suggestion of using the window method rather than the contraction mapping.

References

1. Y. Ben-Asher, D. Gordon, and A. Schuster. Optimal simulations in reconfigurable arrays. Technical Report TR #716, Department of Computer Science, Technion - IIT, February 1992.

2. Y. Ben-Asher, D. Peleg, R. Ramaswami, and A. Schuster. The power of reconfiguration. *Journal of Parallel and Distributed Computing*, 13(2), October 1991. Special issue on Massively Parallel Computation.

3. Y. Ben-Asher, D. Peleg, and A. Schuster. The complexity of reconfiguring networks models. In *Proc. of the Israel Symposium on the Theory of Computing and Systems*, May 1992.

4. E. Hao, P.D. MacKenzie, and Q.F. Stout. Selection in $O(\log n)$ time on the reconfigurable mesh. In *4th Symp. on Frontiers of Massively parallel computing*, 1992.

5. J. Jang, H. Park, and V.K. Prasanna. An optimal multiplication algorithm on reconfigurable mesh. In *Proc. Symp. on Parallel and Distributed Processing*, pages 381–391, 1992.

6. J. Jang and V.K. Prasanna. A fast sorting algorithm on higher dimensional reconfigurable mesh. In *26th Conf. on Information Sciences and Systems*, 1992.

7. J. Jang and V.K. Prasanna. An optimal sorting algorithm on reconfigurable mesh. In *Proc. Inter. Parallel Processing Symp.*, pages 130–137, March 1992.

8. M. Kaufmann, J.F. Sibeyn, and R. Raman. Randomized routing on meshes with buses. In *proc. 1st European Symp. on Algorithms*, September 1993.

9. F.T. Leighton. *Introduction to parallel algorithms and architectures*. Morgan Kaufmann publishers, 1991.

10. H. Li and M. Maresca. Polymorphic-torus network. *IEEE Transactions on Computing*, 38(9):1345–1351, 1989.

11. M. Maresca and H. Li. Connection autonomy in SIMD computers: A VLSI implementation. *Journal of Parallel and Distributed Computing*, 7(2):302–320, 1989.

12. M. Maresca and H. Li. Virtual parallelism support in reconfigurable processor arrays. unpublished manuscript, 1993.

13. R. Miller, V.K. Prasanna-Kumar, D.I. Reisis, and Q.F. Stout. Parallel computations on reconfigurable meshes. Technical Report TR IRIS#229, Dept. of Computer Science, University of Southern California, 1987. To appear in IEEE Transactions on Computers.

14. M. Nigam and S. Sahni. Sorting n numbers on $n \times n$ reconfigurable meshes with buses. In *Proc. of Intl. Parallel Processing Symposium*, pages 174–181, April 1993.

15. H. Park and V.K. Prasanna. $O(1)$ time optimal algorithms for some arithmetic operations on reconfigurable mesh. Unpublished Manuscript, July 1992.

16. S. Rajasekaran. Mesh connected computers with fixed and reconfigurable buses: packet routing, sorting and selection. In *proc. 1st European Symp. on Algorithms*, September 1993.

17. B. Wang and G. Chen. Constant time algorithms for the transitive closure and some related graph problems on processor arrays with reconfigurable bus systems. *IEEE Transactions on Parallel and Distributed Systems*, 1(4):500–507, 1990.

18. C.C. Weems, R. Deepak, D.B. Shu, and G. Nash. Reconfiguration in the low and intermediate levels of the image understanding architecture. Technical Report COINS TR 90-10, University of Massachusetts at Amherst, February 1990.

Optimal Upward Planarity Testing
of Single-Source Digraphs[*]

(Extended Abstract)

Paola Bertolazzi[†]   *Giuseppe Di Battista*[‡]   *Carlo Mannino*[§]

Roberto Tamassia[¶]

Abstract: A directed graph is upward planar if it has a planar drawing such that all the edges are monotone with respect to the vertical direction. Testing upward planarity and constructing upward planar drawings is important for displaying hierarchical network structures, which frequently arise in software engineering, project management, and visual languages. In this paper we investigate upward planarity testing of single-source digraphs: we provide a new combinatorial characterization of upward planarity, and give an optimal algorithm for upward planarity testing. Our algorithm tests whether a single-source digraph with n vertices is upward planar in $O(n)$ sequential time, and in $O(\log n)$ time on a CRCW PRAM with $n \log \log n / \log n$ processors, using $O(n)$ space. The algorithm also constructs an upward planar drawing if the test is successful. The previous best result is an $O(n^2)$-time algorithm by Hutton and Lubiw. No efficient parallel algorithms for upward planarity testing were previously known.

1 Introduction

Upward planarity of directed graphs is an important issue in the area of graph drawing and has been extensively investigated. A digraph is upward planar if it has a planar upward drawing, i.e. a planar drawing such that all the edges are monotone with respect to the vertical direction. Planarity and acyclicity are necessary but not sufficient conditions for upward planarity.

---

[*]Research supported in part by the National Science Foundation under grant CCR-9007851, by the U.S. Army Research Office under grant DAAL03-91-G-0035, by the NATO Scientific Affairs Division under collaborative research grant 911016, by the Progetto Finalizzato Sistemi Informatici e Calcolo Parallelo of the Italian National Research Council, and by the Esprit BRA of the European Community – ALCOM Contract 7141.

[†]IASI-CNR Viale Manzoni, 30 – 00185 Roma, Italy bertola@iasi.rm.cnr.it

[‡]Dip. di Informatica e Sistemistica Università di Roma "La Sapienza" Via Salaria, 113 – 00198 Roma, Italy dibattista@iasi.rm.cnr.it

[§]Dip. di Statistica Università di Roma "La Sapienza" P.le A. Moro, 5 – 00185 Roma, Italy mannino@iasi.rm.cnr.it

[¶]Dept. of Computer Science Brown University Providence, RI 02912–1910 rt@cs.brown.edu

Testing upward planarity and constructing upward planar drawings is important for displaying hierarchical network structures. Key domains of application include software engineering, project management, and visual languages. Especially significant in a number of applications are single-source digraphs (in the following sT-digraphs), such as subroutine-call graphs, is-a hierarchies (i.e. subset relationships between classes of objects), and organization charts. Also, upward planarity of sT-digraphs has deep combinatorial implications in the theory of ordered sets. Namely, upward planar sT-digraphs have bounded poset dimension, so that their transitive closure can be compactly represented. A survey on algorithms for planarity testing and graph drawing can be found in [25, 10].

Combinatorial results on upward planarity for covering digraphs of lattices were first given in [17, 19]. Further results on the interplay between upward planarity and ordered sets can be found in [28, 23]. Lempel, Even, and Cederbaum [18] relate the planarity of biconnected undirected graphs to the upward planarity of st-digraphs. A combinatorial characterization of upward planar digraphs is provided in [16, 5]. Di Battista, Tamassia, and Tollis [5, 8] give algorithms for constructing upward planar drawings of st-digraphs, and investigate area bounds and symmetry display. Tamassia and Vitter [26] show that the above drawing algorithms can be efficiently parallelized. Upward planar drawings of trees and series-parallel digraphs are studied in [22, 24, 3, 9] and [2], respectively. In [4] it is shown that for the special case of bipartite digraphs, upward planarity is equivalent to planarity. In [1] a polynomial time algorithm is given for testing upward planarity of triconnected digraphs. Concerning sT-digraphs, Thomassen [27] characterizes upward planarity in terms of forbidden circuits. Hutton and Lubiw [14] use Thomassen's result and a decomposition scheme to test upward planarity of an n-vertex sT-digraph in $O(n^2)$ time.

In this paper we investigate upward planarity testing of sT-digraphs. Our main results are summarized as follows: (1) We provide a new combinatorial characterizations of upward planarity within a given embedding in terms of a forest embedded in the face-vertex incidence graph. (2) We reduce the upward planarity testing problem to the one of finding a suitable orientation of a tree. (3) We show that the above combinatorial results yield an optimal $O(n)$-time upward planarity testing algorithm for sT-digraphs. The algorithm also constructs an upward planar drawing if the test is successful. Our algorithm improves over the previous best result [14] by an $O(n)$ factor in the time complexity. (4) We efficiently parallelize the above algorithm to achieve $O(\log n)$ time on a CRCW PRAM with $n \log \log n / \log n$ processors. Hence, we provide the first efficient parallel algorithm for upward planarity testing. Our parallel time complexity is the same as the one of the best parallel algorithm for planarity testing [20, 21]. (5) We give an optimal parallel algorithm to test acyclicity of a planar n-vertex sT-digraph in $O(\log n)$ time with $n / \log n$ processors on an EREW PRAM.

The rest of this paper is organized as follows. Section 2 contains preliminary definitions and results. The problem of testing upward planarity for a fixed embedding is investigated in Section 3. A combinatorial characterization of upward planarity in sT-digraphs is given in Section 4. The complete upward planarity testing algorithm for sT-digraphs is presented in Section 5.

2 Preliminaries

A drawing of a graph maps each vertex to a distinct point of the plane and each edge (u, v) to a simple Jordan curve with endpoints u and v. A drawing is planar if no two edges intersect, except, possibly, at common endpoints. A graph is planar if it has a planar drawing. Two planar drawings of a planar graph G are equivalent if, for each vertex v, they have the same circular clockwise sequence of edges incident on v. Hence, the planar drawings of G are partitioned into equivalence classes. Each such class is called an embedding of G. An embedded planar graph is a planar graph with a prescribed embedding. A triconnected planar graph has a unique embedding, up to a reflection. A planar drawing divides the plane into topologically connected regions delimited by circuits, called faces. The external face is the boundary of the unbounded region. Two drawings with the same embedding have the same faces. Hence, one can speak about the faces of an embedding.

Let G be a digraph (i.e., a directed graph). A *source* (*sink*) of G is a vertex without incoming (outgoing) edges. An *internal vertex* of G has both incoming and outgoing edges. Let f be a face of planar drawing (or embedding) of a digraph. A *source-switch* (*sink-switch*) of f is a source (sink) of f. Note that a *source-switch* (*sink-switch*) is not necessarily a source (sink) of G.

An *upward* drawing of a digraph is such that all the edges are represented by directed curves increasing monotonically in the vertical direction. A digraph has an upward drawing if and only if it is acyclic. A digraph is *upward planar* if it admits a planar upward drawing. Note that a planar acyclic digraph does not necessarily have a planar upward drawing. An upward planar digraph also admits a planar upward straight-line drawing [16, 5]. A planar *st*-digraph is a planar digraph with exactly one source s and one sink t, connected by edge (s, t). A digraph is upward planar if and only if it is a subgraph of a planar *st*-digraph [16, 5]. If a digraph has a single source, then it is upward planar if and only if its biconnected components are upward planar [14].

Now we summarize SPQR-trees (more details in [6, 7]). SPQR-trees are closely related to the classical decomposition of biconnected graphs into triconnected components [12].

Let G be a biconnected graph. A *split pair* of G is either a separation-pair or a pair of adjacent vertices. A *split component* of a split pair $\{u, v\}$ is either an edge (u, v) or a maximal subgraph C of G such that $\{u, v\}$ is not a split pair of C. Let $\{s, t\}$ be a split pair of G. A *maximal split pair* $\{u, v\}$ of G with respect to $\{s, t\}$ is such that for any other split pair $\{u', v'\}$, vertices u, v, and t are in the same split component.

Let $e(s, t)$ be an edge of G, called *reference edge*. The *SPQR-tree* T of G with respect to e describes a recursive decomposition of G induced by its split pairs. Tree T is a rooted ordered tree whose nodes are of four types: S, P, Q, and R. Each node μ of T has an associated biconnected multigraph, called the *skeleton* of μ, and denoted by $skeleton(\mu)$. Also, it is associated with an edge of the skeleton of the parent ν of μ, called the *virtual edge* of μ in $skeleton(\nu)$. Tree T is recursively defined as follows. *Trivial Case:* If G consists of exactly two parallel edges between s and t, then T consists of a single Q-node whose skeleton is G itself. *Parallel Case:* If the split pair $\{s,t\}$ has at least three split components $G_1, \cdots, G_k$ ($k \geq 3$), the root of T is a P-node μ. Graph $skeleton(\mu)$ consists of k parallel edges between s and t, denoted $e_1, \cdots, e_k$, with $e_1 = e$. *Series Case:* Otherwise, the split pair $\{s,t\}$ has exactly two split components, one of them is the reference edge e, and we denote with G' the other split component. If G' has cutvertices $c_1, \cdots, c_{k-1}$ ($k \geq 2$) that partition G into its blocks $G_1, \cdots, G_k$, in this order from s to t, the root of T is an S-node μ. Graph $skeleton(\mu)$ is the cycle $e_0, e_1, \cdots, e_k$, where $e_0 = e$, $c_0 = s$, $c_k = t$, and e_i connects c_{i-1} with c_i ($i = 1 \cdots k$). *Rigid Case:* If none of the above cases applies, let $\{s_1, t_1\}, \cdots, \{s_k, t_k\}$ be the maximal split pairs of G with respect to $\{s,t\}$ ($k \geq 1$), and for $i = 1, \cdots, k$, let G_i be the union of all the split components of $\{s_i, t_i\}$ but the one containing the reference edge e. The root of T is an R-node μ. Graph $skeleton(\mu)$ is obtained from G by replacing each subgraph G_i with the edge e_i between s_i and t_i.

Except for the trivial case, μ has children $\mu_1, \cdots, \mu_k$ in this order, such that μ_i is the root of the SPQR-tree of graph $G_i \cup e_i$ with respect to reference edge e_i ($i = 1, \cdots, k$). The tree so obtained has a Q-node associated with each edge of G, except the reference edge e. We complete the SPQR-tree by adding another Q-node, representing the reference edge e, and making it the parent of μ so that it becomes the root. Observe that we are defining SPQR-trees of graphs, however the same definition can be applied to digraphs. See an example in Fig. 1.

Edge e_i is the *virtual edge* of node μ_i in $skeleton(\mu)$ and of node μ in $skeleton(\mu_i)$. Graph G_i is called the *pertinent graph* of node μ_i, and of edge e_i.

Let μ be a node of T. We have: (1) if μ is an R-node, then $skeleton(\mu)$ is a triconnected graph; (2) if μ is an S-node, then $skeleton(\mu)$ is a cycle; (3) if μ is a P-node, then $skeleton(\mu)$ is a triconnected multigraph consisting of a bundle of multiple edges; (4) if μ is a Q-node, then $skeleton(\mu)$ is a biconnected multigraph consisting of two multiple edges.

The skeletons of the nodes of T are homeomorphic to subgraphs of G. The SPQR-trees of G with respect to different reference edges are isomorphic and are obtained one from the other by selecting a different Q-node as the root. Hence, we can define the *unrooted SPQR-tree* of G without ambiguity.

The SPQR-tree T of a graph G with n vertices and m edges has m Q-nodes and $O(n)$ S-, P-, and R-nodes. Also, the total number of vertices of the skeletons stored at the nodes of T is $O(n)$.

A graph G is planar if and only if the skeletons of all the nodes of the SPQR-tree T of G are planar. An SPQR-tree T rooted at a given Q-node represents all the planar drawing of G having the reference edge (associated to the Q-node at the root) on the external face (see Fig. 1). Namely, such drawings can be constructed by the following recursive procedure: (1) Construct a drawing of the skeleton of the root ρ with the reference edge of the parent of ρ on the external face. (2) For each child μ of ρ: let e be the virtual edge of μ in $skeleton(\rho)$, and let H be the pertinent graph of μ plus edge e; recursively draw H with the reference edge e on the external face; in $skeleton(\rho)$, replace virtual edge e with the the above drawing of H minus edge e.

3 Embedded Digraphs

In this section we give a new combinatorial characterization of upward planarity for planar sT-digraphs with a given embedding. This characterization yields an optimal algorithm for testing whether an embedded planar sT-digraph has an upward planar drawing that preserves the embedding.

Let Γ be a planar upward drawing of an sT-digraph G. We say that a sink t of G is *assigned* to a face f of Γ if the region of the plane bounded by face f in Γ is to the left of the leftmost incoming edge of t and to the right of the rightmost incoming edge of t. Informally, t is assigned to f if it "penetrates" into f.

Since G has a unique source, the following properties can be easily derived:

Fact 1 *The source s of G is the bottommost vertex of Γ.*

Fact 2 *For the external face h of Γ, all the sink-switches are sinks of G and are assigned to h.*

Fact 3 *For each internal face f, at most one sink-switch (the topmost vertex of f in Γ) is not a sink of G and all but one sink switches are assigned to f.*

We shall also use the following result about cycles in planar sT-digraphs:

Lemma 1 *Let G and G' be planar sT-digraphs such that G' is obtained from G by means of one of the following operations: (1) adding a new vertex v and a new edge (u, v) or (v, u), connecting v to a vertex u of G; (2) adding a directed edge between the source and a sink on the same face in some embedding of G; (3) adding a directed edge between two sink-switches that are on the same face in some embedding of G. Then G is acyclic if and only if G' is acyclic.*

Sketch of Proof: The acyclicity is trivially preserved by the first two operations. Regarding the third operation, consider an embedding of G with the source s on the external face, and assume, for a contradiction, that G is acyclic and adding the edge (t', t'') between sink-switches t' and t'' of face f causes the resulting graph G' to have a cycle γ. Cycle γ must consist of edge (t', t'') and a directed path π' in G from t'' to t'. Let v be the neighbor of t'' in f inside γ, and let π'' be a directed path from the source s of G to v. Since s is external to cycle γ, path π'' must have at least a vertex in common with path π'. Let u be the last vertex of π'' that is also on π'. We have that G has a cycle consisting of edge (v, t''), the subpath of π' from t'' to u, and the subpath of π'' from u to v, a contradiction. □

Given an embedded planar sT-digraph G, the *face-sink graph* F of G is the incidence graph of the faces and the sink-switches of G (see Figs. 2.a). Namely: (1) The vertices of F are the faces and the sink-switches of G. (2) Graph F has an edge (f, v) if v is a sink-switch on face f.

Theorem 1 *Let G be an embedded planar sT-digraph and h a face of G. Digraph G has an upward planar drawing that preserves the embedding and with external face h if and only if all the following conditions are satisfied:*

1. the face-sink graph F is a forest;

2. *a tree T of F has no internal vertices of G, while the remaining trees have exactly one internal vertex;*

3. *h is in tree T; and*

4. *the source of G is in the boundary of h.*

Also, if G has such a drawing, then an embedded planar st-digraph G′ containing G as an embedded subgraph is obtained as follows:

1. *root tree T at h and each remaining tree of F at its (unique) internal vertex;*

2. *orient F by directing edges towards the roots;*

3. *prune the leaves from every tree of F; and*

4. *add the resulting forest $\hat{F}$ and the edge (s, h) to G.*

Sketch of Proof: *Only If.* Let Γ be any planar upward drawing of G that preserves the original planar embedding and has external face h. By Fact 1, Condition 4 is verified. Orient the face-sink graph F of G by directing edge (v, f) from v to f if v is a sink assigned to face f, and from f to v otherwise. By Facts 2–3, each vertex of F has at most one outgoing edge. Specifically, each internal face and each sink has exactly one outgoing edge, while each internal vertex and the external face have no outgoing edges. Now, label the vertices of F as follows: the label of a sink-switch is its y-coordinate in Γ; the label of an internal face f is $y(v) - \epsilon$, where v is the sink-switch not assigned to f, and ϵ is a suitably small positive real value; and the label of the external face h is $+\infty$. Since Γ is an upward drawing, the edges of F are directed by increasing labels. We conclude that F is a forest of sink-trees. One tree is rooted at h, while the other trees are rooted at internal vertices. This shows Conditions 1–2. Condition 3 follows from Fact 1. Additional geometric considerations show that adding $\hat{F}$ to G yields a planar st-digraph.

If. We show that, if F satisfies the conditions of the theorem, then G is a subgraph of a planar st-digraph $G′$, which is obtained as the union of G and $\hat{F}$. This implies that G is upward planar. Planarity is preserved since a star is inserted in each face. Also, $G′$ has exactly one source (s) and one sink (h) connected by a directed edge. It remains to be shown that $G′$ is acyclic. By the construction of $G′$ and Lemma 1, we have that $G′$ is acyclic if and only if G is acyclic. Assume, for a contradiction, that G is not acyclic. Let γ be a cycle of G that does not enclose any other cycle. Note that the source s must be outside γ. If γ is a face of G, then F has an isolated vertex associated with face γ, a contradiction. Otherwise (γ is not a face of G), the subgraph $\hat{F}′$ of $\hat{F}$ enclosed by γ consists of a forest of trees, each with exactly one internal vertex. Let H be the digraph obtained from the subgraph of G enclosed by γ by removing the edges of γ, and adding a new vertex $s′$ together with edges from $s′$ to all the vertices of γ. By our choice of cycle γ, H is a planar sT-digraph. Adding $\hat{F}′$ to H yields a planar sT-digraph without sinks, and hence a digraph with cycles. By Lemma 1, H must also have cycles, again a contradiction. $\square$

Fig. 2 illustrates Theorem 1. The following algorithm tests whether an embedded planar sT-digraph G is upward planar and reports all the faces of G that can be external in an upward planar drawing of G with the given embedding.

Algorithm *Embedded-Test:* (1) Construct the face-sink graph F of G. (2) Check Conditions 1 and 2 of Theorem 1. If these conditions are not verified, then return "not-upward-planar" and stop. (3) Report the set of faces of G that contain vertex s in their boundary and are associated with nodes of tree T. If such set of faces is empty, then return "not-upward-planar" else return "upward-planar".

For the example of Fig. 2, Algorithm *Embedded-Test* returns "upward-planar" and reports two faces.

Theorem 2 *Let G be an embedded planar sT-digraph with n vertices. Algorithm* Embedded-Test *determines whether G has an upward planar drawing that preserves the embedding, and reports all the admissible external faces. It runs in $O(n)$ sequential time and in $O(\log n)$ time on a CRCW PRAM with $n \cdot \alpha(n) / \log n$ processors, using $O(n)$ space.*

4   Upward Planarity and SPQR-Trees

Let G be a biconnected sT-digraph. In this section we give a combinatorial characterization of the upward planarity of G using SPQR-trees.

A digraph is said to be *expanded* if every internal vertex has exactly one incoming edge or one outgoing edge. The *expansion* of a digraph is obtained by replacing each internal vertex v with two new vertices v_1 and v_2, which inherit the incoming and outgoing edges of v, respectively, and the edge (v_1, v_2). Clearly, a digraph is acyclic if and only if its expansion is acyclic.

Lemma 2 *A digraph is upward planar if and only if its expansion is upward planar.*

We call *peak* a digraph consisting of two directed edges (a, t) and (b, t). Let G be a planar sT-digraph, and T the corresponding SPQR-unrooted tree. The *directed skeleton* of a node μ of T is the digraph obtained from the skeleton of μ by replacing each virtual edge $e = (u, v)$ with a *directed virtual edge*, which is a directed edge or a peak. Let K be the pertinent digraph of e, and let $H = G - K$, and $K^0 = K - \{u, v\}$. We distinguish the following cases:

1. u and v are sources of K: e is replaced with a peak.

2. u is a source of K, and v is a sink of K:

   (a) $s \notin K^0$: e is replaced with a directed edge (u, v).

   (b) $s \in K^0$: e is replaced with a peak.

3. u is a source of K and v is an internal vertex of K:

   (a) v is a source of H and $s \in H$: e is replaced with a directed edge (u, v).

   (b) Otherwise: e is replaced with a peak.

4. u and v are not sources of K:

   (a) u is a source of H: e is replaced with a directed edge (u, v).

   (b) u is not a source of H: e is replaced with a directed edge (v, u).

Examples of directed skeletons are shown in Fig. 3.

Given planar biconnected expanded sT-digraphs G and H, we say that H is a *minor* of G if it can be obtained by a sequence of vertex deletions, edge deletions, and edge contractions, such that all the intermediate digraphs are planar expanded biconnected sT-digraphs.

Lemma 3 *Let G be a planar expanded sT-digraph, and H a minor of G. If G has a planar upward drawing, then H has a planar upward drawing.*

Lemma 4 *The directed skeletons of the nodes of T are minors of G.*

The main result of this section is summarized in the following theorem:

Theorem 3 *A biconnected acyclic sT-digraph G is upward planar if and only if there is a rooting of the SPQR-tree T of the expansion of G at a reference edge containing the source such that the directed skeleton of each node μ of T has a planar upward drawing with reference directed virtual edge on the external face.*

Sketch of Proof: *Only If.* It follows from Lemmas 2–4.

If. We use two levels of induction: on the number of nodes of T, and on the number of children of each node of T. At the first level of induction we show that for each node μ of T, the digraph obtained by replacing the directed virtual edges in the directed skeleton of (μ) associated with the children of μ with the corresponding pertinent digraphs is a biconnected upward planar expanded sT-digraph. The inductive step shows that, if this property holds for the children of μ, it also holds for μ. At the second level of induction, we perform the above replacements one by one, and show that each intermediate digraph is a biconnected upward planar expanded sT-digraph. We combine in a bottom-up fashion the embeddings of the directed skeletons and the corresponding face-sink forests (see Theorem 1). We show that combining face-sink forests results in a sequence of "prune" and "graft" operations . As a side effect, this construction also yields a planar st-digraph that includes G, and hence a planar upward drawing of G.
□

As an example of Theorem 3, consider the digraph of Fig. 1 and Fig. 3. Such a digraph does not have an upward drawing. In fact, any choice of reference edge consistent with Theorem 3 causes a rooting of T in which the directed virtual edge e of the directed skeleton of μ_2 is forced to be on the external face and no upward drawing of the directed skeleton of μ_2 exists with such property.

5 Algorithm for General Single-Source Digraphs

Let G be a biconnected sT-digraph. In this section we present an algorithm for testing whether G is upward planar.

Algorithm *Test:* (1) Construct the expansion G' of G. (2) Test whether G' is planar. If G' is not planar, then return "not-upward-planar" and stop, else, construct an embedding for G'. (3) Test whether G' is acyclic. If G' is not acyclic, then return "not-upward-planar" and stop. (4) Construct the SPQR-tree

T of G' and the skeletons of its nodes. (5) For each virtual edge e of a skeleton, classify each endpoint of e as a source, sink, or internal vertex in the pertinent digraph of e. Also, determine if the pertinent digraph of e contains the source. (6) For each node μ of T, compute the directed skeleton of μ. (7) For each R-node μ of T: (7.a) Test whether the directed skeleton of μ is upward planar by means of algorithm *Embedded-Test*. If *Embedded-Test* returns "not-upward-planar", then return "not-upward-planar" and stop. (7.b) Mark the virtual edges of the skeleton of μ whose endpoints are on the external face in some upward drawing of the directed skeleton of μ. (7.c) For each unmarked virtual edge e of the skeleton of μ, constrain the tree edge associated with e to be directed towards μ. (7.d) If the source is not in *skeleton*(μ), let ν be the node neighbor of μ whose pertinent digraph contains the source, and constrain the tree edge (μ, ν) to be directed towards ν. (8) Determine whether T can be rooted at a Q-node in such a way that orienting edges from children to parents satisfies the constraints of Steps 7.c–7.d. If such a rooting exists then return "upward-planar", else return "not-upward-planar".

For sT-digraphs that are not biconnected we apply the above algorithm to each biconnected component.

Theorem 4 *Upward planarity testing of a sT-digraph with n vertices can be done in $O(n)$ time using $O(n)$ space.*

Sketch of Proof: All the steps take $O(n)$ time. Steps 1 and 3 can be trivially performed in $O(n)$ time. Planarity testing in Step 2 can also be done in $O(n)$ time [13]. The construction of the SPQR-tree and the skeletons of its nodes (Step 4) takes time $O(n)$ using a variation of the algorithm of [12]. The preprocessing of Step 5 consists essentially of a visit of T, and can be done in $O(n)$ time. Let n_μ be the number of vertices of the skeleton of μ. The information collected in Step 5 allows to perform Step 6 in $O(n)$ time and Step 7.d in $O(n_\mu)$ time. By Theorem 2, Step 7.a takes $O(n_\mu)$ time. The output of Step 7.a allows to perform Steps 7.b and 7.c in $O(n_\mu)$ time. Since $\sum_\mu n_\mu = O(n)$, the total complexity of Step 7 is $O(n)$. Finally, Step 8 consists of a visit of T and takes $O(n)$ time. $\square$

To parallelize Algorithm *Test*, we need an efficient way of testing in parallel whether a planar sT-digraph with n vertices is acyclic. We can use the algorithm of [15], which runs in $O(\log^3 n)$ time on a CRCW PRAM with n processors. However, the particular structure of planar sT-digraphs allows us to perform this test optimally. The following characterization is inspired by some ideas in [15].

Let G be an embedded expanded planar sT-digraph. The *clockwise subgraph* of G is obtained by taking the first incoming edge of each internal vertex, in the clockwise order. The *counterclockwise subgraph* of G is similarly obtained by taking the first incoming edge of each internal vertex, in the counterclockwise order. Such subgraphs of G have all vertices with indegree 1 or 0.

Theorem 5 *An embedded expanded sT-digraph G is acyclic if and only if both the clockwise and counterclockwise subgraphs of G are acyclic.*

Sketch of Proof: The only-if part is trivial. For the if part, assume for a contradiction that G is not acyclic, and consider an arbitrary drawing of G with the prescribed embedding and with the source on the external face. We will show the existence of a cycle in either the clockwise or counterclockwise subgraph. Let γ be a cycle of G that does not enclose any other cycle. Since the source of G must be outside γ, all the edges incident on vertices of γ and inside γ must be outgoing edges. Hence, γ is contained in the clockwise or counterclockwise subgraph, depending on whether it is a clockwise or counterclockwise cycle. $\square$

The structure of each connected component of the clockwise (counterclockwise) subgraphs is either a source-tree, or a set of source trees with their roots connected in a directed cycle; so, we can test if such subgraphs are acyclic by standard parallel techniques. Since expansion preserves acyclicity, we have:

Theorem 6 *Given an embedded planar sT-digraph G with n vertices, one can test if G is acyclic in $O(\log n)$ time with $n/\log n$ processors on an EREW PRAM.*

Finally, by applying the result of Theorem 6 and various parallel techniques (in particular [20, 21, 11]) we can efficiently parallelize Algorithm *Test*:

Theorem 7 *Upward planarity testing of a sT-digraph with n vertices can be done in in $O(\log n)$ time on a CRCW PRAM with $n \log \log n/\log n$ processors, using $O(n)$ space.*

Theorem 8 *Algorithm Test can be extended so that it constructs a planar upward drawing such that the edges are represented with polygonal lines if the digraph is upward planar. The complexity bounds stay unchanged.*

References

[1] P. Bertolazzi and G. Di Battista, "On Upward Drawing Testing of Triconnected Digraphs," Proc. ACM Symp. on Computational Geometry, pp. 272-280, 1991.

[2] P. Bertolazzi, R.F. Cohen, G. Di Battista, R. Tamassia, and I.G. Tollis, "How to Draw a Series-Parallel Digraph," Proc. SWAT, Lecture Notes in Computer Science, vol. 621, pp. 272-283, 1992.

[3] P. Crescenzi, G. Di Battista, and A. Piperno, "A Note on Optimal Area Algorithms for Upward Drawings of Binary Trees," Technical Report RAP.11.91, Dip. di Informatica e Sistemistica, Università degli Studi di Roma La Sapienza, 1991.

[4] G. Di Battista, W.-P. Liu, and I. Rival, "Bipartite Graphs, Upward Drawings, and Planarity," Information Processing Letters, vol. 36, pp. 317-322, 1990.

[5] G. Di Battista and R. Tamassia, "Algorithms for Plane Representations of Acyclic Digraphs," Theoretical Computer Science, vol. 61, pp. 175-198, 1988.

[6] G. Di Battista and R. Tamassia, "On-Line Graph Algorithms with SPQR-Trees," Automata, Languages and Programming (Proc. 17th ICALP), Lecture Notes in Computer Science, vol. 442, pp. 598-611, 1990.

[7] G. Di Battista and R. Tamassia, "On-Line Maintenance of Triconnected Components with SPOR-Trees," Technical Report CS-92-40, Dept. of Computer Science, Brown Univ., 1992.

[8] G. Di Battista, R. Tamassia, and I.G. Tollis, "Area Requirement and Symmetry Display of Planar Upward Drawings," Discrete and Computational Geometry, vol. 7, no. 4, pp. 381-401, 1992.

[9] P. Eades, T. Lin, and X. Lin, "Two Tree Drawing Conventions," Technical Report No. 174, Key Centre for Software Technology, Department of Computer Science, The Univ. of Queensland, 1990.

[10] P. Eades and R. Tamassia, "Algorithms for Automatic Graph Drawing: An Annotated Bibliography," Technical Report CS-89-09, Dept. of Computer Science, Brown Univ., 1989.

[11] D. Fussell, V. Ramachandran, and R. Thurimella, "Finding Triconnected Components by Local Replacements," Automata, Languages and Programming (Proc. 16th ICALP), Lecture Notes in Computer Science, vol. 372, pp. 379-393, 1989.

[12] J. Hopcroft and R.E. Tarjan, "Dividing a Graph into Triconnected Components," SIAM J. Computing, vol. 2, no. 3, pp. 135-158, 1973.

[13] J. Hopcroft and R.E. Tarjan, "Efficient Planarity Testing," J. ACM, vol. 21, no. 4, pp. 549-568, 1974.

[14] M.D. Hutton and A. Lubiw, "Upward Planar Drawing of Single Source Acyclic Digraphs," Proc. ACM-SIAM Symp. on Discrete Algorithms, pp. 203-211, 1991.

[15] M.-Y. Kao and G.E. Shannon, "Local Reorientations, Global Order, and Planar Topology," Proc. 30th IEEE Symp. on Foundations of Computer Science, pp. 286-296, 1989

[16] D. Kelly, "Fundamentals of Planar Ordered Sets," Discrete Mathematics, vol. 63, pp. 197-216, 1987.

[17] D. Kelly and I. Rival, "Planar Lattices," Canadian J. Mathematics, vol. 27, no. 3, pp. 636-665, 1975.

[18] A. Lempel, S. Even, and I. Cederbaum, "An Algorithm for Planarity Testing of Graphs," in Theory of Graphs, Int. Symposium (Rome, 1966), P. Rosenstiehl, ed., pp. 215-232, Gordon and Breach, New York, 1967.

[19] C. Platt, "Planar Lattices and Planar Graphs," J. Combinatorial Theory, Series B, vol. 21, pp. 30-39, 1976.

[20] V. Ramachandran and J.H. Reif, "An Optimal Parallel Algorithm for Graph Planarity," Proc. IEEE Symp. on Foundations of Computer Science, 1989.

[21] V. Ramachandran and J.H. Reif, "Planarity Testing in Parallel," Technical Report TR-90-15, Dept. of Computer Sciences, Univ. of Texas at Austin, 1990.

[22] E. Reingold and J. Tilford, "Tidier Drawing of Trees," IEEE Trans. on Software Engineering, vol. SE-7, no. 2, pp. 223- 228, 1981.

[23] I. Rival, "Graphical Data Structures for Ordered Sets," in Algorithms and Order, ed. I. Rival, pp. 3-31, Kluwer Academic Publishers, 1989.

[24] K.J. Supowit and E.M. Reingold, "The Complexity of Drawing Trees Nicely," Acta Informatica, vol. 18, pp. 377-392, 1983.

[25] R. Tamassia, G. Di Battista, and C. Batini, "Automatic Graph Drawing and Readability of Diagrams," IEEE Transactions on Systems, Man and Cybernetics, vol. SMC-18, no. 1, pp. 61-79, 1988.

[26] R. Tamassia and J.S. Vitter, "Parallel Transitive Closure and Point Location in Planar Structures," SIAM J. Computing, vol. 20, no. 4, pp. 708-725, 1991.

[27] C. Thomassen, "Planar Acyclic Oriented Graphs," Order, vol. 5, no. 4, 1989.

[28] W.T. Trotter and J. Moore, "The Dimension of Planar Posets," J. Combinatorial Theory, Series B, vol. 22, pp. 54-67, 1977.

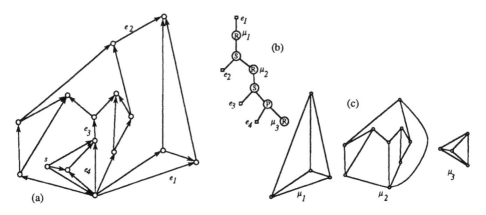

Figure 1: (a) A planar biconnected digraph G. (b) SPQR-tree of G. (c) Skeletons of the R-nodes.

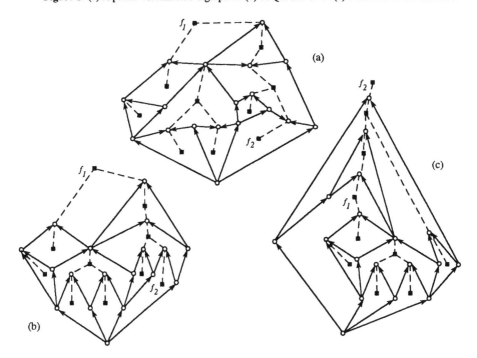

Figure 2: (a) An embedded planar single-source digraph G and its face-sink graph. (b-c) Upward drawings of G that preserve the embedding with different external faces.

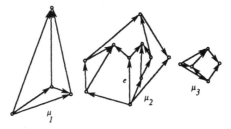

Figure 3: The directed skeletons of the R-nodes of the digraph of Fig. 1.

On Bufferless Routing of Variable-Length Messages in Leveled Networks

(Extended Abstract)

S. N. Bhatt[1], G. Bilardi[2], G. Pucci[2,3],
A. Ranade[4], A. L. Rosenberg[5], E. J. Schwabe[6]

[1] Bell Communications Research, Morristown NJ, USA
[2] Dip. di Elettronica e Informatica, Università di Padova, Padova, ITALY
[3] International Computer Science Institute, Berkeley CA, USA
[4] Computer Science Division, University of California Berkeley CA, USA
[5] Dept. of Computer Science, University of Massachusetts, Amherst MA, USA
[6] Dept. of EECS, Northwestern University, Evanston IL, USA

Abstract. We study the most general communication paradigm on a multiprocessor, wherein each processor has a distinct message (of possibly distinct lengths) for each other processor. We study this paradigm, which we call *chatting*, on multiprocessors that do not allow messages once dispatched ever to be delayed on their routes. By insisting on *oblivious* routes for messages, we convert the communication problem to a pure scheduling problem. We introduce the notion of a *virtual chatting schedule*, and we show how efficient chatting schedules can often be produced from efficient virtual chatting schedules. We present a number of strategies for producing efficient virtual chatting schedules on a variety of network topologies.

1 Introduction

We study a paradigm of interprocessor communication within a network of processors, called *all-to-all personalized communication*, or, for short, *chatting*. This paradigm is characterized by the network's Processing Elements (PEs) sending (possibly) distinct messages of (possibly) distinct lengths to any subset of the other PEs. This modality of communication is computationally more demanding than its more commonly studied relatives, where the communication pattern and/or the message length are typically restricted.

We study how to obtain time-efficient schedules for the transmission of messages in a chatting operation within the context of the following conventions.

Communication in rounds. Our processor networks operate in alternating phases of local (or, intra-PE) computation and global (or, inter-PE) communication.

Message integrity. Each message in a chatting operation travels through the network as a contiguous stream of atomic units, called *flits*. This convention, which minimizes the amount of addressing information that must travel throughout the network along with the messages, is also a characteristic of *wormhole* routing [4, 7, 12].

Oblivious Routing. Each message in a chatting operation travels through the network along a fixed path determined solely by the source and destination PEs of the message. This convention, which contrasts with typical studies of packet routing (see, e.g., [15]) and wormhole routing, reduces our message routing problems to pure scheduling problems.

Bufferless communication. The PEs of our networks have neither buffers nor queues to store messages being relayed through them; hence messages, once dispatched, are never delayed en route. This convention is shared with the regimen of *hot potato* routing [6, 11]; it appears also in the study of *scattering* and *gathering* general messages in general networks, in [2].

The confluence of all of these conventions — in particular, our focus on communication in rounds, on oblivious routing of messages, and on general patterns of passing arbitrarily many messages of arbitrary lengths — makes the object of our study rather different from other studies of communication in processor networks.

Section 2 introduces the formal setting for our study. We develop there the notion of a chatting schedule for a set of messages and define its duration — the quantity we strive to minimize. In Section 3, we introduce the notion of a virtual chatting schedule and show that, for the class of "forward-leveled" networks, one can convert an efficient virtual chatting schedule (which are often easier to develop than are efficient chatting schedules) into an almost equally efficient chatting schedule. The remaining sections build on the transformation technique of Section 3; they focus on designing efficient virtual chatting schedules which will later be converted to efficient "ordinary" chatting schedules with almost equal efficiency. Section 4 is devoted to showing how to use known solutions to the so-called rectangle compaction problem to create virtual chatting schedules for linear array networks, that are within a small constant factor of optimal. In Section 5, we study how to devise efficient virtual chatting schedules that are based on the lengths of the messages, and apply this strategy to two-dimensional meshes. Section 6 explores an approach to message scheduling based on decomposing the host network by suitable cuts. This approach is then shown to yield efficient chatting schedules for tree networks. Due to space limitations, the proofs of the theorems are omitted from this extended abstract. They can be found in [1].

2 The Problem Formalized

Processor Networks. As is customary, we identify a processor network with a directed graph $G = (V, E)$ whose nodes V represent the network's PEs and whose arcs E represent its communication links. We also associate with each processor network G an *atlas* that designates a path $\rho(u, v)$ from every node $u \in V$ to every node $v \in V$ that is reachable from u; any message from node u to node v is routed along path $\rho(u, v)$. As a result, all message passing is *oblivious* in the sense that the route of a message is determined entirely by the message's source and destination nodes.

Our processor networks operate in a *pulsed* fashion: a network alternates *computation phases*, in which (in parallel) each PE performs some computation, and *communication phases*, in which messages flow among the PEs. The networks operate synchronously, at least during the communication phases. The PEs in a processor network $G = (V, E)$ *do not have message buffers*. In particular, this means that if, at time t, a PE $u \in V$ receives a flit that is not destined for it, then u sends that flit out toward its destination at time $t + 1$.

The Chatting Problem. A *message* is a sequence $M = \langle m_0, m_1, \ldots, m_{\ell-1} \rangle$ of flits; we call ℓ the *length* of message M. An *addressed message in processor network* $G = (V, E)$ is a message M, together with a designated source PE $u \in V$ and a designated destination PE $v \in V$ (hence with a designated path $\rho(u, v)$ in G). A *chatting problem for processor network* $G = (V, E)$ is a set of addressed messages in G. This paper is devoted to studying algorithms that solve a chatting problem by scheduling the transmission of all messages so that they get from their designated sources to their designated destinations honoring the oblivious, bufferless communication regimen we are studying. We also insist that messages travel as *indivisible units*, i.e., that at any moment, all flits of an addressed message occupy a contiguous path of arcs in G, with one message-flit per path-arc.

Chatting Schedules. A *chatting schedule* for a chatting problem $\mathcal{M}$ on a processor network G is a function τ that associates an integer (time) with each flit m of an addressed message $M \in \mathcal{M}$ and each arc e on the designated path from the source to the destination of M. The interpretation is that flit m traverses arc e at step $\tau(m, e)$ of the communication phase. Our assumptions about bufferless transmission of indivisible messages translate into the following constraints:

- *Conflict-free Arcs.* At most one flit traverses any given arc at any given time:

$$[\tau(m, e) = \tau(m', e)] \Rightarrow [m = m']. \tag{1}$$

- *Bufferless nodes.* Once transmitted from its source PE, a flit m moves "forward" one arc at every step: if $\langle e_0, e_1, \ldots, e_{d-1} \rangle$ is the designated path for the message that m belongs to, then, for $k = 0, 1, \ldots, d-1$,

$$\tau(m, e_k) = \tau(m, e_0) + k. \tag{2}$$

- *Message integrity.* Flits of the same message $\langle m_0, m_1, \ldots, m_{\ell-1} \rangle$ traverse a given arc e at consecutive times: for $h = 0, 1, \ldots, \ell - 1$,

$$\tau(m_h, e) = \tau(m_0, e) + h. \tag{3}$$

We call a chatting schedule that honors these restrictions *admissible.*

The *duration* $T(\tau)$ of a chatting schedule τ for a chatting problem $\mathcal{M}$ is the amount of time it takes to deliver all messages in $\mathcal{M}$; formally:

$$T(\tau) = \max_{m,e}\{\tau(m, e)\} - \min_{m,e}\{\tau(m, e)\} + 1, \tag{4}$$

where the minimization and maximization are over all flits m in $\mathcal{M}$ and all arcs e that occur in the designated paths for the addressed messages in $\mathcal{M}$. An objective is to devise techniques for constructing admissible chatting schedules with (close to) minimum duration.

Certain simple parameters of a chatting problem $\mathcal{M}$ yield straightforward, but useful, lower bounds on the duration $T(\tau)$ of any chatting schedule τ for $\mathcal{M}$. Define:

$\ell(M)$	length of message $M \in \mathcal{M}$
$d(M)$	length of the routing path of $M \in \mathcal{M}$
$C(e)$	number of flits in $\mathcal{M}$ traversing arc $e \in E$
$L(\mathcal{M}) = \max\{\ell(M) : M \in \mathcal{M}\}$	maximum message length in $\mathcal{M}$
$D(\mathcal{M}) = \max\{d(M) : M \in \mathcal{M}\}$	maximum routing-path length for any $M \in \mathcal{M}$
$Q(\mathcal{M}) = \max\{\ell(M) + d(M) - 1 : M \in \mathcal{M}\}$	longest transit time for any $M \in \mathcal{M}$
$C(\mathcal{M}) = \max\{C(e) : e \in E\}$	maximum arc congestion for problem $\mathcal{M}$

Proposition 1. *For any chatting schedule τ for a chatting problem $\mathcal{M}$,*

$$T(\tau) \geq \max(C(\mathcal{M}), Q(\mathcal{M})).$$

Unidirectional Linear Arrays: A Running Example. To illustrate the various notions we have introduced, it is helpful to refer repeatedly to a simple specific network. We consider the *unidirectional linear array* (ULA), whose node set is $V = \{0, 1, \ldots, N - 1\}$ and whose arc set is $E = \{(0, 1), (1, 2), \ldots, (N - 2, N - 1)\}$.

A chatting schedule τ for a ULA admits the following convenient geometric representation, that originates in [2]. Consider the positive quadrant of the integer plane as the processor-time plane for a ULA: abscissa x corresponds to the ULA-arc $\eta_x \overset{\text{def}}{=} (x, x + 1)$, while ordinate t corresponds to discrete time-step t. Given a chatting schedule τ for a chatting problem $\mathcal{M}$, say that integer point (x, t) in the plane is *marked* if, for some flit m within $\mathcal{M}$, $\tau(m, \eta_x) = t$; point (x, t) is *unmarked* otherwise. Because of Conditions (2) and (3) on τ, the marked region corresponding to any single message $M \in \mathcal{M}$ is a *parallelogram* with two sides parallel to the vertical (time) axis and the other two parallel to the main bisector (see Figure 1).

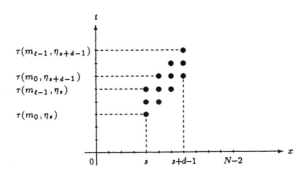

Fig. 1. Graphical representation of a chatting schedule τ for one message.

Condition (1) guarantees — in fact, is equivalent to — the condition that parallelograms corresponding to distinct messages do not overlap. This geometric interpretation of message transmission in a ULA reduces the problem of finding a minimum-duration chatting schedule for a chatting problem on a ULA to the following geometric problem.

Parallelogram Compaction.

Given: A set of parallelograms in the integer plane (with sides as described above), each having a fixed projection on the horizontal (ULA) axis but being free to slide up or down in the vertical (time) direction,

Find: A pairwise nonoverlapping placement of the parallelograms that minimizes the difference between the maximum and minimum ordinates of any of their points.

3   Virtual Chatting Schedules

In this section, we introduce *virtual* chatting schedules, and we study their relation with "ordinary" chatting schedules. We show, by example, that devising an efficient virtual chatting schedule for a problem is sometimes easier than devising an efficient chatting schedule. Moreover, we specify a class of networks for which one can convert an efficient virtual chatting schedule into an efficient chatting schedule.

The Notion of Virtual Schedule. A *virtual chatting schedule* for a chatting problem $\mathcal{M}$ on a processor network G is a function σ that associates an integer (time) with each flit m of an addressed message $M \in \mathcal{M}$ and each arc e on the designated path from the source to the destination of M. Every virtual chatting schedule σ must satisfy certain constraints; however, in this case the constraints do not necessarily correspond to either architectural or algorithmic features.

– *Arc conflict-free.* At most one flit traverses a given arc at any given time:

$$[\sigma(m, e) = \sigma(m', e)] \Rightarrow [m = m']. \tag{5}$$

– *Message "transmission" and integrity.* Every flit of a message $M = \langle m_0, m_1, \ldots, m_{\ell-1} \rangle$ "traverses" all the arcs in M's designated path $\langle e_0, e_1, \ldots, e_{d-1} \rangle$ at the same time. Formally, for all $h = 1, 2, \ldots, \ell - 1$ and all $k = 1, 2, \ldots, d - 1$,

$$\sigma(m_h, e_k) = \sigma(m_0, e_0) + h. \tag{6}$$

The *duration* $S(\sigma)$ of a virtual chatting schedule σ is the quantity

$$S(\sigma) = \max_{m,e}\{\sigma(m, e)\} - \min_{m,e}\{\sigma(m, e)\} + 1, \tag{7}$$

where the minimization and maximization are over all flits m in the chatting problem $\mathcal{M}$ and all arcs e that occur in the designated paths for the addressed messages in $\mathcal{M}$. There are also simple lower bounds on the duration of virtual chatting schedules.

Proposition 2. *For any virtual chatting schedule σ for a chatting problem $\mathcal{M}$,*

$$S(\sigma) \geq \max(C(\mathcal{M}), L(\mathcal{M})).$$

Unidirectional Linear Arrays: A Running Example. A virtual chatting schedule for a ULA admits a geometric representation similar to the one developed for a chatting schedule in Section 2. In the virtual scenario, the marked region associated with a given message is an *isothetic rectangle,* (see Figure 2).

The problem of finding a virtual chatting schedule of minimum duration reduces to the following purely geometric problem.

Rectangle Compaction.

Given: A set of isothetic rectangles in the integer plane, each having a fixed projection on the horizontal (ULA) axis but being free to slide up or down in the vertical (time) direction,

Find: A pairwise nonoverlapping placement of the rectangles that minimizes the difference between the maximum and minimum ordinates of any of their points.

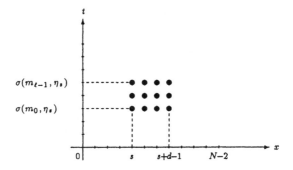

Fig. 2. Graphical representation of a virtual chatting schedule σ for one message.

Comparing Ordinary and Virtual Chatting Schedules. For *forward-leveled networks,* there is an intimate relationship between ordinary and virtual chatting schedules.

A *forward-leveled network (FL-network,* for short) is one whose underlying directed graph $G = (V, E)$ is *forward-leveled,* in the following sense. The node-set V of G admits a partition $V = V_0 + V_1 + \cdots + V_H$ in such a way that, for every arc $e \in E$, there is an i such that $e \in V_i \times V_{i+1}$; in this case, we say that level$(e) = i$. In the partition of V, we call block V_i the ith *level* of G. Note that in an FL-network, for each chatting path $\langle e_0, e_1, \ldots, e_{d-1} \rangle$, level$(e_k) =$ level$(e_{k-1}) + 1$, for $k = 1, 2, \ldots, d - 1$.

While FL-networks form a narrow class of networks, the results that we obtain for this class can be applied to broader classes of networks in a variety of ways. One instance of this would be the decomposition of a *bidirectional* linear array into two ULAs. Such indirect use of a restricted class of networks to implement communication primitives in broader classes can be found also in [2, 10, 13].

The LS Transformation. Let $\mathcal{M}$ be a chatting problem on the FL-network G. Consider functions τ and σ, each of which associates an integer with each flit m of an addressed message $M \in \mathcal{M}$ and each arc e on the designated path from the source to the destination of M. Call any such function a *pseudo-schedule* for $\mathcal{M}$. If τ and σ are related by the equations

$$\tau(m, e) = \sigma(m, e) + \text{level}(e) \tag{8}$$

for all flits m and arcs e, then we say that they are *LS-related.*

Theorem 3. *Let $\mathcal{M}$ be a chatting problem on the $(H + 1)$-level FL-network G; let τ and σ be pseudo-schedules for $\mathcal{M}$ that are LS-related.*
(a) The pseudo-schedule τ is a chatting schedule for $\mathcal{M}$ if and only if the pseudo-schedule σ is a virtual chatting schedule for $\mathcal{M}$.
(b) If τ and σ are, respectively, a chatting schedule and a virtual chatting schedule for $\mathcal{M}$, then

$$|T(\tau) - S(\sigma)| \leq H - 1. \tag{9}$$

Next we present two global transformations, one that produces an ordinary chatting schedule from a given virtual one, the other that performs the converse transformation. Both of these transformations have better performance guarantees than are given by inequality (9) for the LS-transformation.

The VO Transformation. Let σ be a virtual chatting schedule of duration $S(\sigma)$, for the chatting problem $\mathcal{M}$ on the FL-network G. Letting $\lambda(e) = \text{level}(e) - (\text{level}(e) \bmod S(\sigma))$ for each arc e of G, define the *VO-derived pseudo-schedule* τ_σ as follows. In order to evaluate $\tau_\sigma(m, e)$ for a given flit m belonging to some message $M \in \mathcal{M}$ and a given arc e of G, one looks at the first flit, call it m_0, of message M and at the first arc, call it e_0, of the designated path of message M. One then assigns

$$\tau_\sigma(m, e) \stackrel{\text{def}}{=} \begin{cases} \sigma(m, e) + \text{level}(e) - \lambda(e_0) & \text{if } \sigma(m_0, e_0) \leq S(\sigma) - (\text{level}(e_0) \bmod S(\sigma)) \\ \sigma(m, e) + \text{level}(e) - \lambda(e_0) - S(\sigma) & \text{otherwise.} \end{cases}$$
(10)

Theorem 4. (VO-derived chatting schedules)
Let σ be a virtual chatting schedule of duration $S(\sigma)$, for the chatting problem $\mathcal{M}$ on the FL-network G; let τ_σ be the pseudo-schedule that is VO-derived from σ. Then τ_σ is a chatting schedule for $\mathcal{M}$ of duration

$$T(\tau_\sigma) < S(\sigma) + Q(\mathcal{M}).$$
(11)

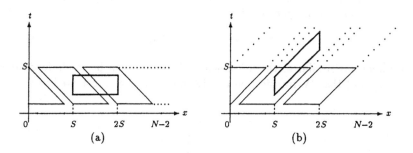

Fig. 3. Graphical representation of transformation (10) for a ULA. Continuous, dotted and thick lines denote, respectively, base regions, extended regions and messages in (a), and their images in (b).

The VO-transformation and Theorem 4 admit an intuitively appealing geometric interpretation on an n-node ULA, which is depicted schematically in Figure 3. Call the region of the virtual schedule plane at or below time $S = S(\sigma)$ and extending rightward $n-1$ units the *problem region*. We partition the problem region into disjoint *base regions*. The ith base region, for each $i \geq 0$, is a parallelogram (slanting from southeast to northwest) whose corners are $(Si, 1)$, $(S(i+1) - 1, 1)$, $(S(i-1) + 1, S)$ and (Si, S). Each message M in the chatting problem is assigned a *home region*, namely, that base region that contains the southwest corner of the rectangle that represents M's (virtual) journey through the ULA; this corner. Superimposed on this structure, each base region R is associated to the right with an *extended region* that is the trapezoid of corners $(S(i + 1) - 1, 1)$, (Si, S), $(N - 2, 1)$, and $(N - 2, S)$. The chatting schedule τ is obtained from the virtual chatting schedule σ by applying the transformation specified in (10). One can view this transformation as separate applications of the LS transformation to each pair consisting of a base region and its associated extended region, *with respect to suitably translated axes*.

The OV Transformation. The second transformation derives virtual chatting schedules from ordinary ones. In conjunction with Theorem 4, it will yield an important bound (14) on the quality of chatting schedules produced by the VO transformation.

Let τ be a chatting schedule of duration $T(\tau)$, for the chatting problem $\mathcal{M}$ on the FL-network G. Define the *OV-derived pseudo-schedule* σ_τ as follows. In order to evaluate $\sigma_\tau(m, e)$ for a given flit m belonging to some message $M \in \mathcal{M}$ and a given arc e of G, one looks at the first flit, call it m_0, of message M and at the first arc, call it e_0, of the designated path of message M. One then assigns

$$\sigma_\tau(m, e) \stackrel{\text{def}}{=} \begin{cases} \tau(m, e_0) - (\text{level}(e_0) \bmod T(\tau)) & \text{if } \tau(m_0, e_0) > \text{level}(e_0) \bmod T(\tau) \\ \tau(m, e_0) - (\text{level}(e_0) \bmod T(\tau)) + T(\tau) & \text{otherwise.} \end{cases}$$

$$(12)$$

Theorem 5. (OV-derived virtual chatting schedules)
Let τ be a chatting schedule of duration $T(\tau)$, for the chatting problem $\mathcal{M}$ on the FL-network G; let σ_τ be the pseudo-schedule that is OV-derived from τ. Then σ_τ is a virtual chatting schedule for $\mathcal{M}$ of duration

$$S(\sigma_\tau) < T(\tau) + L(\mathcal{M}).$$

$$(13)$$

Considerations similar to those following Theorem 4 can be made to develop a graphical view of the OV-transformation (12) on a ULA, as illustrated in Figure 4.

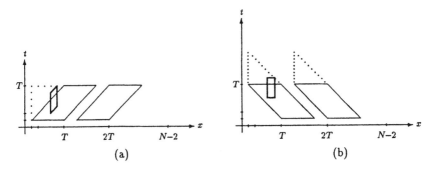

(a)   (b)

Fig. 4. Graphical representation of transformation (12) for a ULA. Continuous, dotted and thick lines denote, respectively, base regions, extended regions and messages in (a), and their images in (b).

Theorems 4 and 5 yield the following relation between the duration $T_{\text{opt}}(\mathcal{M})$ of the optimal ordinary chatting schedule τ_{opt} for a given chatting problem $\mathcal{M}$ and the duration $T_{\text{opt}}^{(VO)}(\mathcal{M})$ of the VO-derived chatting schedule $\tau_{\sigma_{\text{opt}}}$ that is derived from the optimal virtual chatting schedule σ_{opt} for problem $\mathcal{M}$.

Corollary 6.

$$T_{\text{opt}}^{(VO)}(\mathcal{M}) < 2T_{\text{opt}}(\mathcal{M}) + L(\mathcal{M}) - 1. \tag{14}$$

4 Virtual Schedules via Rectangle Compaction

We now develop a strategy for producing chatting schedules for ULAs. Our approach first derives a virtual chatting schedule σ for a given chatting problem $\mathcal{M}$ and then employs the VO transformation (10) to obtain the desired chatting schedule. The virtual schedule σ is obtained by solving the rectangle compaction problem that corresponds to problem $\mathcal{M}$, as in Section 3. Using known results about a problem that is equivalent to rectangle compaction, we obtain, in linear time, a virtual chatting schedule that is within a small constant factor of optimal. The quality of the derived chatting schedule can then be estimated via the bound (14).

Unit-Length Messages. ($L(\mathcal{M}) = 1$). In this case, the rectangle compaction problem reduces to the problem of arranging the collection of segments derived from message in $\mathcal{M}$ on the minimum number of *tracks* (that is, lines parallel to the x-axis at integer coordinates) so that no two segments overlap. Via a reduction to node-coloring of interval graphs [8] we obtain:

Theorem 7. *The rectangle compaction problem that represents a chatting problem $\mathcal{M}$ with $L(\mathcal{M}) = 1$ can be solved in (low-degree) polynomial time. In the optimal solution, the segments are allocated to precisely $C(\mathcal{M})$ tracks.*

Corollary 8. *In (low-degree) polynomial time, one can find, for any chatting problem $\mathcal{M}$ with unit-length messages, a chatting schedule τ of duration $T(\tau) < C(\mathcal{M}) + Q(\mathcal{M}) \leq 2T_{\text{opt}}(\mathcal{M})$.*

Arbitrary-Length Messages. When a chatting problem has arbitrary message lengths, the reduction of optimal virtual scheduling to optimal rectangle compaction still affords one an efficient avenue to efficient chatting schedules, but not as easily as in the unit-length case. The decision version of the general rectangle compaction problem can be easily shown to be NP-complete, being equivalent to the *Dynamic Storage Allocation* problem defined in [8] and shown NP-complete in [5].

By results of Doenhardt and Lengauer [5], and Kierstead [9] we have:

Theorem 9. **(a)** [5] *The rectangle compaction problem for a chatting problem with arbitrary-length messages is NP-hard.*
(b) [9] *Given a chatting problem $\mathcal{M}$, there exists a polynomial-time approximation algorithm which yields a placement of the rectangles on at most $6C(\mathcal{M})$ tracks.*

By applying Theorem 4 to the virtual chatting schedule obtained by solving the rectangle compaction problem on the base of Theorem 9, we obtain:

Corollary 10. *In (low-degree) polynomial time, one can find, for any chatting problem $\mathcal{M}$, a chatting schedule τ of duration $T(\tau) < 6C(\mathcal{M}) + Q(\mathcal{M}) \leq 7T_{\text{opt}}(\mathcal{M})$.*

5 Virtual Schedules via Problem Decomposition

This section is devoted to a technique for deriving moderately efficient virtual chatting schedules for chatting problems by decomposing each problem $\mathcal{M}$ into (roughly) $\log L(\mathcal{M})$ problems on the basis of message length. The strategy produces a virtual chatting schedule that is within a factor of $O(\log L(\mathcal{M}))$ of optimal.

Consider an indexed family $\mathbf{M} = \{\mathcal{M}_\ell\}_{\ell \in N}$ of *related, uniform* chatting problems for G as follows. Each $\mathcal{M}_\ell$ consists exclusively of messages of length ℓ, and $\mathcal{M}_\ell$ has a message from u to v if and only if any other chatting problem $\mathcal{M}_{\ell'} \in \mathbf{M}$ has a message from u to v. The following theorem shows that finding a virtual chatting schedule for any one $\mathcal{M}_\ell \in \mathbf{M}$ is essentially equivalent to finding a virtual schedule for every problem in $\mathbf{M}$.

Theorem 11. *Given a family* $\mathbf{M} = \{\mathcal{M}_\ell\}_{\ell \in N}$ *of related, uniform chatting problems on the FL-network* G:
(a) *One can transform any virtual chatting schedule* σ_1 *for the problem* $\mathcal{M}_1$ *into a virtual chatting schedule* σ_ℓ *for the problem* $\mathcal{M}_\ell$ *of duration* $S(\sigma_\ell) = \ell S(\sigma_1)$.
(b) *One can transform any virtual chatting schedule* σ_ℓ *for the problem* $\mathcal{M}_\ell$ *into a virtual chatting schedule* σ_1 *for the p oblem* $\mathcal{M}_1$ *of duration* $S(\sigma_1) = \lfloor S(\sigma_\ell)/\ell \rfloor$.
(c) *The durations* $S_{\mathrm{opt}}(\mathcal{M}_1)$ *and* $S_{\mathrm{opt}}(\mathcal{M}_\ell)$ *of the optimal virtual chatting schedules for problems* $\mathcal{M}_1$ *and* $\mathcal{M}_\ell$, *respectively, stand in the relation*

$$S_{\mathrm{opt}}(\mathcal{M}_\ell) = \ell S_{\mathrm{opt}}(\mathcal{M}_1).$$

We now build on the single-message-length strategy implicit in Theorem 11(a) to devise a strategy for arbitrary chatting problems. Roughly speaking, this general strategy partitions an arbitrary problem into a sequence of single-message-length problems which are then solved sequentially. Although the virtual chatting schedules produced by the general strategy cannot deviate from optimality by more than a factor of $\log L(\mathcal{M})$, these schedules probably do deviate by that much in general. In the following, denote by $S_{\mathrm{opt}}(\mathcal{M})$ the duration of the optimal virtual chatting schedule for the chatting problem $\mathcal{M}$.

Theorem 12. *Let us be given an FL-network* G *and a procedure* P *that produces, for any uniform chatting problem* $\mathcal{M}'$ *on* G, *a virtual chatting schedule* $\sigma_{\mathcal{M}'}$ *whose duration is within the factor* α *of* $S_{\mathrm{opt}}(\mathcal{M}')$. *One can transform procedure* P *into a procedure* P' *that produces, for any chatting problem* $\mathcal{M}$ *for* G, *a virtual chatting schedule* $\sigma_{\mathcal{M}}$ *of duration*

$$S(\sigma_{\mathcal{M}}) \leq 2\alpha(\lceil \log L(\mathcal{M}) \rceil + 1)S_{\mathrm{opt}}(\mathcal{M}). \tag{15}$$

Two-Dimensional Meshes. The $N \times N$ *Eastward-Southward mesh network* (*ESM*, for short) has node-set $V = \{(i,j) : 0 \leq i, j \leq N - 1\}$ and arcs connecting each node (i,j) to node $(i + 1, j)$ providing that $i < N - 1$, and to node $(i, j + 1)$, providing that $j < N - 1$. One verifies easily that the $N \times N$ ESM is an FL-network: for $\ell = 0, 1, \ldots, 2N - 2$, the ℓth level comprises the set $V_\ell = \{(i,j) : i + j = \ell\}$. In conformance with the metaphor implicit in the name of ESMs, we say that the rows of an ESM run eastward, while the columns run southward. Just as ULAs are useful intermediate structures for handling chatting problems on linear arrays (Section 3), ESMs are useful for handling chatting problems on full meshes. Specifically, a "full mesh" can be obtained by superposing four appropriately rotated ESMs. This approach yields the following Theorem.

Theorem 13. *Let $\mathcal{M}$ be a chatting problem with single-flit messages on the $N \times N$ ESM. In time $O(|\mathcal{M}|C(\mathcal{M}))$, one can produce a virtual chatting schedule σ for $\mathcal{M}$, of duration $S(\sigma) \leq 2C(\mathcal{M})$.*

As shown in the full paper, schedule σ above is obtained by efficiently determining, for any message set $\mathcal{M}$, at most $2C(\mathcal{M})$ classes of non overlapping message paths. Theorem 13, together with Theorems 12 and 4, yields the following.

Corollary 14. *Let $\mathcal{M}$ be a chatting problem on the $N \times N$ ESM. In time $O(|\mathcal{M}|C(\mathcal{M}))$, one can produce a chatting schedule τ for $\mathcal{M}$, of duration*

$$T(\tau) \leq 4\left(\lceil \log L(\mathcal{M}) \rceil + 1\right) C(\mathcal{M}) + L(\mathcal{M}) + 2N.$$

6 Virtual Schedules via Network Cuts

The strategy of this final section produces chatting schedules by decomposing the given chatting problem with respect to a suitable set of *cuts* of the host FL-network G. A *cutset* of the network G is a set A of arcs whose removal partitions G into two subnetworks that are *disjoint* in the sense every path in G that originates in one subnetwork and terminates in the other crosses some arc in A.

One can derive a (virtual) chatting schedule for a chatting problem $\mathcal{M}$ for G by successively decomposing G via cuts. First we partition G into subnetworks by removing the arcs of some cutset. This partitions the set $\mathcal{M}$ of messages into two disjoint set of messages that are each entirely contained in some subnetwork. We can recursively apply this cutting procedure to obtain (virtual) chatting schedules for these message sets, finally scheduling the messages that cross the cut. This strategy is demonstrably useful for certain networks that have small cuts into large subnetworks, as illustrated by next example.

Bidirectional Trees. A *bidirectional tree* (*BT*, for short) is a complete tree wherein any two adjacent nodes u and v are connected by two opposing arcs, (u, v) and (v, u). Given any pair of nodes x and y of a BT, we designate the (unique) shortest path from x to y as the path along which all messages having source x and destination y are routed. Clearly, a BT is not an FL-network; however, any set of designated paths that traverse a given arc of the BT do form an FL-subnetwork of the tree; hence, for such a subnetwork, one can use the VO transformation of (10) to transform an efficient virtual chatting schedule to an efficient ordinary chatting schedule. Using separation techniques [14], we establish:

Theorem 15. *Let $\mathcal{M}$ be a chatting problem for a BT of n nodes and maximum node-degree $\delta > 1$. One can efficiently produce a chatting schedule τ for $\mathcal{M}$, of duration $T(\tau) \leq 2(C(\mathcal{M}) + Q(\mathcal{M}))\lceil \delta \log n \rceil.$*

Acknowledgments This research was supported in part by the Istituto Trentino di Cultura, through the Leonardo Fibonacci Institute, in Trento, Italy. Further research support for S. N. Bhatt by NSF Grant CCR-88-07426, by NSF/DARPA Grant CCR-89-08285, and by Air Force Grant AFOSR-89-0382; for G. Bilardi by CNR and the Italian Ministry of Research; for G. Pucci by the CNR project "Sistemi Informatici e Calcolo Parallelo"; for A. Ranade by AFOSR Grant F49620-87-C-0041; and for A. L. Rosenberg by NSF Grant CCR-90-13184. We gratefully acknowledge constructive discussions with Franco P. Preparata.

References

1. S.N. Bhatt, G. Bilardi, G. Pucci, A. Ranade, A.L. Rosenberg, E.J. Schwabe (1993): *On Bufferless Routing on Variable-Length Messages in Leveled Networks.* Tech. Rep., University of Massachusetts, Amherst, MA.

2. S.N. Bhatt, G. Pucci, A. Ranade, A.L. Rosenberg (1992): Scattering and Gathering Messages in Networks of Processors. *IEEE Trans. Comput.,* to appear. See also *Advanced Research in VLSI and Parallel Systems 1992* (T. Knight and J. Savage, eds.) 318-332.

3. A. Borodin and J.E. Hopcroft (1985): Routing, merging, and sorting on parallel models of computation. *J. Comp. Syst. Sci. 30,* 130-145.

4. W.J. Dally and C.L. Seitz (1987): Deadlock-free message routing in multiprocessor interconnection networks. *IEEE Trans. Comp., C-36,* 547-553.

5. G. Doenhardt and T. Lengauer: Algorithmic Aspects of One-Dimensional Layout Compaction. *IEEE Trans. Comp. Aided Design.*

6. U. Feige and P. Raghavan (1992): Exact analysis of hot-potato routing. *33rd IEEE Symp. on Foundations of Computer Science,* 553-562.

7. S. Felperin, P. Raghavan, E. Upfal (1992): A theory of wormhole routing in parallel computers.. *33rd IEEE Symp. on Foundations of Computer Science,* 563-572.

8. H.A. Kierstead (1988): The linearity of first-fit coloring of interval graphs. *SIAM J. Discr. Math. 1,* 526-530.

9. H.A. Kierstead (1991): A polynomial time approximation algorithm for Dynamic Storage Allocation. *Discr. Math. 88,* 231-237.

10. G. Kortsarz and D. Peleg (1992): Approximation algorithms for minimum time broadcast. *Theory of Computing and Systems (ISTCS '92). Lecture Notes in Computer Science 601,* Springer-Verlag, N.Y., pp. 67-78.

11. D.H. Lawrie and D.A. Padua (1984): Analysis of message switching with shuffle-exchanges in multiprocessors. In *Interconnection Networks,* IEEE Computer Soc. Press, N.Y.

12. D.H. Linder and J.C. Harden (1991): An adaptive and fault tolerant wormhole routing strategy for k-ary n-cubes. *IEEE Trans. Comp. 40,* 2-12.

13. D. Peleg and J.D. Ullman (1989): An optimal synchronizer for the hypercube. *SIAM J. Comput. 18,* 740-747.

14. L.G. Valiant (1981): Universality considerations in VLSI circuits. *IEEE Trans. Comp., C-30,* 135-140.

15. L.G. Valiant (1982): A scheme for fast parallel communication. *SIAM J. Comput. 11,* 350-361.

16. L.G. Valiant (1989): Bulk-synchronous parallel computers. In *Parallel Processing and Artificial Intelligence* (M. Reeve and S.E. Zenith, eds.) Wiley, N. Y., pp. 15-22.

Saving Comparisons in the Crochemore-Perrin String Matching Algorithm

Dany Breslauer*

Institut National de Recherche en Informatique
et en Automatique
B.P. 105, 78153 Le Chesnay Cedex, France

Abstract

Crochemore and Perrin discovered an elegant linear-time constant-space string matching algorithm that makes at most $2n - m$ symbol comparison. This paper shows how to modify their algorithm to use fewer comparisons.

Given any fixed $\epsilon > 0$, the new algorithm takes linear time, uses constant space and makes at most $n + \lfloor \frac{1+\epsilon}{2}(n - m) \rfloor$ symbol comparisons. If $O(\log m)$ space is available, then the algorithm makes at most $n + \lfloor \frac{1}{2}(n - m) \rfloor$ symbol comparisons. The pattern preprocessing step also takes linear time and uses constant space.

These are the first string matching algorithms that make fewer than $2n - m$ symbol comparisons and use sub-linear space.

1 Introduction

String matching is the problem of finding all occurrences of a short string $\mathcal{P}[1..m]$ that is called *a pattern* in a longer string $T[1..n]$ that is called *a text*. In this paper we study the exact comparison complexity of the string matching problem. We assume that the only access the algorithms have to the input strings is by pairwise symbol comparisons that result in equal or unequal answers.

Several algorithms solve the string matching problem in linear time. Most known perhaps is the algorithm of Knuth, Morris and Pratt [19] that makes $2n - m$ comparisons in the worst case. A variant of the Boyer-Moore [3] algorithm that was designed by Apostolico and Giancarlo [1] also makes $2n - m$ comparisons. The original Boyer-Moore algorithm makes about $3n$ comparisons as shown recently by Cole [6]. All these algorithms work in two steps: in the first step the pattern is preprocessed and some information is stored and used later in a text processing step. Our bounds do not account for comparisons

---

*Partially supported by the IBM Graduate Fellowship while studying at Columbia University in New York and by the European Research Consortium for Informatics and Mathematics postdoctoral fellowship while staying at the Centrum voor Wiskunde en Informatica in Amsterdam, The Netherlands. Part of this work was done while visiting at Università de L'Aquila, L'Aquila, Italy in summer 1991.

that are made in the pattern preprocessing step that can compare even all pairs of pattern symbols.

Research on the exact number of comparisons required to solve the string matching problem has been stimulated by Colussi's [9] discovery of an algorithm that makes at most $n + \frac{1}{2}(n - m)$ comparisons. This bound was improved by Galil and Giancarlo [14], Breslauer and Galil [4] and most recently by Cole and Hariharan [7] who show that the string matching problem can be solved using at most $n + \frac{8}{3m}(n - m)$ comparisons[1]. Lower bounds given by Galil and Giancarlo [13] and Cole et al. [8] still leave a small gap between the lower and upper bounds.

The computation model considered in this paper consists of random-access read-only input registers, random-access write-only output registers and a limited number of auxiliary random-access read-write data registers. The number of bits per data register is bounded by some constant times the logarithm of $n + m$. The term *space* in this model refers to the number of auxiliary data registers used. Namely, a constant-space algorithm can use only a constant number of auxiliary registers.

The algorithms mentioned above use $O(m)$ auxiliary memory registers. However, the naive approach to string matching can find all occurrences of the pattern in the text in $O(nm)$ time using only constant auxiliary space. Galil and Seiferas [17] were the first to discover a linear-time constant-space string matching algorithm, disproving conjectures about a time-space tradeoff [2, 16].

Crochemore and Perrin [11] discovered a simple linear-time constant-space string matching algorithm that makes at most $2n - m$ comparisons. Crochemore and Rytter [12] show how to reduce the number of comparisons made by the Galil-Seiferas [17] algorithm by a better choice of parameters. Crochemore [10] gives another constant-space string matching algorithm. The comparison bounds achieved by Galil and Seiferas [17], Crochemore and Rytter [12] and by Crochemore [10] are larger than $2n - m$.

This paper focuses on the number of comparisons required by constant-space string matching algorithms. It is shown that for any fixed $\epsilon > 0$, there exists a linear-time constant-space string matching algorithm that makes at most $n + \lfloor \frac{1+\epsilon}{2}(n - m) \rfloor$ comparisons. Our results are developed in three steps:

1. The Crochemore-Perrin string matching algorithm is modified to use the periodicity structure of the pattern in order to record some pattern suffixes that occur in the text. The modified algorithm takes linear time and uses $O(m)$ auxiliary space. It makes at most $n + \lfloor \frac{\min(\pi_1, m - \pi_1)}{m}(n - m) \rfloor \leq n + \lfloor \frac{1}{2}(n - m) \rfloor$ comparisons, where π_1 denotes the period length of the pattern.

2. The periodicity structure of the pattern that is used in the modified algorithm can be stored in $\lceil \log_{\frac{3}{2}} m \rceil$ memory registers. Thus, the algorithm can be implemented using $O(\log m)$ auxiliary memory registers.

3. If only $c \geq 1$ registers are available to store the periodicity structure of the pattern,

---

[1] All the string matching algorithm that are mentioned take linear time. The pattern preprocessing steps which are not accounted in the bounds take $O(m^2)$ time in Cole and Hariharan's algorithm and linear time in the other algorithms.

then a hybrid between the original Crochemore-Perrin algorithm and the modified algorithm makes at most $n + \lfloor \frac{1}{2-(\frac{4}{3})^{c-1}}(n-m) \rfloor$ comparisons.

This establishes that there exists a linear-time constant-space string matching algorithm that makes fewer than $2n - m$ comparisons.

The pattern preprocessing step of the new algorithms can be implemented in linear time using a constant number of auxiliary memory registers except the registers that store the portion of the periodicity structure of the pattern which is used in the text processing step.

We proceed with the definitions of periods and their basic properties in Section 2. Section 3 overviews the original Crochemore-Perrin algorithm and Section 4 presents the modified algorithm. Section 5 gives more properties of periods which are used in Section 6 to save space. The pattern preprocessing step is discussed in Section 7.

2 Properties of Strings

This sections gives some basic definitions and properties of strings.

Definition 2.1 *A string $S[1..k]$ has a* period *of length π if $S[i] = S[i+\pi]$, for $i = 1, \cdots, k - \pi$.*

We define the set $\Pi^{S[1..k]} = \{\pi_i^S | 0 = \pi_0^S < \pi_1^S < \cdots < \pi_p^S = k\}$ to be the set of all periods of a string $S[1..k]$. π_1^S, the smallest non-zero period of $S[1..k]$ is called *the period* of S. We use the terms *period* and *period length* synonymously.

A *substring* or a *factor* of a string $S[1..k]$ is a contiguous block of symbols $S[i..j]$. A *factorization* of $S[1..k]$ is a way to break S into few factors. We only consider factorizations of a string into two factors: a *prefix* $S[1..l]$ and a *suffix* $S[l+1..k]$. Such a factorization is said to be *non-trivial* if neither of the two factors is equal to the empty string. Note that a factorization can be represented by a single integer which is the position at which the string is partitioned.

Definition 2.2 *Given a factorization $(S[1..l], S[l+1..k])$, a* local period *of the factorization is defined as a non-empty string that is consistent with both sides of the factorization. Namely, a string that matches the prefix $S[1..l]$ aligned at its end and also matches suffix $S[l+1..k]$ aligned at its start. The shortest local period of a factorization is called the* local period. *See Figure 1 for an example.*

Definition 2.3 *A non-trivial factorization of a string $S[1..k]$ is called a* critical factorization *if the local period of the factorization is of the same length as the period of $S[1..k]$.*

The following theorem states that critical factorizations always exist. It is the basis for the Crochemore-Perrin string matching algorithm.

Theorem 2.4 *(The Critical Factorization Theorem, Cesari and Vincent [5, 20]) Let π_1^S be the period length of a string $S[1..k]$. Then, if we consider any $\pi_1^S - 1$ consecutive non-trivial factorizations, at least one is a critical factorization.*

$$a \mid b\,a\,a\,a\,b\,a \qquad\qquad a\,b \mid a\,a\,a\,b\,a \qquad\qquad a\,b\,a \mid a\,a\,b\,a$$
$$b\,a \quad b\,a \qquad\qquad\qquad a\,a\,a\,b \quad a\,a\,a\,b \qquad\qquad\quad a \quad a$$
$$(a) \qquad\qquad\qquad\qquad (b) \qquad\qquad\qquad\qquad (c)$$

Figure 1: The local periods of the first three non-trivial factorizations of 'abaaaba'. Note that in some cases the local period can overflow to either side; this happens when the local period is longer than either of the two factors. The factorization (b) is a critical factorization.

3 The Crochemore-Perrin Algorithm

Crochemore and Perrin [11] used the Critical Factorization Theorem to obtain a simple and elegant linear-time constant-space string matching algorithm. The pattern prepro-cessing step of their algorithm, which is discusses in Section 7, also takes linear time and uses constant space. In the rest of this section we assume that the period length of the pattern and a critical factorization $(\mathcal{P}[1..\chi],\ \mathcal{P}[\chi+1..m])$ of the pattern, such that $\chi < \pi_1^P$, are given. We describe a somewhat simplified version of the Crochemore-Perrin algorithm.

The Crochemore-Perrin string matching algorithm tries to match the pattern aligned starting at a certain text position. It compares symbols starting from the middle of the pattern and tries first to match the pattern suffix $\mathcal{P}[\chi+1..m]$. Only then, after this suffix was discovered in the text, the algorithm tries to match the pattern prefix $\mathcal{P}[1..\chi]$ that was skipped.

Lemma 3.1 *(Crochemore and Perrin [11]) Let $(\mathcal{P}[1..\chi],\mathcal{P}[\chi+1..m])$ be a critical fac-torization of the pattern and let $\rho \le \max(\chi, m-\chi)$ be the length of a local period of this factorization. Then ρ is a multiple of π_1^P, the period length of the pattern.*

Theorem 3.2 *(Crochemore and Perrin [11]) There exist a constant-space linear-time string matching algorithm that makes at most $2n-m$ comparisons.*

Proof: The Crochemore-Perrin algorithm is given is Figure 2. We prove it correctness and show that is makes at most $2n-m$ symbol comparisons.

The algorithm aligns the pattern starting at some text position σ and tries to match the pattern suffix $\mathcal{P}[\chi+1..m]$ with the text symbols that are aligned with it. Initially $\sigma = 1$, and later σ is incremented if there are mismatches or if occurrences of the pattern are discovered. The algorithm maintains the invariant that $T[\sigma+\chi..\theta-1] = \mathcal{P}[\chi+1..\theta-\sigma]$, where θ is the text position that is compared next. There are two conditions in which the while loop that tries to match the pattern suffix terminates.

1. **Mismatch:** If the loop that matches the pattern suffix terminated with $\theta < \sigma + m$, then there was a mismatch $T[\theta] \ne \mathcal{P}[\theta-\sigma+1]$. Clearly, there can be no occurrence of the pattern starting at text position σ.

   Assume that an occurrence of the pattern starts at text position $\overline{\sigma}$, $\sigma < \overline{\sigma} \le \theta - \chi$. Then, the critical factorization $(\mathcal{P}[1..\chi],\ \mathcal{P}[\chi+1..m])$ must have a local period of length $\overline{\sigma} - \sigma$. See Figure 3.

Since $\overline{\sigma} - \sigma \leq m - \chi$, by Lemma 3.1, $\overline{\sigma} - \sigma$ is a multiple of π_1^P. But then, by the definition of a period, $\mathcal{P}[\theta - \sigma + 1] = \mathcal{P}[\theta - \overline{\sigma} + 1]$ and $\mathcal{T}[\theta] \neq \mathcal{P}[\theta - \overline{\sigma} + 1]$. Therefore, there can be no occurrence of the pattern starting at text position $\overline{\sigma}$ and thus, the smallest text position at which an occurrence of the pattern may start is $\theta - \chi + 1$.

The algorithm proceeds by setting $\sigma = \theta - \chi + 1$.

- π_1^P is the period length of the pattern $\mathcal{P}[1..m]$.
- $(\mathcal{P}[1..\chi], \mathcal{P}[\chi + 1..m])$ is a given critical factorization, such that $\chi < \pi_1^P$.
- σ is the current text position that the pattern is aligned with.
- θ is the current text position we have to compare.

```
σ = 1
θ = 1 + χ
while σ ≤ n − m + 1 do
            − Try to match the pattern suffix.
            − '&&' is the conditional and operator.
      while θ < σ + m  &&  T[θ] = P[θ − σ + 1] do
            θ = θ + 1
      if θ < σ + m then   − If there was a mismatch.
            θ = θ + 1
            σ = θ − χ
      else        − The pattern suffix P[χ + 1..m] was matched.
                  − It remains to match the prefix P[1..χ].
                  − The original algorithm compares the symbols in the next statement
                  − from right to left. However, any order can be used.
            if T[σ..σ + χ − 1] = P[1..χ] then
                  Report an occurrence of the pattern starting at text position σ.
            σ = σ + π₁ᴾ
            if σ + χ > θ then
                  θ = σ + χ
      end
end
```

Figure 2: The Crochemore-Perrin algorithm.

2. **Match:** If the loop terminated with $\theta = \sigma + m$, then an occurrence of the pattern suffix $\mathcal{P}[\chi + 1..m]$ was discovered at text position $\sigma + \chi$. The algorithm proceeds to match the pattern prefix $\mathcal{P}[1..\chi]$ that was skipped. If an occurrence of this pattern prefix is discovered, the algorithm can report an occurrence of the complete pattern starting at text position σ.

In any case, the pattern is shifted ahead with respect to the text by π_1^P positions since an occurrence of the pattern at any text position $\overline{\sigma}$, such that $\sigma < \overline{\sigma} < \sigma + \pi_1^P$, would imply that the critical factorization $(\mathcal{P}[1..\chi], \mathcal{P}[\chi + 1..m])$ has a local period whose length is smaller than π_1^P. Note that if after incrementing σ by π_1^P, $\sigma + \chi < \theta$, then $\mathcal{T}[\sigma..\theta - 1] = \mathcal{P}[1..\theta - \sigma]$ and in particular $\mathcal{T}[\sigma + \chi..\theta - 1] = \mathcal{P}[\chi + 1..\theta - \sigma]$. Therefore, the invariant is already maintained and there is no need to go back and compare parts of the pattern and the text that were compared earlier.

It remains to count the number of comparisons made by the algorithm. There are at most $n - \chi$ comparisons made in the loop that matches the pattern suffix since θ is

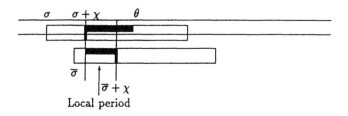

Local period

Figure 3: Applying critical factorizations. If $T[\sigma + \chi..\theta - 1] = \mathcal{P}[\chi + 1..\theta - \sigma]$ and there is an occurrence of the pattern at text position $\overline{\sigma}$, $\sigma < \overline{\sigma} \leq \theta - \chi$, then the factorization $(\mathcal{P}[1..\chi], \mathcal{P}[\chi + 1..m])$ has a local period of length $\overline{\sigma} - \sigma$.

incremented after each comparison is made and initially $\theta = \chi + 1$. The second comparison statement that matches the pattern prefix makes at most χ comparisons each time it is reached. But then, σ is incremented by π_1^P and $\chi < \pi_1^P$. Thus, there are at most $n - m + \chi$ comparisons made by this statement throughout the execution of the algorithm and the total number of comparisons is at most $2n - m$. $\square$

4   Saving Comparisons

The Crochemore-Perrin algorithm is oblivious in the sense that it sometimes "forgets" comparisons that it made and repeats them later. In this section we show how to avoid some of the repeated comparisons. The obvious implementation of the suggested algorithm uses $O(m)$ memory registers to store the periods of the pattern. Section 6 shows how to reduce the space requirements.

Theorem 4.1 *The Crochemore-Perrin string matching algorithm can be modified in such a way that it takes linear-time and makes at most $n + \lfloor \frac{\max(\pi_1^P, m - \pi_1^P)}{m}(n - m) \rfloor$ comparisons.*

Proof: The modified Crochemore-Perrin algorithm is given is Figure 4. The main observation in the modified algorithm is that when the original Crochemore-Perrin algorithm tries to match the pattern prefix $\mathcal{P}[1..\chi]$, this prefix might overlap the pattern suffix $\mathcal{P}[\chi + 1..m]$ that was previously discovered in the text. It is possible to avoid repeating some comparisons by keeping track of suffix-prefix overlaps. For this purpose, the modified algorithm keeps an additional index τ which holds the text position immediately after the last discovered pattern suffix $\mathcal{P}[\chi + 1..m]$.

In addition to the invariant $T[\sigma + \chi..\theta - 1] = \mathcal{P}[\chi + 1..\theta - \sigma]$ that was maintained in the algorithm that was given in the previous section, the modified algorithm maintains a second invariant: If $\sigma < \tau$, then $T[\sigma..\tau - 1] = \mathcal{P}[1..\tau - \sigma]$. Namely, if there would be an occurrence of the pattern starting at text position σ and this occurrence overlapped the last discovered pattern suffix $\mathcal{P}[\chi + 1..m]$, then the overlapping parts must be are identical. Therefore, if $\sigma < \tau$, then it suffices to compare $T[\tau..\sigma_{+\chi} - 1]$ to $\mathcal{P}[\tau - \sigma..\chi]$ to check if $T[\sigma..\sigma + \chi - 1] = \mathcal{P}[1..\chi]$.

Note that suffix-prefix overlaps correspond to periods since $\mathcal{P}[\pi+1..m] = \mathcal{P}[1..m-\pi]$ if and only if $\pi \in \Pi^{\mathcal{P}}$. The second invariant is clearly maintained after the pattern suffix $\mathcal{P}[\chi+1..m]$ is discovered in the text and the pattern is shifted ahead by $\pi_1^{\mathcal{P}}$ positions. The algorithm makes sure that this invariant is maintained each time that a mismatch is encountered by shifting the pattern further ahead until it is maintained, if necessary. The correctness of the algorithm follows similarly to Theorem 3.2. We show that the algorithm makes at most $n + \lfloor \frac{\max(\pi_1^{\mathcal{P}}, m - \pi_1^{\mathcal{P}})}{m}(n-m) \rfloor$ comparisons and takes linear time.

- $\pi_1^{\mathcal{P}}$ is the period length of the pattern $\mathcal{P}[1..m]$.
- $(\mathcal{P}[1..\chi], \mathcal{P}[\chi+1..m])$ is a given critical factorization, such that $\chi < \pi_1^{\mathcal{P}}$.
- σ is the current text position that the pattern is aligned with.
- θ is the current text position we have to compare.
- τ is the text position immediately after the last discovered pattern suffix $\mathcal{P}[\chi+1..m]$.
- The algorithm does not compare text symbols at positions that are smaller than τ.

```
σ = 1
θ = 1 + χ
τ = 0
while σ ≤ n − m + 1 do
          - Try to match the pattern suffix.
        - '&&' is the conditional and operator.
      while θ < σ + m  &&  T[θ] = P[θ − σ + 1] do
            θ = θ + 1
      if θ < σ + m then   - If there was a mismatch.
            θ = θ + 1
            σ = θ − χ
            if σ < τ then   - Maintain the invariant T[σ..τ − 1] = P[1..τ − σ].
                  σ = min{τ − m + π|π ∈ Π^P and τ − m + π ≥ σ}
                  if σ + χ > θ then
                        θ = σ + χ
            end
      else       - The pattern suffix P[χ + 1..m] was matched.
                 - It remains to match the prefix P[1..χ].
            α = max(σ, τ)
            if T[α..σ + χ − 1] = P[α − σ + 1..χ] then
                  Report an occurrence of the pattern starting at text position σ.
            σ = σ + π_1^P
            τ = θ
            if σ + χ > θ then
                  θ = σ + χ
      end
end
```

Figure 4: The modified Crochemore-Perrin algorithm.

Partition the execution of the algorithm into phases. A phase ends after the algorithm has found an occurrence of the pattern suffix $\mathcal{P}[\chi+1..m]$ in the text and it tried to match the pattern prefix $\mathcal{P}[1..\chi]$, or when the end of the text is reached. The following phase starts immediately after the algorithm has shifted the pattern ahead with respect to the text by $\pi_1^{\mathcal{P}}$ positions. Let σ^{ϕ} denote the value of σ, the text position that the pattern is aligned with, at the end of the phase number ϕ. Then, in the first phase $\sigma^1 \geq 1$ and in the last phase $\sigma^l \leq n - m + 1$. In phase number ϕ, $2 \leq \phi \leq l$, $\sigma^{\phi-1} + \pi_1^P \leq \sigma^\phi$. Note that during phase number ϕ, $\tau = \sigma^{\phi-1} + m$ is the text position position immediately after the last discovered pattern suffix $\mathcal{P}[\chi + 1..m]$.

We use a simple policy of charging comparisons to text symbols: each comparison is charged to the text symbol that is compared and the charge might be later transferred to a smaller text position. Using this charging policy it is clear that at the beginning of phase number ϕ all text positions that are larger than or equal to τ are not charged with any comparison.

The charges are transferred as follows. The comparisons that were charged during phase number ϕ to text positions between τ and $\sigma^\phi + \chi$ are transferred χ positions back. Note that the number of these comparisons is bounded by $\sigma^\phi - \sigma^{\phi-1} - \pi_1^P$ and only the $m - \pi_1^P$ text positions that are larger than or equal to $\sigma^{\phi-1} + \pi_1^P$ might be charged with a second comparison. This charge transfer has the advantage that all text symbols at positions between $\max(\sigma^\phi, \tau)$ and $\sigma^\phi + \chi$ do not have a comparison charged to them. Each of these text positions are charged with at most one comparison when the algorithm tries to match the pattern prefix $\mathcal{P}[1..\chi]$.

Clearly, a second comparison might be charged to a text position only when the charges are transferred. We obtain an upper bound on the number of text symbols that are charged with a second comparison in phase number ϕ by bounding the ratio between the number of these symbols to $\sigma^\phi - \sigma^{\phi-1}$, the number of positions by which the pattern was shifted in phase number ϕ. If this ratio can be bounded by a constant c in all phases, then there are at most $\lfloor c(\sigma^\phi - \sigma^{\phi-1}) \rfloor$ text symbols charged with a second comparison in phase number ϕ and the total number of text symbols charged with two comparisons is bounded by $\lfloor c \sum_{i=2}^{l} (\sigma^\phi - \sigma^{\phi-1}) \rfloor \leq \lfloor c(n-m) \rfloor$.

There are two cases:

1. There are at most $\sigma^\phi - \sigma^{\phi-1} - \pi_1^P$ text positions charged with a second comparison in phase number ϕ, but in any case no more than $m - \pi_1^P$. The ratio $\frac{\min(\sigma^\phi - \sigma^{\phi-1} - \pi_1^P, m - \pi_1^P)}{\sigma^\phi - \sigma^{\phi-1}}$ is maximized for $\sigma^\phi - \sigma^{\phi-1} = m$ and is bounded by $\frac{m - \pi}{m}$.

2. If $m - \pi_1^P > \pi_1^P$, then it is possible to achieve better bounds. If $\phi^\phi = \phi^{\phi-1} + \pi_1^P$, then there are clearly no text symbols charged with a second comparison in phase number ϕ since there were no charges transferred.

   Otherwise, there was at least one mismatch while the algorithm was trying to match the pattern suffix $\mathcal{P}[\chi + 1..m]$. Since $\chi < \pi_1^P$, we have that $\sigma^\phi - \sigma^{\phi-1} > m - \pi_1^P$. The number of text symbols charged with a second comparison in phase number ϕ is bounded by $\sigma^\phi - \sigma^{\phi-1} + \pi_1^P - m$ and only text symbols that are larger than or equal to $\tau - \pi_1^P$ might be charged with a second comparison. Thus, in any case there are not more than π_1^P such symbols. The ratio $\frac{\min(\sigma^\phi - \sigma^{\phi-1} + \pi_1^P - m, \pi_1^P)}{\sigma^\phi - \sigma^{\phi-1}}$ is maximized for $\sigma^\phi - \sigma^{\phi-1} = m$ and is bounded by $\frac{\pi_1^P}{m}$.

Therefore, the total number of comparisons made by the modified algorithm is bounded by $n + \lfloor \frac{\min(\pi_1^P, m - \pi_1^P)}{m}(n-m) \rfloor$.

It remains to show that the algorithm takes linear time. The only part which might take longer is the search for the smallest period length of the pattern which is larger than or equal to $\sigma - \tau + m$ when $\sigma < \tau$. It is possible to precompute a table in the preprocessing step that would provide this information in a single step. In Theorem 6.1 we show how this step can be implemented without precomputing such a table. □

5 The Periodicity Structure

The following lemma shows that the periodicity structure of a string can be represented economically. Note that 0 and k are always period lengths of a string $S[1..k]$ and do not have to be specified by the representation.

Lemma 5.1 *Given a string $S[1..k]$, it is possible to represent all period lengths $\pi_i^S \in \Pi^S$, such that $\pi_i^S \leq k - \lfloor \frac{1}{2}(\frac{2}{3})^{c-1}k \rfloor$, by specifying only $c \geq 1$ period lengths. Furthermore, it is possible to compute this representation from the periods of $S[1..k]$ in linear time and using constant space while the periods are given in an increasing order, and it is also possible to generate the periods from this representation in an increasing order in time that is linear in the number of generated periods and using constant space.*

Corollary 5.2 *The set $\Pi^{S[1..k]}$ of all periods of a string $S[1..k]$ can be represented by specifying only $\lceil \log_{\frac{3}{2}} k \rceil$ periods.*

Remark. The compact representation of periods of a string is not new. Galil and Seiferas [15] used similar arguments in a variant of the Knuth-Morris-Pratt string matching algorithm that uses only $O(\log m)$ space. Guibas and Odlyzko [18] characterized all possible periodicity structures of a string of length k and showed that there are $k^{\Theta(\log k)}$ such structures, independent of the alphabet size. Thus, any encoding of the periodicity structure requires $\Omega(\log^2 k)$ bits and our representation can not be uniformly improved by more than a constant multiplicative factor.

6 Saving Space

This section shows how to use the economic representation of the periodicity structure of the pattern in the modified Chrochemore-Perrin algorithm that was given in Section 4.

Theorem 6.1 *The modified Crochemore-Perrin algorithm from Section 4 can be implemented in linear-time using only $O(\log m)$ auxiliary memory registers.*

Proof: The algorithm uses constant space except for storing of the periods of the pattern. By Corollary 5.2, the periods can be represented in $O(\log m)$ memory registers. By Lemma 5.1, the periods can be generated from this representation in an increasing order, in time that is linear in the number of periods generated and using constant space.

The periods are used only in one place in the algorithm where the smallest period of the pattern that is larger than or equal to $\sigma - \tau + m$ is needed. But σ only increases during the execution of the algorithm, so as long that τ is fixed, the periods that are needed also increase and can be found by scanning the periods in an increasing order. The time is clearly bounded by the amount of increase of σ, and therefore is linear.

However, τ increases each time an occurrence of the pattern suffix $\mathcal{P}[\chi + 1..m]$ is discovered in the text. In this case the algorithm returns to generate the periods in an increasing order starting from the smallest period. Note, that in this case $\tau = \sigma + m - \pi_1^{\mathcal{P}}$, the algorithm will need only periods that are larger than $\pi_1^{\mathcal{P}}$, and the time to generate the periods will be bounded by the amount of increase of σ. Thus, the algorithm still takes linear time. $\square$

If only constant space is available, then a part of the periodicity structure of the pattern can still be stored. The resulting algorithm is a hybrid between the Crochemore-Perrin algorithm given in Section 3 and the modified algorithm from Section 4.

Theorem 6.2 *If $c \geq 1$ registers are available to store the periodicity structure of the pattern, then the modified Crochemore-Perrin algorithm can be implemented in linear time and constant space. It makes at most $n + \lfloor \frac{1}{2-(\frac{2}{3})^{c-1}}(n-m) \rfloor$ comparisons.*

Proof: Since the period $\pi_1^{\mathcal{P}}$ is used in the original algorithm, the number of registers used to store the periodicity structure is larger than one. By Lemma 5.1 all period lengths $\pi_\alpha^{\mathcal{P}} \leq m - \lfloor \frac{1}{2}(\frac{2}{3})^{c-1}m \rfloor$ can be represented by $c \geq 1$ registers.

Recall the proof of Theorem 4.1. In phase number ϕ, if $\sigma^\phi - \sigma^{\phi-1} \leq m - \lfloor \frac{1}{2}(\frac{2}{3})^{c-1}m \rfloor$, then the algorithm can proceed as in Theorem 6.1. The problem arises if $\sigma < \tau$ and $\sigma^\phi - \sigma^{\phi-1} > m - \lfloor \frac{1}{2}(\frac{2}{3})^{c-1}m \rfloor$. Since the algorithm cannot maintain the invariant that $T[\sigma..\tau - 1] = \mathcal{P}[1..\tau - \sigma]$ it will behave as the original Crochemore-Perrin algorithm of Section 3 and compare the complete prefix $\mathcal{P}[1..\chi]$ of the pattern if necessary.

This may cause second charges to $\min(\pi_1^{\mathcal{P}}, m - \pi_1^{\mathcal{P}})$ text symbols while $m \geq \sigma^\phi - \sigma^{\phi-1} > m - \lfloor \frac{1}{2}(\frac{2}{3})^{c-1}m \rfloor$. Thus in phase number ϕ, the ratio between the number of text symbols that are charged with a second comparison to $\sigma^\phi - \sigma^{\phi-1}$ is bounded by,

$$\frac{\min(\pi_1^{\mathcal{P}}, m - \pi_1^{\mathcal{P}})}{m - \lfloor \frac{1}{2}(\frac{2}{3})^{c-1}m \rfloor} \leq \frac{\frac{1}{2}}{1 - \frac{1}{2}(\frac{2}{3})^{c-1}} = \frac{1}{2 - (\frac{2}{3})^{c-1}}$$

establishing the claimed bound. $\square$

7   The Pattern Preprocessing

The pattern preprocessing step of the Crochemore and Perrin algorithm takes linear time, uses constant space and make at most $5m$ symbol comparisons. However, it uses order comparisons that may result in less-than, equal-to, or greater-than answers. This preprocessing is not sufficient for our purpose since it does not find all the periods of the pattern. In fact, if the period of the pattern is longer than half of the pattern length, then the Crochemore-Perrin pattern preprocessing step does not compute it at all.

Theorem 7.1 *The pattern preprocessing step of the algorithms presented in this paper takes linear time and uses constant space. It uses order comparisons to find a critical factorization of the pattern.*

Proof: The preprocessing consists of two parts:

1. A critical factorization of the pattern is computed by Crochemore and Perrin's pattern preprocessing algorithm. This computation requires the use of order comparisons.

2. Galil and Seiferas [17] and Crochemore and Rytter [12] show that their linear-time constant-space string matching algorithms can find all overhanging occurrences of the pattern in the text and therefore find all period lengths of the pattern.

   These algorithms find the periods in an increasing order of their length as required in Lemma 5.1. The construction of the economic representation of the periods proceeds as the periods are found and does not require any additional symbol comparisons. It takes linear time and uses constant space, except for the registers which are used to store the representation.

The number of comparisons made is linear with a constant that is not very large. □

8 Acknowledgments

I am in debt to Alberto Apostolico for discussions that led to the results given in this paper and for comments on early versions of the paper. Uri Zwick provided several suggestions that helped to improve the presentation.

References

[1] A. Apostolico and R. Giancarlo. The Boyer-Moore-Galil string searching strategies revisited. *SIAM J. Comput.*, 15(1):98–105, 1986.

[2] A. B. Borodin, M. J. Fischer, D. G. Kirkpatrick, N. A. Lynch, and M. Tompa. A time-space tradeoff for sorting on non-oblivious machines. In *Proc. 20th IEEE Symp. on Foundations of Computer Science*, pages 294–301, 1979.

[3] R.S. Boyer and J.S. Moore. A fast string searching algorithm. *Comm. of the ACM*, 20:762–772, 1977.

[4] D. Breslauer and Z. Galil. Efficient Comparison Based String Matching. *J. Complexity*, 1993. To appear.

[5] Y. Cesari and M. Vincent. Une caractérisation des mots periodiques. *C.R. Acad. Sci. Paris*, 286(A):1175–1177, 1978.

[6] R. Cole. Tight bounds on the complexity of the Boyer-Moore pattern matching algorithm. In *Proc. 2nd ACM-SIAM Symp. on Discrete Algorithms*, pages 224–233, 1991.

[7] R. Cole and R. Hariharan. Tighter Bounds on The Exact Complexity of String Matching. In *Proc. 33rd IEEE Symp. on Foundations of Computer Science*, pages 600–609, 1992.

[8] R. Cole, R. Hariharan, M.S. Paterson, and U. Zwick. Which patterns are hard to find. In *Proc. 2nd Israeli Symp. on Theoretical Computer Science*, pages 59–68, 1993.

[9] L. Colussi. Correctness and efficiency of string matching algorithms. *Inform. and Control*, 95:225–251, 1991.

[10] M. Crochemore. String-matching on ordered alphabets. *Theoret. Comput. Sci.*, 92:33–47, 1992.

[11] M. Crochemore and D. Perrin. Two-way string-matching. *J. Assoc. Comput. Mach.*, 38(3):651–675, 1991.

[12] M. Crochemore and W. Rytter. Periodic Prefixes in Texts. In R. Capocelli, A. De Santis, and U. Vaccaro, editors, *Proc. of the Sequences '91 Workshop: "Sequences II: Methods in Communication, Security and Computer Science"*, pages 153–165. Springer-Verlag, 1993.

[13] Z. Galil and R. Giancarlo. On the exact complexity of string matching: lower bounds. *SIAM J. Comput.*, 20(6):1008–1020, 1991.

[14] Z. Galil and R. Giancarlo. The exact complexity of string matching: upper bounds. *SIAM J. Comput.*, 21(3):407–437, 1992.

[15] Z. Galil and J. Seiferas. Saving space in fast string-matching. *SIAM J. Comput.*, 2:417–438, 1980.

[16] Z. Galil and J. Seiferas. Linear-time string-matching using only a fixed number of local storage locations. *Theoret. Comput. Sci.*, 13:331–336, 1981.

[17] Z. Galil and J. Seiferas. Time-space-optimal string matching. *J. Comput. System Sci.*, 26:280–294, 1983.

[18] L. Guibas and A. M. Odlyzko. Periods in strings. *Journal of Combinatorial Theory, Series A*, 30:19–42, 1981.

[19] D.E. Knuth, J.H. Morris, and V.R. Pratt. Fast pattern matching in strings. *SIAM J. Comput.*, 6:322–350, 1977.

[20] M. Lothaire. *Combinatorics on Words*. Addison-Wesley, Reading, MA., U.S.A., 1983.

Unambiguity of Extended Regular Expressions in SGML Document Grammars

Anne Brüggemann-Klein[*]

Abstract

In the Standard Generalized Markup Language (SGML), document types are defined by context-free grammars in an extended Backus-Naur form. The right-hand side of a production is called a *content model*. Content models are extended regular expressions that have to be unambiguous in the sense that "an element ... that occurs in the document instance must be able to satisfy only one primitive content token without looking ahead in the document instance." In this paper, we present a linear-time algorithm that decides whether a given content model is unambiguous.

A similar result has previously been obtained not for content models but for the smaller class of standard regular expressions. It relies on the fact that the languages of marked regular expressions are local — a property that does not hold any more for content models that contain the new &-operator. Therefore, it is necessary to develop new techniques for content models.

Besides solving an interesting problem in formal language theory, our results are relevant for developers of SGML systems. In fact, our definitions are causing changes to the revised edition of the SGML standard, and the algorithm to test content models for unambiguity has been implemented in an SGML parser.

1 Introduction

The Standard Generalized Markup Language (SGML) is an ISO standard that provides a syntactic meta-language for the definition of textual markup systems, which are used to indicate the structure of documents so that they can be electronically typeset, searched, and communicated [ISO86, Bar89, Gol90].

In SGML, document types are defined by context-free grammars in an extended Backus-Naur form. The right-hand side of a production is called a *content model*.

---
[*]Universität–GH–Paderborn, Fachbereich für Mathematik und Informatik, Warburger Straße 100 33098 Paderborn, Germany.
E-mail: brueggem@uni-paderborn.de.

Content models are similar to standard regular expressions; they are built from symbols in an alphabet Σ with unary operators ?, *, and + and binary operators +, ·, and &. The language $L(E)$ represented by a content model E is defined inductively:

$$L(a) = \{a\} \text{ for } a \in \Sigma \qquad L(FG) = \{vw \mid v \in L(F),\ w \in L(G)\}$$
$$L(F + G) = L(F) \cup L(G) \qquad L(F\ \&\ G) = L(FG + GF)$$
$$L(F?) = L(F) \cup \{\epsilon\} \qquad L(F^*) = \{v_1 \ldots v_n \mid n \geq 0,\ v_1, \ldots, v_n \in L(F)\}$$
$$L(F^+) = \{v_1 \ldots v_n \mid n \geq 1,\ v_1, \ldots, v_n \in L(F)\}$$

Note that neither $\emptyset$ nor ϵ are syntactic constituents of content models. It is not hard to see that the languages denoted by content models are exactly the regular languages that are different from $\emptyset$ and $\{\epsilon\}$.

In an SGML document grammar, only those content models are allowed that are *unambiguous* in the sense of Clause 11.2.4.3 of the standard, as cited in the abstract. In other words, only such content models are valid that enable us to uniquely determine which appearance of a symbol in a content model matches the next symbol in an input word without looking beyond that symbol in the input word. For example, $((a + b)^*a)?$ is ambiguous, whereas $(b^*a)^*$ is unambiguous.

This definition gives rise to the following question, both of theoretical interest and of relevance for systems supporting SGML: Given a content model E, how can we decide whether E is unambiguous? This question has already been answered not for content models, but for the smaller class of standard regular expressions [BKW92, BK92a, BK92b]. (As usual, a *standard regular expression* is built from ϵ, $\emptyset$, and symbols in Σ with the unary operator * and the binary operators · and +.) In this paper, we give a rigorous definition of unambiguity for content models, and we present an optimal-time algorithm to test content models for unambiguity.

The definition of unambiguity is based on the concept of marking a content model; that is, assigning different subscripts to different occurrences of the same symbol. For example, $(a?\&b)a^+$ can be marked as $(a_1?\&b_1)a_2^+$. Now, each word denoted by the content model corresponds to at least one sequence of subscripted symbols. In our example, $baaa$ corresponds to $b_1a_1a_2a_2$ and to $b_1a_2a_2a_2$. If we mark a standard regular expression, the language L of the marked expression is local [Pin92]: If $u_1xu_2, v_1xv_2 \in L$ for some words u_1, u_2, v_1, and v_2 of subscripted symbols and some subscripted symbol x, then $u_1xv_2, v_1xu_2 \in L$. This observation leads to a decision algorithm for unambiguity of standard regular expressions [BK92a, BK92b]. However, this algorithm is not capable of dealing with content models containing &-operators because they do in general not have the property of locality. In our example above, the possible continuations of the word b_1a_1 are a_2^n, $n \geq 1$, but the possible continuations of the word a_1 are $b_1a_2^n$, $n \geq 1$. Clark [Cla92] has proposed a modification of our algorithm [BK92a, BK92b] that also works for content models with &-operators.

Yet he has not given a proof *why* this algorithm works. We describe this approach in Section 3, give a formal proof of correctness, and optimize the running time. Our results imply also that the unambiguity test that is implemented in the SGML parser *sgmls* [Cla92] is correct.

2 Definitions

In this paper, we consider *extended regular expressions* built from ϵ and symbols in Σ with unary operators ?, *, and $^+$ and binary operators $+$, $\cdot$, and $\&$. Since $\emptyset$ can be eliminated as a syntactic constituent without disturbing unambiguity, we do not allow it in extended regular expressions. From now on, we use the term *expression* for extended regular expressions.

A *marked* expression E is an expression over Σ', the alphabet of subscripted symbols, such that each subscripted symbol occurs at most once in E. For a subscripted symbol x, let $\chi(x)$ denote the underlying symbol in Σ. We use uppercase letters from E through J as variables for expressions and for marked expressions, a, b, and c for symbols in Σ, x, y, and z for subscripted symbols in Σ', also called positions, and u, v, and w for words over Σ or over Σ'.

Note that, given a *marking* E' of expression E, the words of $L(E)$ can be obtained from the words of $L(E')$ by dropping the subscripts. In general, several words in $L(E')$ may correspond to a single word in $L(E)$. For example, let $E = a?(a+b)^*$ and $E' = a_1?(a_2+b_3)^*$. Then, aaa in $L(E)$ corresponds to $a_1a_2a_2$ and to $a_2a_2a_2$ in $L(E')$. We can now give a concise definition of what the SGML standard calls unambiguous.

Definition 2.1 Let E' be a marking of the expression E. Then, E' is *unambiguous* if and only if for all words u, v, and w over Σ' and all symbols x, y in Σ' holds: If uxv and uyw are in $L(E')$ and if $\chi(x) = \chi(y)$, then $x = y$. The expression E is *unambiguous* if and only if E' is unambiguous.

It is not hard to see that this definition is independent of the marking chosen for E.

3 The decision algorithm for unambiguity

In this section, we compute, given a marked expression E, all pairs of positions in E that *compete* in the sense of Definition 3.1 below. Then, E is unambiguous if and only if any two competing positions of E have different underlying symbols in Σ.

Definition 3.1 Let E be a marked expression. Two positions x and y *compete* in E if and only if there are words u, v, and w such that uxv and uyw are in $L(E)$.

Definition 3.2 For a marked expression E, we define

$sym(E) = \{y \mid \text{there are words } v \text{ and } w \text{ over } \Sigma' \text{ such that } vyw \in L(E)\}$,

$first(E) = \{y \mid \text{there is a word } w \text{ over } \Sigma' \text{ such that } yw \in L(E)\}$,

$last(E) = \{y \mid \text{there is a word } w \text{ over } \Sigma' \text{ such that } wy \in L(E)\}$, and

$follow(E, x) = \{y \mid \text{there are words } v \text{ and } w \text{ over } \Pi \text{ such that } vxyw \in L(E)\}$,

for each $x \in sym(E)$.

It is not hard to see that, for a marked standard regular expression E, positions x and y of E compete if and only if $x, y \in first(E)$ or if there is a position z of E such that $x, y \in follow(E, z)$. This fact is due to the principle of locality. However, if E contains an &-operator, there may be positions x, y, and z such that $x, y \in follow(E, z)$ yet x and y do not compete. Consider $E = (I \,\&\, J)H$, $z \in last(J)$, $x \in first(I)$, $y \in first(H)$. Then we have $x, y \in follow(E, z)$. Yet for any prefix uz of a word in $L(E)$, either u has not yet satisfied I and, thus, uz may be continued with x but (assuming $\epsilon \notin L(I)$) not with y, or u has already satisfied I and thus, uz may be continued with y but not with x. Therefore, x and y do not compete. Clark [Cla92] has proposed to consider a subset $follow^-(E, z)$ of $follow(E, z)$ that in this case no longer contains x. In Theorem A, we characterize competing positions using Clark's definition.

Definition 3.3 For a marked expression E, we define $follow^-(E, x)$ for x in $sym(E)$ by induction on E as follows.

$E = x$: $follow^-(E, x) = \emptyset$.

$E = F + G$:
$$follow^-(E, x) = \begin{cases} follow^-(F, x) & \text{if } x \in sym(F), \\ follow^-(G, x) & \text{if } x \in sym(G). \end{cases}$$

$E = FG$:
$$follow^-(E, x) = \begin{cases} follow^-(F, x) & \text{if } x \in sym(F), x \notin last(F), \\ follow^-(F, x) \cup first(G) & \text{if } x \in last(F), \\ follow^-(G, x) & \text{if } x \in sym(G). \end{cases}$$

$E = F \,\&\, G$:
$$follow^-(E, x) = \begin{cases} follow^-(F, x) & \text{if } x \in sym(F), x \notin last(F) \\ & \text{or if } x \in last(F), \epsilon \notin L(G), \\ follow^-(F, x) \cup first(G) & \text{if } x \in last(F), \epsilon \in L(G), \\ follow^-(G, x) & \text{if } x \in sym(G), x \notin last(G) \\ & \text{or if } x \in last(G), \epsilon \notin L(F), \\ follow^-(G, x) \cup first(F) & \text{if } x \in last(G), \epsilon \in L(F). \end{cases}$$

$E = F?$: $follow^-(E, x) = follow^-(F, x)$.

$E = F^*, F^+$:
$$follow^-(E, x) = \begin{cases} follow^-(F, x) & \text{if } x \in sym(F), x \notin last(F), \\ follow^-(F, x) \cup first(F) & \text{if } x \in last(F). \end{cases}$$

Theorem A *Let E be a marked expression. Then, positions x and y of E compete if and only if one of the following three conditions holds:*

1. $x, y \in first(E)$,

2. *there is a position z of E such that $x, y \in follow^-(E, z)$, or*

3. *there is a subexpression $I \& J$ or $J \& I$ of E and a position $z \in last(I)$ such that $x \in follow^-(I, z)$ and $y \in first(J)$ or $y \in follow^-(I, z)$ and $x \in first(J)$.*

Theorem B *A marked expression E is unambiguous if and only if it satisfies the following three conditions:*

1. *If $x, y \in first(E)$ and $\chi(x) = \chi(y)$, then $x = y$.*

2. *If $x, y \in follow^-(E, z)$ and $\chi(x) = \chi(y)$, then $x = y$.*

3. *If $I \& J$ or $J \& I$ is a subexpression of E and $z \in last(I)$ such that $x \in follow^-(I, z)$ and $y \in first(J)$, then $\chi(x) \neq \chi(y)$.*

Theorem A immediately implies Theorem B. We prove now Theorem A. As an auxiliary result we use Lemma 3.2, which establishes a weak form of locality for extended expressions. Four types of arguments that are valid only in the context of marked expressions repeat themselves throughout the proofs. We present these arguments first in Lemma 3.1 below. For a set S of symbols, the S-prefix (S-suffix) of a word is its longest prefix (suffix) that consists only of symbols in S.

Lemma 3.1 *Let E be a marked expression.*

1. *Let $E = F + G$ or $E = G + F$. If w in $L(E)$ contains a symbol in $sym(F)$, then $w \in L(F)$ and $w \notin L(G)$.*

2. *Let $E = FG$. If $uxv \in L(E)$ and $x \in sym(F)$, we factorize v as $\dot{v}\ddot{v}$ where $\dot{v}$ is the $sym(F)$-prefix of v; then $ux\dot{v} \in L(F)$ and $\ddot{v} \in L(G)$. Analogously, if $x \in sym(G)$, we factorize u as $\dot{u}\ddot{u}$ where $\ddot{u}$ is the $sym(G)$-suffix of u; then $\dot{u} \in L(F)$ and $\ddot{u}xv \in L(G)$.*

3. *Let $E = F \& G$ or $E = G \& F$. If $uxv \in L(E)$ and $x \in sym(F)$, we factorize u as $\dot{u}\ddot{u}$ and v as $\dot{v}\ddot{v}$ where $\ddot{u}$ is the $sym(F)$-suffix of u and $\dot{v}$ is the $sym(F)$-prefix of v; then $\ddot{u}x\dot{v} \in L(F)$, one of $\dot{u}$ and $\ddot{v}$ is in $L(G)$ and the other one is the empty word.*

4. *Let $E = H^*$ or $E = H^+$ and $H = F + G$, $H = G + F$, $H = FG$, $H = GF$, $H = F \& G$, or $H = G \& F$. If $uzv \in L(E)$ and $z \in sym(F)$, then $\dot{u}z\dot{v} \in L(F^*)$ where $\dot{u}$ is the $sym(F)$-suffix of u and $\dot{v}$ is the $sym(F)$-prefix of v. If H is a concatenation and $\epsilon \notin L(G)$, then even $\dot{u}z\dot{v} \in L(F)$.*

Lemma 3.2 *Let E be a marked expression and $z \in last(E)$. Then $x \in follow^-(E, z)$ if and only if $uz, uzxv \in L(E)$ for some u and v.*

PROOF   The proof is by induction on the size of E.

$E = \epsilon$:   The precondition that $z \in last(E)$ cannot be satisfied.

$E = z$:   Neither $x \in follow^-(E, z)$ nor $uzxv \in L(E)$.

$E = F + G$:   Without loss of generality, $z \in sym(F)$; that is, $z \in last(F)$ by Lemma 3.1. By definition, $follow^-(E, z) = follow^-(F, z)$. Furthermore, $uz, uzxv \in L(E)$ if and only if $uz, uzxv \in L(F)$, by Lemma 3.1. An application of the induction hypothesis to F completes the proof.

$E = FG$:   Since $z \in last(E)$, the language of E is not empty, and neither are the languages of F nor of G. We divide the proof into three cases. First, let $z \in sym(G)$; that is, $z \in last(G)$, by Lemma 3.1. Then $follow^-(E, z) = follow^-(G, z)$. Finally, by Lemma 3.1, $uz, uzxv \in L(E)$ for some u and v if and only if $\dot{u}z, \dot{u}zxv \in L(G)$ for some $\dot{u}$ and v, because $L(F) \neq \emptyset$. An application of the induction hypothesis to G completes the proof for this case.

Next, let $z \in sym(F)$ and $x \in sym(G)$. Then $x \in follow^-(E, z)$ if and only if $z \in last(F)$ and $x \in first(G)$; but this means there are u and v such that $uz \in L(F)$ and $xv \in L(G)$, or, equivalently, $uz, uzxv \in L(E)$, by Lemma 3.1. Hence, in this case we have a direct proof that does not resort to the induction hypothesis.

Finally, let $z, x \in sym(F)$; in particular, $z \in last(F)$ by Lemma 3.1. Then $x \in follow^-(E, z)$ if and only if $x \in follow^-(F, z)$. Furthermore, $uz, uzxv \in L(E)$ for some u and v if and only if $uz, uzx\dot{v} \in L(F)$ for some u and $\dot{v}$, because $L(G) \neq \emptyset$. An application of the induction hypothesis to F completes the proof.

$E = F \& G$:   As in the previous case, neither $L(F)$ nor $L(G)$ are empty. Without loss of generality, $z \in sym(F)$; that is, $z \in last(F)$, by Lemma 3.1. If $x \in sym(F)$, then $x \in follow^-(E, z)$ if and only if $x \in follow^-(F, z)$; furthermore, $uz, uzxv \in L(E)$ for some u and v if and only if $\dot{u}z, \dot{u}zx\dot{v} \in L(F)$ for some $\dot{u}$ and $\dot{v}$, by Lemma 3.1; hence, the claim follows by the induction hypothesis for F.

On the other hand, if $x \in sym(G)$, then $x \in follow^-(E, z)$ if and only if $z \in last(F)$ $x \in first(G)$, and $\epsilon \in L(G)$; but this means there are u and v such that $uz \in L(F)$, $xw \in L(G)$, and still $\epsilon \in L(G)$, or, equivalently, $uz, uzxw \in L(E)$, by Lemma 3.1.

$E = F?$:   This case is proved by a simple application of the induction hypothesis to F.

$E = H^*$ or $E = H^+$:   To make sure that the induction hypothesis can be applied, we observe that $last(E) = last(H)$ and, hence, $z \in last(H)$. Now we prove the two implications of the lemma separately.

First, let $x \in follow^-(E, z)$. We wish to prove that $uz, uzxv \in L(E)$ for some u and v. If $x \in follow^-(H, z)$, we only have to apply the induction hypothesis to H. On the other hand, if $x \notin follow^-(H, z)$, then $x \in first(H)$; therefore, $uz, xv \in L(H)$ for some u and v; that is, $uz, uzxv \in L(E)$.

Second, let $uz, uzxv \in L(E)$. To prove that $x \in follow^-(E, z)$, we carry out a case analysis that depends on the structure of H. Since $uz \in L(E)$, it cannot happen that $H = \epsilon$. If $H = y$, then $z = x = y$ and, hence, $x \in follow^-(E, z)$. If $H = F?$, $H = F^*$, or $H = F^+$, the proof is an easy application of the induction hypothesis to F^*, which is smaller than E. We demonstrate now the case when $H = F + G$, $H = FG$, or $H = F \& G$. Without loss of generality, $z \in sym(F)$; that is, $z \in last(F)$. We consider two cases. If $x \in sym(G)$, then $uzxv \in L(E)$ implies that $z \in last(F)$ and $x \in first(G)$; hence, $x \in follow^-(E, z)$. On the other hand, if $x \in sym(F)$, we consider the $sym(F)$-suffix $\dot{u}$ of u and the $sym(F)$-prefix $\dot{v}$ of v; then $\dot{u}z, \dot{u}zx\dot{v} \in L(F^*)$, by Lemma 3.2; the induction hypothesis for F^* implies that $x \in follow^-(F^*, z)$; then either $x \in follow^-(E^*, z)$, and we are done, or H is a concatenation and $\epsilon \notin L(G)$; in the latter case, however, $\dot{u}z, \dot{u}zx\dot{v} \in L(F)$ even, and an application of the induction hypothesis to F completes the proof.

□

We are now ready to prove the left-to-right direction of Theorem A. If the two positions x and y compete, then $x, y \in first(E)$ or there are u, v, w, and z such that $uzxv, uzyw \in L(E)$. In the latter case, there are three possibilities: First, $x, y \in follow^-(E, z)$; second, exactly one of x and y is in $follow^-(E, z)$; or, third, $x, y \notin follow^-(E, z)$. In the first case we are done; the second case is dealt with in Lemma 3.3 and the third case in Lemmas 3.4 and 3.5 below.

Lemma 3.3 *Let E be a marked expression. If $uzxv, uzyw \in L(E)$, $x \in follow^-(E, z)$, and $y \notin follow^-(E, z)$, then $x \in follow^-(F, z)$ and $y \in first(G)$ for some subexpression $F \& G$ or $G \& F$ of E where $z \in last(F)$.*

PROOF The proof is by induction on the size of E.

$E = \epsilon$ or $E = x$: In these cases, the preconditions of the lemma are not satisfied.

$E = F + G$: By Lemma 3.1 and the definition of $follow^-(E, z)$, all preconditions for an application of the induction hypothesis to F respectively to G are satisfied, depending on whether $z \in sym(F)$ or $z \in sym(G)$.

$E = FG$: If $z \in sym(G)$, we can, by Lemma 3.1, reduce the proof to an application of the induction hypothesis to G. If $z \in sym(F)$, an application of Lemma 3.1 to $uzyw$ yields $y \in sym(F)$, since $y \notin follow^-(E, z)$. By Lemma 3.2, $x \in sym(F)$ as well, since otherwise $uz, uzy\dot{w} \in L(F)$, yet $y \notin follow^-(F, z)$. we are now prepared for an application of the induction hypothesis to F, since, by Lemma 3.1, $uzx\dot{v}, uzy\dot{w} \in L(F)$ for some $\dot{v}$ and $\dot{w}$ and $x \in follow^-(F, z)$, $y \notin follow^-(F, z)$.

$E = F \& G$: Without loss of generality, $z \in sym(F)$. First we consider the case when $\epsilon \notin L(G)$; in particular, $x \in follow^-(F, z)$. If $y \in sym(G)$, then $z \in last(F)$ and $y \in first(G)$ and the claim is proved. On the other hand, if $y \in sym(F)$, then $y \notin follow^-(F, z)$ and $\dot{u}z x \dot{v}, \dot{u}z y \dot{w} \in L(F)$ for some $\dot{u}$, $\dot{v}$, and $\dot{w}$, by Lemma 3.1 applied to $uz x v$ and $uz yw$; hence, we can now apply the induction hypothesis to F.

Next we consider the case when $\epsilon \in L(G)$. Then $y \in sym(F)$, because $y \notin follow^-(E, z)$. As in the case when $E = FG$, Lemma 3.2 implies that $x \in sym(F)$; that is, $x \in follow^-(F, z)$; by Lemma 3.1, $\dot{u}z x \dot{v}, \dot{u}z y \dot{w} \in L(F)$ for some $\dot{u}$, $\dot{v}$, and $\dot{w}$. We can now apply the induction hypothesis to F again.

$E = H^*$ or $E = H^+$: As in the proof of Lemma 3.2, we carry out a case analysis that depends on the structure of H. In the cases when $H = \epsilon$ or $H = x$, the preconditions of the lemma are not satisfied. If $H = F?$, $H = F^*$, or $H = F^+$, the proof is an easy application of the induction hypothesis to F^*, which is smaller than E. We demonstrate now the cases when $H = F + G$, $H = FG$, or $H = F \& G$. Then $y \in sym(F)$, since $y \notin follow^-(E, z)$. If $H = FG$ and $\epsilon \notin L(G)$, then $x \in sym(G)$ implies that $\dot{u}z, \dot{u}z y \dot{w} \in L(F)$ for some $\dot{u}$ and $\dot{w}$, whereas $y \notin follow^-(F, z)$, in contradiction to Lemma 3.2; thus, $x \in sym(F)$; that is, $\dot{u}z x \dot{v}, \dot{u}z y \dot{w} \in L(F)$ for some $\dot{u}$, $\dot{v}$ and $\dot{w}$, $x \in follow^-(F, z)$, and $y \notin follow^-(F, z)$; hence, in this case, an application of the induction hypothesis to F completes the proof. In all other cases, $x \in sym(G)$ implies that $\dot{u}z$, $\dot{u}z y \dot{w} \in L(F^*)$ for some $\dot{u}$ and $\dot{w}$. Yet it contradicts Lemma 3.2 that $y \notin follow^-(F^*, z)$. Thus, $x \in sym(F)$; that is, $\dot{u}z x \dot{v}, \dot{u}z y \dot{w} \in L(F^*)$ for some $\dot{u}$, $\dot{v}$ and $\dot{w}$, $x \in follow^-(F^*, z)$, and $y \notin follow^-(F^*, z)$; now an application of the induction hypothesis to F^* completes the proof.

$\square$

Lemma 3.4 *Let E be a marked expression. If $uz x v, uz yw \in L(E)$, $x, y \notin follow^-(E, z)$, then $x, y \in first(H)$ for some subexpression H of E.*

PROOF The proof is by induction on the size of E.

$E = \epsilon$ or $E = x$: The precondition that $uz x v \in L(E)$ is not satisfied.

$E = F + G$: By Lemma 3.1, the proof is reduced to the induction hypothesis for F or G, depending on whether $z \in sym(F)$ or $z \in sym(G)$.

$E = FG$: If $z \in sym(G)$, then $\dot{u}z x v, \dot{u}z yw \in L(G)$ for some $\dot{u}$ and $x, y \notin follow^-(G, z)$. On the other hand, if $z \in sym(F)$, then $x, y \in sym(F)$, since $x, y \notin follow^-(E, z)$; therefore, $uz x \dot{v}, uz y \dot{w} \in L(F)$ for some $\dot{v}$ and $\dot{w}$ and $x, y \notin follow^-(F, z)$. We can now apply the induction hypothesis to F and G, respectively.

$E = F \& G$: Without loss of generality, let $z \in sym(F)$. First, if $x, y \in sym(F)$ as well, then $\dot{u}z x \dot{v}, \dot{u}z y \dot{w} \in L(F)$ for some $\dot{u}$, $\dot{v}$ and $\dot{w}$ and $x, y \notin follow^-(F, z)$; we can then apply the induction hypothesis to F. Second, if $x, y \in sym(G)$, then

$x, y \in first(G)$ and the claim is proved. Finally, we show by contradiction that either both or none of x and y must belong to $sym(F)$; without loss of generality, we assume $x \in sym(F)$ and $y \in sym(G)$. Then, by Lemma 3.2, $uzx\dot{v}, uz \in L(F)$ for some $\dot{v}$; by Lemma 3.2, $x \in follow^-(F, z)$, in contradiction to the assumption that $x \notin follow^-(E, z)$.

$E = H^*$ or $E = H^+$: Again we carry out a case analysis that depends on the structure of H. If $H = \epsilon$ or $H = x$, the preconditions are not satisfied. If $H = F?$, $H = F^*$, or $H = F^+$, the proof is an easy application of the induction hypothesis to F^*, which is smaller than E. We demonstrate now the cases when $H = F + G$, $H = FG$, or $H = F \& G$. Without loss of generality, $z \in sym(F)$. Then $x, y \in sym(F)$, because $x, y \notin follow^-(E, z)$. In the subcase when $H = F \& G$ and $\epsilon \notin L(G)$, we have $x, y \notin follow^-(F, z)$ and $\dot{u}zx\dot{v}, \dot{u}zy\dot{w} \in L(F)$ for some $\dot{u}$, $\dot{v}$ and $\dot{w}$. In all other subcases, $x, y \notin follow^-(F^*, z)$ and $\dot{u}zx\dot{v}, \dot{u}zy\dot{w} \in L(F^*)$. We can complete the proof by applying the induction hypothesis to F in the former case and to F^* in the latter case.

$\square$

Lemma 3.5 *Let E be a marked expression and H be a subexpression E. Then $first(H) \subseteq first(E)$ or $first(H) \subseteq follow^-(E, z)$ for some z in $sym(E)$.*

The proof is a straightforward induction on E that we omit.

Finally, we prove the right-to-left direction of Theorem A. We wish to show that each of the three conditions in the lemma implies that x and y compete in E. Obviously, this is correct for the first condition. For the second condition, we carry out a structural induction on E. So we assume that $z \in sym(E)$ and $x, y \in follow^-(E, z)$. We have to demonstrate that x and y compete in E.

$E = \epsilon$ or $E = z$: The preconditions are not satisfied.

$E = F + G$: This case is an easy application of the induction hypothesis to F respectively to G, depending on whether $z \in sym(F)$ or $z \in sym(G)$.

$E = FG$ or $E = F \& G$: Since $z \in sym(E)$, neither $L(F)$ nor $L(G)$ are empty. Without loss of generality, $z \in sym(F)$. If $x, y \in sym(F)$ as well, then $x, y \in follow^-(F, z)$ and we can apply the induction hypothesis to F; since $L(G) \neq \emptyset$, this completes the proof. If $x, y \in sym(G)$, then $x, y \in first(G)$; that is, $xv, yw \in L(G)$ for some v and w; since $L(F) \neq \emptyset$, x and y compete in E. Finally, we assume without loss of generality that $x \in sym(F)$ and $y \in sym(G)$; then $x, y \in follow^-(E, z)$ implies that $z \in last(F)$, $x \in follow^-(F, z)$, and $y \in first(G)$; that is, by Lemma 3.2, $uz, uzxv \in L(F)$ and $yw \in L(G)$ for some u, v and w; hence, $uzxvyw, uzyw \in L(E)$.

$E = F^*$ or $E = F^+$: If $x, y \in follow^-(F, z)$, then we have only to apply the induction hypothesis to F. If $x, y \notin follow^-(F, z)$, then $x, y \in follow^-(E, z)$ implies that $x, y \in first(F)$ and we are done. Finally, we assume without loss of generality that $x \in follow^-(F, z)$ and $y \notin follow^-(F, z)$. The fact that $y \in$

$follow^-(E, z)$ implies that $z \in last(F)$ and $y \in first(F)$; since $x \in follow^-(F, z)$, we can apply Lemma 3.2 and get $uz, uzxv \in L(F)$ for some u and v; furthermore, $yw \in L(F)$ for some w; hence, $uzxv, uzyw \in L(E)$.

Finally, we turn to the third condition. If $E = F \& G$ or $E = G \& F$ and $z \in last(F)$, $x \in follow^-(F, z)$, and $y \in first(G)$, then, as in the induction above, $uz, uzxv \in L(F)$ and $yw \in L(G)$ for some u, v, and w, and, thus, $uzxvyw, uzyw \in L(E)$; hence, x and y compete in E. We complete the proof with the following observation:

Lemma 3.6 *If x and y compete in a subexpression H of a marked expression E, then they compete in E as well.*

The proof of this observation is a straightforward induction on E that we omit.

Theorem C *For a fixed-size alphabet, it can be decided, for an expression E, in time linear in the size of E whether E is unambiguous.*

PROOF   We only sketch the proof for the case when E is marked. To test E for unambiguity, we compute $first(E)$, $last(E)$, and $follow^-(E, x)$ for x in $sym(E)$ bottom up from the subexpressions of E. Definition 3.3 gives the equations necessary to compute $follow^-$. We are especially interested in the case when all the set unions in Definition 3.3 are disjoint; that is, E is in *star normal form* as defined below. In this case, we can partition the computation of $follow^-(E, x)$ into constant-time steps such that each step computes a new element in $follow^-(E, x)$ for some x in $sym(E)$. During this computation, we also monitor conditions 2 and 3 of Theorem B. Therefore, we can detect any unambiguity in E before more than time linear in the size of E has been spent. The next theorem justifies the assumption that E be in star normal form. It is a generalization of an earlier result for standard regular expressions [BK92a, BK92b].   □

Definition 3.4 Let E' be a marking of the expression E. Then, E' is in *star normal form* if and only if for each subexpression H^* or H^+ of E' the condition

$$follow^-(H, last(H)) \cap first(H) = \emptyset$$

holds. E is in *star normal form* if and only if E' is in star normal form.

Theorem D *Given a marked expression E, we can compute in time linear in the size of E an expression E^* in star normal form such that E is unambiguous if and only if E^* is unambiguous. If E contains no &-operator, then E^* does not contain one either; in this case, even $L(E) = L(E^*)$.*

4   Conclusions

We have presented an optimal-time algorithm to test SGML content models for unambiguity. This paper clarifies the SGML standard and entails a proof of correctness for the unambiguity test in Clark's parser *sgmls*. Our work is currently reviewed by DIN and ISO for the revised edition of the SGML standard.

The "non-local" &-operator, that can occur in content models but not in standard regular expressions, has required a new approach to unambiguity testing. It is also quite powerful from the languages point of view: Applying our characterization of unambiguous regular languages [BKW92], it is not hard to see that the language L_0 of the unambiguous content model $(a \& b? \& c?)^*$ cannot be denoted by any unambiguous standard regular expression. Incidentally, L_0 can neither be denoted by any unambiguous content model in star normal form [BK93]. While we have solved the semantic problem for unambiguous standard regular expressions in an earlier paper [BKW92], it is an open problem to characterize the regular languages that can be denoted by unambiguous content models.

5 Acknowledgement

In January 1992, Derick Wood's and mine work on unambiguous standard regular expressions and its implementation by James Clark in his SGML parser *sgmls* kindled a discussion on unambiguous content models in the usenet newsgroup comp.text.sgml. The contributions of James Clark and Erik Naggum stimulated the work in this paper and are gratefully acknowledged.

References

[Bar89]   D. Barron. Why use SGML? *Electronic Publishing — Origination, Dissemination and Design*, 2(1):3–24, April 1989.

[BK92a]   A. Brüggemann-Klein. Regular expressions into finite automata. In I. Simon, editor, *Latin '92*, pages 87–98, Berlin, 1992. Springer-Verlag. Lecture Notes in Computer Science 583.

[BK92b]   A. Brüggemann-Klein. Regular expressions into finite automata, 1992. To appear in Theoretical Computer Science.

[BK93]   A. Brüggemann-Klein. Formal models in document processing. Habilitationsschrift. Submitted to the Faculty of Mathematics at the University of Freiburg, 1993.

[BKW92] A. Brüggemann-Klein and D. Wood. Deterministic regular languages. In A. Finkel and M. Jantzen, editors, *STACS 92*, pages 173–184, Berlin, 1992. Springer-Verlag. Lecture Notes in Computer Science 577.

[Cla92]   J. Clark, 1992. Source code for SGMLS. Available by anonymous ftp from ftp.uu.net and sgml1.ex.ac.uk.

[Gol90] C. F. Goldfarb. *The SGML Handbook*. Clarendon Press, Oxford, 1990.

[ISO86] ISO 8879: Information processing — Text and office systems — Standard Generalized Markup Language (SGML), October 1986. International Organization for Standardization.

[Pin92] J.-E. Pin. Local languages and the Berry-Sethi algorithm. Unpublished Manuscript, 1992.

ON THE DIRECT SUM CONJECTURE
IN THE STRAIGHT LINE MODEL

Nader H. Bshouty

Department of Computer Science
University of Calgary
2500 University Drive N.W.
Calgary, Alberta, Canada T2N 1N4
E-mail: bshouty@cpsc.ucalgary.ca

ABSTRACT

We prove that if a quadratic system satisfies the direct sum conjecture strongly in the quadratic algorithm model, then it satisfies the
direct sum conjecture strongly in the straight line algorithm model.
Therefore, if the strong direct sum conjecture is true for the quadratic
algorithm model then it is also true for the straight line algorithm
model.

1. INTRODUCTION

Let F be a field and let $x = (x_1, \cdots, x_n)^T$ be a vector of indeterminates. Let $Q^x = \{x^T A_1 x, \cdots, x^T A_m x\}$ be a vector of quadratic forms on $x_1, \cdots, x_n$ over F where each A_i is a $n \times n$ matrix with entries from F. A *straight line algorithm* that computes Q^x is a sequence of rational functions $\sigma_1, \cdots, \sigma_L$ where

1) For every $1 \le j \le L$, we have $\sigma_j = w_{j,1} \circ_j w_{j,2}$ where $\circ_j \in \{\times, \div\}$ and

$$w_{j,1}, w_{j,2} \in \left(F + \sum_{i=1}^{n} F x_i + \sum_{i=1}^{j-1} F \sigma_i \right) \backslash F,$$

or $\circ_j = \div$, $w_{j,1} \in F$ and $w_{j,2} \in (F + \sum_{i=1}^{n} F x_i + \sum_{i=1}^{j-1} F \sigma_i) \backslash F$,

2) and we have

$$Q^x \subseteq F + \sum_{i=1}^{n} F x_i + \sum_{i=1}^{L} F \sigma_i.$$

We call the operation $\circ$ in 1) a *non-scalar* $\circ$. Therefore, in this model we count only non-scalar multiplications/divisions.

The minimal L is denoted by $L(Q^x)$ or $L_F(Q^x)$ and is called the *multiplicative complexity* of Q^x.

When we compute Q^x by an algorithm $\sigma_1, \cdots, \sigma_\mu$ with $\sigma_j = w_{j,1} \times w_{j,2}$ and $w_{j,1}, w_{j,2} \in \sum_{i=1}^{n} F x_i$ then we call the algorithm a *quadratic algorithm*. The minimal μ is denoted by $\mu(Q^x)$ or $\mu_F(Q^x)$ and is called the quadratic complexity of Q^x. In [12], Strassen proved that for infinite fields F

$$L(Q^x) = \mu(Q^x). \tag{1}$$

Let $(y_1, \cdots, y_n)$ be a vector of new indeterminates and

$$Q_1^x = \{x^T A_1 x, \cdots, x^T A_{m_1} x\}, \qquad Q_2^y = \{y^T A_{m_1+1} y, \cdots, y^T A_{m_2} y\}$$

be two sets of quadratic forms. It is obvious that

$$\mu(Q_1^x \cup Q_2^y) \le \mu(Q_1^x) + \mu(Q_2^y). \tag{2}$$

Fiduccia- Zalcstein [8], Strassen [12] and Winograd [13], conjecture that for any two sets of quadratic forms Q_1^x, Q_2^y

$$\mu(Q_1^x \cup Q_2^y) = \mu(Q_1^x) + \mu(Q_2^y), \tag{3}$$

and that every minimal quadratic algorithm $\sigma_1, \cdots, \sigma_\mu$ for $(Q_1^x \cup Q_2^y)$ can be separated into two minimal algorithms

$$s_1 = (\sigma_i)_{i \in I}, \qquad s_2 = (\sigma_i)_{i \in J}, \tag{4}$$

where $I \cup J = \{1, \cdots, \mu\}$, $I \cap J = \emptyset$, and s_1 and s_2 are minimal quadratic algorithms for Q_1^x and Q_2^y, respectively. The set $Q_1^x \cup Q_2^y$ is called the *direct sum* of Q_1^x and Q_2^y.

When (3) is satisfied for Q_1^x and Q_2^y then we say that Q_1^x and Q_2^y satisfy the *direct sum conjecture* in the model of quadratic algorithms. We define DSC_{QA} or $DSC_{QA}(F)$ to be the set of all pairs (Q_1^x, Q_2^y) such that Q_1^x and Q_2^y satisfy the direct sum conjecture in the quadratic algorithm model. When (4) is satisfied for Q_1^x and Q_2^y then we say that Q_1^x and Q_2^y satisfy the *direct sum conjecture strongly* in the model of quadratic algorithms. We define $DSCS_{QA}$ or $DSCS_{QA}(F)$ to be the set of all pairs (Q_1^x, Q_2^y) such that Q_1^x and Q_2^y satisfy the direct sum conjecture strongly in the quadratic algorithm model.

Similarly, we define the classes DSC_{SLA} and $DSCS_{SLA}$ for the straight line model. It is obvious that

$$DSC_M \subseteq DSCS_M \tag{5}$$

for every model of computation M. By the results of Strassen in [12], for infinite fields we also have

$$DSC_{SLA} = DSC_{QA}.$$

In this paper we prove

Theorem 1. *For infinite fields F we have*

$$DSCS_{SLA} = DSCS_{QA}.$$

That is, if Q_1^x and Q_2^y satisfy the direct sum conjecture strongly in the model of quadratic algorithms then they satisfy the direct sum conjecture strongly in the straight line model.

For finite fields we prove

Theorem 2. *Let F be a finite field. Let Q_1^x, Q_2^y be sets of quadratic forms. If there exists an infinite extension field $E \supseteq F$ of F such that*

(i) $\mu_E(Q_1^x) = \mu_F(Q_1^x)$ *and* $\mu_E(Q_2^y) = \mu_F(Q_2^y)$

(ii) $(Q_1^x, Q_2^y) \in DSCS_{QA}(E)$,

then $(Q_1^x, Q_2^y) \in DSCS_{SLA}(F)$.

Using the above theorems we classify all the minimal straight line algorithms for multiplying two polynomials modulo a squarfree polynomial and for computing some direct sums of quadratic systems.

2. PRELIMINARY RESULTS

In this section we give the preliminary results needed to prove Theorem 1

Let $x = (x_1, \cdots, x_n), y = (y_1, \cdots, y_n)$,

$$Q_1^x = \{x^T A_1 x, \cdots, x^T A_{m_1} x\} \quad \text{and} \quad Q_2^y = \{y^T A_{m_1+1} y, \cdots, y^T A_{m_2} y\}.$$

In [3], we proved

Lemma 1. *We have $(Q_1^x, Q_2^y) \in DSCS_{QA}$ if and only if every minimal quadratic algorithm $\sigma_1, \cdots, \sigma_\mu$ for $Q_1^x \cup Q_2^y$ satisfies $\sigma_i \in F[x]$ or $\sigma_i \in F[y]$ for every $0 \leq i \leq \mu$.* ∎

In this section we shall prove the "if" direction of this lemma for straight line algorithms. The "only if" direction will be proved in section 3.

Lemma 2. *We have $(Q_1^x, Q_2^y) \in DSCS_{SLA}$ if every minimal straight line algorithm $\sigma_1, \cdots, \sigma_\mu$ for $Q_1^x \cup Q_2^y$ satisfies $\sigma_i \in F(x)$ or $\sigma_i \in F(y)$ for every $0 \leq i \leq \mu$.*

Proof. Let $\sigma_1, \cdots, \sigma_L$ be a minimal straight line algorithm for $Q_1^x \cup Q_2^y$. Let

$$s_1 = \{\sigma_i\}_{i \in I} = \{\sigma_i | \sigma_i \in F(x)\} \quad \text{and} \quad s_2 = \{\sigma_i\}_{i \in J} = \{\sigma_i | \sigma_i \in F(y)\}.$$

Obviously, s_1 and s_2 are straight line algorithms for Q_1^x and Q_2^y, respectively, and therefore $|I| \geq L(Q_1^x)$ and $|J| \geq L(Q_2^y)$. If s_1 is not minimal then $|I| > L(Q_1^x)$ and

$$L(Q_1^x \cup Q_2^y) = |I| + |J| > L(Q_1^x) + L(Q_2^y),$$

which contradicts (2) with (1). ∎

Definition 1. We denote the ring of *formal power series* in the indeterminates $x_1, \cdots, x_n$ by $F[[x]]$. Let $f \in F[[x]]$ be a power series. Then we can write $f = f^{(0)} + f^{(1)} + \cdots$ where

$$f^{(l)} = \sum_{i_1 + \cdots + i_n = l} a_{i_1, \cdots, i_n} x_1^{i_1} \cdots x_n^{i_n}.$$

We call $f^{(l)}$ the *l-homogeneous* part of f. For any rational function $r = p/q$ for $p, q \in F[x]$ where $q(0) \neq 0$, we correspond the power series $f_r = pq^{-1} \in F[[x]]$. We shall write $r^{(0)}, r^{(1)}, \cdots$ for $f_r^{(0)}, f_r^{(1)}, \cdots$, respectively. It is easy to show that $r^{(0)} = r(0)$ and $r^{(1)} = \sum_{i=1}^n \left(\frac{\partial r}{\partial x_i} \Big|_{x=0} x_i \right)$.

We say that the straight line algorithm $\sigma_1, \cdots, \sigma_L$ where $\sigma_i = w_{i,1} \circ_i w_{i,2}$ is *pure* if for all $i = 1, \cdots, L$, we have $w_{i,1}(0)$ and $w_{i,2}(0)$ are not zero. It is well known that when $w_{i,2}(0) \neq 0$, then

$$\sigma_i^{(2)} \in F\overline{w}_{i,1}^{(1)} w_{i,2}^{(1)} + F w_{i,1}^{(2)} + F w_{i,2}^{(2)} \tag{6}$$

where

$$\overline{w}_{i,1}^{(1)} = \begin{cases} w_{i,1}^{(1)} & \circ_i = \times \\ w_{i,2}^{(0)} w_{i,1}^{(1)} - w_{i,1}^{(0)} w_{i,2}^{(1)} & \circ_i = \div \end{cases}, \qquad i = 1, \cdots, L. \tag{7}$$

Lemma 3. *(Strassen) Let $\sigma_1, \cdots, \sigma_L$ be a minimal pure algorithm that computes Q^x where $\sigma_i = w_{i,1} \circ_i w_{i,2}$. Then the algorithm*

$$\overline{w}_{1,1}^{(1)} w_{1,2}^{(1)}, \cdots, \overline{w}_{L,1}^{(1)} w_{L,2}^{(1)},$$

where $\overline{w}_{i,1}^{(1)}$ is as defined in (7), is a minimal quadratic algorithm for Q^x.

Proof. We shall first prove by induction that

$$\sum_{i=1}^{j} F\sigma_i^{(2)} \subseteq \sum_{i=1}^{j} \left(F\overline{w}_{i,1}^{(1)} w_{i,2}^{(1)} \right). \tag{8}$$

and

$$\{w_{j,1}^{(2)}, w_{j,2}^{(2)}\} \subseteq \sum_{i=1}^{j-1} \left(F\overline{w}_{i,1}^{(1)} w_{i,2}^{(1)} \right). \tag{9}$$

Since $w_{1,1}, w_{1,2} \in F + \sum_{i=1}^{n} Fx_i$ we have $w_{1,1}^{(2)}, w_{1,2}^{(2)} = 0$ and by (6), $F\sigma_1^{(2)} = F\overline{w}_{1,1}^{(1)} w_{1,2}^{(1)}$. This implies (8) and (9) for $j = 1$. Since $\sigma_j = w_{j,1} \circ_j w_{j,2}$ and (by the definition of straight line algorithm)

$$w_{j,1}, w_{j,2} \in F + \sum_{i=1}^{n} Fx_i + \sum_{i=1}^{j-1} F\sigma_i \tag{10}$$

by (6) we have

$$\sigma_j^{(2)} \in F\overline{w}_{j,1}^{(1)} w_{j,2}^{(1)} + Fw_{j,1}^{(2)} + Fw_{j,2}^{(2)}, \tag{11}$$

By (10) and (11),

$$w_{j,1}^{(2)}, w_{j,2}^{(2)} \in \sum_{i=1}^{j-1} F\sigma_i^{(2)} \subseteq \sum_{i=1}^{j-1} F\overline{w}_{i,1}^{(1)} w_{i,2}^{(1)} + Fw_{i,1}^{(2)} + Fw_{i,2}^{(2)}, \tag{12}$$

and by the induction hypothesis (9) for $j - 1, j - 2, \cdots$ we have

$$\sum_{i=1}^{j-1} F\overline{w}_{i,1}^{(1)} w_{i,2}^{(1)} + Fw_{i,1}^{(2)} + Fw_{i,2}^{(2)} \subseteq \sum_{i=1}^{j-1} F\overline{w}_{i,1}^{(1)} w_{i,2}^{(1)} \tag{13}$$

which with (12) implies (9) for j. Now (11) with (9) for j implies (8) for j. This complete the proof of (8) and (9).

Now since by (8),

$$Q^x \in \sum_{i=1}^{L} F\sigma_i^{(2)} \subseteq \sum_{i=1}^{L} F\overline{w}_{i,1}^{(1)} w_{i,2}^{(1)},$$

the quadratic algorithm $\overline{w}_{1,1}^{(1)} w_{1,2}^{(1)}, \cdots, \overline{w}_{L,1}^{(1)} w_{L,2}^{(1)}$ computes Q^x.

Furthermore, the algorithm is minimal because by (1) we have $L(Q^x) = \mu(Q^x)$. ∎

We now give some properties of straight line and quadratic algorithms.

Lemma 4. *Let* $\sigma_1, \cdots, \sigma_{L(Q^x)}$ *be a pure minimal algorithm in* $F(x)$ *that computes* Q^x *where* $\sigma_i = w_{i,1} \circ_i w_{i,2}$. *Then* $\overline{w}_{i,1}^{(1)}, w_{i,2}^{(1)} \neq 0$ *for* $i = 1, \cdots, L$.

Proof. If $w_{i_0,2}^{(1)} = 0$, then $\overline{w}_{i_0,1}^{(1)} w_{i_0,2}^{(1)} = 0$ so we can delete $\overline{w}_{i_0,1}^{(1)} w_{i_0,2}^{(1)}$ from the quadratic algorithm $\overline{w}_{1,1}^{(1)} w_{1,2}^{(1)}, \cdots, \overline{w}_{L(Q^x),1}^{(1)} w_{L(Q^x),2}^{(1)}$ and obtain $\mu(Q^x) \leq L(Q^x) - 1$, in contradiction to (1). ■

Lemma 5. *If $\sigma_1, \cdots, \sigma_L$ is an algorithm for Q^x, then for any $\psi \in F^n$, the algorithm $\sigma_1(x + \psi), \cdots, \sigma_L(x + \psi)$ is an algorithm for Q^x.*

Proof. Since $\sigma_1(x + \psi), \cdots, \sigma_L(x + \psi)$ computes

$$(x+\psi)^T A_i(x+\psi) = x^T A_i x + \psi^T A_i x + x^T A_i \psi + \psi^T A_i \psi \in x^T A_i x + \sum_{i=1}^{n} F x_i + F,$$

the result follows. ■

The following properties are needed for the proof of the theorems.

Lemma 6. *Let F be an infinite field. Let $w_{1,1} \circ_1 w_{1,2}, \cdots, w_{L,1} \circ_L w_{L,2}$ be a minimal straight line algorithm for $Q_1^x \cup Q_2^y$. If $w_{i_0,2} \in F(x,y) \backslash (F(x) \cup F(y))$ then there exist $\psi_1, \psi_2 \in F^n$ such that for $g_{i,j}(x,y) = w_{i,j}(x + \psi_1, y + \psi_2)$ we have:*

(i) The algorithm $g_{1,1} \circ_1 g_{1,2}, \cdots, g_{L,1} \circ_L g_{L,2}$ is a pure algorithm for $Q_1^x \cup Q_2^y$.

(ii) $g_{i_0,2}^{(1)} \in F[x,y] \backslash (F[x] \cup F[y])$.

Proof. We have $g_{i,j}(0,0) = w_{i,j}(\psi_1, \psi_2)$ and

$$g_{i_0,2}^{(1)} = \left.\frac{\partial g_{i_0,2}}{\partial x_1}\right|_{\substack{x=0 \\ y=0}} x_1 + \cdots + \left.\frac{\partial g_{i_0,2}}{\partial x_n}\right|_{\substack{x=0 \\ y=0}} x_n + \left.\frac{\partial g_{i_0,2}}{\partial y_1}\right|_{\substack{x=0 \\ y=0}} y_1 + \cdots + \left.\frac{\partial g_{i_0,2}}{\partial y_n}\right|_{\substack{x=0 \\ y=0}} y_n$$

$$= \left.\frac{\partial w_{i_0,2}}{\partial x_1}\right|_{\substack{x=\psi_1 \\ y=\psi_2}} x_1 + \cdots +$$

$$\left.\frac{\partial w_{i_0,2}}{\partial x_n}\right|_{\substack{x=\psi_1 \\ y=\psi_2}} x_n + \left.\frac{\partial w_{i_0,2}}{\partial y_1}\right|_{\substack{x=\psi_1 \\ y=\psi_2}} y_1 + \cdots + \left.\frac{\partial w_{i_0,2}}{\partial y_n}\right|_{\substack{x=\psi_1 \\ y=\psi_2}} y_n$$

Therefore (i) and (ii) cannot be satisfied if and only if for every $(\psi_1, \psi_2) \in F^{2n}$ we have

$$(\forall i) \frac{\partial w_{i_0,2}(\psi_1, \psi_2)}{\partial x_i} = 0, \; or \; (\forall i) \frac{\partial w_{i_0,2}(\psi_1, \psi_2)}{\partial y_i} = 0, \; or$$

$$w_{1,1}(\psi_1, \psi_2) = 0, \; or \; \cdots \; or \; w_{L,2}(\psi_1, \psi_2) = 0. \tag{14}$$

Let $u = (u_1, \cdots, u_n)$ be new indeterminates. Since F is infinite field, (14) holds if and only if

$$\left(\sum_{i=1}^{n} \frac{\partial w_{i_0,2}}{\partial x_i} u_i \right) \left(\sum_{i=1}^{n} \frac{\partial w_{i_0,2}}{\partial y_i} u_i \right) \left(\prod_{i=0,\cdots,k;j=1,2} w_{i,j} \right) \equiv 0$$

and this implies that (a) $(\forall i) \frac{\partial w_{i_0,2}}{\partial x_i} \equiv 0$ which means that $w_{i_0,2} \in F(y)$ or (b) $(\forall i) \frac{\partial w_{i_0,2}}{\partial y_i} \equiv 0$ which means that $w_{i_0,2} \in F(x)$ or (c) one of $w_{1,1}, \cdots, w_{L,2}$ is zero. In all cases we have a contradiction. ∎

Lemma 7. *Let F be an infinite field. Let $w_{1,1} \circ_1 w_{1,2}, \cdots, w_{L,1} \circ_L w_{L,2}$ be a minimal straight line algorithm for $Q_1^x \cup Q_2^y$. Let $w_{i_0,2} \in F(x)\backslash F$ and $w_{i_0,1} \in (F(x) + F(y))\backslash F(x)$. There exist $\psi_1, \psi_2 \in F^n$ such that for $g_{i,j}(x,y) = w_{i,j}(x + \psi_1, y + \psi_2)$, we have*

(i) The algorithm $g_{1,1} \circ_1 g_{1,2}, \cdots, g_{L,1} \circ_L g_{L,2}$ is a pure algorithm for $Q_1^x \cup Q_2^y$.

(ii) $g_{i_0,1}^{(0)} g_{i_0,2}^{(1)} - g_{i_0,1}^{(1)} g_{i_0,2}^{(0)} \in F[x,y]\backslash(F[x] \cup F[y])$.

Proof. We have $g_{i,j}(0,0) = w_{i,j}(\psi_1, \psi_2)$ and

$$g_{i_0,1}^{(0)} g_{i_0,2}^{(1)} - g_{i_0,1}^{(1)} g_{i_0,2}^{(0)}$$

$$= \sum_{i=1}^{n} \left(g_{i_0,1} \frac{\partial g_{i_0,2}}{\partial x_i} - g_{i_0,2} \frac{\partial g_{i_0,1}}{\partial x_i} \right) \Big|_{\substack{x=0 \\ y=0}} x_i + \sum_{i=1}^{n} \left(g_{i_0,1} \frac{\partial g_{i_0,2}}{\partial y_i} - g_{i_0,2} \frac{\partial g_{i_0,1}}{\partial y_i} \right) \Big|_{\substack{x=0 \\ y=0}} y_i$$

$$= \sum_{i=1}^{n} \left(w_{i_0,2} \frac{\partial w_{i_0,1}}{\partial x_i} - w_{i_0,1} \frac{\partial w_{i_0,2}}{\partial x_i} \right) \Big|_{\substack{x=\psi_1 \\ y=\psi_2}} x_i + \sum_{i=1}^{n} w_{i_0,2} \frac{\partial w_{i_0,1}}{\partial y_i} \Big|_{\substack{x=\psi_1 \\ y=\psi_2}} y_i.$$

Therefore (i) and (ii) do not hold if and only if

$$F_1 F_2 F_3 \equiv$$

$$\left(\sum_{i=1}^{n} w_{i_0,2} \frac{\partial w_{i_0,1}}{\partial y_i} u_i \right) \left(\sum_{i=1}^{n} \left(w_{i_0,2} \frac{\partial w_{i_0,1}}{\partial x_i} - w_{i_0,1} \frac{\partial w_{i_0,2}}{\partial x_i} \right) u_i \right) \left(\prod_{i,j} w_{i,j} \right) \equiv 0.$$

Since $F_3 \not\equiv 0$ we have $F_1 \equiv 0$ or $F_2 \equiv 0$. If $F_1 \equiv 0$ then $(\forall i) \frac{\partial w_{i_0,1}}{\partial y_i} \equiv 0$ which implies that $w_{i_0,1} \in F(x)$. This is a contradiction.

If $F_2 \equiv 0$ then $(\forall i) \frac{\partial(w_{i_0,1}/w_{i_0,2})}{\partial x_i} \equiv 0$ which implies that $w_{i_0,1}/w_{i_0,2} \in F(y)$ and therefore $w_{i_0,1} \equiv w_{i_0,2}(x)l(y)$. Since $w_{i_0,1} \in (F(x) + F(y))\backslash F(x)$ we have $l(y) \in F$ and therefore $w_{i_0,1} \in F(x)$. This is again a contradiction. ∎

3. PROOF OF THEOREMS 1 AND 2

In this section we prove Theorems 1 and 2.

Theorem 1. *For infinite fields F we have*

$$DSCS_{SLA} = DSCS_{QA}.$$

I.e., if Q_1^x and Q_2^y satisfy the direct sum conjecture strongly in the model of quadratic algorithms then they satisfy the direct sum conjecture strongly in the straight line model.

Proof. Let $(Q_1^x, Q_2^y) \in DSCS_{QA}$, where

$$Q_1^x = \{x^T A_1 x, \cdots, x^T A_{m_1} x\} \text{ and } Q_2^y = \{y^T A_{m_1+1} y, \cdots, y^T A_{m_2} y\}.$$

Let $\sigma_1, \cdots, \sigma_L$ be a minimal straight line algorithm for $Q_1^x \cup Q_2^y$. By Lemma 5, if we substitute $x + \psi_1$ and $y + \psi_2$ for x and y, respectively, we obtain a new minimal straight line algorithm for $Q_1^x \cup Q_2^y$.

If $(Q_1^x, Q_2^y) \notin DSCS_{SLA}$, then by Lemma 2 there exists a σ_i such that $\sigma_i \in F(x, y) \backslash (F(x) \cup F(y))$. Let $\sigma_{i_0} = w_{i_0,1} \circ w_{i_0,2}$ be the first element in the algorithm such that $\sigma_{i_0} \in F(x, y) \backslash (F(x) \cup F(y))$.

If $w_{i_0,2} \in F(x, y) \backslash (F(x) \cup F(y))$, then by Lemma 6, there exist $\psi_1, \psi_2 \in F^n$ such that $w_{i,j}(\psi_1, \psi_2) \neq 0$ for $i = 1, \cdots, L$, $j = 1, 2$ (the algorithm is pure) and $(w_{i,2}(x + \psi_1, y + \psi_2))^{(1)} \in F[x, y] \backslash (F[x] \cup F[y])$. Then substituting $x + \psi_1$ and $y + \psi_2$ for x and y in the algorithm we obtain, by Lemma 5, a new algorithm $\sigma_1' = w_{1,1}' \circ w_{1,2}', \cdots, \sigma_L' = w_{L,2}' \circ w_{L,2}'$ for $Q_1^x \cup Q_2^y$. Now, by Lemma 3, since the algorithm is pure, we have $\overline{w'}_{1,1}^{(1)} w_{1,2}'^{(1)}, \cdots, \overline{w'}_{L,1}^{(1)} w_{L,2}'^{(1)}$ is a minimal quadratic algorithm for $Q_1^x \cup Q_2^y$. And since

$$w_{i_0,2}'^{(1)} = (w_{i_0,2}(x + \psi_1, y + \psi_2))^{(1)} \in F[x, y] \backslash (F[x] \cup F[y]),$$

we have a contradiction to Lemma 1. This contradiction implies that $w_{i_0,2} \in F(x)$ or $w_{i_0,2} \in F(y)$. Assume (without loss of generality) that

$$w_{i_0,2} \in F(x). \tag{15}$$

Since $\sigma_{i_0} \in F(x,y) \backslash (F(x) \cup F(y))$ and $w_{i_0,1} \in F + \sum_{i=1}^{n} Fx_i + \sum_{i=1}^{n} Fy_i + \sum_{i=1}^{i_0-1} \sigma_i$ and since $\sigma_i \in F(x)$ or $\sigma_i \in F(y)$ for $i < i_0$, we have

$$w_{i_0,1} \in (F(x) + F(y)) \backslash F(x). \tag{16}$$

If $\circ_{i_0} = \times$, then as before it can be easily shown that there exists a substitution $x + \psi_1$ and $y + \psi_2$ for x and y, respectively, such that $w_{i,j}(\psi_1, \psi_2) \neq 0$ for $i = 1, \cdots, L$, $j = 1, 2$ and

$$(\overline{w}_{i_0,1}(x + \psi_1, y + \psi_2))^{(1)} = (w_{i_0,1}(x + \psi_1, y + \psi_2))^{(1)} \in F[x,y] \backslash F[x].$$

Since $w_{i_0,2}(x + \psi_1, y + \psi_2) \in F(x)$ we have, by Lemma 4, $(w_{i_0,2}(x + \psi_1, y + \psi_2))^{(1)} \in F[x] \backslash F$, and then

$$(\overline{w}_{i_0,1}(x + \psi_1, y + \psi_2))^{(1)}(w_{i_0,2}(x + \psi_1, y + \psi_2))^{(1)} \in F[x,y] \backslash (F[x] \cup F[y]).$$

As before we have a contradiction to Lemma 1.

If $\circ_i = \div$ then, by (15), (16) and Lemma 7, there exist $\psi_1, \psi_2 \in F^n$ such that $w_{i,j}(\psi_1, \psi_2) \neq 0$ for $i = 1, \cdots, L$, $j = 1, 2$ and

$$(\overline{w}_{i_0,2}(x + \psi_1, y + \psi_2))^{(1)} =$$

$$(w_{i_0,1}(x + \psi_1, y + \psi_2))^{(0)}(w_{i_0,2}(x + \psi_1, y + \psi_2))^{(1)} -$$

$$(w_{i_0,1}(x + \psi_1, y + \psi_2))^{(1)}(w_{i_0,2}(x + \psi_1, y + \psi_2))^{(0)} \in$$

$$F[x,y] \backslash (F[x] \cup F[y]).$$

Again, we have a contradiction to Lemma 1. ∎

We now complete the second direction of Lemma 2.

Lemma 8. We have $(Q_1^x, Q_2^y) \in DSCS_{SLA}$ if and only if every minimal straight line algorithm $\sigma_1, \cdots, \sigma_\mu$ for $Q_1^x \cup Q_2^y$ satisfies $\sigma_i \in F(x)$ or $\sigma_i \in F(y)$ for every $0 \leq i \leq \mu$.

Proof. The "if" direction was proved in Lemma 2. Let $(Q_1^x, Q_2^y) \in DSCS_{SLA}$ and assume that $\sigma_1, \cdots, \sigma_\mu$ is a minimal straight line algorithm such that $\sigma_{i_0} \in F(x,y) \backslash (F(x) \cup F(y))$. Since $(Q_1^x, Q_2^y) \in DSCS_{SLA}$, we also have $(Q_1^x, Q_2^y) \in DSCS_{QA}$. Now we treat the algorithm the same as in the proof of Theorem 1 and obtain a minimal quadratic algorithm $\overline{\sigma}_1, \cdots, \overline{\sigma}_L$ with $\overline{\sigma}_{i_0} \in F[x,y] \backslash (F[x] \cup F[y])$ which contradicts Lemma 1. ∎

Let F be a field and E be an extension of F. Let $\mu_E(Q_1^x)$ denote the quadratic complexity of Q_1^x over the field E, i.e., the minimal length of the quadratic algorithm over E that computes Q_1^x. It is obvious that $\mu_E(Q_1^x) \leq \mu_F(Q_1^x)$. For some sets of quadratic forms it is known that this inequality is strict. Similar to this definition we can define L_E, L_F, $DSCS_{QA}(F)$, $DSCS_{SLA}(E)$, etc.

For finite fields we prove in the full paper

Theorem 2. *Let F be a finite field. Let Q_1^x and Q_2^y be sets of quadratic forms. If there exists an infinite extension field $E \supseteq F$ such that*

(i) $\mu_E(Q_1^x) = \mu_F(Q_1^x)$ and $\mu_E(Q_1^y) = \mu_F(Q_1^y)$

(ii) $(Q_1^x, Q_2^y) \in DSCS_{QA}(E)$,

then $(Q_1^x, Q_2^y) \in DSCS_{SLA}(F)$.

Acknowlegement : I wish to thank Michael Kaminski for a number of helpful comments.

REFERENCES

[1] A. Averbuch, Z. Galil, S. Winograd, Classification of all the minimal bilinear algorithms for computing the coefficient of the product of two polynomials modulo a polynomial in the algebra $G[u]/ < u^n >$.

[2] A. Averbuch, Z. Galil, S. Winograd, Classification of all the minimal bilinear algorithms for computing the coefficient of the product of two polynomials modulo a polynomial in the algebra $G[u]/ < Q(u)^l >$ $l > 1$, *Theoretical Computer Science* **58** (1988), 17-56.

[3] N. H. Bshouty, On the extended direct sum conjecture, Proceedings 21st Annual ACM Symposium on Theory of Computing, (May 1989).

[4] E. Feig, On systems of bilinear forms whose minimal division-free algorithms are all bilinear, *Journal of Algorithms*, **2** , (1981), 261-281.

[5] E. Feig, Certain systems of bilinear forms whose minimal algorithms are all quadratic, *Journal of Algorithms,* **4** , (1983), 137-149.

[6] A. Fellmann, Optimal algorithms for finite dimensional simply generated algebras, *Lecture Notes in Computer Science,* (1986).

[7] E. Feig, S. Winograd, On the direct sum conjecture, *Linear Algebra and Its Application,* **63** (1984), 193-219.

[8] C.M. Feduccia, Y. Zalcstein, Algebras having linear multiplicative complexity, *J. ACM,* **24** (1977), 311-331.

[9] H. F. Groote, Characterization of division algebras of minimal rank and the structure of their algorithm varieties, *SIAM J. Comput.* **12** (1983), 101-117.

[10] H. G. Groote, Lectures on the complexity of bilinear problems. *LN Comput. Sci.* **245** , Springer, Berline 1987.

[11] J. Ja' Ja', J. Takche, On the validity of the direct sum conjecture, *SIAM. J.Comput,* **15, 4,** (1986), 1004-1020.

[12] V. Strassen, Vermeidung von Divisionen, *J. Reine Angew. Math.* **264** (1973), 184-202.

[13] S. Winograd, Some Bilinear Forms Whose Multiplicative Complexity Depends on the Field Constants, *Math. System Theory,* **10** (1976/77), 169-180.

[14] S. Winograd, On multiplication in algebraic extension field, *Theoret. Comput. Sci.,* **8** (1979), 359-377.

Combine and Conquer:
a General Technique for Dynamic Algorithms*
(Extended Abstract)

Robert F. Cohen[1] and Roberto Tamassia[2]

[1] Department of Computer Science, University of Newcastle, University Drive, Callaghan, New South Wales 2308, Australia.
[2] Department of Computer Science, Brown University, Providence, RI 02912–1910, USA.

Abstract. We present a general technique for dynamizing a significant class of problems whose underlying structure is a computation graph embedded in a tree. This class of problems includes the evaluation of linear expressions over k-tuples from a semiring with binary and unary operators, attribute grammars with linear dependencies, point location in binary space partitions, compaction of slicing floorplans, graph drawing, generalized heaps, and a variety of optimization problems in bounded tree-width graphs. For problems in this class, we support a complete repertory of dynamic operations in logarithmic time using linear space.

1 Introduction

The development of dynamic algorithms and data structures is a challenging area of research that has attracted increasing interest in the last years. While considerable progress has been made in dynamic computational geometry (see, e.g., the survey paper [5]), considerably fewer results exist on the dynamization of graph algorithms. A crucial development in the area of dynamic geometric searching has been the identification of general methods that apply to a large class of problems. In particular, the techniques developed by Overmars et al. (summarized in [23]) for the class of decomposable search problems constitute a fundamental contribution toward the dynamization of large classes of geometric problems. Our research is motivated by the lack of general techniques in the area of dynamic graph algorithms.

Our approach is motivated by the observation that a number of dynamic graph algorithms [10,11,14,15,17,18,21,25,33], developed mostly for connectivity problems, appear to be based on the following fundamental idea: Decompose a graph into subgraphs with limited overlap, and represent such a decomposition by means of a tree so that dynamic operations on the graph are reflected into corresponding dynamic tree operations, which are in turn supported by variations of the link-cut trees of Sleator and Tarjan [30].

We associate values, called *attributes*, with the nodes, paths, and subtrees of our trees. Path attributes form a *path attribute system*, if they are maintained in constant time under path concatenation. Additionally, attributes form a *tree attribute system*

---
* Research supported in part by the National Science Foundation under grant CCR-9007851, by the U.S. Army Research Office under grant DAAL03-91-G-0035, and by the Office of Naval Research and the Defense Advanced Research Projects Agency under contract N00014-91-J-4052, ARPA order 8225.

if the tree attributes of the tail of a path Π are determined in constant time from the path attributes of Π. A large class of problems can be modeled by tree attribute systems, including the evaluation of linear expressions over k-tuples from a semiring with binary and unary operators, attribute grammars with linear dependencies, point location in binary space partitions, compaction of slicing floorplans, graph drawing, generalized heaps, and a variety of optimization problems in bounded tree-width graphs. For problems in this class, we show how to support a complete repertory of dynamic operations in logarithmic time using linear space.

We also introduce a new data structure called a *linear attribute grammar*. An *attribute grammar* [22] is a tree-based expression where the values a node μ are calculated from the values at the parent, siblings, and/or the children of μ. A linear attribute grammar, is an attribute grammar where all dependencies are linear. Attribute grammars are much studied and have applications in areas such as language based editors [3,28], and VLSI design [20]. Incremental algorithms for attribute grammars have been presented [19,27] which give a $O(a \cdot t)$ running time per operation where a is the number of attributes affected by a change and t is the time to make a single change. In general, $a = O(n)$ for an attribute grammar of size n. Incremental evaluation of general circuits is studied in [1].

Related work has been very recently done by Frederickson [16], who provides an alternative technique for the dynamic evaluation of arithmetic expressions matching the performance of [8], and by Eppstein et al. [12], who present a technique for speeding up dynamic graph algorithms.

Our contributions can be summarized as follows: We provide a framework for maintaining attribute systems on trees in a fully dynamic environment. Our technique extends and generalizes the dynamic trees of [30] and the dynamic expression trees of [8]. Our technique also extends and generalizes also the work on order-decomposable search problems by Overmars [23]. We show that given a semiring $\mathcal{S}$, a set of linear expressions with binary and unary operators over $\mathcal{S}^k$ can be dynamically maintained in a fully dynamic environment using linear space and logarithmic time per operation. We show that a linear attribute grammar can be dynamically maintained in a fully dynamic environment using linear space and logarithmic time per operation. Linear attribute grammars can be used as the data structure for dynamic algorithms for several problems in graph drawing [6] and bounded tree-width graphs [7]. Full details of this work can be found in [9].

The rest of this extended abstract is organized as follows. Section 2 describes fully dynamic algorithms to maintain attributes on paths and trees. Separate algorithms are presented for trees of bounded and unbounded degrees. Section 3.1 applies the results of the previous section to present a fully dynamic algorithm for maintaining the solutions of linear expressions. Section 3.2 presents a fully dynamic algorithm for linear attribute grammars. Finally, section 4 presents other applications including two types of generalized heaps.

2 Attribute Systems on Paths and Trees

In this section we introduce the concept of path attribute systems and tree attribute systems, which significantly extend and generalize dynamic trees [30]. We begin by discussing dynamic algorithms on paths, and then show that trees can be maintained as a collection of paths.

2.1 Path Attribute Systems

Paths are directed. The first and last nodes of a path Π are called the *head* and *tail* of Π, respectively, and are denoted with $head(\Pi)$ and $tail(\Pi)$. The *reversed path* $\overline{\Pi}$ of Π is the path obtained by reversing all edges of Π.

A *node attribute* N is a function on nodes. The values N can take are arbitrary, but can be stored in $O(1)$ space. A *path attribute* P is a function on paths. The value of $P(\Pi)$ for a path Π is dependent on the values of $N(\mu)$ for each node μ on Π, and on the order of the nodes of Π. The value of $P(\Pi)$ can be stored in $O(1)$ space. If the values of a node attribute N are taken from a monoid, then N is said to be *globally updatable*, otherwise N is *locally updatable*. We define $\mathcal{N}_G(\mu)$ and $\mathcal{N}_L(\mu)$ to be the set of globally and locally updatable node attributes of μ. A *node attribute set* $\mathcal{N}$ is a finite collection of node attributes $N_1, \ldots, N_r$. Similarly, a *path attribute set* $\mathcal{P}$ is a finite collection of path attributes $P_1, \ldots, P_s$. For a node μ, the *value* of $\mathcal{N}(\mu)$ is the vector $(N_1(\mu), \ldots, N_r(\mu))$. For a path Π, the value of $\mathcal{P}(\Pi)$ is the vector $(P_1(\Pi), \ldots, P_s(\Pi))$. We consider the node attribute sets $\mathcal{N}(head(\Pi))$ and $\mathcal{N}(tail(\Pi))$ to be included in path attribute set $\mathcal{P}(\Pi)$.

Suppose $\mathcal{N}$ is a node attribute set, $\mathcal{P}$ is a path attribute set, and $F : \mathcal{P} \times \mathcal{P} \to \mathcal{P}$ is a function. The triple $(\mathcal{N}, \mathcal{P}, F)$ is a *path attribute system* Q if for any path Π that is the concatenation of paths Π' and Π'', $\mathcal{P}(\Pi)$ can be determined in $O(1)$ time as $F(\mathcal{P}(\Pi'), \mathcal{P}(\Pi''))$. Function F is called the *concatenation function* of Q.

A decomposable search problem π on a set D, locates a distinguished element $x = \pi(D)$ with the conditions of given any partition of D into subsets D' and D'', we can determine in constant time if x is in D' or D''. A number of methods are presented in [23] to maintain the solution for a decomposable search problem. We extend this notion to paths as follows: Given a path attribute system $Q = (\mathcal{N}, \mathcal{P}, F)$, a *path-selection query* Q for Q maps a path Π and a query argument q into a node $\mu = Q(\Pi, q)$ of Π. Suppose path Π is the concatenation of Π' and Π''. A *path-selection function* $S(\Pi, q)$ for Q, determines in $O(1)$ time whether μ is in Π' or Π'' from q and the values $\mathcal{P}(\Pi')$ and $\mathcal{P}(\Pi'')$. We show that we can perform a selection query given the associated selection function.

We consider a rather general dynamic environment for a set of paths equipped with a path attribute system Q and a collection S of path-selection functions. Update operations include: *splitting, concatenating,* and *reversing paths*; and *updating node attributes*. Query operations include: *evaluating* node and path attributes, and *computing path-selection functions*.

We allow both "local" and "global" updates of node attributes. A *local update* consists of changing the value $N(\mu)$ of a single node μ. A *global update* can be performed on a globally updatable node attribute N with values taken from a monoid with operator $\odot$, and consists of applying an incremental change δ to the attribute of every node in a path Π. Formally, a global update sets $N(\mu) = N(\mu) \odot \delta$ for every node μ of Π, where we assume that $x \odot y$ can be computed in $O(1)$ time. If a path attribute P depends on a globally updatable node attribute N, we require that the new value of $P(\Pi)$ can be computed in $O(1)$ time given the variation δ of N in Π.

Our data structure consists of representing each path Π as a balanced binary

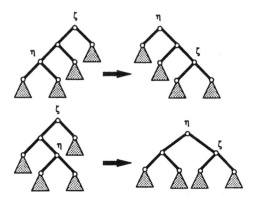

Fig. 1. The rotations performed in operation $splaystep(\eta)$.

tree $\mathcal{B}_\Pi$, called the *path-tree* of Π, and identifying Π by the root ζ of $\mathcal{B}_\Pi$. Each leaf λ of $\mathcal{B}_\Pi$ represents a node of Π, and the left-to-right order of the leaves of $\mathcal{B}_\Pi$ corresponds to the head-to-tail order of the nodes of Π. We store at λ the values $\mathcal{N}(\lambda)$. For a globally updatable node attribute N_i of $\mathcal{N}$, the value stored at λ is not the actual value of N_i, but a "partial value" that can be used to compute $N_i(\lambda)$, as will become apparent later.

Each internal node η of $\mathcal{B}_\Pi$ represents the subpath $\Pi(\eta)$ of Π associated with the leaves in the subtree of η. Note that we identify path $\Pi(\eta)$ with η. We store at η pointers $head(\eta)$ and $tail(\eta)$ to the head and tail of $\Pi(\eta)$. Additionally, we store at node η the two values $\mathcal{P}(\Pi(\eta))$ and $\mathcal{P}(\overline{\Pi(\eta)})$. Notice that we can consider the pointers $head$ and $tail$ as part of path attribute set $\mathcal{P}$. For the path attributes that depend on globally updatable node attributes, we store partial values. We also keep in η a vector $\Delta(\eta)$ such that for any globally updatable node attribute N_i of $\mathcal{N}$ the i-th component $\Delta_i(\eta)$ represents a change to the value of N_i for each node μ on $\Pi(\eta)$. (If N_j is a locally updatable node attribute, then we conventionally set $\Delta_j(\eta) = 0$.) Therefore, for any node μ of Π, $N_i(\mu)$ is the combination (using operation $\odot_i$, the monoid operator for values of N_i) of the "partial value" of N_i stored at μ and of the "offsets" $\Delta_i(\eta)$ of all the nodes η on the path from μ to the root ζ of $\mathcal{B}_\Pi$. Similarly, $\mathcal{P}(\mu)$ can be computed from the partial value of $\mathcal{P}$ stored at μ and the "offsets" $\Delta(\eta)$ of all the nodes η on the path from μ to ζ in $\mathcal{B}_\Pi$.

As an example, we show how to compute a path selection query. For full details of the implementation of all operations see [9]. We need the following operation from [31] in our implementation of $find(\Pi, S, q)$:

- $splaystep(node\ \eta)$ — For binary tree $\mathcal{B}$ with root ζ and node η a grandchild of ζ, restructure $\mathcal{B}$ such that the relative order of the leaves of $\mathcal{B}$ remain fixed, and every node in the subtree rooted at η has its depth reduced by one or two. (See Fig. 1.)

Operation $splaystep$ is implemented with a constant number of rotations, and hence takes $O(1)$ time. Operation $find(\zeta)$ is implemented as follows, starting at node ζ and repeating until a value is returned: if μ is a child or grandchild of ζ then return μ. Otherwise, determine the subtree rooted at the grandchild η of ζ which contains μ. Do *splaystep* at η and recur. After μ is found, we undo all the *splaysteps* to restore the balance of the path-tree. This takes $O(d_\mu)$ time where d_μ is the depth of the returned node μ.

Theorem 1. *Let $Q = (\mathcal{P}, \mathcal{N}, F)$ be a path attribute system and $\mathcal{S}$ a collection of path-selection functions for Q. There is a fully dynamic data structure for maintaining Q and $\mathcal{S}$ over a set of paths with the following performance: A path of length ℓ uses $O(\ell)$ space, and the operations of evaluating node and path attributes, computing path-selection functions, splitting, concatenating, and reversing paths, and locally and globally updating node attributes, take each time $O(\log \ell)$, where ℓ is the length of the paths involved.*

2.2 Tree Attribute Systems

We now discuss a general class of dynamic algorithms on trees. We consider rooted, ordered, unbounded degree trees with edges directed from the child to parent. Hence, a path in a tree is always directed from a descendant to an ancestor.

Rooted tree T has *bounded degree* if there is a constant d such that no node of T has more than d children, otherwise it is of *unbounded degree*. Our algorithm supports both bounded and unbounded degree trees. In this extended abstract we discuss the algorithm for bounded degree trees. The extension to unbounded degree trees is described in [9]. Tree T is *ordered* if for each node μ the left to right order of the children of μ is fixed. We develop our dynamic algorithm for trees in the same manner as for paths. We introduce the concept of tree attribute systems, and show a large collection of dynamic operations on a tree storing a tree attribute system.

A node attribute set $\mathcal{N}$ for a tree T consists of a finite collection of node attributes for the nodes of T. Similarly, a path attribute set $\mathcal{P}$ for T consists of a fixed collection of path attributes for the paths of T. A *tree attribute* R is a function on trees such that, for a tree T, $R(T)$ depends on the values $\mathcal{N}(\mu)$ for each node μ on T, on the parent-child relationships between nodes of T, and on the order of the children of each node of T. We assume that $R(T)$ can be stored in $O(1)$ space. A *tree attribute set* $\mathcal{R}$ is a fixed collection of tree attributes. For a node μ of tree T, $\mathcal{R}(\mu)$ denotes the tree attribute set of the subtree of T rooted at μ.

Suppose $\mathcal{N}$ is a node attribute set, $\mathcal{P}$ is a path attribute set, $\mathcal{R}$ is a tree attribute set, and F is a concatenation function. Given a bounded degree tree T and a path Π of T, we denote with $\tilde{\mathcal{N}}$ the extended node attribute set that, for a node μ of Π, consists of $\mathcal{N}(\mu)$, $\mathcal{R}(\mu')$ for each child μ' of μ not on Π, and the ordering of the children of μ. We say that the 4-tuple $\mathcal{Z} = (\mathcal{N}, \mathcal{P}, \mathcal{R}, F)$ is a *tree attribute system* if: (*i*) the triple $(\tilde{\mathcal{N}}, \mathcal{P}, F)$ is a path attribute system for the set of paths of T (where concatenations are restricted to paths Π' and Π'' in T such that $head(\Pi'')$ is the parent of $tail(\Pi')$ in T); and (*ii*) the value of $\mathcal{R}(head(\Pi))$ can be determined in constant time from the values $\mathcal{P}(\Pi)$, $\tilde{\mathcal{N}}(head(\Pi))$, and $\tilde{\mathcal{N}}(tail(\Pi))$.

We extend the concept of selection to trees. We define a tree selection function, which, if available, allows us to quickly find a distinguished node of tree T. Given a tree attribute system $\mathcal{Z} = (\mathcal{N}, \mathcal{P}, \mathcal{R}, F)$, a *tree-selection query* Q for $\mathcal{Z}$ maps tree T and a query argument q into a node $\mu = Q(T, q)$ of T. Suppose path Π is the concatenation of Π' and Π'' in T such that $tail(\Pi)$ is the root of T. Suppose node

μ' is the node closest to $tail(\Pi)$ such that μ is a descendant of μ'. A *tree-selection function* $S(\Pi, q)$ for Q is a path-selection function to find node μ'.

We allow trees to be merged and separated by *linking* and *cutting*. We also provide three types of global restructuring of trees: reflecting, everting and cycling. *Reflecting* tree T at node μ consists of reversing the left-to-right order of the children of μ and their descendants. Suppose node ν is the parent of μ. *Cycling* ν at μ consists of cyclicly permuting the children of ν such that node μ is the rightmost child of ν. *Everting* T at μ consists of reversing the parent-child relationships between all nodes on the path Π from μ to the root of T. Each node of Π keeps an *eversion rule* which describes changes to the order of the children of μ. Possibilities include: *simple* — the left and right-trees of μ remain unchanged; *steady* — the left and right-trees of μ are exchanged; and *mirrored* — the left and right-trees of μ are reflected.

We consider a general dynamic environment for a set of trees equipped with a tree attribute system $\mathcal{Z} = (\mathcal{N}, \mathcal{P}, \mathcal{R}, F)$ and a collection S of path-selection and tree-selection functions. Update operations include: *linking* and *cutting* trees (by adding/removing an edge between the root of one tree and a node of the other tree); *everting* a tree (by reversing the parent-child relations along the path from a node μ to the root, so that μ becomes the new root); and *updating the node attributes* (either locally or globally along a path). Query operations include *evaluating* node, path, and tree *attributes*, and *computing path- and tree-selection queries*.

As in [30], we represent an n-node tree T as a collection of disjoint paths. We partition the edges of T into *solid* or *dashed* such that at most one solid edge is incoming into a node. Therefore, every node is in exactly one maximal path of solid edges (of length 0 or more), called a *solid path* of T. The partition of edges into solid and and dashed is obtained by means of the following *size invariant*. Let $size(\mu)$ denote the number of nodes in the subtree of T rooted at node μ. An edge (μ, ν) of T is solid if $size(\mu) > size(\nu)/2$. If the partition satisfies the size invariant, then every path of T has $O(\log n)$ dashed edges. In order to achieve the logarithmic time per dynamic operation, we we use biased search trees [2] to represent the path-trees.

We demonstrate how to perform operation *TreeFind*. The implementation of the remaining path operations is given in [9]. Consider tree selection function S. We implement the tree selection query using S as follows. If ν is not the tail of its solid path, we split the path containing ν at $Parent(\nu)$ so ν then becomes the tail of its solid path Π. We repeat the following two steps until the node μ we are searching for is found.

1. Perform a path selection query using S on Π. The node ν returned by this query is the deepest node on Π that is an ancestor of μ.

2. Find the the child ν' of ν that is an ancestor of μ by repeatedly computing path selection functions. If no such child exists, we set $\mu = \nu$ and stop, else let Π be the solid path ending at ν'.

After μ is found, we undo the restructuring to restore the path trees.

Due to the implementation of the solid paths by biased search trees, the execution times of the selection queries performed in Step 1 form a telescoping sum, for a total of $O(\log n)$ time. Each path selection function computed in Step 2 takes constant time, so that each execution of Step 2 takes time $O(d) = O(1)$. Since there are $O(\log n)$ solid paths along a root-to-leaf path, the total time for all iterations of Step 2 is $O(\log n)$.

Theorem 2. *Let $Z = (\mathcal{N}, \mathcal{P}, \mathcal{R}, F)$ be a tree attribute system, and S be a collection of path-selection and tree-selection functions. There is a fully dynamic data structure for maintaining Z and S over a set of trees such that a tree of size n uses $O(n)$ space, and the operations of evaluating node and tree attributes, computing path and tree-selection functions, linking, cutting, everting, and reflecting trees, and locally and globally updating node attributes, take each $O(\log n)$ time, where n is the size of the trees involved.*

3 Expression Trees

3.1 Linear Expression Trees

In this section, we consider the dynamic evaluation of linear expressions, which extends and generalizes the results of of [8,16]. We show how to solve this problem using a tree attribute system. Let S be a semiring with binary operators $\oplus$ and $\otimes$, such that $\otimes$ is distributive with respect to $\oplus$. We assume that elements of S can be stored in $O(1)$ space and binary operations can be performed in $O(1)$ time. The extension to the general case is straightforward. We consider operations in S^r, the space of r-tuples of S for some fixed r. Let $x = (x_1, \ldots, x_r)$ and $y = (y_1, \ldots, y_r) \in S^r$. A *binary linear operator* $\odot : S^r \times S^r \to S^r$ is defined by

$$(x \odot y)_i = x^T A_i y \oplus b_i^T x \oplus c_i^T y \oplus d_i$$

for $i = 1, \cdots, r$, where $A_i \in S^r \times S^r$, $b_i, c_i \in S^r$, and $d_i \in S$. A *unary linear operator* $\nabla : S^r \to S^r$ is defined by

$$\nabla x = Ax \oplus b$$

where $A \in S^r \times S^r$, and $b \in S^r$. (In the above definitions we have used standard linear algebra notation, and convention that xy denotes $x \otimes y$.) Note that, given a binary linear operator $\odot$, for any fixed $\bar{y}$, $x \odot \bar{y}$ is a unary linear operator. Also, given a unary linear operator ∇, there exists a binary linear operator $\odot$ such that $\nabla x = x \odot 0$.

A *linear expression* $\mathcal{E}$ over S^r is an expression with variables in S^r involving binary and unary linear operators. A linear expression is represented by a tree T called a linear expression tree, such that the internal nodes are the operators and the leaves are the variables of $\mathcal{E}$ (see Fig. 2). We associate a binary linear operator with each internal node of T. We maintain a tree attribute system to find the values of subexpressions of T.

Lemma 3. *Suppose $\mathcal{E}$ is a linear expression. The dependence of the value y of $\mathcal{E}$ on a single variable x can be expressed by a unary linear operator $y = Ax \oplus b$.*

By Lemma 3, we call *transfer pair* a (matrix, vector) pair (A, b) that characterizes the dependency of a linear expression on a variable. In particular, given a solid path Π of T, we denote with $P(\Pi)$ the transfer pair associated with the dependency of the value of $tail(\Pi)$ from the value of $head(\Pi)$.

Lemma 4. *Let $\mathcal{E}$ be a linear expression represented by a tree T, and let $x(\mu)$ be the value of the subexpression given by the subtree of T rooted at μ. Then there exists a*

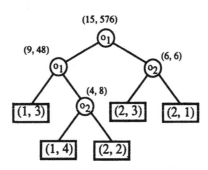

Fig. 2. A linear expression tree with two binary linear operators. For $x = (x_1, x_2)$ and $y = (y_1, y_2)$ operators are defined to be: $o_1(x, y) = (x_1 + y_2, 2y_1 y_2)$ and $o_2(x, y) = (2x_1 + x_1 y_2, x_2 y_1)$. The unary operators associated with the edges are the identity operators.

tree attribute system that supports tree attribute $x(\mu)$ by means of the transfer pairs of the solid paths of T, which form a path attribute system.

Theorem 5. *Let S be a semiring and r an integer constant. There is a fully dynamic data structure for maintaining a set of expressions with (unary and binary) linear operators over S^r such that a linear attribute grammar of size n uses $O(n)$ space, and the operations of evaluating, everting, composing, and decomposing expressions, and updating variables, take each time $O(\log n)$, where n is the size of the linear expression trees involved in the operation.*

3.2 Linear Attribute Grammars

In this section, we consider the dynamic evaluation of linear attribute grammars and show that this problem can also be solved using a tree attribute system.

An *attribute grammar* T is a rooted, ordered tree such that each node μ of T stores an r-tuple of attributes $x(\mu) = (x_1(\mu), \ldots, x_r(\mu))$. *Synthesized* attributes are calculated from the attributes of the children of μ. *Inherited* attributes are calculated from the attributes of the parent of μ, and the synthesized attributes of μ and its siblings. A *binary attribute grammar* is an attribute grammar on a binary tree. We consider attribute grammars where the values of each attribute are taken from a semiring S.

Attribute grammars [22] are much studied and have applications in areas such as language based editors [3,28], and VLSI design [20]. Incremental algorithms for attribute grammars have been presented [19,27] which give a $O(a \cdot t)$ running time per operation where a is the number of attributes affected by a change and t is the time to make a single change. In general, $a = O(n)$ for an attribute grammar of size n. Incremental evaluation of general circuits is studied in [1].

The *precedence graph* for an attribute grammar T is a digraph consisting of a vertex for each attribute $x_i(\mu)$, and a directed edge from $x_i(\mu)$ to $x_j(\nu)$ whenever the value $x_i(\mu)$ is used in the calculation of $x_j(\nu)$. A *binary linear attribute grammar* T is a binary attribute grammar such that: (i) The value of all attributes are taken from a semiring S. (ii) Each node μ of T has an associated linear operator to calculate the value $x(\mu)$ from the values of the attributes stored at the parent, children, and siblings of μ. (iii) The precedence graph of G is acyclic. (iv) The dependence of the value an attribute $x_i(\mu)$ on the value of any other attribute $x_j(\nu)$ is linear. Therefore, we can express the dependencies between attributes in the form $x_i(\mu) = a x_j(\nu) + b$ where a and b are members of S. Hence, a (binary) linear expression tree is a (binary) linear attribute grammar without inherited attributes.

Suppose $x = (x_1, \ldots, x_h)$ is a vector. The power vector $power(x)$ is a vector with 2^h entries $(1, x_1, \ldots, x_h, x_1 x_2, \ldots, x_{h-1} x_h, \ldots, x_1 \cdots x_h)$.

Lemma 6. *Suppose G is a precedence graph such that all dependencies between attributes are linear. Consider attributes $y_0, \ldots, y_f$ of G. The dependence of the value of y_0 on the values of $y_1, \ldots, y_f$ can be expressed as $y_0 = a \cdot \mathrm{power}((y_1, \ldots, y_f))$ for some vector a with 2^f components.*

We associate with each node μ in linear attribute grammar T a directed acyclic graph known as $summarygraph(\mu)$ which describes the dependencies if the synthesized attributes of μ on the inherited attributes of μ. Similarly, We associate with each path Π of T a directed, acyclic graph $summarygraph(\Pi)$ which describes the dependencies of the synthesized attributes of $tail(\Pi)$ and the inherited attributes of $head(\Pi)$ on the synthesized attributes of $head(\Pi)$ and the inherited attributes of $tail(\Pi)$. In the following lemma, the vector a of Lemma 6 is used to describe the dependencies in $summarygraph(\Pi)$ and $summarygraph(\mu)$.

Lemma 7. *Let T be a linear attribute grammar, $R(\mu)$ be the graph summarygraph(μ), $N(\mu)$ be the bidirectional binary operator $\odot$, and $P(\Pi)$ be the directed graph summarygraph(Π). Then there exists a tree attribute system that supports tree attribute $R(\mu)$ by means of the path attribute $P(\Pi)$.*

Theorem 8. *Let S be a semiring and r an integer. There is a fully dynamic data structure for maintaining a set of linear attribute grammars over S^r such that a linear attribute grammar of size n uses $O(n)$ space, and the operations of evaluating, everting, composing, and decomposing expressions, updating variables, and exchanging operators, take each time $O(\log n)$, where n is the size of the linear attribute grammars involved in the operation.*

4 Applications

Our techniques have many applications to important problems in data structures, graph algorithms, computational geometry, and VLSI layout. A number of existing data structures can be expressed as tree attribute systems. This includes dynamic trees [30], dynamic expression trees [8], and edge ordered trees [13]. In [7], linear attribute grammars are used to develop fully dynamic algorithms for a large number of optimization problems on bounded tree-width graphs. In [6], linear attribute grammars are used to maintain drawings of series-parallel graphs and trees. In the rest of this section, we present the following applications: two generalizations of heaps, *summation heaps* and *blocking heaps*, point location in binary space partitions, and compaction of slicing floorplans. Further applications of our techniques are presented in the full paper.

4.1 Blocking Heaps

Blocking heaps are a generalization of heaps. They can be used in dynamic algorithms for bounded tree-width graphs [7]. Suppose S is a totally-ordered set. We consider operations in S^r, the space of r-tuples of S for some fixed r. Let $x = (x_1, \ldots, x_r)$ and $y = (y_1, \ldots, y_r) \in S^r$. A *binary minimization operator* $\odot : S^r \times S^r \to S^r$ is defined such that each component $(x \odot y)_i$ calculates the minimum of some subset of the components of x and some subset of the components of y. Given a component μ_i at node μ, a *source* of μ_i is a component λ_j at leaf λ such that reducing the value of λ_j, reduces the value of μ_i. A *blocking heap* is a linear expression tree of binary minimization operators which also supports the query operation of returning a source for each of the components of μ.

We can represent a blocking heap as a linear expression tree and find sources using a selection function.

Theorem 9. *There exists a fully dynamic $O(n)$-space data structure for maintaining a collection of blocking heaps of total size n, such that a query or update operation takes time $O(\log n)$.*

A summation heap is an expression tree over operators $+$ and min where queries consist of determining the node representing the minimum (or maximum) value subexpression. It is easy to see that we can represent a summation heap as a blocking heap.

Theorem 10. *There exists a fully dynamic $O(n)$ space data structure for maintaining a collection of summation heaps of total size n, such that a query or update operation takes time $O(\log n)$.*

4.2 Point Location in Binary Space Partitions

A binary space partition is a binary tree where each node corresponds to a connected region of the space (or plane) and to a geometric object that bipartitions such a region. E.g., a node corresponds to a trapezoid and to a segment that divides it into two trapezoids (for details see [24]). Hence, a binary space partition is a planar subdivision whose regions are associated with the leaves of the tree. A point location query determines the region containing a given query point. If the discrimination of a point with respect to the dividing object (i.e., figuring out which of the two children regions contains the point) can be done in polylog time (usually $O(1)$ time), then one can do point location in polylog time. A variety of geometric problems can be efficiently solved by point location in binary space partitions. Hence, previous work has addressed the problem of constructing balanced binary space partitions [26,29]. We show that point location can be efficiently performed in general (unbalanced) partitions in a fully dynamic environment. A specialized version of this technique was introduced in [4].

We can represent a binary space partition as an unbalanced tree and perform point location using selection functions.

Theorem 11. *There is a fully dynamic data structure for maintaining a collection of binary space partitions such that a binary space partition of size n requires $O(n)$ space, point location queries take $O(q_1(n) \cdot \log n)$ time, and update operations take*

$O(q_1(n) \cdot \log n)$ *time, where $q_1(n)$ is the time required to perform point location in a single region and $q_2(n)$ is the time required to build a single partition.*

4.3 Slicing Floorplan Compaction

Linear attribute grammars have immediate application to the problem of compacting *slicing floorplans*, a layout technique widely used in VLSI (see, e.g., [32]). A slicing floorplan is either a rectangle (called basic rectangle), or is the union of two slicing floorplans that share a horizontal side (called horizontal slice) or a vertical side (called vertical slice). Each basic rectangle r has a minimum width w_r and a minimum height h_r, which describe the dimensions of a circuit module to be placed inside r.

An important problem in VLSI layout is determining the location of basic rectangles while minimizing the area of the slicing floorplan, subject to the above constraints on the height and width of the basic rectangles. (For full details, see [32].) This problem can be solved using a linear attribute grammar T. Leaves of T represent basic rectangles, and internal nodes represent horizontal or vertical slices. By implementing the parse tree of a slicing floorplan as a tree attribute system, we get:

Theorem 12. *There exists a fully dynamic $O(n)$-space data structure for maintaining a collection of slicing floorplans of total size n, such that a query or update operation takes $O(\log n)$ time.*

References

[1] B. Alpern, R. Hoover, B. Rosen, P. Sweeney, and F.K. Zadeck, "Incremental Evaluation of Computational Circuits," *Proc. ACM-SIAM Symp. on Discrete Algorithms* (1990), 32–42.

[2] S.W. Bent, D.D. Sleator, and R.E. Tarjan, "Biased Search Trees," *SIAM J. Computing* 14 (1985), 545–568.

[3] G. M. Beshers and R. H. Campbell, "Maintained and Constructor Attributes," *Proc. ACM Symp. on Language Issues in Programming Environments, ACM SIGPLAN Notices* (1985), 34–42.

[4] Y.-J. Chiang, F.P. Preparata, and R. Tamassia, "A Unified Approach to Dynamic Point Location, Ray Shooting and Shortest Paths in Planar Maps," *Proc. ACM-SIAM Symp. on Discrete Algorithms* (1993).

[5] Y.-J. Chiang and R. Tamassia, "Dynamic Algorithms in Computational Geometry," *IEEE Proc., Special Issue on Computational Geometry* 80 (1992), 362–381.

[6] R. F. Cohen, G. Di Battista, R. Tamassia, I.G. Tollis, and P. Bertolazzi, "A Framework for Dynamic Graph Drawing," *Proc. ACM Symp. on Computational Geometry* (1992), 261–270.

[7] R.F. Cohen, S. Sairam, R. Tamassia, and J.S. Vitter, "Dynamic Algorithms for Optimization Problems in Bounded Tree-Width Graphs," *Proc. 3rd Integer Programming and Combinatorial Optimization Conference* (1993).

[8] R.F. Cohen and R. Tamassia, "Dynamic Expression Trees and their Applications," *Proc. ACM-SIAM Symp. on Discrete Algorithms* (1991), 52–61.

[9] R.F. Cohen and R. Tamassia, "Combine and Conquer," Dept. Computer Science, Brown Univ., Technical Report CS-92-19, 1992.

[10] G. Di Battista and R. Tamassia, "Incremental Planarity Testing," *Proc. 30th IEEE Symp. on Foundations of Computer Science* (1989), 436–441.

[11] G. Di Battista and R. Tamassia, "On-Line Graph Algorithms with SPQR-Trees," *Automata, Languages and Programming (Proc. 17th ICALP), Lecture Notes in Computer Science* 442 (1990), 598–611.

[12] D. Eppstein, Z. Galil, G.F. Italiano, and A. Nissenzweig, "Sparsification — A technique for speeding up dynamic graph algorithms," *Proc. IEEE Symp. on Foundations of Computer Science* (1992).

[13] D. Eppstein, G.F. Italiano, R. Tamassia, R.E. Tarjan, J. Westbrook, and M. Yung, "Maintenance of a Minimum Spanning Forest in a Dynamic Plane Graph," *J. of Algorithms* 13 (1992), 33–54.

[14] G.N. Frederickson, "Data Structures for On-Line Updating of Minimum Spanning Trees, with Applications," *SIAM J. Computing* 14 (1985), 781–798.

[15] G.N. Frederickson, "Ambivalent Data Structures for Dynamic 2-Edge-Connectivity and k Smallest Spanning Trees," *Proc. 32th IEEE Symp. on Foundations of Computer Science* (1991).

[16] G.N. Frederickson, "A Data Structure for Dynamically Maintaining Rooted Trees.," *Proc. ACM-SIAM Symp. on Discrete Algorithms* (1993).

[17] Z. Galil and G.F. Italiano, "Fully Dynamic Algorithms for Edge-Connectivity Problems," *Proc. 23th ACM Symp. on Theory of Computing* (1991), 317–327.

[18] Z. Galil and G.F. Italiano, "Maintaining Biconnected Components of Dynamic Planar Graphs," *Automata, Languages and Programming (Proc. 18th ICALP), Lecture Notes in Computer Science* (1991).

[19] L. G. Jones, "Incremental Compaction of Flat Symbolic IC Layouts," Department of Computer Science, University of Illinois, Urbana, Illinois, Technical Report No. UIUCDCS-R-87-1386, 1987.

[20] L. G. Jones and J. Simon, "Hierarchical VLSI Design Systems Based on Attribute Grammars," *Proc. 13th ACM Symp. on Principles of Programming Languages* (1986), 58–69.

[21] A. Kanevsky, R. Tamassia, J. Chen, and G. Di Battista, "On-Line Maintenance of the Four-Connected Components of a Graph," *Proc. 32th IEEE Symp. on Foundations of Computer Science* (1991), 793–801.

[22] D. E. Knuth, "Semantics of Context-Free Languages," *Mathematical Systems Theory* 2 (1968), 127–145.

[23] M. Overmars, "The Design of Dynamic Data Structures," *Lecture Notes in Computer Science* 156 (1983).

[24] M.S. Patterson and F.F. Yao, "Optimal Binary Space Partitions for Orthoganal Objects," *Proc. 1st ACM-SIAM Symp. on Discrete Algorithms* (1990), 100–106.

[25] J.A. La Poutre, "Dynamic Graph Algorithms and Data Structures," Dept. of Computer Science, University of Utrechet, Utrechet, Ph.D. Thesis, 1991.

[26] F.P. Preparata, "A New Approach to Planar Point Location," *SIAM J. Computing* 10 (1981), 473–483.

[27] T.W. Reps, *Generating Language-Based Environments*, The MIT Press, 1984.

[28] T.W. Reps and T. Teitelbaum, *The Synthesizer Generator*, Springer-Verlag, 1989.

[29] C. Schwarz, M. Smid, and J. Snoeyink, "An Optimal Algorithm for the On-line Closest-Pair Problem," *Proc. ACM Symp. on Computational Geometry* (1992).

[30] D.D. Sleator and R.E. Tarjan, "A Data Structure for Dynamic Trees," *J. Computer Systems Sciences* 24 (1983), 362–381.

[31] D.D. Sleator and R.E. Tarjan, "Self-Adjusting Binary Search Trees," *J. ACM* 32 (1985), 652–686.

[32] L. Stockmeyer, "Optimal Orientation of Cells in Slicing Floorplan Design," *Information and Control* 57 (1983), 91–101.

[33] J. Westbrook and R.E. Tarjan, "Maintaining Bridge-Connected and Biconnected Components On-Line," *Algorithmica* 7 (1992), 433–464.

Optimal CREW-PRAM Algorithms for Direct Dominance Problems*

Amitava Datta[†]   Anil Maheshwari[‡]   Jörg-Rüdiger Sack[§]

Abstract

We present optimal parallel solutions to direct dominance problems for planar point sets. Our algorithms are deterministic and designed to run on the concurrent read exclusive write parallel random-access machine (CREW PRAM). In particular, we provide algorithms for counting the number of points that are directly dominated by each point of a planar point set, and for reporting these point sets. The counting algorithm runs in $O(\log n)$ time using $O(n)$ processors; the reporting algorithm runs in $O(\log n)$ time using $O(n+k/\log n)$ processors, where k is the size of the output. The total work of each algorithm matches the respective sequential lower bound. As an application of our results, we present an algorithm for the maximum empty rectangle problem, which is work optimal in the expected case.

1 Introduction

Let $P = \{p_1, p_2, ..., p_n\}$ be a planar point set of n points $p_i=(x_i, y_i)$, $i=1,...,n$. A point p_i is said to *dominate* a point p_j, if both $x_i \geq x_j$ and $y_i \geq y_j$ and $p_i \neq p_j$. The dominance problem is to enumerate all dominances of a given point set.

Dominance problems are directly related to well studied geometric problems like range searching, finding maximal elements and maximal layers, computing a largest area empty rectangle in a point set and interval/rectangle intersection problems (see e.g. [21]).

The 2-dimensional dominance problem is equivalent to the interval enclosure problem which asks for all enclosures of a given line interval set (see [15]). The enclosure problem and hence the dominance problems have several practical applications in computer aided design systems for VLSI circuits [15].

Optimal sequential algorithms for counting and reporting the dominances for each point of a given point set P are discussed in [21]. The time complexities are $O(n \log n)$ and $O(n \log n + k)$, respectively, where k denotes the total number of dominance pairs.

In the *direct dominance* reporting problem, all non-redundant dominances in a set are to be reported. Since the dominance relation is transitive, in the worst case, $\Theta(n^2)$ such redundant dominances are a possible output.

---

*This research was supported in part by the ESPRIT Basic Research Actions Program, under contract No. 7141 (project ALCOM II).

[†]Max-Planck Institut für Informatik, Im Stadtwald, W-6600 Saarbrücken, Germany; e-mail: datta@mpi-sb.mpg.de

[‡]CSC Group, TIFR, Homi Bhabha Road, Bombay 400 005, India; e-mail: manil@tifrvax.tifr.res.in

[§]School of Computer Science, Carleton University, Ottawa, Ontario K1S 5B6, Canada; e-mail: sack@scs.carleton.ca. This research was in part supported by Natural Sciences and Engineering Council of Canada.

Dominance problems have been studied for the CREW-PRAM (for details on this model see e.g., [16, 17]). Atallah et al. [3] solved the two-set dominance counting problem, in optimal $O(\log n)$ time using $O(n)$ processors, where n is the total number of points in the given sets. In this problem, given two point sets A and B, all pairs (a, b) are to be counted where $a \in A$ dominates $b \in B$. In the reporting mode of the problem, all dominance pairs are to be enumerated. Goodrich [14] solved this problem in $O(\log n)$ time using $O(n/\log n + k)$ processors, where k is the total number of dominance pairs. Note that the algorithm of Goodrich computes the value of k on the fly and the processor allocation is explicit and global, see [14] for the details of the computation model.

A point $p_i \in P$ *directly dominates* a point $p_j \in P$ if p_i dominates p_j and there exists no other point $p_k \in P$ such that p_i dominates p_k and p_k dominates p_j. It is easy to see that the dominance relation is the transitive closure of the direct dominance relation. The set of all points directly dominated by p_i is denoted by $DOM(p_i)$. In P, points which are not dominated by any other point are called *maximal*; the set of all maximal points is denoted by $MAX(P)$. The *direct dominance counting* problem is to compute the numbers $|DOM(p_i)|$, for every point $p_i \in P$; where $||$ denotes the cardinality of a set. The *direct dominance reporting* problem is to report the sets $DOM(p_i)$, i.e., to list the elements, for every point $p_i \in P$.

In the sequential model of computation, the dominance counting and reporting problems have tight $\Theta(n \log n)$ and $\Theta(n \log n + k)$ time bounds, respectively, where k is the total number of direct dominance pairs in P [15]. Note that k is at least $n - 1$, but it can be as large as $\Omega(n^2)$. Güting et al. [15] and Overmars et al. [20] have studied the direct dominance problem in the context of rectangular visibility. Gewali *et al.* [13] use rectangular visibility for solving a covering problem.

Chen and Friesen [8] presented a parallel algorithm for solving the direct dominance reporting problem which takes $O(\log n + j)$ time and uses $O(n)$ processors on the CREW PRAM, where j denotes the maximum of the number of direct dominances reported by a single point in the set. Because one point can directly dominate all other points in the set, in the worst case the run-time of their algorithm is $\Omega(n)$.

In this paper, we present optimal parallel algorithms for the problems of direct dominance counting and reporting. As an application of our results, we give an algorithm to find the maximum empty rectangle inside a planar point set; the resulting algorithm is work optimal in the expected case. All our parallel algorithms run on the CREW PRAM variety of parallel model of computation. Our direct dominance counting algorithm runs in optimal $O(\log n)$ time using $O(n)$ processors. The algorithm for the direct dominance reporting problem runs in $O(\log n)$ time, uses $O(n + k/\log n)$ processors, and $O(n \log n)$ space, where k is the size of the output. The total work, i.e., the processor-time product is $O(n \log n + k)$ and thus matches the tight sequential lower bound known for this problem. Our algorithm uses an output sensitive number of processors. The output size i.e., k, is not known in advance. We compute the size of the output on the fly and allocate the required number of processors in the spirit of [14]. The processors are 'spawned' depending on the output size. When new processors are allocated, a global array of pointers is created. A processor knows the exact location from where it should start working by accessing this array. For the details of computation model, see Goodrich [14].

We use the following standard tools in parallel algorithms: Parallel merge-sort [6], lowest common ancestor in a tree [22], tree operations including the Euler tour technique

[23], centroid level of nodes in a tree [9], parallel prefix, list ranking and doubling, described e.g., in [16, 17].

The remainder of the paper is organized as follows. In Section 2, we give some preliminaries. In Section 3, we present a data structure crucial to our computation of direct dominances. The algorithms for the direct dominance counting and reporting problems are presented in Section 4. The algorithm for the maximum empty rectangle problem is discussed in Section 5.

2 Preliminaries

Let p_i be a point in a given planar point set P. We define the set $Window(p_i)$ as $\{p_j | p_j \in P \text{ and } x_j > x_i, y_j < y_i\}$ which is the lower right quadrant of an orthogonal coordinate system centered at p_i. We denote by $MAX(Window(p_i))$ the points which are maximal in $Window(p_i)$. We define a data structure, denoted by $MF(P)$, on P, by keeping a pointer for each $p_i \in P$ to the highest y-coordinate point in its set $MAX(Window(p_i))$. In the following lemmas, we state properties of this data structure as relevant in the context of this paper.

Lemma 2.1 *For any input point set P, MF(P) is a directed forest.*

Proof: Omitted in this extended abstract. □

We refer to this directed forest as the *Maximal Forest* for the set P, and denote it by *MF(P)*.

Lemma 2.2 *There exists a directed path from a point p_i to a point p_j in $MF(P)$, if and only if (i) $x_j > x_i$ and $y_j < y_i$, and (ii) there is no point p_k such that $x_k > x_i, x_k > x_j$, and $y_i > y_k > y_j$.*

Proof: 'only if': If there is a directed path from p_i to p_j, the first condition follows from the construction of the forest. We now prove that, if there is a directed path from p_i to p_j, the second condition is also true. The proof is by contradiction. Suppose, a point p_k satisfying the condition (ii) exists. Then there are two possibilities. Either there is a directed path from p_i to p_k or there is no such path. In the first case, no directed path exists from p_k to p_j, because $x_j < x_k$. Suppose, there is a directed path from p_i to p_j. This path cannot pass through p_k. We traverse backwards towards p_i along the two paths from p_j and p_k. These two paths must diverge from either p_i or from some other point p_l below p_i. This contradicts the fact that every point has only one outgoing pointer. In the second case (i.e., there is no directed path from p_i to p_k), the directed path from p_i is already attached to some other point p_m such that $y_m > y_k$ and $x_m > x_k$. This implies that $x_m > x_j$ and hence no directed path can exist from p_m to p_j. The proof that no directed path from p_i to p_j exists, can be completed as in the previous case.

'if': We now prove the lemma in the other direction. Suppose, no point p_k exists satisfying (i) and (ii). This implies that all the points in between p_i and p_j (according to y-coordinate) have an x-coordinate less than p_j. Consider a point p_l having the least y-coordinate among all these points. Then a directed edge exists from p_l to p_j. Continuing this construction backwards, a directed path from p_i to p_j is shown to exist. □

Corollary 2.3 *Assume that p_i, p_j and p_k are three points for which $x_i < x_j < x_k$ and $y_i > y_j > y_k$ and that there are directed paths from p_i to p_j and from p_j to p_k. Then the unique directed path from p_i to p_k visists p_j.*

Proof: Follows from Lemma 2.1. □

Suppose all points in the set P are placed in the leaves of a complete binary tree T sorted by increasing x-coordinate from left to right. Consider a node $u \in T$ and let its left and right children be v and w, respectively. Let p_i be a point in the subtree rooted at w. We want to determine the points in the subtree rooted at v which are directly dominated by p_i. Let p_j be the point having the highest y-coordinate among all points directly dominated by p_i and lying in the subtree rooted at w. (In case there are several points with the same y-coordinate, we choose the one with the least x-coordinate among them.) Define $H(p_i)$ and $H(p_j)$ to be horizontal lines through p_i and p_j, respectively. Let $Strip(p_i, p_j)$ be the set of points in the node v whose y-coordinates lie between $H(p_i)$ and $H(p_j)$.

Lemma 2.4 *Any point in v which is directly dominated by p_i belongs to the set $MAX(Strip(p_i, p_j))$.*

Consider the maximal forest at v, i.e., $MF(v)$. The two points just below the horizontal lines $H(p_i)$ and $H(p_j)$ in $MF(v)$ are called $below(p_i)$ and $below(p_j)$, respectively. We denote the *lowest common ancestor* of the points $below(p_i)$ and $below(p_j)$ by $lca(below(p_i), below(p_j))$.

Lemma 2.5 *If $below(p_i)$ and $below(p_j)$ are in the same connected component in $MF(v)$, then the points on $MAX(Strip(p_i, p_j))$ are the vertices in the directed path in $MF(v)$ from $below(p_i)$ to the point just before $lca(below(p_i), below(p_j))$. If $below(p_i)$ and $below(p_j)$ are in the different components in $MF(v)$, then the points on $MAX(Strip(p_i, p_j))$ are the vertices in the path from $below(p_i)$ to the root of the component of $MF(v)$ containing it.*

Proof: We first consider the case when the points $below(p_i)$ and $below(p_j)$ are in the same connected component $MF_i(v)$ of $MF(v)$. Suppose, the lowest common ancestor of $below(p_i)$ and $below(p_j)$ i.e., $lca(below(p_i), below(p_j))$ is the node p_k. It is easy to see that there is a directed path from $below(p_i)$ to p_k. Suppose, p_l is the node just before p_k on this path. We claim that the points directly dominated by p_i in the subtree rooted at v are the points from $below(p_i)$ to p_l (including these two points). Note that, p_k is below the horizontal line $H(p_j)$. This follows from the fact that p_k is an ancestor of $below(p_j)$ in the directed tree $MF_i(v)$. To prove our claim, we show that p_l is above the line $H(p_j)$. This will prove that the points in $MAX(Strip(p_i, p_j))$ are the points from p_i to p_l.

We show this by contradiction. Assume that, p_l is below the line $H(p_j)$. Then two possibilities arise: p_l is either higher than $below(p_j)$ or lower than $below(p_j)$. In the first case, the definition of $below(p_j)$ is contradicted. In other words, p_l would be $below(p_j)$. So, we need to consider only the second case, i.e., p_l has y-coordinate less than that of $below(p_j)$. From Corollary 2.3, follows that there is a directed path from $below(p_j)$ to p_k, this path visits p_l. Furthermore, p_l lies on the directed path from $below(p_i)$ to p_k. This implies that p_l is the lowest common ancestor of $below(p_i)$ and $below(p_j)$, which is a contradiction.

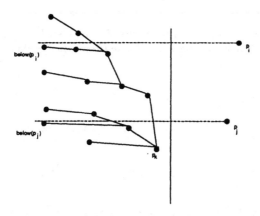

Figure 1: Illustration for Lemma 2.5.

Suppose, $below(p_i)$ and $below(p_j)$ are in different components. Then we observe that points which are maximal in $Window(p_i)$ start at $below(p_i)$ and lie on a directed path upto the root of its component. Any other point in $Strip(p_i, p_j)$ is dominated by some point on this path. □

3 Parallel Algorithm for Constructing the Data Structure

In this section, we describe a parallel algorithm for constructing a data structure common to our counting and reporting algorithms. The algorithm for constructing this data structure consists of four phases, as given in Algorithm 1. The computation carried out in each phase is discussed in detail in the following. We assume for simplicity that the number of points in the input set P is $n = 2^k$, for some $k \geq 1$.

Phase 1 : Sort the points in P according to increasing x-coordinates and store them in the leaves of a complete binary tree T from the left to the right. This can be performed in $O(\log n)$ time using n processors by the algorithm of Cole [6].

Phase 2 : (This phase is based on Cole's parallel-merge sort algorithm [6].) Sort the points according to decreasing y-coordinate. At each internal node u of T, maintain a list of points, sorted by y-coordinate, in the subtree rooted at u. Compute the maximal forest, $MF(u)$, of u for the points in the subtree rooted at u. In addition to this, establish cross-ranking pointers from a child node to its parent in T to facilitate search in Phase 4.

The generic merge step is performed as follows. Let v and w be the left and the right child of node u in T, respectively. Further, assume that the maximal forests for the nodes v and w have already been computed. We state the procedure to compute $MF(u)$ from $MF(v)$ and $MF(w)$. Denote by $Y(u)$, $Y(v)$ and $Y(w)$ the arrays of elements, sorted by decreasing y-coordinate and corresponding to the nodes u,v and w, respectively.

---

1. Place the points of the set P as the leaves of a complete binary tree T so that the leaves are sorted by increasing x-coordinate (from left to right).

2. Perform a bottom-up sort of the points according to increasing y-coordinate and for each node u of T, compute the maximal forest $MF(u)$ of the set of points in the subtree rooted at u.

3. For each node $u \in T$ and each $p_i \in MF(u)$, compute the number of nodes in the directed path from p_i to the root of the component of $MF(u)$ to which p_i belongs.

4. For each node $u \in T$, preprocess the maximal forest $MF(u)$ for answering lowest common ancestor queries.

Algorithm 1: The main steps in constructing the data structure

Observe that the merge step does not alter $MF(w)$. Consider now an element $p_i \in Y(v)$ and denote its rank in $Y(w)$ by $rank(p_i)$. Let $succ(p_i)$ denote the parent of p_i in $MF(v)$. Observe that if $rank(p_i) = rank(succ(p_i))$, then both p_i and $succ(p_i)$ have been ranked between the same pair of elements in $Y(w)$. Since p_i has a higher y-coordinate than $succ(p_i)$, the successor pointer of p_i remains unchanged. If, on the other hand, $rank(p_i) > rank(succ(p_i))$, then a point p_j in $Y(w)$ exists which lies between p_i and $succ(p_i)$. In this case, p_j becomes the new successor of p_i in the merged set. It is easy to see that the above computation can be performed in $O(\log n)$ time using $O(n)$ processors for the entire tree.

In another step of this phase a pointer is established from an element $p_i \in Y(u)$ to the element just below p_i in the array $Y(v)$ if $p_i \notin Y(v)$. Otherwise, i.e., if $p_i \in Y(v)$, the pointer points to p_i. Similar links are set up between the elements of $Y(u)$ and $Y(w)$. Recall that the algorithm of Cole [6] cross ranks the elements in $Y(v)$ and $Y(w)$ through the elements of $Y(u)$. In other words, a point $p_i \in Y(v)$ is first ranked in $Y(u)$ and each element of $Y(u)$ is ranked in $Y(w)$. So, in effect, the rank of p_i in $Y(w)$ is known. In Cole's algorithm, these pointers from $Y(v)$ to $Y(u)$ and from $Y(u)$ to $Y(w)$ are not maintained throughout the execution of the algorithm. Our algorithm, instead, will maintain these pointers. Using these pointers, in $O(1)$ time we can access from an element $p_i \in Y(u)$ the corresponding element in $Y(v)$, or $Y(w)$, or the element immediately below p_i in $Y(v)$ or $Y(w)$. We refer to such a pointer from $p_i \in Y(u)$ to $Y(v)$ and $Y(w)$ by $l\_link(p_i)$ and $r\_link(p_i)$, respectively. There are only two pointers for each element of the array $Y(u)$. So, the space requirement does not exceed that of the size of the array $Y(u)$. This concludes the description of Phase 2 of the algorithm. It can be seen that this phase requires $O(\log n)$ time using $O(n)$ processors.

Phase 3 : Consider a node u of T and its maximal forest $MF(u)$. Recall that at the end of Phase 2, every point $p_i \in u$ knows its successor in $MF(u)$. We want to compute the number of nodes (denoted as $count(p_i)$) of $MF(u)$ in the directed path from p_i. This is equivalent to computing the vertex level of each node. It can be computed by the list ranking algorithm of [9, 23]. Let n_i be the number of points in the subtree rooted at u. Assign $O(n_i/\log n)$ processors for the array $Y(u)$ at the node u. These processors can compute the rank of every element in $Y(u)$ in $O(\log n_i)$ time. Furthermore, this computation can be performed simultaneously at all nodes of T. So, this step can be performed in $O(\log n)$ time. Note that each point p_i in T appears in $O(\log n)$ nodes along the path from the root to the leaf containing p_i. So, the total number of processors required is $O(n \log n / \log n)$ i.e., $O(n)$. Hence, Phase 3 can be performed in $O(\log n)$ time using $O(n)$ processors.

Phase 4 : We invoke the algorithm of Schieber and Vishkin [22] to prepare all maximal forests $MF(u)$ for answering the *lowest common ancestor* queries. Each $MF(u)$ is preprocessed in $O(\log |MF(u)|)$ time using $O(|MF(u)|/\log |MF(u)|)$ processors. This preprocessing is done at all nodes of the tree simultaneously in $O(\log n)$ time using $O(n)$ processors. After preprocessing, lowest common ancestor of two points $p_i, p_j \in MF(u)$, is computed in $O(1)$ time by a uniprocessor. We summarize the properties of the data structure in the following lemma.

Theorem 3.1 *In $O(\log n)$ time using $O(n)$ processors, a data structure is constructed on the tree T allowing for any node $u \in T$, the following operations to be performed in $O(1)$ time by a single processor*
(i) Given a point $p_i \in u$, locate p_i or the point just below p_i among the children of u in $O(1)$ time.
(ii) Given a point $p_i \in u$, compute $count(p_i)$, i.e., the number of points on the path from p_i to the root of the component of $MF(u)$ containing p_i, in $O(1)$ time.
(iii) Given two points $p_i, p_j \in MF(u)$, compute the lowest common ancestor of p_i and p_j,i.e., $lca(p_i, p_j)$ in $O(1)$ time.

4 Algorithms for the Direct Dominance Counting and Reporting Problems

In this section, we present our optimal parallel algorithms for both direct dominance problems. To solve these problems we use the data structure developed in the previous section.

4.1 Direct Dominance Counting

Consider a path of p_i in T from a leaf node containing p_i to the root of T and denote it by $path(p_i)$. Our algorithm for the direct dominance counting problem traverses $path(p_i)$ from the leaf node containing p_i to the root of T. During this traversal it counts the direct dominances of p_i at each node in the path. Let u be a node of T belonging to $path(p_i)$. Two cases arise depending on whether the left child v of u or the right child w belongs to $path(p_i)$.

First, we assume that $v \in path(p_i)$. Point p_i does not dominate any point in the subtree rooted at w since all points in this subtree have a larger x-coordinate than p_i. Thus the sets of points which p_i directly dominates in the subtrees rooted at u and at v are equal.

Now consider the case that $w \in path(p_i)$. Point p_i directly dominates zero or more points of the set of points in the subtree rooted at v. Let p_j be the point having the largest y-coordinate in this set. Let $below(p_i)$ and $below(p_j)$ be defined as in Lemma 2.5. The points $below(p_i)$ and $below(p_j)$ can be located among the points in the subtree rooted at v by using the l_link pointers of Phase 2 of Algorithm 1. Count the number of points directly dominates by p_i in the subtree rooted at v, by using Lemma 2.5, as follows. First compute the lowest common ancestor node of $below(p_i)$ and $below(p_j)$ in $MF(v)$ by the algorithm of Schieber and Vishkin [22]. Let $lca(below(p_i), below(p_j))$ be the lowest common ancestor node. Using the preprocessing of Phase 4 of Algorithm 1, compute the number of nodes between $below(p_i)$ and $lca(below(p_i), below(p_j))$. Hence for each node u in $path(p_i)$, report the number of nodes directly dominated by p_i can be reported. Since, for any point $p_i \in P$, $path(p_i)$ has $O(\log n)$ nodes, a single processor can compute $|DOM(p_i)|$ in $O(\log n)$ time. Hence for all points p_i, $|DOM(p_i)|$ is reported in $O(\log n)$ time using $O(n)$ processors.

This concludes the description of the algorithm for the direct dominance counting problem. The space requirement is $O(n \log n)$ because each point appears in $O(\log n)$ nodes along $path(p_i)$. We summarize the results in the following theorem.

Theorem 4.1 *The direct dominance counting problem can be solved in optimal $O(\log n)$ time using $O(n)$ processors and $O(n \log n)$ space on the CREW PRAM.*

Proof: For a point p_i, the points it directly dominates have x-coordinates less than that of p_i. All such points appear in the left child of the nodes along $path(p_i)$. By counting all direct dominances in these nodes, the total count $|DOM(p_i)|$ is obtained. Furthermore, from Lemma 2.5 it follows that no point is missed which is directly dominated by p_i at such a node. □

Corollary 4.2 *Given a set of n points P in the plane, a data structure can be computed in $O(\log n)$ time using $O(n)$ processors and $O(n \log n)$ space on the CREW PRAM, such that a single processor can report in $O(\log n)$ time the number of points directly dominated in P by a given query point.*

4.2 Direct Dominance Reporting

Now we present a parallel algorithm for reporting the set $DOM(p_i)$ for all $p_i \in P$. As in the algorithm for the counting problem, we compute $path(p_i)$. Let u and w be in $path(p_i)$ as before. Further assume that we have reported the set of points which are directly dominated by p_i among the set of points in the subtree rooted at w. From Lemma 2.5 follows that the set of points in the subtree rooted at v which are directly dominated by p_i are the vertices on the path from $below(p_i)$ to the point just above $lca(below(p_i), below(p_j))$ in $MF(v)$. Thus the problem reduces to reporting this path in $MF(v)$. In the following, we solve the more general problem on reporting paths in binary trees efficiently and then show that the direct dominance reporting problem can be solved using a solution to the general problem.

A parallel algorithm (see Algorithm 2) for preprocessing an n-node rooted binary tree B such that the path between the two query nodes a and b in B can be reported efficiently is presented next. Let $size(v)$ be the number of vertices in the subtree rooted at v. The *centroid level* of a vertex v is given by $clevel(v) = \lceil \log_2(size(v)) \rceil$ [9]. Observe that for any node $v \in B$, there is at most one child of v which has the same centroid level as that of v.

Using Algorithm 2, we report the vertices in the path between two query nodes a and b (denoted as $path(a, b)$) in B as follows. First compute the lowest common ancestor node of a and b in B and let it be c. Now the problem reduces to reporting the paths $path(a, c)$ and $path(b, c)$. For simplicity assume that b is an ancestor of a. If $clevel(a) = clevel(b)$ then both a and b belong to the same array and the vertices in $path(a, b)$ are the elements

1. Using the algorithm of Schieber and Vishkin [22], preprocess B for answering lowest common ancestor queries.

2. Perform an Euler Tour of B and compute $size(v)$ for each node v of B by the algorithm of Tarjan and Vishkin [23].

3. For each node v of B, compute $clevel(v)$.

4. Partition the set of vertices of B into maximal disjoint paths, where each vertex v on a maximal path has the same value of $clevel(v)$. Store the vertices on a path in an array. This step can be performed by the algorithm of [9, 16].

5. Establish pointers from nodes of B to the indices in the array where they appear and vice versa.

6. For each node v in B, establish a pointer between v and the furthest ancestor of v in BT which has the same centroid level as $clevel(v)$. If v and the parent of v have different centroid levels, then we set up a pointer from v to the parent of v. These pointers can be computed by the algorithm of [16].

Algorithm 2: Algorithm for preprocessing a binary tree for reporting paths

in the array between a and b. The elements in the array between a and b can be reported in $O(\log n)$ time using $O(|path(a,b)|/\log n)$ processors.

Consider the case when $clevel(a) \neq clevel(b)$. Note that there are at most $O(\log n)$ distinct centroid levels in any $path(a,b)$. Hence a processor can identify the arrays which contains the vertices along $path(a,b)$ in $O(\log n)$ time. Using the pointers computed in Steps 3 and 4 of Algorithm 2 the two indices in each of these arrays can be located such that the elements between them are in $path(a,b)$. Thus the problem reduces to reporting elements between these two indices in each array. Allocate a total of $O(|path(a,b)|/\log n)$ processors to report the elements in $path(a,b)$. Each processor reports at most $O(\log n)$ nodes. The ith processor finds in $O(\log n)$ time the array and the indices of the elements in the array which it needs to report. Now processors have sufficient information to report the elements in $path(a,b)$ in $O(\log n)$ time. Hence, using the above data structure, the vertices in $path(a,b)$ can be reported in $O(\log n)$ time using $O(|path(a,b)|/\log n)$ processors.

Theorem 4.3 *An n−node binary tree can be preprocessed in $O(\log n)$ time using $O(n)$ space and $O(n/\log n)$ processors, such that the vertices in the path between two query nodes a and b, $path(a,b)$, can be reported in $O(\log n)$ time using $O(|path(a,b)|/\log n)$ processors on the CREW PRAM.*

Proof: The correctness of the algorithm follows from the property of $clevel(v)$ for each $v \in B$. Now we analyze the complexity of the algorithm. Each step of Algorithm 2 runs in $O(\log n)$ time using $O(n/\log n)$ processors [9, 16, 22, 23]. The preprocessing algorithm runs within the claimed complexity bounds, since there can be at most $O(\log n)$ different centroid levels on any $path(a,b)$ in B. The data structure required is of linear size since each step of Algorithm 2 requires a linear space. □

Using the above theorem, we design an optimal parallel algorithm for reporting the set of points of P which are directly dominated by $p_i \in P$. Recall that the remaining subproblem to solve was to report the vertices on the path from $below(p_i)$ to the point just above $lca(below(p_i), below(p_j))$ in $MF(v)$. If $MF(v)$ is a forest, then allocate a dummy root and this transforms $MF(v)$ to a tree. For the remaining discussion assume that that $MF(v)$ is a tree. Since $MF(v)$ need not be binary tree we transform $MF(v)$ to a binary tree using the optimal algorithm by [16]. Now using Theorem 4.3, we have an optimal procedure for reporting paths in $MF(v)$.

We analyze the complexity of the above algorithm. Each $MF(v)$ is preprocessed in $O(\log |MF(v)|)$ time using $O(|MF(v)|/\log |MF(v)|)$ processors to efficiently report the paths between two query nodes. Since each point of $p_i \in P$ is stored in $O(\log n)$ nodes, the total number of vertices summed over $MF(v)$ for all v is $O(n \log n)$. Hence the preprocessing algorithm runs in $O(\log n)$ time using $O(n)$ processors and it requires $O(n \log n)$ space. The algorithm for the direct dominance reporting requires an output sensitive number of processors. The algorithm computes the size of the output on the fly and allocates processors accordingly. We summarize the results in the following theorem.

Theorem 4.4 *The direct dominance reporting problem can be solved in optimal $O(\log n)$ time using $O(n + K/\log n)$ processors and $O(n \log n)$ space, where $K = \Sigma_{p_i \in P} |DOM(p_i)|$, on the CREW PRAM.*

Corollary 4.5 *Given a set of n points P in the plane, a data structure can be computed in $O(\log n)$ time using $O(n)$ processors and $O(n \log n)$ space on the CREW PRAM, such that $O(max\{1, K/\log n\})$ processors can report in $O(\log n)$ time the points directly dominated in P by a query point, where K is the number of points directly dominated in P by the query point.*

5 Fast parallel algorithm for the maximum empty rectangle problem

The *Maximum Empty Rectangle (MER)* problem is the following. Given an n-point set P inside a bounding isothetic rectangle BR, the problem is to find the maximum area/perimeter isothetic rectangle R such that R lies completely inside BR and R does not include in its interior any point from the set P. This problem has been extensively studied in recent years [2, 4, 5, 7, 11, 18, 19].

Aggarwal and Suri [2] have proved a sequential lower bound of $\Omega(n \log n)$ for this problem. The best known sequential algorithm runs in $O(n \log^2 n)$ time for the area problem and in $O(n \log n)$ time for the perimeter problem [2]. There are several algorithms which solve this problem by enumerating all *restricted rectangles* (rectangles whose sides are supported by the sides of BR or points from the set P and which are empty) in the point set P. The complexity of such algorithms is $O(n \log n + K)$ [5, 11, 19], where K is the number of restricted rectangles for a problem instance. It has been proved [18], that K is $O(n^2)$ and $O(n \log n)$ in the worst and expected cases respectively. So, these algorithms are optimal in the expected case.

In the domain of PRAM, this problem has been studied in [1, 12]. On the EREW PRAM, this problem was solved in $O(\log n)$ time using $O(n^2/\log n)$ processors [12].

The best known algorithm for this problem on the CREW PRAM is by Aggarwal et al [1]. They solve this problem in $O(\log^2 n \log \log n)$ time using $O(n \log n / \log \log n)$ processors. We present a parallel algorithm which matches the performance of the sequential algorithms in [5, 11, 19]. Our algorithm is work optimal in the expected case. We solve this problem using $O(n + K / \log n)$ processors and $O(\log n)$ time, where K is the number of restricted rectangles for a problem instance. A rectangle is called *restricted* if its sides either pass through the points of the set P or are flush with the sides of the bounding rectangle BR.

The complexity of computing the maximum empty rectangle is bounded by the complexity of the direct dominance reporting problem [4]. Using our parallel direct dominance reporting algorithm we get the following result (the details are found in the full version of this paper).

Theorem 5.1 *The maximum empty rectangle problem can be solved in $O(\log n)$ time using $O(n + K / \log n)$ processors on the CREW PRAM, where K is the number of restricted rectangles for a problem instance.*

Acknowledgement: The authors thank Kurt Mehlhorn for providing the excellent environment that allowed this research to be carried out.

References

[1] A. Aggarwal, D. Kravets, J. Park and S. Sen. *Parallel searching in generalized Monge arrays with applications.* Proc. 2nd ACM Symp. on Parallel Algorithms and Architectures, 1990, pp. 259-268.

[2] A. Aggarwal and S. Suri. *Fast algorithms for computing the largest empty rectangle.* Proc. 3rd Annual ACM Symp. on Comp. Geom., 1987, pp. 278-290.

[3] M. Atallah, R. Cole and M. Goodrich. *Cascading divide-and-conquer : a technique for designing parallel algorithms.* SIAM J. Computing, **18** (1989), pp. 499-532.

[4] M. Atallah and G. Fredrickson. *A note on finding the maximum empty rectangle.* Discrete Applied Mathematics. **13** (1986) pp. 87-91.

[5] M. Atallah and S. R. Kosaraju. *An efficient algorithm for maxdominance, with applications.* Algorithmica 4 (1989) pp. 221-236.

[6] R. Cole. *Parallel merge sort.* SIAM J. Computing, **17**, (1988), pp. 770-785.

[7] B. Chazelle, R. Drysdale and D. T. Lee. *Computing the largest empty rectangle.* SIAM J. Computing 15 (1986), pp. 300-315.

[8] I. W. Chen and D. K. Friesen. *Parallel algorithms for some dominance problems based on a CREW PRAM.* Proc. 2nd International Symposium on Algorithms, LNCS 557, 1991, pp. 375-384.

[9] R. Cole and U. Vishkin. *The accelerated centroid decomposition technique for optimal parallel tree evaluation in logarithmic time.* Algorithmica, **3** (1988), pp. 329-346.

[10] R. Cole and U. Vishkin. *Approximate parallel scheduling, Part I: The basic technique with applications to optimal parallel list ranking in logarithmic time.* SIAM J. Computing, **17** (1988), pp. 128-142.

[11] A. Datta. *Efficient algorithms for the largest rectangle problem.* Information Sciences, **64** (1992), pp. 121-141.

[12] A. Datta and K. Krithivasan. *Efficient algorithms for the maximum empty rectangle problem in shared memory and other architectures.* Proc. 1990 International Conference on Parallel Processing, 1990, **III**, pp. 344-345.

[13] L. Gewali, M. Keil, S. Ntafos, *On Covering Orthogonal Polygons with Star-Shaped Polygons.* Information Sciences, **65** (1992), pp. 45-63.

[14] M. T. Goodrich. *Intersecting line segments in parallel with an output-sensitive number of processors.* SIAM J. Computing, **20**, (1991), pp. 737-755.

[15] R. Güting, O. Nurmi and T. Ottmann. *Fast algorithms for direct enclosures and direct dominances.* J. Algorithms **10** (1989), pp. 170-186.

[16] J. JáJá. *An Introduction to Parallel Algorithms.* Addison-Wesley, 1992.

[17] R. M. Karp and V. Ramachandran, *Parallel Algorithms for Shared-Memory Machines*, Handbook of Theoretical Computer Science, Ed. J. van Leeuwen, Vol 1, Elsevier Science Publishers B.V, 1990.

[18] A. Namaad, W. Hsu and D. T. Lee. *On maximum empty rectangle problem.* Discrete Applied Mathematics **8** (1984), pp. 267-277.

[19] M. Orlowski. *A new algorithm for the largest empty rectangle problem.* Algorithmica **5** (1990), pp. 65-73.

[20] M. Overmars and D. Wood. *On rectangular visibility.* J. of Algorithms **9** (1988), pp. 372-390.

[21] F. P. Preparata and M. I. Shamos. *Computational Geometry: an Introduction.* Springer-Verlag, New York, 1985.

[22] B. Schieber and U. Vishkin. *On finding lowest common ancestors: Simplification and Parallelization.* SIAM. J. Computing, **17** (1988), pp. 1253-1262.

[23] R. E. Tarjan and U. Vishkin. *Finding biconnected components and computing tree functions in logarithmic parallel time.* SIAM J. Computing, **14** (1985), pp. 862-874.

Trekking in the Alps Without Freezing or Getting Tired*

Mark de Berg[1] and Marc van Kreveld[2]

[1] Department of Computer Science, Utrecht University,
P.O.Box 80.089, 3508 TB Utrecht, the Netherlands.
[2] School of Computer Science, McGill University,
3480 University St., Montréal, Québec, Canada H3A 2A7.

Abstract. For a polyhedral terrain F with n vertices, the concept of height level map is defined. This concept has several useful properties for paths that have certain height restrictions. The height level map is used to store F, such that for any two query points, one can decide whether there exists a path on F between the two points whose height decreases monotonically. More generally, one can compute the minimum height difference along any path between the two points. It is also possible to decide, given two query points and a height, whether there is a path that stays below this height. Although the height level map has quadratic worst case complexity, it is stored implicitly using only linear storage. The query time for all the above queries is $O(\log n)$, and the structure can be built in $O(n \log n)$ time. A path with the desired property can also be reported in additional time that is linear in the description size of the path.

1 Introduction

Polyhedral terrains (mountain landscapes) are important concepts in several application areas of computational geometry, including Geographic Information Systems. For that reason they are well-studied in computational geometry, and efficient algorithms have been given for e.g. hidden surface removal, ray shooting, and intersections of terrains [1, 4, 6, 9, 11]. The problem of computing shortest paths in the Euclidean metric on polyhedral terrains has also received attention. See for instance Sharir and Shorr [12] and Chen and Han [5], who solve the shortest path problem on the surface of a simple polyhedron in $O(n^2)$ time. For a fixed target, they obtain a data structure such that the shortest path from a given query point to the target can be found in $O(\log n)$ time.

However, the Euclidean length is not the only important feature when walking in terrains. Indeed, the shortest path is probably not advisable if it involves

* This research was performed when the second author visited the first author at Utrecht University. The research of the first author is supported by the Dutch Organization for Scientific Research (N.W.O.) and by ESPRIT Basic Research Action 7141 (project ALCOM II: *Algorithms and Complexity*). The research of the second author is supported by an NSERC international fellowship.

going to great height, with the risk of freezing to death. Furthermore, it is well known that walking in mountainous terrain can be very tiresome, so it is wise to take a path that is as level as possible. These considerations have led us to study the following questions for a polyhedral terrain F:

1. Given two points s and t in F and a height z, is there a path between them that stays below height z?
2. Given two points s and t in F, is there a path between them that stays on one height?
3. Given two points s and t in F, is there a path between them whose height is monotonically decreasing?
4. Given two points s and t in F, what is the minimum summed height difference of any path between them?

Observe that question 2 is a special case of question 3, which, in turn, can be answered using question 4. For all of the above questions, we develop an optimal data structure that requires linear storage and answers queries in $O(\log n)$ time. An actual path can also be found in additional time proportional to the description size of the path; in this extended abstract, however, we restrict ourselves to the decision problems.

It turns out that in these questions the saddle vertices play an important role. (A vertex is called a saddle vertex it has four incident edges that are going up, down, up, and down, when they are visited in cylic order.) Imagine drawing for each saddle vertex all paths of constant height emanating from it. Project these paths onto the xy-plane and add the projections of all peak vertices and valley vertices (i.e. local maxima and minima). We call the resulting planar map the *height level map* of the terrain, and we prove that this map contains all the information that is necessary to answer the above questions. Unfortunately, the complexity of the height level map can be quadratic. Therefore our data structure is an implicit representation of the map that uses only linear storage; point location queries in the map can still be answered in logarithmic time.

The remainder of this paper is organized as follows. In Section 2 we give the necessary definitions. Properties of the height level map are proved in Section 3. In Section 4 we present the data structure that underlies the solution to all four query problems. The actual query algorithms are given in Section 5. In Section 6 the concluding remarks are given.

2 Preliminaries

Let F^* be a polygonal terrain defined over the entire xy-plane. We shall restrict our attention to the part of this terrain defined on some rectangular area of the xy-plane. We denote this part by F and we define n to be the number of vertices of F, including vertices that arise because edges of F^* are clipped. The boundary of F is denoted by ∂F. Vertices on ∂F are called *boundary vertices*; the remaining vertices are called *interior vertices*. We denote the height of a point

p on F (that is, its z-coordinate) by $h(p)$, and we assume that no two vertices of F have the same height. (We believe that this restriction is not necessary; we are currently studying the technical details that are involved in removing the restriction.) Because we can triangulate every face of the terrain in linear time [3] we may assume without loss of generality that every face of F is a triangle. (If the faces are not necessarily simple polygons, i.e. if they are allowed to have holes, then we need $O(n \log n)$ time to triangulate them, but this will also not influence our final bounds.) With a slight abuse of notation, we denote by F both the (clipped) polyhedral terrain in 3-space, and its projection.[3]

For a vertex v of F, we denote by $N_1(v), \ldots, N_i(v)$ the cyclic sequence of neighbor vertices of v, ordered counterclockwise, where i is the degree of v in F. If v is a boundary vertex then we order its neighbors such that $N_1(v)$ and $N_i(v)$ are also a boundary vertices. A vertex v is a *local maximum* if all neighbors of v have smaller height than v. A vertex v is a *local minimum* if all neighbors of v have greater height than v. A vertex that is a local maximum or local minimum is a *local extremum*. An interior vertex v is a *saddle vertex* if v has two higher neighbors and two lower neighbors which alternate around v. More precisely, there are four neighbors $N_{i_1}(v), N_{i_2}(v), N_{i_3}(v), N_{i_4}(v)$ of v with $1 \leq i_1 < i_2 < i_3 < i_4 \leq i$, such that $h(v) > h(N_{i_1}(v))$ and $h(v) < h(N_{i_2}(v))$ and $h(v) > h(N_{i_3}(v))$ and $h(v) < h(N_{i_4}(v))$, or such that $h(v) < h(N_{i_1}(v))$ and $h(v) > h(N_{i_2}(v))$ and $h(v) < h(N_{i_3}(v))$ and $h(v) > h(N_{i_4}(v))$. A boundary vertex v is a saddle vertex if there are three neighbors $N_{i_1}(v), N_{i_2}(v), N_{i_3}(v)$ of v with $1 \leq i_1 < i_2 < i_3 \leq i$, such that $h(v) > h(N_{i_1}(v))$ and $h(v) < h(N_{i_2}(v))$ and $h(v) > h(N_{i_3}(v))$, or $h(v) < h(N_{i_1}(v))$ and $h(v) > h(N_{i_2}(v))$ and $h(v) < h(N_{i_3}(v))$. A vertex that is a local extremum or a saddle vertex is a *special vertex*. For a height h, the *h-map of F* is defined to be the intersection of F with the horizontal plane $z = h$. Any h-map may contain cycles and paths. By non-degeneracy, it contains at most one vertex of F. Moreover, there is at most one point shared by two or more cycles or paths. If there is such a point, then it must be a saddle vertex v of F, and $h = h(v)$. If two cycles share a saddle vertex v of F then one can be contained in the other, but this need not be the case. Note that any path (which is not a cycle) starting at v must end at the boundary of F. Let v be a saddle vertex of F, and consider the $h(v)$-map of F. We call the cycles and paths that contain v the *main component* M_v. Finally, we define the *height level map* of F to be the planar map consisting of the main components of all saddle vertices of F, to which are added all local extrema, see Figure 1. It appears that the height level map contains all the necessary information to solve the four above problems efficiently. Since any h-map has linear size, it follows that the height level map has size $O(n^2)$. Unfortunately, there exist polyhedral terrains for which the height level map has quadratic size. Therefore, we devise a data structure that stores the height level map implicitly: using only linear storage, we can find the region of the height level map that contains a given query point in $O(\log n)$ time.

---

[3] Here and in the sequel, projection always means orthogonal projection onto the xy-plane.

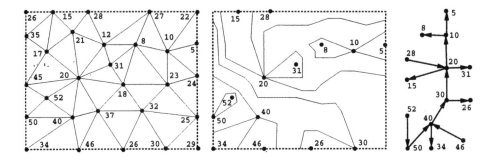

Fig. 1. A polyhedral terrain, its height level map and the graph G_F on the special vertices.

A second structure we use is a directed graph G_F of which the nodes correspond to the saddle vertices and local extrema of F. For convenience, for any special vertex v of F, we denote the corresponding node in G_F also by v. There is a directed arc (v, w) in G_F if and only if $h(v) > h(w)$, and there is a region in the height level map with v and w on the boundary of that region. We prove later that the undirected version T_F of G_F is actually a tree, and that the degree of a node v in G_F is the number of regions incident to the saddle vertex v. The local extrema correspond to the leaves of G_F. Furthermore, every region of the height level map corresponds to one edge of G_F, and vice versa.

3 The height level map

In this section we prove several important properties of the height level map and the corresponding graphs G_F and T_F.

Lemma 1. *Any region in the height level map is incident to exactly two special vertices, which appear in different components of the boundary of that region (a local extremum is also considered a boundary of the region in which it lies).*

Proof. Let R be any region of the height level map. Let v_{min} be the vertex inside or on the boundary of R with minimum height. If v_{min} is a local minimum, then v_{min} is a special vertex by definition. We claim that if v_{min} is not a local minimum then it must be on the boundary of R and, hence, a saddle vertex. Indeed, if v_{min} is interior to R but not a local minimum then v_{min} must have some edges going down and crossing the boundary of R. But then v_{min} is higher than this part of the boundary of R and the corresponding saddle vertex. Similarly, we can show that the highest vertex v_{max} of R is a local maximum or saddle vertex.

Suppose that there exists a third special vertex w incident to R. Assume w.l.o.g. that w is locally maximal inside R, that is, w is a local maximum if we

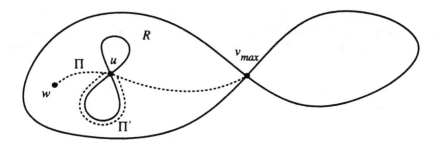

Fig. 2. Illustration of the proof of Lemma 1.

restrict our attention to the points in R. Let Π be a path in R from w to v_{max} such that the minimum height on Π is maximal. Since v_{max} and w are both local maxima in R, path Π must contain a lowest point u distinct from w and v_{max}. This point must be a saddle vertex, otherwise we obtain a contradiction with the maximality of the lowest point on Π. Now consider the main component M_u. If we can 'walk around' on M_u, as in Figure 2, then we also obtain a contradiction with the maximality of the height of Π, because we can push Π a small distance ε away from M_v to obtain a higher path Π'. But if we cannot walk around, then v_{max} and w are not incident to the same region.

Corollary 2. T_F *is a tree.*

Lemma 3. *Two points s and t on F can be connected with a path on one height if and only if $h(s) = h(t)$ and s and t lie in the same region of the height level map.*

Proof. Suppose there is a constant height path between s and t. Clearly, such a path cannot cross an edge of some main component and, hence, s and t are in the same region.

Now suppose for a contradiction that s and t lie in the same region R and $h(s) = h(t)$, but they cannot be connected with a constant height path. Assume w.l.o.g. that there is a path in R between s and t such that for the lowest point u on the path we have $h(u) < h(s)$. Let Π be such a path where $h(u)$ is maximal. As in the proof of Lemma 1, we can now show that u must be a saddle vertex and derive a contradiction.

For any region R of the height level map, the *lower boundary* is the component of the boundary of R with height smaller than the interior of R. The *lower boundary vertex* of R is the saddle vertex or local minimum on the lower boundary of R. The lower boundary and lower boundary vertex of a point p are the lower boundary and lower boundary vertex, respectively, of the region R in which it lies. Similarly, we define the *higher boundary* and the *higher boundary vertex* of R and of p. For any point p, we denote the lower boundary vertex by $lbv(p)$ and the higher boundary vertex by $hbv(p)$. If p lies on the boundary of a region in R, then $lbv(p)$ and $hbv(p)$ are both the saddle vertex that also lies on that boundary.

Lemma 4. *Two points s and t on F with $h(s) > h(t)$ can be connected with a path of monotonically decreasing height if and only if s and t lie in the same region of the height level map, or there exists a directed path in G_F from the node corresponding to $lbv(s)$ to the node corresponding to $hbv(t)$.*

Proof. We first show that any directed arc (u, v) in G_F implies the existence of a monotone decreasing path from the special vertex u to the special vertex v on F. First, observe that there exists a region R for which u is the upper boundary vertex and v is the lower boundary vertex. Since the region R does not contain any local minima except v, we can find a monotonically decreasing path inside R from u to the main component of v. From such an intersection point, follow the main component using a constant height path until it reaches v. This path is clearly monotone and decreasing. For the same reasons, there is a monotone decreasing path from any point inside R to the lower boundary vertex, and a monotone increasing path to the higher boundary vertex. If s and t lie in the same component, then a decreasing path starting at s to the lower boundary vertex will contain some point in R on height $h(t)$. By the previous lemma, there is a constant height path between this point and t.

On the other hand, suppose that there exists a monotonically decreasing path Π from s to t. Let $R_1, \ldots, R_k$ be the sequence of regions in the height level map that Π crosses. Since for every region R_i, there is an arc from the higher boundary vertex to the lower boundary vertex, it follows that Π is represented in the graph G_F.

Using a similar argument we can show the following.

Lemma 5. *Let s and t be two points on F. If s and t lie in the same region of the height level map then the minimum height difference on any path from s to t is $|h(s) - h(t)|$. Otherwise the minimum height difference is the minimum, over all choices $v \in \{lbv(s), hbv(s)\}$, $w \in \{lbv(t), hbv(t)\}$, of the quantity $|h(s) - h(v)| + W(v, w) + |h(w) - h(t)|$, where $W(v, w)$ is the sum of the height differences over the edges of the path from v to w in T_F.*

Lemma 6. *Two points s and t on F can be connected with a path that has height at most z, with $z \geq h(s)$ and $z \geq h(t)$, if and only if s and t lie in the same region of the height level map, or there exists a path in T_F from the node corresponding to $lbv(s)$ to the node corresponding to $lbv(t)$ which does not contain any nodes with height greater than z.*

4   The data structure

In this section we describe the data structure for point location in the height level map. For simplicity, we assume that every node in the tree T_F has constant degree, or, in other words, that every main component consists of a constant number of distinct paths and cycles. (This restriction is not difficult to remove, as we show in the full paper.)

The search tree $\mathcal{T}$. Before we describe the search tree, recall that the arcs of the tree T_F correspond one-to-one with the regions of the height level map. The *hbv* and *lbv* of that region are the two nodes incident to that arc.

We construct a constant degree search tree $\mathcal{T}$ as follows. Let δ be the root of $\mathcal{T}$, and let it correspond to the whole polyhedral terrain F. Find a node u in the graph T_F such that the removal of u from T_F gives c subgraphs[4] with at most $\lfloor n/2 \rfloor$ nodes each. We add a copy of node u back to all c subgraphs, so that every arc in T_F appears in exactly one subgraph. Therefore, the subgraphs correspond exactly to the cells into which F is partitioned by the main component of u. We associate the saddle vertex u with δ (but we should not store the main component explicitly, since it may have large complexity). For every subgraph that arises in the above way (or, equivalently, every cell defined by M_u) we make a child node γ of δ which is the root of a recursively defined subtree for that subgraph (or, that cell). If the subgraph consists of one single arc, then γ is a leaf of $\mathcal{T}$. The leaf γ represents the region of the height level map that corresponds to the single arc. We store $lbv(R)$ and $hbv(R)$—the two nodes incident to the arc—explicitly at γ. See Figure 3 for an example of T_F and the search tree $\mathcal{T}$. The tree $\mathcal{T}$ has linear size and it has depth $O(\log n)$.

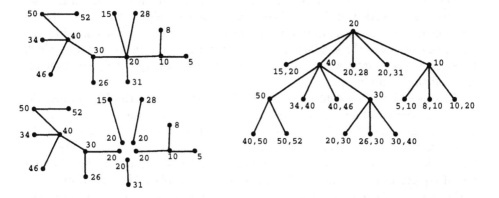

Fig. 3. The graph T_F of Figure 1, its partitioning into subgraphs and the search tree $\mathcal{T}$. Note that there is precisely one leaf for every region of the height level map.

---

[4] Even though T_F is a tree, we will speak of *subgraphs* of T_F to avoid confusion with subtrees of $\mathcal{T}$.

The conceptual query in $\mathcal{T}$. Let q be a point on F. We wish to know the region of the height level map that contains it. In particular, we wish to know $lbv(q)$ and $hbv(q)$. The search with point q starts at the root δ of $\mathcal{T}$. Each of the children corresponds to a cell in the plane separated by the paths and cycles of the main component corresponding to δ. We use an oracle $\omega(q, \delta)$ to determine in which of the cells q lies. Then the search continues recursively in the subtree corresponding to that cell, until a leaf is reached. At the leaf, $hbv(q)$ and $lbv(q)$ are stored. The query time is $O(\log n)$ times the time needed for the oracle.

The oracle for decisions in $\mathcal{T}$. We now explain how to implement the oracle $\omega(q, \delta)$. Thus, for a given node δ and query point q, we want to determine in which cell q lies with respect to the cycles and paths corresponding to δ. First, we need a planar point location structure on the xy-projection of the terrain F. This structure also stores for every vertex v of F its hbv and lbv, which we have precomputed. (How this precomputation is done is discussed later). Second, we assign an arbitrary node of T_F to be the root, and we preprocess T_F for $O(1)$ time lowest common ancestor queries [8].

At the start of the query with q, we determine the triangle t_q of F that contains q. We assume to simplify the discription that q is contained in the interior of t_q. Let v_1 be the vertex of triangle t_q with maximal height, and let v_2 be the vertex of t_q with minimal height. We have precomputed $lbv(v_1)$, $hbv(v_1)$, $lbv(v_2)$ and $hbv(v_2)$, which correspond to nodes in the tree T_F. We let $w_1 = hbv(v_1)$ and $w_2 = lbv(v_2)$. See Figure 4(i).

Let $\gamma_1, \ldots, \gamma_c$ be the children of δ in $\mathcal{T}$. Let us recall our convention that we denote special vertices in F and their corresponding nodes in T_F (or $\mathcal{T}$) by the same identifier. Also recall that a subtree of δ corresponds to a cell defined by the main component of the saddle vertex δ. To determine the subtree of δ in which to continue the search, we should determine which subgraph of T_F corresponds to a cell that contains q.

To this end we first determine which of the subtrees of δ (in $\mathcal{T}$) contains w_1, as follows. See also Figure 4(ii). Note that only one of the children γ_i of δ in $\mathcal{T}$ is not a descendant of δ in T_F. We can decide which child this is by computing the lowest common ancestor in T_F of each γ_i with δ; the unique γ_j (γ_2 in Figure 4) for which the answer is not δ is the one that is not a descendant in T_F. We first test if w_1 lies in the subtree of $\mathcal{T}$ rooted at this γ_j; this is the case if and only if the lowest common ancestor of δ and w_1 is not δ. If w_1 does not lie in the subtree of γ_j then we test the other children of δ: w_1 lies in the subtree of $\mathcal{T}$ rooted at γ_i if and only if the lowest common ancestor of w_1 and γ_i is not δ. (This last test is correct because we already know that w_1 is a descendant of δ in T_F.)

In the same way we compute the subtree of $\mathcal{T}$ that contains w_2. If w_1 and w_2 lie in the same subtree, then q must lie in that subtree as well and we have solved the oracle. (If either one of w_1 or w_2 is δ, then q must lie in the subtree chosen by the other of w_1 and w_2.) The reason that q must lie in that subtree is the following. The regions corresponding to the subtrees are separated by curves

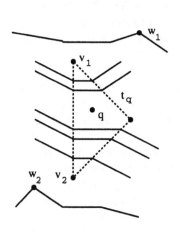 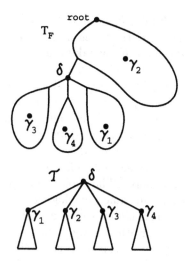

Fig. 4. (i): the triangle t_q in a part of the height level map, and the vertices w_1 and w_2. (ii): deciding in which subtree below δ to continue by means of lowest common ancestor queries in T_F.

of the main component M_δ that contains the saddle vertex δ. Recall that M_δ contains edges on one height only. Therefore, the triangle t_q of F intersects M_δ in at most one edge. Since w_1 and w_2 lie in the same region of M_δ, so must the whole triangle t_q.

On the other hand, if w_1 and w_2 lie in different subtrees, then t_q intersects M_δ and we do not know yet which of the two subtrees contains q. There are two candidates: the subtree that contains w_1 and the subtree that contains w_2. However, we know the height of q, the height of the saddle vertex δ and the heights of w_1 and of w_2. We have $h(w_1) > h(\delta) > h(w_2)$ and also $h(w_1) > h(q) > h(w_2)$. If $h(q) > h(\delta)$, then the search with q should continue in the same subtree of $\mathcal{T}$ as w_1. If $h(q) < h(\delta)$, then the search with q should continue in the same subtree of $\mathcal{T}$ as w_2. If $h(q) = h(\delta)$, then q lies on the main component M_δ.

The preprocessing. Suppose we know for every vertex of the terrain its *lbv* and *hbv*. From this we can construct the graphs G_F and T_F in linear time, and the search tree $\mathcal{T}$ in $O(n \log n)$ time. The preprocessing of T_F for lowest common ancestor queries requires $O(n)$ time. Furthermore, we preprocess the xy-projection of F in linear time for planar point location [7, 10], so that for each query point p, we can find the triangle that contains it efficiently. So the whole preprocessing requires $O(n \log n)$ time if we can compute the *lbv*'s and *hbv*'s. Below we sketch how to compute for each vertex v of F the lowest vertex w of F which lies higher than v, and which can be reached with a monotone increasing path. From this the *lbv*'s and *hbv*'s can easily be computed.

Find a value *mid* such that at most $\lceil n/2 \rceil$ vertices of F have height greater (smaller) than *mid* and no vertex of F has height equal to *mid*. Intuitively, we will split F by taking a plane $z = mid$ and deleting all edges that intersect the plane. The deleted edges are either 'simple cycles of edges' incident to a cycle of triangles, or 'simple paths of edges' incident to a path of triangles and which separate the terrain. For every component, we add edges to form a valid terrain again. This comes down to adding an edge between the highest vertex of a component and any other vertex of that component, so that the terrain is again triangulated. (It is possible that these new edges will intersect existing edges when embedded as straight line segments, but this is of no importance to us here.) The choice of the highest vertex assures that all other vertices still have the same next higher vertex in the next terrain. The highest vertex finds the next higher vertex during the merge step among the endpoints of the deleted edges. Both the divide and the merge step take linear time, so the divide-and-conquer algorithm runs in $O(n \log n)$ time. A detailed description and proof of correctness can be found in the full paper.

5   The height query algorithms

In this section we describe the query algorithms to the four problems stated in the introduction of this paper. The solutions are based on the properties of the height level map that were given in Section 3, and they make use of the data structures described in the previous section.

Constant-height paths. Given two points s and t with equal height, we wish to determine whether there exists a path between them that stays on the height $h(s) = h(t)$. By Lemma 1, such a path exists if and only if s and t lie in the same region of the height level map. Therefore, to solve the query, we determine $hbv(s)$, $lbv(s)$, $hbv(t)$ and $lbv(t)$. Since the hbv and lbv of any point determines uniquely the region of the height level map that contains it, we conclude that there is a constant height path between s and t if and only if $hbv(s) = hbv(t)$ and $lbv(s) = lbv(t)$.

Theorem 7. *Let F be a polyhedral terrain with n vertices with distinct heights. F can be preprocessed in $O(n \log n)$ time into a data structure of size $O(n)$, such that for any two query points, one can determine in $O(\log n)$ time whether there exists a path between them that stays on one height (assuming a RAM model of computation).*

The other queries. We now discuss monotone paths, minimum height difference paths and paths below a given query height z. The solutions to these problems turn out to be very similar. Recall that in the tree T_F, the arcs correspond one-to-one to regions of the height level map, and the endpoints of each arc are the hbv and lbv of the corresponding region. Let s and t be two query points that lie in different regions, and let a_s and a_t be the arcs in T_F corresponding to these regions. It is not difficult to see that any path Π connecting s and t in the height level map crosses the regions that correspond to the arcs between a_s and a_t in T_F. The nodes between a_s and a_t in T_F are the saddle vertices whose main components must be crossed by Π. It follows that we can test any of the above three requirements by considering only the heights of the saddle vertices on the path in T_F between v and w (see also Lemmas 3 and 4).

We use an extra data structure, or in fact an adaptation of T_F, to compute so-called weighted subpath queries in T_F. If T_F is a weighted tree with n nodes, then T_F can be preprocessed in linear time so that for any two nodes v, w in T_F, the sum of the weights of the arcs between v and w can be computed in $O(\alpha(n))$ time [2]. In fact, the function *sum* can be replaced by any other which, together with the weights, forms a commutative semigroup (for instance, the function *maximum*).

To solve the minimum total height difference problem on F, we construct the tree $\mathcal{T}$ of the previous section, and we let any arc in T_F be weighted by the height difference of its endpoints. We preprocess T_F for weighted subpath queries [2]. Given two query points s and t, we search in $\mathcal{T}$ to obtain $s_1 = hbv(s)$, $s_2 = lbv(s)$, $t_1 = hbv(t)$ and $t_2 = lbv(t)$. Then we determine the weights of the subpaths of s_1 and t_1, of s_1 and t_2, of s_2 and t_1, and of s_2 and t_2. Assume w.l.o.g. that the minimum weight W is obtained by s_1 and t_1. Then the answer to the query is $W + |h(s_1) - h(s)| + |h(t_1) - h(t)|$. Note that this also solves the monotone path problem: we simply check whether the total weight is equal to the absolute height difference $|h(s) - h(t)|$ of s and t.

To solve the paths below height z problem, we take a similar approach. However, we weight the arcs of T_F with the maximal height of its two endpoints. Furthermore, we do not compute the sum of the weights of a subpath, but the maximum. With the above notation, we need only test the weight (maximum height) of the subpath between s_2 and t_2 in T_F. If the weight is at most z, then there exists a path between s and t with height at most z.

Theorem 8. *Let F be a polyhedral terrain with n vertices with distinct heights. F can be preprocessed in $O(n \log n)$ time into a data structure of size $O(n)$, such that for any two query points one can determine in $O(\log n)$ time whether there exists a path between them that is monotone. It is also possible to compute in $O(\log n)$ time the minimum height difference of any path between the two query points, and to decide, given two query points and a query height z, in $O(\log n)$ time if there is a path that stays below height z. (The bounds hold in a RAM model of computation).*

6 Conclusions

This paper dealt with several path problems in polyhedral terrains, in which the height of paths played the most important role. We captured the structure of a polyhedral terrain by means of a height level map, and showed that it contains all necessary infomation. Then we showed how the height level map could be stored implicitly for point location, using only linear storage and $O(n \log n)$ preprocessing time. Path queries between any two points on the terrain can be answered in $O(\log n)$ time. It would be interesting to combine the work of this paper with the results from [5, 12], to obtain paths that are good both with respect to height difference and to Euclidean length. Furthermore, in some applications it is required that very steep parts of the terrain are avoided. This is also well worth studying.

References

1. Bern, M., D. Dobkin, D. Eppstein, and R. Grossman, Visibility with a Moving Point of View, *Proc. 1st ACM-SIAM Symp. on Discrete Algorithms* (1990), pp. 107–117.
2. Chazelle, B., Computing on a Free Tree via Complexity-Perserving Mappings, *Algorithmica* **3** (1987), pp. 337–361.
3. Chazelle, B., Triangulating a Simple Polygon in Linear Time, *Discrete Comput. Geom.* **6** (1991), pp. 485–524.
4. Chazelle, B., H. Edelsbrunner, L. J. Guibas, and M. Sharir, Lines in Space: Combinatorics, Algorithms and Applications, *Proc. 21nd ACM Symp. on the Theory of Computing* (1989), pp. 382–393.
5. Chen, J., and Y. Han, Shortest Paths on a Polyhedron, *Proc. 6th ACM Symp. on Comp. Geometry* (1990), pp. 360–369.
6. Cole, R., and M. Sharir, Visibility Problems for Polyhedral Terrains, *J. Symb. Computation* **7** (1989), pp. 11–30.
7. Edelsbrunner, H., L. J. Guibas, and J. Stolfi, Optimal Point Location in a Monotone Subdivision, *SIAM J. Computing* **15** (1986), pp. 317–340.
8. Harel, D., and R. E. Tarjan, Fast Algorithms for Finding Nearest Common Ancestors, *SIAM J. Computing* **13** (1984), pp. 338–355.
9. Katz, M.J., M.H. Overmars and M. Sharir, Efficient Hidden Surface Removal for Objects with Small Union Size, *Computational Geometry: Theory and Applications* **2** (1992), pp. 223–234.
10. Kirkpatrick, D. G., Optimal Search in Planar Subdivisions, *SIAM J. Computing* **12** (1983), pp. 28–35.
11. Reif, J., and S. Sen, An Efficient Output-Sensitive Hidden Surface Removal Algorithm and its Parallelization, *Proc. 4th ACM Symp. on Comp. Geometry*, 1988, pp. 193–200.
12. Sharir, M., and A. Schorr, On Shortest Paths in Polyhedral Spaces, *SIAM J. Comput.* **15** (1986), pp. 193–215.

Dog Bites Postman: Point Location in the Moving Voronoi Diagram and Related Problems*

Olivier Devillers and Mordecai Golin

INRIA, B.P.93,
06902 Sophia-Antipolis cedex (France),
E-mail: olivier.devillers@sophia.inria.fr.

Hong Kong Univ. of Science & Technology,
Clear Water Bay, Kowloon, (Hong Kong),
E-mail: golin@cs.ust.hk.

Abstract. We discuss two variations of the two-dimensional post-office problem that arise when the post-offices are replaced by n postmen moving with constant velocities. The first variation addresses the question: given a point q_0 and time t_0 who is the nearest postman to q_0 at time t_0? We present a randomized incremental data structure that answers the query in expected $O(\log^2 n)$ time. The second variation views a query point as a dog searching for a postman to bite and finds the postman that a dog running with speed v_0 could reach first. We show that if the dog is quicker than all of the postmen then the data structure developed for the first problem permits us to solve the second one in $O(\log^2 n)$ time as well.

1 Introduction

In this paper we study the problem of point location in the Voronoi diagram of n moving planar sites. The input sites of the static Voronoi diagram are frequently considered as post-offices; by analogy we think of moving sites as postmen making their appointed rounds. We asssume that each postman moves with constant velocity:

$$p_i(t) = q_i + v_i t, \qquad i = 1, \ldots, n$$

where $p_i(t)$ is the location of the postman at time t, $q_i \in I\!\!R^2$ its location at time 0, and $v_i \in I\!\!R^2$ its velocity. As in the static case we want to preprocess the postmen so as to easily answer the question "given a query point q_0 at time t_0 who is the closest postman?"

In the static case the meaning of "closest" was quite clear. The closest post-office was the nearest post-office, which was also the one that could be reached quickest. In the postman problem we must distinguish between the two different types of closeness. Because the postman are moving the nearest postman at time t_0 might not be the postman that can be reached quickest. We therefore want

---

* *This work was partially supported by the ESPRIT Basic Research Actions 7141 (AL-COMII) and 6546 (PROMotion).*

to be able to answer the following two different types of queries: ($|p - q|$ is the Euclidean distance between p and q)

(1) Moving-Voronoi queries: Given a customer at location $q_0 \in I\!\!R^2$ find the nearest postman at time $t_0 \in I\!\!R$. That is, given q_0, t_0, return i such that

$$|p_i(t_0) - q_0| \leq |p_j(t_0) - q_0|, \quad j = 1. \ldots, n.$$

(2) Dog-Bites-Postman queries: The query inputs are $q_0 \in I\!\!R^2$, $t_0 \in I\!\!R$, and $v_0 > 0$. They specify a dog located at q_0 at time t_0 capable of running at maximum speed v_0. The dog is mean; it wants to catch and bite a postman. It's also impatient; it wants to bite a postman as soon as possible. The problem here is to find the postman the dog can reach quickest. Set

$$t_j = \min\{t \geq t_0 : (t_j - t_0)v_0 = |p_j(t) - q_0|\}, \quad j = 1, \ldots, n$$

to be the first time that the dog can catch postman j. Then the query returns i such that $t_i \leq t_j$, $j = 1, \ldots, n$.

A series of recent articles [Roo91], [Roo90], [AR92], [RN92], [FL91] study the Voronoi diagram of moving points. These papers examine the changes in the Delaunay triangulation of moving points (corresponding to topological changes in the Voronoi diagram) and how to detect them but do not address the problem of point location. [GMR91] give an $O(n^3)$ upper bound on the number of such topological changes, and a nearly cubic bound for more general motions.

The problem of answering general dog-type queries does not seem to have been directly addressed previously. (See though, the paper on Voronoi diagrams in rivers [Sug92] which discusses a problem which is equivalent to the special case in which all of the postmen are moving with the same velocity.)

Our approach to solving the moving-Voronoi problem is to view it as a point-location problem in a three-dimensional, (space, time), cell structure. We construct the point-location data-structure by modifying a randomized-incremental procedure originally devised by Guibas, Knuth and Sharir [GKS92] for solving the static Voronoi diagram problem. The GKS result for static sites is:

Theorem 1. *The radial Voronoi diagram of n sites inserted in random order can be maintained in $O(\log n)$ expected update time and $O(n)$ expected space. Any fixed query point is located in $O(\log^2 n)$ expected time (the expectation is taken over the random order of insertion).*

Our modification of the GKS result will construct a data structure which permits finding the nearest postman to q_0 at time t_0 in expected $O(\log^2 n)$ time where the expectation is only taken over the order in which the postmen were inserted into the data structure and not over the values of q_0, t_0.

Our approach to solving the "dog" problem is to show that, if the dog is faster than all of the postmen, i.e., $v_0 > |v_i|$, $i = 1, \ldots, n$, then the dog problem can be transformed to the moving-Voronoi one. The transformation is highly dependent upon the speed v_0.

The structure of the paper is as follows. In section 2 we quickly review the basic ideas of the GKS algorithm and data-structure. In section 3 we discuss how to solve the moving-Voronoi problem. In section 4 we define a bijection that transforms the three dimensional cell structure describing the moving-Voronoi problem into the structure that describes the dog problem and then show how to use this bijection to solve the dog problem.

2   The GKS Algorithm

We quickly review the GKS algorithm and its analysis (see [GKS92] for details) since we will have need of it in the next section. This algorithm uses a common trick in the design of randomized algorithm: the use of a structure storing the history of the construction [BT86, Sch91]. The algorithm presented here is slightly modified from the generic one that appears in [GKS92] in that it utilizes pointers to travel from a triangle to the exterior triangles that destroy it but, as noticed by Guibas *et al.*, this modification does not affect the complexity. We utilize the modified algorithm because it is easier to generalize to the moving point case.

GKS maintain a triangulation of the Voronoi diagram. Each Voronoi cell is triangulated, with edges drawn from the site defining a cell to all of the vertices of the cell (Figure 1). We call this the *radial Voronoi diagram*. A fundamental observation is that each (bounded) triangle depends on exactly four sites (or three sites for unbounded triangles). For example, in Figure 1, the sites p, q, r and s define the shaded triangle ptu (and also qtu), where t and u are the Voronoi vertices which are the centers of the empty disks circumscribing Delaunay triangles pqr and pqs.

When a new site k is inserted in the radial Voronoi diagram, two types of triangles are created. The interior type triangles are the ones that triangulate the Voronoi cell associated with site k and the exterior type triangles are triangles that border on the cell of site k Previously existing triangles that intersect a newly created one are said to be *destroyed* by the insertion of site k.

The location structure   The structure built during the insertion of $p_1, \ldots, p_n$ can be represented as n successive parallel horizontal planes; each plane contains the triangles created by the insertion of site p_i.

If a triangle is destroyed by the insertion of site p then it is contained within the union of the Voronoi cell of p and one or two exterior triangles. Links are created from the destroyed triangle to the corresponding exterior triangles and from the destroyed triangle to the site p. The interior triangles created by the insertion of p are organized around p in a search structure sorted in polar order; this structure is called a *polar sorted list*.

To locate a query point q in the radial Voronoi diagram we walk through all the triangles on successive levels containing q. If a triangle T contains q, either T is still a triangle of the final diagram and we are done, or T is destroyed by the insertion of some site p. In this second case q is first tested against the exterior triangles pointed to by T. If q is not inside these triangles then q is located in one of the interior triangles around p; this triangle can be found by performing a

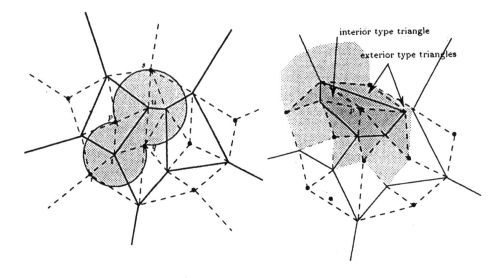

Fig. 1. The triangulated Voronoi diagram, and the triangles created by a site insertion

binary search in the polar sorted list. In both cases the next triangle containing q in the succession of radial Voronoi diagrams has been found and the search can continue.

It is not difficult to maintain the adjacency relationships between triangles in the radial Voronoi diagram. Notice that these relations are not necessarily between triangles in the same plane and may change when new sites are inserted.

Update algorithm When a new point is added, it is first located in the structure and found to be in some triangle T. The current radial triangulation is then explored using the adjacency relationships to find all the destroyed triangles. A new horizontal plane is then created, and all new triangles (exterior and interior) are computed from the destroyed ones and linked to them. Using the adjacency relations it is possible to find the new interior triangles directly sorted in polar order around the new points. Finally, the adjacency relations are updated.

Analysis This algorithm can be analyzed using randomized techniques, i.e. backward analysis [Che85, Sei91, Dev92]. An exhaustive analysis of the algorithm is not done here, the interested reader can refer to the original Guibas *et al.* paper or to the full version of this one.

The main hypothesis is that the sites are inserted in a random order; each of the $n!$ possible insertion sequence are therefore equally likely.

The fundamental observation is that, during the location of a query point q

the Delaunay triangle T containing q at stage k is stored on level k only if k is one of the four sites defining T. Because the sites are inserted in a random order this occurs with probability $\frac{4}{k}$. Summing over all levels k we find that the expected number of triangles visited during a location is $O(\log n)$ and the location time is $O(\log^2 n)$ since a location in a polar sorted list can be required to deduce a visited triangle from the previous one.

The analysis of the update time is done similarly and yields a logarithmic time for inserting a new site.

3 Randomized incremental construction of the moving Voronoi diagram

We now address the problem of moving points. A site is a two-dimensional point moving with constant velocity: $p_i(t) = (x(t), y(t)) = q_i + v_i t$, $i = 1, \ldots, n$, where $q_i \in \mathbb{R}^2$ is the point's location at time 0, and $v_i \in \mathbb{R}^2$ is its velocity. For the sake of simplicity we assume that the velocity is constant but the technique that we introduce can be used in many cases where the motion is not constant, e.g., $p_i(t)$ could have components which are polynomials in t. At time t, the moving sites define a radial Voronoi diagram. As t increases this diagram changes.

It is quite natural to visualize this problem in the 3D space (x, y, t). A moving site is a non-horizontal line (or a curve of degree k if the components of $p_i(t)$ are polynomials of degree k). As t increases the vertices and edges of the radial Voronoi diagram define curves and surfaces. The moving diagram is in fact a subdivision of 3-dimensional space. A region (cell) of this subdivision corresponds to the space swept by a triangle of the moving radial Voronoi diagram between the time when it appears and the time when it disappears. The moving triangle is defined by 4 moving sites (as in the non moving case), and the appearance and the disappearance of the triangle correspond to topological changes in the 2D radial Voronoi diagram, i.e., the disappearance and the appearance of another site in a disk circumscribing three of the four sites. To summarize, a region of the 3D moving radial Voronoi diagram is defined by at most 6 moving sites (see Figure 2). In some cases, fewer sites defines a region when the moving triangle is unbounded or when the site which caused the triangle to appear or disappear is one of the four sites defining the moving triangle.

This section modifies the GKS algorithm to work on the moving Voronoi diagram and proves the following theorem.

Theorem 2. *The moving radial Voronoi diagram of a set S of n moving sites inserted in random order can be maintained in $O(f_S(n)/n + \log n)$ expected insertion time and $O(f_S(n))$ expected space where $f_S(r)$ is the expected size of the moving diagram for a random sample of size r of S. Any fixed query point is located in $O(\log^2 n)$ expected time (the expectation is taken over the random order of insertion).*

The complexity depends on f_S since the size of the moving Voronoi diagram can vary between $\Omega(n)$ and $O(n^3)$. So, our algorithm is partially output sensitive.

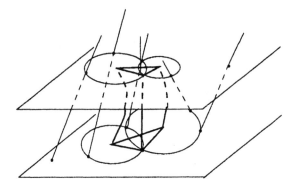

Fig. 2. A region of the 3D moving diagram

More exactly, it is not sensitive to the size of the output but to the expected size of intermediate results, a phenomenon which is found in the analysis of other semi-dynamic randomized algorithms [BDS+92]. Furthermore, no tight upper bound for the size of the moving Voronoi diagram are known. It is not even known if it can be larger than $O(n^2)$ [GMR91].

3.1 Algorithm

The location structure This section describes the structure used for locating a query point $q = (x_q, y_q, t_q)$ in the 3D space and solving the question "at given time t_q, what is the nearest site $p_i(t_q)$ to the point (x_q, y_q)?". As in Section 2 the structure can be viewed as n successive 3D spaces, each containing the regions created by the insertion of the corresponding moving site.

Suppose that region (cell) R which contains point q in the 3D moving Voronoi diagram of sites $p_1, \ldots, p_{k-1}$ no longer exists in the diagram of sites $p_1, \ldots, p_k$. Then we say that R is *destroyed* by the insertion of site k. In such a case R is contained in the union of some new regions at level k. Some of the regions in this collection are called interior ($p(t)$ supports an edge of the region) while others are exterior. These regions will be stored in a way which will enable us to easily determine which of them contains q.

Exterior regions Let $R(t)$ denote a cut at fixed time t of region R. This is a triangle in the radial Voronoi diagram of $p_1(t), \ldots, p_{k-1}(t)$. $R(t)$ overlaps at most two exterior triangles in the radial diagram of $p_1(t), \ldots, p_k(t)$. The regions these triangles belong to will be known as exterior regions. As t changes these exterior regions will remain the same except at certain discrete values of t. Thus the set of new exterior regions overlapping R can be sorted in the time direction, and the new exterior region containing q, if it exists, can be determined by a simple binary search in $O(\log n)$ time. (We use here the fact that there are no more than $O(n^3)$ such regions [GMR91].)

Interior regions The set of all interior regions caused by the insertion of site k is exactly the set of all the regions which have the line $p_k(t)$ as a supporting edge. Note that a region can not be both interior and exterior.

As in the non moving case, if the query q does not belong to an exterior region, we must examine the whole set of interior regions to determine which contains q. In the non-moving case, the interior triangles were sorted according to the angle θ around the new inserted site.

In this case, a two dimensional structure is used: the first dimension is the time t and the second is the polar angle θ_t around $p(t)$. The whole set of interior triangles created by the insertion of $p(t)$ defines a monotone subdivision of this 2D space (t, θ_t). To find q we map it to (t_q, θ_q) (where θ_q is the angle between $p(t)q$ with the x-axis), and this mapped point is located in the above subdivision using any static point location algorithm [EGS86, CS89, Pre90].

Update algorithm The insertion of a new moving site $p(t)$ is done as follows. Some position of the site, for example $p(0)$, is located as above. Then all the regions destroyed by the insertion of $p(t)$ are found using the adjacency relationships between regions. These regions are either those traversed by the line $p(t)$ or are neighbors of the preceding ones. Since each region has a constant number of neighbors, all relevant regions are found in output sensitive time. Then a new level corresponding to the new site is created in the structure. All new regions are computed from the destroyed ones; they are organized in the location structure described above comprising some sorted lists and a planar monotone subdivision. Notice that these structures are static and any location structure can be used, randomized [CS89] or not [EGS86]. The adjacency relations must then be updated.

3.2 Analysis

The analysis uses standard tools of randomized approaches [CS89, BDS+92]. In this extended abstract we only sketch the basic ideas; the complete analysis is left for the full paper.

Location of a given point The location algorithm follows the sequence of regions containing the query point q through the n levels. Given a region containing q the next region is determined in $O(\log n)$ time by solving a planar location query for interior regions and searching in a sorted list for exterior regions.

The number of regions in the sequence is expected $O(\log n)$ as in Section 2. This result follows from the fact that a region is defined by at most 6 sites so a given region (the one containing q) is created by the insertion of the kth site with probability $O(\frac{6}{k})$.

The expected time complexity of the location of any query point is therefore $O(\log^2 n)$, where the expectation is taken over the insertion order of the sites.

Insertion of a site Standard techniques of analyzing randomized algorithms can be used. After the insertion of the kth moving site, the moving Voronoi diagram has $f_S(k)$ regions (expected) and, since a region is defined by at most 6 sites, the number of regions created by the insertion of the kth site is less than $6\frac{f_S(k)}{k}$. Summing over k yields the result.

If the postmen travel with non-constant velocity the algorithm can, in many cases, be generalized without difficulty. The subdivision in space (t, θ_t) is still a monotone subdivision (the degree of the edges may change) and the function f_S can be different, but the general techniques still work.

4 The dog and postmen problem

We now consider dog queries. Imagine that the moving sites are postmen. A query asks "a dog capable of running at maximum speed v_d is put outside at time t_d and location (x_d, y_d). It will run and bite a postman. Which postman can it reach in a minimum time?"

These types of queries differ from the "moving Voronoi" type addressed in the previous section. It is possible that the dog might be able to reach a postman further away (that is travelling towards it) quicker than it can reach a nearby one (that is travelling away from it). The answer to the query depends strongly on v_d.

In what follows we will assume that $v_d > |v_i|$, $i = 1, \ldots, n$, i.e., the dog is faster than all of the postman. This ensures that the dog will always be able to catch at least one postman. Later, we will discuss what happens if the dog is not as fast as some of the postmen. Note that as we let v_d approach infinity the postman that can be reached quickest becomes the nearest postman.

We will prove the following theorem:

Theorem 3. *The postman that any fixed query dog can reach quickest can be found in $O(\log^2 n)$ expected time, with the same space and preprocessing time as in Theorem 2.*

To do this we show that if the dog is faster than all of the postmen then there is an efficient way to transform a dog query into a moving-Voronoi query.

In the previous section we constructed a point-location structure for the moving Voronoi diagram. We denote this diagram by $\mathcal{M}$. In $\mathcal{M}$ vertices are points that are the centers of empty circles (on the horizontal plane through the point) that contain four sites on their boundaries. A point is on an edge of $\mathcal{M}$ if it is the center of an empty circle with three sites on the boundary and on a face if there are two. Finally, a point is in some region associated with postman p_i if p_i is on the boundary of an empty circle with the point at its center.

We now return to the dog problem. Given a query (x_d, y_d, t_d, v_0), it is clear that the dog can reach any point of the 3D space in a vertical cone of apex (x_d, y_d, t_d) and angle $\arctan v_0$, and the answer to the query is the lowest intersection of this cone with one of the moving sites (see Figure 3). The different

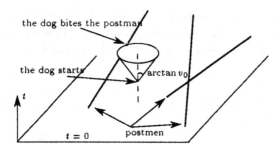

Fig. 3. A query dog

answers to this query split the space in different regions where the answer remain the same. We denote the three dimensional diagram that this splitting induces by $\mathcal{D}$. In a region of the diagram the quickest reachable postman is the same, on a face of $\mathcal{D}$ there are two postmen that can be reached in the same time, on an edge of $\mathcal{D}$ three postmen, and from a vertex of $\mathcal{D}$ four postmen (assuming general position of the sites).

Note that $\mathcal{D}$ is dependent upon v_0 which we are assuming to be fixed. Actually $\mathcal{D}$ is exactly the Voronoi diagram of the lines corresponding to the moving sites for the convex distance associated to a vertical cone of angle arctan v_0 (bounded by an horizontal plane).

Let $z = (x_z, y_z, t_z)$ be a point whose nearest neighbor at time t_z is postman p_i; z is in some region $R \in \mathcal{M}$ that is associated with postman p_i. Set $\bar{z} = (x_z, y_z)$ and let $d(z) = |p_i(t_z) - \bar{z}|$ be the distance from the nearest neighbor to z. Now set $q = \left(x_z, y_z, t_z - \frac{d(z)}{v_0}\right)$. We claim that a dog at q reaches postman p_i at least as quickly as it can reach any other postman and it reaches p_i in time $d(z)/v_0$.

A dog at q can reach postman p_i in time $d(z)/v_0$ by travelling directly towards the point $p_i(t_z)$. Now suppose that the dog can reach some other postman quicker. That means there is some postman p_j and $0 < \delta < d(z)/v_0$, such that

$$\left| p_j \left(t_z - \frac{d(z)}{v_0} + \delta \right) - \bar{z} \right| = \delta v_0.$$

Using the fact that $v_0 > |v_j|$ we find that

$$|p_j(t_z) - \bar{z}| \leq \left| p_j(t_z) - p_j \left(t_z - \frac{d(z)}{v_0} + \delta \right) \right| + \left| p_j \left(t_z - \frac{d(z)}{v_0} + \delta \right) - \bar{z} \right|$$

$$= |v_j| \left(\frac{d(z)}{v_0} - \delta \right) + v_0 \delta \; < \; v_0 \frac{d(z)}{v_0} = d(z).$$

But this implies that postman p_j is closer than distance $d(z)$ to point $\bar{z}$ at time t_z which leads to a contradiction. A dog at q therefore reaches postman p_i at least as quickly as it reaches any other postman.

Now suppose that z is on some face of $\mathcal{M}$. This means that there are two postmen p_i, p_j which are both nearest to $\bar{z}$ at time t_z. From the above discussion this implies that a dog at $q = \left(x_z, y_z, t_z - \frac{d(z)}{v_0}\right)$ can reach these two postmen equally quickly and can not reach any other postman quicker. This means that q is on a face of $\mathcal{D}$. Similarly we can show that if z is on an edge or is a vertex of $\mathcal{M}$ then q is respectively on a corresponding edge or is a vertex of $\mathcal{D}$. This leads us to define the following transformation.

For $z = (x_z, y_z, t_z)$ let $d(z)$ be the distance between z and the nearest postman at time t_z and set

$$\rho(z) = \left(x_z, y_z, t_z - \frac{d(z)}{v_0}\right).$$

We have already seen that ρ maps points on vertices, edges and faces of $\mathcal{M}$ to points on vertices, edges and faces of $\mathcal{D}$. This mapping satisfies a host of nice properties which are listed in the next theorem (the proof of which is not given in this extended abstract).

Theorem 4. *For $z = (x_z, y_z, t_z)$ let $\rho(z)$ be defined as above. Then ρ is a continuous bijection from $\mathbb{R}^3$ to $\mathbb{R}^3$ that is also a bijection of vertices, edges, faces and regions of $\mathcal{M}$ into vertices, edges faces and regions of $\mathcal{D}$. Furthermore, p_i is the nearest postman to the points in a region $R \in \mathcal{M}$ if and only if p_i is the postman that a dog in region $\rho(R) \in \mathcal{D}$ can reach quickest.*

What this theorem says is that $\mathcal{D}$ is topologically equivalent to the moving Voronoi diagram. Intuitively we deform and pull each face of $\mathcal{M}$ downward in the past direction to get the corresponding face of $\mathcal{D}$.

Note that since ρ is a bijection its inverse $h \equiv \rho^{-1}$ exists. If $q = (x_q, y_q, t_q)$ and $f(z)$ is the quickest time that it takes for a dog at q to reach a postman then $h(q) = (x_q, y_q, t_q + f(z))$.

Let $\mathcal{M}_i$, $i = 1, \ldots, n$, be the moving Voronoi diagrams created by the insertion of postmen $p_1, p_2, \ldots, p_i$. We say that a region $R \in \mathcal{M}_{i-1}$ is destroyed by the insertion of site i if $R \notin \mathcal{M}_i$. In the last section we saw how to do point location in $\mathcal{M}$ by doing successive point locations in the $\mathcal{M}_i$.

Let $\mathcal{D}_i$, $i = 1, \ldots, n$, be the "dog" diagrams created by the insertion of postmen $p_1, p_2, \ldots, p_i$. Again, we will say that a region $R \in \mathcal{D}_{i-1}$ is destroyed by the insertion of site i if $R \notin \mathcal{D}_i$. Let ρ^i and h^i be the bijections described by Theorem 4 that map the regions of $\mathcal{M}_i$ to the regions of $\mathcal{D}_i$ and vice-versa. We state without proof the following theorem:

Theorem 5. *A region $R \in \mathcal{M}_{i-1}$ is destroyed by the insertion of site i if and only if region $\rho^{i-1}(R) \in \mathcal{D}_{i-1}$ is destroyed by the insertion of site i. Furthermore, if $z \in R \in \mathcal{M}_{i-1}$ and $z \in \tilde{R} \in \mathcal{M}_i$ then $\rho^{i-1}(z) \in \rho^{i-1}(R)$ and $\rho^i(z) \in \rho^i(\tilde{R})$.*

Given theorems 4 and 5 we can solve a dog query by following the path of successive regions in the $\mathcal{D}_i$ that contain q. We do this by following the path of successive regions in the $\mathcal{M}_i$ that contain $h^i(q)$. Suppose that the query point q is located in some region $S \in \mathcal{D}_i$ at the ith level. We can find the next region that contains q on the search path as follows. Look at region $R = h^i(S) \in \mathcal{M}_i$. Find the next site whose insertion destroys R. This is the same site whose insertion destroys S. Let this be site k. We now search on level k to find the (internal or external) region $\tilde{R} \in \mathcal{M}_k$ which contains $h^k(q)$ on level k of the moving Voronoi diagram. The region $\tilde{S} = \rho^k(\tilde{R})$ is the region in $\mathcal{D}_k$ which contains q.

There is a slight complication here in that to calculate $h^k(q)$ we must know the postman that a dog at q can reach quickest. But the postman that q can reach quickest at level k must be either the postman which was associated with region R on all the levels $< k$ (and which is already known) or the new postman, p_k. We can thus check in constant time which of the two the dog at q can reach quickest and then use the answer to calculate $h^k(q)$.

We have just shown that if the dog is faster than all of the postmen then a "dog" query can be solved by solving a corresponding moving Voronoi query. The proof of Theorem 3 then follows from the results of the previous section.

We should point out that the only data structure that we construct is the one that solves moving Voronoi queries. We never actually construct the structure $\mathcal{D}$. In fact the only time that we use the dog speed v_0 in the algorithm is in the calculation of h^k and ρ^k.

5 Conclusion

We have presented a randomized dynamic structure for the moving Voronoi diagram able to answer queries in the Voronoi diagram at any time, in $O(\log^2 n)$ time. If the sites are not moving with constant speed but with other kind of motions, the structure can be generalized. In the case of sites moving with constant speed, the structure can handle more complicated queries: "dog" queries. In this type of query the customer is also moving, with constant speed and faster than the all of the postmen. Since the speed of the dog is part of the query the query is in fact a 4-dimensional one.

If the dog is slower, the problem become more complicated: a postman can be viewed as a line in 3D space of slope $\frac{1}{v}$ where v is the speed of the postman, and a dog can be viewed as the upper half of a cone of slope $\frac{1}{v_0}$. For a fast dog we have $v_0 > v$ and there is exactly one intersection point between the line and the half-cone. For a slow dog there may be zero or two intersection points meaning that the dog might have two ways to reach the postman running at full speed but one way is quicker than the other.

For the same reason the dog and postmen problem is more difficult to generalize to other kind of motions for postmen, because it is not possible to guarantee that there is only one point of intersection between the half cone (dog) and the postman trajectory.

The algorithm presented here is semi-dynamic: postman can be added. A fully dynamic algorithm which also permits deletions can be obtained using ideas due to Schwarzkopf [Sch91]. This yields an $O(\log^3 n)$ query time (see the full paper for details).

The construction is actually sensitive to the expected size of the intermediate outputs, which allow us to claim to its optimality, even if no tight bound are known on the output. We end by noting that the dog diagram can be viewed as a Voronoi diagram of lines using a convex distance function and information concerning bounds for these kind of diagrams is well known to be one of the important open problem in the field.

References

[AR92]   G. Albers and T. Roos. Voronoi diagrams of moving points in higher dimensional spaces. In *Proc. 3rd Scand. Workshop Algorithm Theory*, LNCS 621:399–409. Springer-Verlag, 1992.

[BDS+92]   J.-D. Boissonnat, O. Devillers, R. Schott, M. Teillaud, and M. Yvinec. Applications of random sampling to on-line algorithms in computational geometry. *Discrete Comput. Geom.*, 8:51–71, 1992.

[BT86]   J.-D. Boissonnat and M. Teillaud. A hierarchical representation of objects: the Delaunay tree. In *Proc. ACM Sympos. Comput. Geom.*, 260–268, 1986.

[Che85]   L. P. Chew. Building Voronoi diagrams for convex polygons in linear expected time. Report, Dept. Math. Comput. Sci., Dartmouth College, Hanover, NH, 1985.

[CS89]   K. L. Clarkson and P. W. Shor. Applications of random sampling in computational geometry, II. *Discrete Comput. Geom.*, 4:387–421, 1989.

[Dev92]   O. Devillers. Randomization yields simple $O(n \log^* n)$ algorithms for difficult $\Omega(n)$ problems. *Internat. J. Comput. Geom. Appl.*, 2(1):97–111, 1992.

[EGS86]   H. Edelsbrunner, L. J. Guibas, and J. Stolfi. Optimal point location in a monotone subdivision. *SIAM J. Comput.*, 15:317–340, 1986.

[FL91]   J.-J. Fu and R. C. T. Lee. Voronoi diagrams of moving points in the plane. *Internat. J. Comput. Geom. Appl.*, 1(1):23–32, 1991.

[GKS92]   L. J. Guibas, D. E. Knuth, and M. Sharir. Randomized incremental construction of Delaunay and Voronoi diagrams. *Algorithmica*, 7:381–413, 1992.

[GMR91]   L. Guibas, J. S. B. Mitchell, and T. Roos. Voronoi diagrams of moving points in the plane. In *Proc. Workshop Graph-Theoret. Concepts Comput. Sci.*, LNCS 570:113–125. Springer-Verlag, 1991. (Remark (3), p. 121).

[Pre90]   F. P. Preparata. Planar point location revisited. *Internat. J. Found. Comput. Sci.*, 1:71–86, 1990.

[RN92]   T. Roos and H. Noltemeier. Dynamic Voronoi diagrams in motion planning. In *Proc. 15th IFIP Conf.*, LNCIS 180:102–111. Springer-Verlag, 1992.

[Roo90]   T. Roos. Voronoi diagrams over dynamic scenes. In *Proc. 2nd Canad. Conf. Comput. Geom.*, pages 209–213, 1990.

[Roo91]   T. Roos. Dynamic Voronoi diagrams. Ph.D. Thesis, Bayerische Julius-Maximilians-Univ., Würzburg, Germany, 1991.

[Sch91]   O. Schwarzkopf. Dynamic maintenance of geometric structures made easy. In *Proc. IEEE Sympos. Found. Comput. Sci.*, pages 197–206, 1991.

[Sei91]   R. Seidel. Backwards analysis of randomized geometric algorithms. ALCOM School on efficient algorithms design, Århus, Denmark, 1991.

[Sug92]   K. Sugihara. Voronoi diagrams in a river. *Internat. J. Comput. Geom. Appl.*, 2(1):29–48, 1992.

Parallel Approximation Schemes for problems on planar graphs [*] (Extended Abstract)

J.Díaz[†]   M.J.Serna[†]   J. Torán[†]

Abstract

This paper describes a technique to obtain NC Approximations Schemes for the Maximun Independent Set in planar graphs and related optimization problems.

1 Introduction

The generalized conjecture that no NP-hard problem can be solved in polynomial time has motivated the study of fast approximation algorithms for NP-hard optimization problems. When dealing with polynomial time approximability, a natural question to consider is whether the full power of P is needed or there are approximation algorithms for NP-hard problems that use less resources. In particular we are interested in the study of parallel approximation algorithms, in the sense of NC; using polynomial number of processors and running in polylogarithmic time. Some work has been done on parallel approximation algorithms [AM86, KSS89, Ser90, LN92]. It is even known that some NP-hard problems can be fully approximated in NC [PR87, AM86, KS93].

In the present paper, we present NC approximation schemes for the Maximum Independent Set on planar graphs, and some other related NP-hard optimization problems restricted to planar graphs, showing that they belong to the classes NCAS and NCAS$^\infty$. The class of optimization problems with an NC approximation scheme is defined in the following way. A problem is in NCAS if there is a parallel algorithm such that given as input a pair (x, ϵ), where x is an instance of the problem, and $\epsilon \in \mathbb{R}^+$, produces a solution to the problem within $1 + \epsilon$ of the optimum, running in polylogarithmic time in the length of x and using a number of processors bounded by a polynomial in the length of x. We can also define the asymptotic version NCAS$^\infty$. (Our definitions follow the standard definitions of PTAS and PTAS$^\infty$ [GJ79].

[*]This research was supported by the ESPRIT BRA Program of the EC under contract no. 7141,project ALCOM II.

[†]Departament de Llenguatges i Sistemes, Universitat Politècnica Catalunya, Pau Gargallo 5, 08028-Barcelona

The Maximum Independent Set problem on planar graphs was shown to belong to the class PTAS by Baker in [Bak83]. The idea in her proof is to decompose the graph in k-outerplanar graphs, and then using dynamic programming techniques obtain the MIS for each of the k-outerplanar graphs. Although we follow the same general scheme in our parallel approximation algorithms, the techniques we use are rather different, in particular we have to introduce original methods to represent in parallel the k-outerplanar graphs as binary trees. The dynamic programming part in Bakers proofs is translated into a shunt operation in the binary trees representing the graphs. For any k the algorithm achieves a $k/(k+1)$ performance ratio.

Chrobak and Naor gave an NC algorithm with a linear number of processors to approximate MIS within 1/2 [CN89]. In the same paper they mention the convenience of obtaining a parallel version of Baker's result.

The advantage of our method is that it can be transformed (in the same way as Baker's algorithm) to obtain parallel approximation schemes for a whole set of NP-hard planar graphs problems. In particular we show how the technique can be transformed to approximate Minimum Vertex Cover Minimum Dominating Set and Minimum Edge Dominating Set on planar graphs. We also show that it can be transformed to obtain NC approximations to some polynomial time computable problems, as Maximum Cut, Maximum Even Degree Set and Maximum Matching.

We also obtain exact NC algorithms for these problems as well as for the 3-Colorability and the Graph Bisection problems restricted to k-outerplanar graphs. To our knowledge this is the first time that these problems are proved to be in the class NC.

The algorithms in this paper are designed for the Concurrent Read Exclusive Write PRAM model of computation [KR90].

Due to space reasons in this extended abstract we omit the proof of most of our results. We refer the reader to the full version of the paper.

2 Definitions and Preliminaries

In this section we introduce the notation and basic facts used in the article. We introduce a method to represent a k-outerplanar graph G by a tree. The nodes of this tree will keep enough information to process the graph. For the basic definitions on graphs and planarity, see [Hag90].

Let $\mathcal{G}$ be a planar embedding of G. The *faces* $\mathcal{F}$ of $\mathcal{G}$ are the connected regions of $\mathbb{R}^2 - \mathcal{G}(V \cup E)$. Let f_∞ be the *exterior face*. Two faces of $\mathcal{G}$ are said to be *adjacent* if they share at least one edge. The *exterior boundary* B of $\mathcal{G}$ is the set of edges separating f_∞ from the other faces of $\mathcal{G}$.

It is known that a plane embedding of a planar graph can be constructed in $O(\log^2 n)$ time by using $O(n)$ processors in the CREW PRAM model see [KR86]. Through the remaining of the paper, the term *embedded graph* will denote a planar graph G together with a particular planar embedding of G. Given an embedded graph G, the *quasidual graph* $G^* = (V^*, E^*)$ is the graph with $V^* = \{\mathcal{F} - f_\infty\} \cup B$, and for any two faces $f_1, f_2 \in \mathcal{F} - f_\infty$, $\{f_1, f_2\} \in E^*$ iff both faces are adjacent, and for any face $f \in \mathcal{F} - f_\infty$ and any $e \in B$, $\{f, e\} \in E^*$ iff e separates f and f_∞. Given an embedded graph G, the *face incidence graph* $G^+ = (V^+, E^+)$ is the graph with $V^+ = \mathcal{F}$, and for any two faces $f_1 \neq f_2$, $\{f_1, f_2\} \in E^+$ exactly if f_1 and f_2 have at least one vertex in common. For any $f \in V^+$ let $d(f)$ be the minimum distance from f to the node representing the exterior face in G^+. A vertex is a *level 1* vertex if it is on the exterior face. Let G^i be the graph obtained by deleting all vertices in levels 1 to i, then the vertices on the exterior face of G^i are the *level $i+1$* vertices.

An embedded graph is *k-outerplanar* if it has no vertices of level greater than k (see figure 1). Every embedded graph is k-outerplanar for some k. The terms outerplanar and 1-outerplanar are equivalent.

By a straightforward induction argument on $d(f)$ we get,

Lemma 1 *Given an embedded graph G, for each vertex v of G, the level of vertex v in G is given by $l(v) = 1 + \min\limits_{v \in f} d(f)$.*

The previous lemma give us a way to compute the level of a vertex,

Theorem 1 *Given a embedded graph G, the level of each vertex can be computed in parallel time $O(\log^2 n)$ using $O(n^3)$ processors.*

In order to simplify notation we will use the term *level i subgraph* to denote a connected component of the subgraph induced by the level i vertices (see figure 2). It follows from the definition of levels that every level subgraph is outerplanar. Furthermore, every level $i+1$ subgraph is in a face of a level i graph. A face in a level i subgraph can have inside more than one level $i+1$ graph. If this is the case, we add dummy edges to G to split the face in such a way that each new face contains exactly one level i graph taking care of preserving the planarity.

Let us assume we are given a biconnected outerplanar graph G'. The *face-face tree* representation of G' is a rooted ordered tree that has as leaves the edges in the exterior boundary of G', and constructed in such a way that each internal node x in the tree corresponds to an interior face f_x of G'. In fact for every interior node x of the tree, we can associate two vertices of G', $b_1(x)$ and $b_2(x)$ such that if y denotes the father of x in the tree, then $(b_1(x), b_2(x))$ is the interior edge of G' separating f_x from f_y. Moreover we also can identify x with the portion of the graph G'_x, induced by all nodes encountered in a counterclockwise tour on the exterior face of G', starting at $b_1(x)$ and ending at $b_2(x)$. In the case of the root r, we have $b_1(r) = b_2(r)$, so that $G'_r = G'$. The face-face tree can be obtained from the quasi-dual graph.

Lemma 2 *Given a biconnected outerplanar graph G' a a face-face tree representation can be obtained in $O(\log n)$ parallel time using $O(n^2)$ processors.*

In order to construct the face-face tree of an outerplanar graph G, we first transform G into a biconnected outerplanar graph G', an use the tree representation of G' as tree representation for G. The construction of G' from G is done by the following procedure,

For each cutpoint p of G whose removal gives $k \geq 2$ components, we add k vertices connected as a cycle. (In the case $k=2$, the 2 added vertices just form an edge.) The planar embedding gives a cyclic order of the edges from p, giving also a cyclic ordering of the components. The new vertices will follow this order. After this step each bridge has been converted into four nodes, we connect the end points by two parallel edges. Finally we remove all vertices that were cut points in G with the corresponding edges. (See fig. 5,6)

The constructed G' is a biconnected graph with all new edges on the exterior face, preserving outerplanarity. To keep information about G in G' we will classify the edges of G' in two types; the *virtual* edges the edges joining two vertices in G' that correspond to the same vertex in G, and the *real* edges that are all the remaining edges.

Theorem 2 *Given an outerplanar graph G and the associated biconnected graph G', a tree representation of G can be obtained in parallel time $O(\log n)$ and using $O(n^2)$ processors.*

It remains to convert the face-face tree in a binary tree adding some extra dummy nodes. It can be done in $O(\log n)$ parallel steps using a linear number of processors. The non-dummy nodes of the binary tree representation can be further classified in two types. A node x is said to be *virtual* if $(b_1(x), b_2(x))$ is a virtual edge of G' that is a vertex in G, otherwise x is said to be *real*. Note that the root r of the tree will be classified as virtual node (see figure 7).

We extend the previous procedure to obtain in parallel a tree representation of a connected k-outerplanar graph. Given an k-outerplanar graph G, we convert it into an associated graph G' with the property that each level subgraph is biconnected. We proceed in parallel for each level subgraph, using the procedure described in the previous section.

To construct the face-face tree for the whole graph, we begin by constructing the face-face tree for each level subgraph. As each level i subgraph is inside a face of a level $i-1$ subgraph, we know from where to hang the tree, however it is necessary to choose appropriately the roots of the different face-face trees, so when the tree for the whole graph is constructed, it will have the desired properties. A level vertex v in a subgraph at level i is said to be *consistent* with a vertex u in a subgraph at level $i-1$ (respectively a level $i-1$ edge e) if the two vertices (the vertex and the middle point of the edge) can be joined by a line preserving planarity. A *consistent set of roots* is a selection of pairs (vertex in face), one for each level subgraph, such that after constructing the corresponding face-face trees, each root of a level ($i > 1$) tree is consistent with the edge (vertex) associated to the enclosing face.

Lemma 3 *A consistent set of roots can be constructed in $O(\log n)$ parallel time using $O(n^2)$ processors.*

Once we have a consistent set of roots, we construct the face-face trees. Given a level tree T, let $a_1, \ldots, a_s$ be the exterior cycle of edges at level i going from the first endpoint to the last and $d_1, \ldots, d_q$ be the path (cycle) of level $i-1$ edges in the enclosing face. To each vertex v we associate a vertex in the enclosing face $d(v)$ as follows. Let v_0 be the first vertex of a_1, v_i be the common vertex to a_{i-1} and a_i, and v_{s+1} the last vertex of a_s. Then $d(v_0)$ is the first vertex of d_1 and $d(v_{s+1})$ is the last vertex of d_q. For other i if v_i has no edges to vertices in the enclosing face, $d(v_i) = d(v_{i-1})$, otherwise $d(v_i) = w$ where w is the last visited level $i-1$ vertex to which v_i is connected.

To associate a portion of the graph G to each node x in a level tree, we consider two ordered set of vertices $B_1(x)$ and $B_2(x)$. Let x be a a node in the tree representing a level i subgraph with $u = b_1(x)$ and $v = b_2(x)$ then $B_1(x) = \langle u, d(u), \ldots, d^{i-1}(u) \rangle$ and $B_2(x) = \langle v, d(v), \ldots, d^{i-1}(v) \rangle$. The associated graph G'_x will be the subgraph containing the vertices in $B_1(x)$ and $B_2(x)$, all vertices encountered in a counterclockwise traversal of level subgraphs starting at $B_1(x)$ and ending at $B_2(x)$, all edges with both endpoints in G'_x are included, except any that leaves $B_1(x)$ in a clockwise direction.

To obtain the final tree we have to connect the trees for the different level subgraphs. We construct the tree as follows:

Suppose that x is the node corresponding to a face that have inside a level subgraph having face-face tree T. We connect tree T as the only son of x. Let y be a leaf of T, $u = b_1(y)$ and $v = b_2(y)$. We consider two cases

Case 1 $d(u) \neq d(v)$ let z be the first son of x such that $b_1(z) = d(u)$, z' be the last son of x such that $b_2(z') = d(v)$, and z'' be the first son of x such that $b_2(z'')$ is connected to v. We add a new nodes labeled u as first son of y, the sons of u will be all sons of x from z to z''. The second son of y when $z' \neq z''$ is a node labeled v that has as sons the next sons of x before z', and when $z' = z$ a node labeled $B_2(y)$ that has no son.

Case 2 $d(u) = d(v)$, we add two new nodes labeled $B_1(y), B_2(y)$ as first and second son of y.

Theorem 3 *Given a k-outerplanar graph, we can compute in parallel a face-face tree representation in $O(\log^2 n)$ time and using $O(n^2)$ processors.*

3 Approximating MIS

We describe the procedure to obtain an approximation to the MIS problem on planar graphs.

3.1 Outerplanar graphs

Given a binary face-face tree, we compute the MIS of the corresponding outerplanar graph. We associate to each node x in the tree, a table t_x with four entries; $t_x(a,b)$, with $a, b \in \{0,1\}$. Each entry contains the maximum size of an independent set in G_x depending on which of the associated vertices are or are not forced to be in the independent set. so that 1 (0) means that the corresponding vertex is (is not) in the IS.

For a real leaf x the corresponding table is, $t_x(0,0) = 0$, $t_x(0,1) = t_x(1,0) = 1$ and $t_x(1,1)$ is undefined. If the leaf is virtual, the associated table is, $t_x(0,0) = 0$, $t_x(1,1) = 1$, with $t_x(0,1)$ and $t_x(1,0)$ undefined.

To compute the MIS for the whole outerplanar graph, transvers the tree in a bottom-up way, computing at each interior node of the tree an operation *merge* of tables, as we have three types of internal nodes we need three types of merging. The basic merge of two tables t_x and t_y is defined: for any $a, b \in \{0,1\}$

$$t_{xy}(a,b) = \max_{c \in \{0,1\}} \{t_x(a,c) + t_y(c,b) - c\}.$$

Suppose that z is an internal node with left child x and right child y the table for z will be $t_z = t_{xy}$ when z is a marked node, $t_z = t_{xy}$ except for the entry $(1,1)$ that will be undefined when z is real, and $t_z(0,0) = t_{xy}(0,0)$, $t_z(1,1) = t_{xy}(1,1) - 1$, and $t_z(1,0)$ and $t_z(0,1)$ will be both undefined, when z is virtual.

An induction proof shows that the table obtained for each internal node, corresponds to the maximum size of an IS set for the associated subgraph. Once we have a table for the root, taking the maximum of its entries we get the size of an MIS for the graph G.

Using standard techniques we can define the shunt operation of the tree contraction [KR90].

Theorem 4 *There is a parallel algorithm to compute a maximum independent set for an outerplanar graph that runs in $O(\log^2 n)$ time using $O(n^2)$ processors.*

3.2   k-outerplanar graphs

Finally, let us compute the MIS of a given k-outerplanar graph. For each node x in the tree with i associated vertices, compute a table t_x with 2^{i+1} entries $t_x(a,b)$ with $a, b \in \{0,1\}^i$, each entry contains the maximim size of an IS in G_x depending on which of the vertices in $B_1(x)$, or $B_2(x)$ are forced to be or not to be in the IS.

In order to compute tables for the nodes of the face-face tree, we consider three operations on tables; *table extension*, *table contraction* and *table merging*.

Given a table for a leaf x on level i, and given a level $i+1$ node w, we will extend t_x to obtain a table t_{x+w} for a node with associated set of vertices $\langle w, B_1(x)\rangle$, $\langle w, B_2(x)\rangle$. For any of $(a,b) \in \{0,1\}^i \times \{0,1\}^i$ representing whether each of the vertices in $B_1(x), B_2(x)$ are forced to be in the IS, the value t_{x+w} will be:

$$t_{x+w}(0a, 0b) = t_x(a, b)$$

$$t_{x+w}(1a, 0b) = t_{x+w}(0a, 1b) = \text{undefined}$$

$$t_{x+w}(1a, 1b) = \begin{cases} \text{undefined} & \text{if } w \text{ and a level } i \text{ vertex present} \\ & \text{in the IS are connected} \\ t_x(a,b) + 1 & \text{otherwise} \end{cases}$$

In the reverse sense we define a table contraction operation. Given a table for the root of a level i tree, recall that $B_1(x) = \langle w, B_1'(x)\rangle$ and $B_2(x) = \langle w, B_2'(x)\rangle$, we will contract t_x to obtain a table t_{x-w} for a node with associated set of vertices $B_1'(x)$, $B_2'(x)$.

$$t_{x-w}(a, b) = \max\{t_x(0a, 0b), t_x(1a, 1b)\}$$

Extend the *merging* operation described in 3.1 to deal with sets of more than one vertex, for any $a, b \in \{0,1\}^i$ ($i \leq k$), let $|c|$ denote the number of 1's in c,

$$t_{xy}(a, b) = \max_{c \in \{0,1\}^i} \{t_x(a, c) + t_y(c, b) - |c|\}.$$

Let's analyze the nodes on the face-face tree corresponding to a k-outerplanar graph. We have two kinds of leaves, all leaves in the level 1 tree are leaves in the tree, the initial table for these leaves will be just the initial tables in the outerplanar case. Other leaves are nodes x labeled by a set of vertices $B(x)$, the initial table is defined as follows; $t_x(a, b)$ will be undefined when $a \neq b$ or when both values are the same but there is an edge between two vertices that have to be included in the IS, otherwise the value will be the number of 1's in a.

The operation assigned to a non leaf node x follows the tree construction using the appropriate operation. Again define a shunt operation to apply the tree contraction technique.

As the merging of tables now needs $O(k)$ time and $O(8^k)$ processors, a shunt operation can be done in the same time bounds, but using $O(16^k)$ processors thus final computation over the tree needs $O(k \log n)$ time and $O(16^k n)$ processors. Putting together all the bounds we get

Theorem 5 *There is a parallel algorithm to compute a maximum independent set for a k-outerplanar graph that runs in $O(\log n(k + \log n))$ time using $O(n(16^k + n))$ processors.*

3.3 Planar graphs

The decomposition of a planar graph into k-outerplanar graphs is the same used in [Bak83]. For each i, $0 \leq i \leq k$. Let G_i be the graph obtained by deleting all nodes of G whose levels are congruent to $i \pmod{k + 1}$. Now every connected component of G_i is k-outerplanar.

An independent set for G_i can be computed as the union of independent sets for each component. Furthermore, for some r, $0 \leq r \leq k$ the solution for G_r is at least $k/(k+1)$ as large as the optimal solution for G. Thus the largest of the solutions for the G_i's is an independent set whose size is at least $k/(k+1)$ optimal.

Theorem 6 *Given k there is a parallel approximation for maximum independent set that runs in $O(\log n(k + \log n))$ time, uses $O(n^2(16^k + n))$ processors and achieves a solution at least $k/(k+1)$ optimal.*

Proof.(Sketch)

We can get a representation of the connected components of G_i by getting k copies of the tree representation of G and removing the corresponding parts. Running the k-outerplanar algorithm on each component we obtain a MIS for each component, after just add up the corresponding sizes and take the maximum. □

Thus we get an NC parallel approximation scheme by letting $\epsilon = 1/k$ and an NC asymptotic approximation scheme taking $k = c \log \log n$ for some constant c.

Theorem 7 *The Maximum Independent Set Problem for planar graphs is in NCAS and $NCAS^\infty$*

4 Conclusions and algorithms for other problems

We have shown that for any k there is parallel algorithm that approximate the Maximum Independent Set problem on planar graphs to a factor $\frac{k-1}{k}$, and runs in time $O(\log n(k + \log n))$ using $O(n^2(16^k + n))$ processors. From this follows that the MIS problem for planar graphs is in NCAS and in NCAS$^\infty$.

The techniques to obtain a parallel approximation algorithm for the MIS problem explained in the previous sections can be transformed to obtain approximations for other optimization problems on planar graphs, as well as exact NC algorithms for these problems when restricted to k-outerplanar graphs. In particular, by a simple adaptation of the explained techniques we are able to show that the following problems restricted to planar graphs are in NCAS and NCAS$^\infty$: Minimum Vertex Cover, Minimum Dominating Set, Minimum Edge Dominating Set, Maximum Cut, Maximum Even Degree Set and Maximum Matching. A consequence of the proofs of these results is that there are exact polylogarithmic parallel algorithms for these problems restricted to k outerplanar graphs. Adapting the same technique it can be shown, the 3-colorability and Graph Bisection problems on k outerplanar graphs belong also to the class NC. However, it remains as an open problem to lower the bounds on the number of processors used for the above problems.

Acknowledgment

The authors would like to thank Torben Hagerup for many helpful comments.

References

[AM86]   R.J. Anderson and E.W. Mayr. Approximating P-complete problems. Technical report, Stanford University, 1986.

[Bak83]   B.S. Baker. Approximation algorithms for NP-complete problems on planar graphs. In *24 IEEE Symposium on Foundations of Computer Science*, pages 265–273, 1983.

[CN89]   M. Chrobak and J. Naor. An efficient parallel algorithm for computing large independent set in a planar graph. In *SPAA-89*, pages 379–387. Association for Computing Machinery, 1989.

[GJ79]   M.R. Garey and D.S. Johnson. *Computers and Intractability: A Guide to the Theory of NP-Completeness*. Freeman, San Francisco, 1979.

[Hag90]   T. Hagerup. Planar depth-first search in log n parallel time. *SIAM Journal on Computing*, 19:678–703, 1990.

[KR86] P.H. Klein and J.H. Reif. An efficient parallel algorithm for planarity. In *IEEE Symposium on Foundations of Computer Science*, pages 465–477. IEEE Society, 1986.

[KR90] R. Karp and V. Ramachandran. Parallel algorithms for shared memory machines. In Jan van Leewen, editor, *Handbook of Theoretical Computer Science, Vol. A*, pages 869–942. Elsevier Science Publishers, 1990.

[KS93] P. Klein and C. Stein. A parallel algorithm for approximating the minimum cycle cover. *Algorithmica*, 9:23–31, 1993.

[KSS89] L.M. Kirousis, M.J. Serna, and P. Spirakis. The parallel complexity of the connected subgraph problem. In *30th. IEEE Symposium on Foundations of Computer Science*, pages 446–456. IEEE, 1989.

[LN92] M. Luby and N. Nisan. A parallel approximation algorithm for positive linear programming. Technical report, International Computer Science Institute, Berkeley, 1992.

[PR87] J.G. Peters and L. Rudolph. Parallel approximation schemes for subset sum and knapsack problems. *Acta Informatica*, 24:417–432, 1987.

[Ser90] M.J. Serna. *The parallel approximability of P-complete problems*. PhD thesis, Universitat Politecnica de Catalunya, 1990.

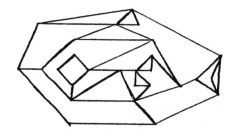

Figure 1: A 3-outerplanar graph G

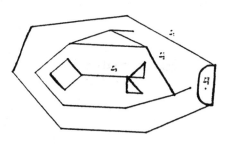

Figure 2: The level subgraphs of graph G

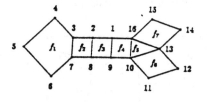

Figure 3: A bi-connected outerplanar graph L

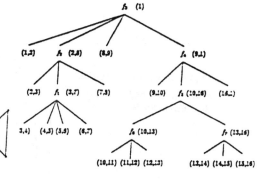

Figure 4: The face-face tree of L

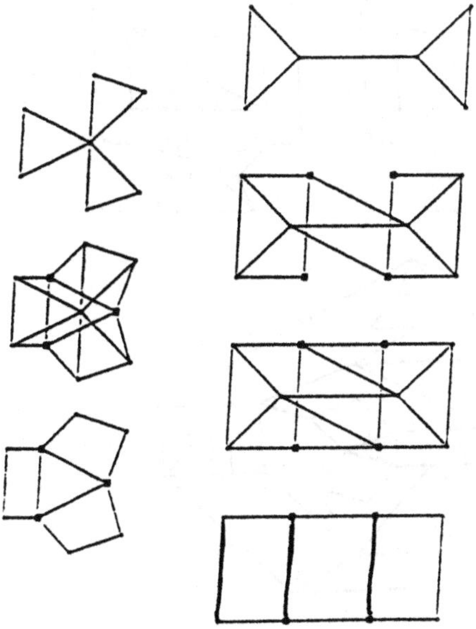

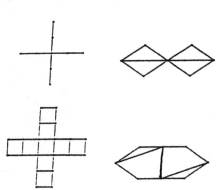

Figure 5: Transformation of a cutpoint with

three components and a bridge

Figure 6: The associated graph for

some outerplanar graphs

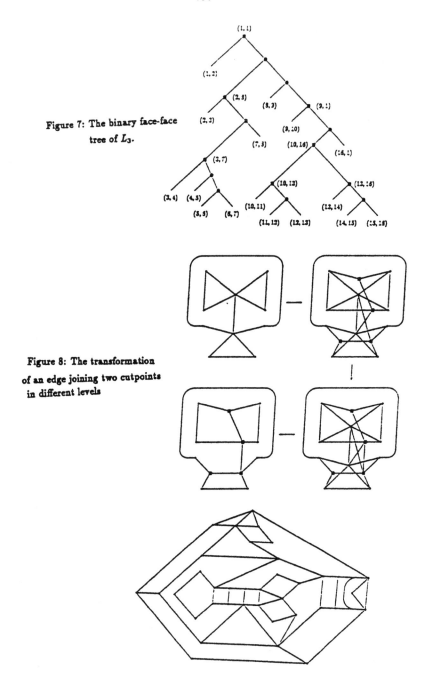

Figure 7: The binary face-face tree of L_3.

Figure 8: The transformation of an edge joining two cutpoints in different levels

Figure 9: The associated graph G'

DNA Physical Mapping: Three Ways Difficult

Michael R. Fellows[1], Michael T. Hallett[1], H. Todd Wareham[2]

[1] Department of Computer Science
University of Victoria
Victoria, British Columbia V8W 3P6, Canada
[2] Department of Computer Science
Memorial University of Newfoundland
St. John's, Newfoundland A1C 5S7, Canada

Abstract. We study the computational complexity of the combinatorial problem *Intervalizing Colored Graphs* (or ICG) that has some applications in DNA physical mapping. The three distinct conceptual frameworks of NP-completeness, algorithmic techniques for bounded treewidth, and parameterized complexity theory are shown to fit together neatly in an integrated complexity analysis of ICG. It is shown that ICG is intractable in three different ways: (1) it is NP-complete, (2) it is hard for the parameterized complexity class $W[1]$ and (3) it is not finite-state for bounded treewidth or pathwidth, and is therefore resistant to the usual algorithm design methodologies. The proofs of these three results are related in interesting ways which suggest useful heuristic connections between the three complexity frameworks for intractable problems of bounded treewidth and pathwidth.

1 Introduction

The focus of this paper is a combinatorial problem (*Intervalizing Colored Graphs*) that models in a straightforward but limited way the determination of contig assemblies in DNA physical mapping.

Definition 1 *An undirected graph $G = (V, E)$ is called an interval graph if there exists an injection $\Phi : V \to \mathcal{I} : u \mapsto I_u$ where $\mathcal{I}$ is a set of intervals of a linearly ordered set (such as the real number line) such that*

$$(u, v) \in E_G \Leftrightarrow |I_u \cap I_v| \neq 0 \qquad I_u, I_v \in \mathcal{I}$$

The set $\mathcal{I}$ is called an interval representation of G.

Intervalizing Colored Graphs (ICG)
Instance: A graph $G = (V, E)$ and a coloring $c : V \to [k]$.
Parameter: k
Question: Is there a supergraph $G' = (V, E')$ of G which is properly colored by c and which is an interval graph?

ICG can be seen to be closely related to the problem of *Triangulating Colored Graphs* (TCG) studied in [4, 13, 14]. For this problem, the question is whether there is a properly colored supergraph which is *chordal*, that is, having no induced cycles of length ≥ 4. (Note that interval graphs are chordal.)

DNA Physical Mapping and ICG

A general problem which has applications at several levels of sequence reconstruction is: Given a set of fragments of a sequence X and a measure of overlap between pairs of sequence fragments in this set, reconstruct the order of these fragments in X. This problem is typically broken into four steps (see [9]):

1. Fragment the sequence X. (This step may be repeated for several identical copies of X.)
2. Determine a set of characteristics for each fragment, termed its *fingerprint* or *signature*.
3. Compute a *similarity* or *overlap* measure between pairs of fragments based on their respective fingerprints.
4. Using the overlap information, assemble the fragments into islands of contiguous fragments, termed *contigs*.

Depending on the nature of the sequence under investigation, there are many ways in which these steps might be accomplished [9].

The fragmentation of a copy of X in step (1) is termed a *digest*. Where X is a piece of DNA, the fragments produced in step (1) can be reproduced in large quantities, and are termed *clones*. Obviously, if two clones originate in step (1) from the same copy of X, then they do not overlap. ICG models the situation where step (1) is applied to k copies of X, i.e., where k digests are performed in creating the clone library. The vertices of the input graph G correspond to the clones created by the k digests, and the edges correspond to pairs of clones that are known to overlap. The goal is to predict further overlaps, and ultimately to reconstruct the sequence X.

2   ICG and Bounded Pathwidth

The starting point is easy; we state it as a theorem only because of its central importance to our discussion. An analog for TCG is proved in [14].

Theorem 1 *If G is a k-colored graph which can be properly intervalized, then the pathwidth of G is bounded by $k-1$ and G can be properly intervalized to a $(k-1)$-path.*

For both ICG and TCG a strong case can be made that there are useful applications for small fixed values of the parameter k (see [13]). Thus we could hope for the kind of complexity we observe for *Min Cut Linear Arrangement*, *Graph Search Number*, *Planar Face Cover*, *Feedback Vertex Set* and a host of other problems that can be solved in linear time for each fixed parameter value. A standard route for such results involves two steps: (1) In time $O(n)$ by the algorithm of Bodlaender [3] we can either determine that the answer is "no" (by an analog of Theorem 1 for the problem), or compute a tree-decomposition of bounded width. (2) Given the tree-decomposition, we can solve the problem in linear time by means of one of the standard algorithm design techniques for bounded treewidth (e.g. [2]). *All* of the problems above are NP-complete, but this implies nothing about the fixed-parameter complexity.

In [1], a graph-theoretic analog of the Myhill-Nerode theorem (of formal language theory) is proved, giving a simple necessary and sufficient condition for the applicability of the standard algorithm design techniques for bounded treewidth and pathwidth that are based on "finite-state" dynamic programming and related ideas. Intuitively, this criteria concerns the amount of information flow across a bounded-size cutset necessary to recogize a graph family F. The relevant cutset size is t for graphs of treewidth (or pathwidth) bounded by t.

Definition 2 *A tree-decomposition (path-decomposition) of a graph $G = (V, E)$ is a tree T together with a collection $\mathcal{T}$ of subsets $S_x \subseteq V$ indexed by the vertices x of T that satisifies (1) for every edge uv of G there is some x such that $\{u, v\} \subseteq S_x$, (2) if y is a vertex on the unique path in T from x to z then $S_x \cap S_z \subseteq S_y$. The width of a tree-decomposition (path-decomposition) is the maximum over the vertices x of the tree T of the decomposition of $|S_x| - 1$. A graph G has treewidth (pathwidth) at most k if there is a tree decomposition of G of width at most k.*

Graphs of bounded treewidth (or pathwidth) can be represented by labeled trees (or paths) where the labels are taken from a finite alphabet and respresent operations for building the graph.

Definition 3 *A t-boundaried graph $G = (V, E, B, f)$ is an ordinary graph $G = (V, E)$ equipped with a distinguished subset $B \subseteq V$ of t vertices, termed the boundary of G, together with a 1:1 map $f : B \to \{1, ..., t\}$ that labels the boundary. The binary operator $\oplus$ on two t-boundaried graphs is defined: $G \oplus H$ is the t-boundaried graph obtained by identifying the ith boundary node of G with the ith boundary node of H, for $i = 1, ..., t$.*

Definition 4 *An n-ary t-boundaried composition operator $\otimes$ is defined by the data (1) a t-boundaried graph $T_\otimes = (V_\otimes, E_\otimes, B_\otimes, f_\otimes)$, (2) injective maps $f_i : \{1, ..., t\} \to V_\otimes$ for $i = 1, ..., n$. For the binary case, if G_i for $i = 1, 2$ is a pair of t-boundaried graphs $G_i = (V_i, E_i, B_i, f_i)$ then $G_1 \otimes G_2$ is defined to be the t-boundaried graph for which the ordinary underlying graph is formed from the disjoint union of G_1, G_2 and $T_\otimes$ by identifying each vertex u of B_i (for $i = 1, 2$) with its image $f_i(u)$ in $V_\otimes$. The boundary set and the labeling for $G_1 \otimes G_2$ is given by $B_\otimes$ and $f_\otimes$.*

Graphs of treewidth at most t can be parsed using a small number of t-ary operators of boundary size t; they can also be parsed with binary operators of size $t + 1$.

Definition 5 *Two t-boundaried graphs X and Y are F-equivalent ($X \sim_F Y$) if and only if for every t-boundaried graph Z, $X \oplus Z \in F \leftrightarrow Y \oplus Z \in F$. The small universe U^t_{small} is the set of all t-boundaried graphs that arise in the* *parsing of graphs of treewidth (pathwidth) at most t. A graph family F is t-cutset regular if and only if $\sim_F$ has finite index on U^t_{small}. A graph family F is t-finite-state if and only if there is a finite-state tree automata (linear automata in the case of pathwidth) that accepts precisely the width t tree-decompositions (path-decompositions) represented symbolically, of graphs in F.*

Theorem [1]. *A graph family F is t-finite state if and only if F is t-cutset regular.*

Using this theorem, we can show that the class of intervalizable k-colored graphs is not t-finite state, for $k \geq 3$ and $t \geq 1$. The method of argument is much as in the application of the Myhill-Nerode theorem to show that a language (for example, $\{a^n b^n : n \geq 0\}$) is not regular. In this more familiar setting, one simply exhibits an infinite set of words S (for example, $S = \{a^i : i \geq 0\}$), and for each pair of words $x_i, x_j \in S$, a witness z_{ij} (for example, $z_{ij} = b^i$) showing that x_i and x_j are not equivalent (in this example, because $x_i z_{ij} = a^i b^i \in L$ and $x_j z_{ij} = a^j b^i \notin L$). Replace "words" with "boundaried graphs" and concatenation with $\oplus$ in this discussion, and you see essentially how the theorem is applied in the graph setting to obtain negative results such as the following.

Theorem 2 *k-ICG is not t-finite-state for all $k \geq 3$ and $t \geq 1$.*

Proof. We sketch the argument for 3-colored graphs and $t = 1$. For $i \geq 1$, let $X_i = (V_i, E_i, B_i, f_i)$ be constructed as follows: $V_i = A \cup B$ where $A = \{a_1, a_2, a_3, a_4\}$ and $B = \{b_k | 1 \leq j \leq 4i + 1\}$; $E_i = \{(a_1, a_2), (a_3, a_4)\} \cup \{(a_1, b_1), (a_2, b_1), (b_{4i+1}, a_3), (b_{4i+1}, a_4)\} \cup \{(b_k, b_{k+1}) \mid 1 \leq k < 4i + 1\}$ $B_i = \{b_{2i+1}\}$; $f_i(b_{2i+1}) = 1$.

Let $c : V_i \to \{1, 2, 3\}$ be the coloring of X_i: $c(a_1) = c(a_3) = 1$, $c(a_2) = c(a_4) = 2$, $c(b_j) = (j + 1 \bmod 2) + 1$ for $1 \leq j \leq 4i + 1$

For $j \geq 1$, let $Y_j = (V'_j, E'_j, B'_j, f'_j)$ be constructed as follows: $V'_j = \{h_k \mid 1 \leq k \leq 8j\} \cup \{d\}$; $E'_j = \{(d, h_1), (d, h_{4k+1})\} \cup \{(h_k, h_{k+1}) | 1 \leq k < 4j\} \cup \{(h_k, h_{k+1}) | 4j + 1 \leq k < 8j\}$; $B'_j = \{d\}$; $f'_j(d) = 1$.

Let $c' : V'_j \to \{1, 2, 3\}$ be the coloring of Y_j: $c'(d) = 1$, $c'(h_k) = 3$ if $k = 1$ (mod 4) or $k = 3$ (mod 4), $c'(h_k) = 1$ if $k = 2$ (mod 4), $c'(h_k) = 2$ if $k = 0$ (mod 4).

It is straightforward to verify that for $i < j$, $X_j \oplus Y_j$ can be intervalized, but $X_i \oplus Y_j$ cannot. An example is shown in Fig. 1. □

Paralleling the results of [13] in the case of TCG, we have the following positive result for the few remaining values of k and t.

Theorem 3 *For all t and for $k \leq 2$, k-ICG is t-finite-state.*

Proof sketch. It can be shown that a 2-colored graph G is intervalizable if and only if G is a properly colored caterpillar, and caterpillarhood can be expressed in second order monadic logic. By the results of [6] the problem is finite-state for all t. □

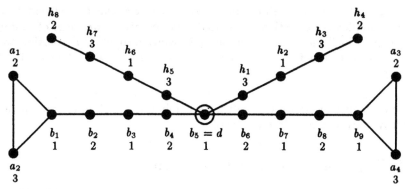

Fig. 1. $X_2 \oplus Y_1$.

The consequence of Theorem 2 is that we cannot demonstrate, for example, that ICG can be solved in linear time for each fixed parameter value *by the standard route*, but the possibility is left open that there may be some other way to accomplish this (or something similar).

In [7] a theory of parameterized computational complexity is developed in order to address such questions. The theory is motivated by the striking qualitative distinction between the complexity behavior of, e.g., *Min Cut Linear Arrangement* or *Feedback Vertex Set*, and the behavior of problems such as *Bandwidth* and *Dominating Set*. While the former can be solved in linear time for each fixed parameter value, for the latter the best known algorithms require time $O(n^{k+1})$ and are essentially based on "brute force." This resembles the qualitative distinction between problems in P and problems which are NP-complete.

Definition 6 *A parameterized problem is a set* $L \subseteq \Sigma^* \times N$.

Definition 7 *We say that a parameterized problem L is fixed-parameter tractable if there is an algorithm to decide membership in L for (x, k) in time $f(k)|x|^\alpha$ where $f : N \to N$ is an arbitrary function and α is a constant.*

In [7] a hierarchy of complexity classes of parameterized problems is defined:

$$FPT \subseteq W[1] \subseteq W[2] \subseteq \cdots \subseteq W[P]$$

FPT denotes the parameterized problems which are fixed-parameter tractable. Many parameterized problems have been identified as complete or hard for various levels of the hierarchy. As with NP-completeness, it is not necessary to understand the details of the theory in order to apply it to a concrete problem. All that is necessary is to find an appropriate combinatorial reduction! In [8] it is shown that *Independent Set* is complete for $W[1]$. A Turing machine characterization of $W[1]$ and a variety of $W[1]$ complete problems are identified in [5].

Definition 8 *A reduction of a parameterized problem L to a parameterized problem L' is an algorithm that transforms (x, k) into $(x', g(k))$ in time $f(k)|x|^\alpha$, so that $(x, k) \in L$ if and only if $(x', g(k)) \in L'$.*

In the next section we show that *Independent Set* can be reduced to ICG. For this complexity reduction, it is necessary to solve a problem of "information transmission" inside of a bounded pathwidth gadget. Our solution to this problem suggests the following heuristic connection between two kinds of intractability.

★ For problems of bounded treewidth, the combinatorics of finite-state intractability and W-hardness are closely related.

3 A Key Combinatorial Reduction

The gadgetry is quite complex, but at the most basic level, the combinatorial mechanism involved is similar to that exploited in the proof of Theorem 3.

Definition 9 *A graph $G = (V, E)$ with vertex coloring $c : V \to Z$ is c-intervalized if there exists a supergraph $H = (V, E')$, $E \subseteq E'$, that is properly colored by c and has an interval representation $\mathcal{I}$. The supergraph H is said to be a c-intervalization of G.*

Definition 10 *We say that an interval $I_v < x$ (I_v is properly less than x) if and only if for all y in $I_v, y < x$. Furthermore $I_v < I_w$ (I_v is properly less than I_w) if and only if for all x in I_v and all y in $I_w, x < y$. If an interval I_v is a proper subinterval of I_w, then we say that I_v lands on I_w.*

Let $G_{IS} = (V, E)$ be a graph with the vertices labeled from $1 \ldots |V| = n$. Let $m = nC(n, 2)$ and $m' = nC(n, 2) - n + 1$, where $C(n, 2) = \binom{n}{2}$. From G_{IS} we create a graph $G_{CGI} = (V', E')$ that has coloring $c : V' \to \{1, \ldots, 3k, block, floor, queen, slide\}$. G_{IS} has four sets of components.

Component 1: Vertex Selection Components (B's and Queen B's) We create k decision components $\mathcal{B}_i$ each of which consists of m' B components where: $V(B) = \{b_1, \ldots, b_5\}$ and $E(B) = \{(b_j, b_{j+1}) \mid 1 \leq j < 5\}$.

Let $B' = B_1 \cup B_2 \cup \ldots \cup B_{m'}$. Within B', identify vertex $b_5 \in B_j$ with vertex $b_1 \in B_{j+1}$ for $1 \leq j < m'$.

Let $\mathcal{B}_i = B' \cup \{\hat{b}_1, \hat{b}_2, \ldots, \hat{b}_{C(n,2)}\} \cup \{\underline{b}_1, \underline{b}_2\}$. We add the following edges to $\mathcal{B}_i$: $\{(\hat{b}_j, b) \mid 1 \leq j \leq C(n, 2), \forall b \in B_{n(j-1)+1}\}$ and $\{(\underline{b}_1, b_1), (\underline{b}_2, b_5) \mid b_1 \in B_1$ of $\mathcal{B}_i$, $b_5 \in B_{m'}$ of $\mathcal{B}_i\}$.

Let $QB = \{B_{nj+1} \in \mathcal{B}_i \mid 1 \leq i \leq k, 1 \leq j \leq C(n, 2)\}$ (the Queen B's).

We color each $\mathcal{B}_i$ as follows: $c(b_1) = c(b_5) = 3(i - 1) + 1$; $c(b_2) = c(b_4) = 3(i - 1) + 2$; $c(\underline{b}_1) = c(\underline{b}_2) = 3(i - 1) + 3$; $c(b_3) = floor$; $c(\hat{b}_j) = queen$ for all j.

Component 2: Complete Edge-Set Representation Component (Flowers and Flower Pots) Our second component $\mathcal{F}$ consists of m F components (flowers) where $V(F) = Q \cup R \cup S \cup T$ where $Q = \{q_j \mid 1 \leq j \leq k + 1\}$, $R = \{r_j \mid 1 \leq j \leq k\}$, $S = \{s_j \mid 1 \leq j \leq k + 1\}$, and $T = \{t_1, t_2\}$.

E_F contains all the edges necessary so that (1) the vertices of Q form a clique, (2) the vertices of R form a clique, (3) the vertices of S form a clique, (4) t_1 is adjacent to every vertex in Q and R, and (5) t_2 is adjacent to every vertex in R and S.

Each vertex of $Q - \{q_{k+1}\}$ and $S - \{s_{k+1}\}$ is assigned a unique color from $\{3j + 2 \mid 0 \le j \le k - 1\}$ and each vertex of R is assigned a unique color from $\{3j + 1 \mid 0 \le j \le k - 1\}$. Additionally, let $c(t_1) = c(t_2) = floor$, $c(q_{k+1}) = c(s_{k+1}) = block$.

Let $\mathcal{F} = F_1 \cup F_2 \cup \ldots \cup F_m$ identifying all the vertices in the set S of F_j with their like-colored vertices in the set Q of F_{j+1}, $1 \le j < m$.

Flower F_j of component $\mathcal{F}$ is shown in Figure ??.

For each decision component $\mathcal{B}_i$, we add edges between $\underline{b}_1$ and every vertex of Q in F_1 and edges between $\underline{b}_2$ and every vertex of S in F_m of $\mathcal{F}$.

Lastly, let $\hat{F} = \{\hat{f}_1, \hat{f}_2, \ldots, \hat{f}_{C(n,2)-1}\}$. We add edges from $\hat{f}_i$ to every vertex in the set S and t_2 of F_{ni}, and from $\hat{f}_i$ to t_1 of F_{ni+1}. Let $c(\hat{f}_i) = slide$.

Note that the number of flowers in this component is equal to the total possible number of edges in graph G_{IS} multiplied by n. Establish a bijection between groups of n contiguous F components and all possible edges in G_{IS} (flower pots). For instance, let us use the bijection in which the pair $\{1, 2\}$ corresponds to $\{F_1, \ldots, F_n\} \subseteq \mathcal{F}$, $\{1, 3\}$ corresponds to $\{F_{n+1}, \ldots, F_{2n}\}$ and $\{n - 1, n\}$ corresponds to $\{F_{nC(n,2)-n}, \ldots, F_m\}$. This bijection will be used in component 4 described below.

Component 3: Intervalization Limits Components Let C_1 and C_2 be two cliques of size $3k + 4$. Every vertex in C_1 (and C_2) is assigned a unique color. We identify the vertices of C_1 which have been assigned colors from the set $\{3i + 2 \mid 0 \le i \le k - 1\} \cup \{block\}$ with their like-colored vertices in the set Q of F_1. Similarly, we identify the vertices of C_2 with their like-colored vertices in the set S of F_m.

Component 4: Edge-Constraint Components (Enforcers) Use the bijection described above for the $\mathcal{F}$ component, and let $F_{p+1}, \ldots, F_{p+n}$ be the corresponding F components for a pair of vertices $\{u, v\}$. If $(u, v) \in E(G_{IS})$ then, we create a component N where $V(N) = \{z_1, \ldots, z_5\}$ and $E(N) = \{(z_i, z_j) \mid 1 \le i \ne j \le 4\} \cup \{(z_i, z_j) \mid 2 \le i \ne j \le 5\}$.

Additionally, we place edges between z_1 and every vertex of R in F_u and we place edges between z_5 and every vertex of R in F_v. For each F_i, $i \ne u$ or v, add a vertex d_i s.t. d_i is adjacent to t_1 and t_2 of F_i. If $(u, v) \notin E(G_{IS})$ then we add such a vertex d to every F component (in the flower pot).

We color these components as follows: $c(z_1) = c(z_5) = slide$; $c(z_2) = queen$; $c(z_3) = floor$; $c(z_4) = c(d_i) = block$ for all i.

A sample construction of a graph G_{CGI} for a given graph G_{IS} is included at the end of this paper. Informally, the chain of proofs given below proceeds as follows: After establishing that the cliques C_1 and C_2 create sets of intervals that contain the rest of the c-intervalization of G_{CGI} (Claim 1), we show that the intervals corresponding to the flowers of $\mathcal{F}$ are ordered in linear fashion between these two limits (Claim 2). Not only must each individual B land in the center of a flower (Claim 3), but the m' B components of each $\mathcal{B}_i$ must land in m' contiguous flowers (Claim 4); moreover, as no two distinct Queen B's can land in the same flower (Claim 5), each B_i effectively selects a vertex in G_{IS} corresponding to the number of the flower on which that B_i initially lands in the first flower pot. As Queen B's cannot land in both of the u- and v-flowers specified by the edge-enforcer for edge $\{u, v\}$ (Claim 6), the k B components must select vertices corresponding to an independent set in G_{IS} (Claim 7).

Let $\alpha = max(\{x \mid x \in \bigcap_{y \in C_1} I_y\})$ and $\omega = min(\{x \mid x \in \bigcap_{y \in C_2} I_y\})$. The proof of the following is straighforward and omitted.

Claim 1 *If $\mathcal{I}$ is a c-intervalization of $G_{CGI} = (V, E)$ using at most $3k + 4$ colors, then for all $v \in V - (C_1 \cup C_2)$ either $\alpha < I_v < \omega$ or $\alpha > I_v > \omega$.*

For the remainder of this section, we assume without loss of generality that $\alpha < \omega$. Let $\beta_i = min(\{x \mid x \in \bigcap_{y \in t_1 \cup Q} I_y\})$, $\gamma_i = max(\{x \mid x \in \bigcap_{y \in t_1 \cup R} I_y\})$, $\delta_i = min(\{x \mid x \in \bigcap_{y \in t_2 \cup R} I_y\})$, $\epsilon_i = max(\{x \mid x \in \bigcap_{y \in t_2 \cup S} I_y\})$ where $t_1, t_2, Q, R, S \in F_i$.

Claim 2 *In any c-intervalization of the graph induced by $\mathcal{F} \cup C_1 \cup C_2$ the following holds for the m F components of $\mathcal{F}$:*

$$\alpha < \beta_1 \leq \gamma_1 < \delta_1 \leq \epsilon_1 < \ldots < \beta_m \leq \gamma_m < \delta_m \leq \epsilon_m < \omega.$$

Proof. We show this by induction of the size of $\mathcal{F}$. Let $\mathcal{F}$ consist of one F component. That $\alpha < \beta_1, \gamma_1, \delta_1, \epsilon_1 < \omega$ follows from Claim 1. Suppose $\gamma_1 < \beta_1$; that is, $max(\{x \mid x \in \bigcap_{y \in t_1 \cup R} I_y\}) < min(\{x \mid x \in \bigcap_{y \in t_1 \cup Q} I_y\})$. Clearly it must be the case that $\gamma_1 < \forall z \in \bigcap_{y \in Q} I_y$ but $Q \subseteq C_1$, which implies that $\gamma_1 < \alpha$. This contradicts Claim 1. A symmetric argument can be given to show that $\delta_1 \not> \epsilon_1$.

Clearly $\delta_1 \neq \gamma_1$, so suppose $\delta_1 < \gamma_1$. Since $c(t_1) = c(t_2)$, δ_1 can not be in the interval $(\beta_1, \ldots, \gamma_1)$ so $\gamma_1 < \beta_1$ and $I_{t_2} < \beta_1$ implying $\epsilon_1 < \beta_1$. Note however that the vertices of Q and S are colored identically. Therefore $\epsilon_1 < \bigcap_{y \in Q} I_y$ but $Q \subseteq C_1$ implies that $\epsilon_1 < \alpha$, again contradicting Claim 1.

Assume that the above property holds for $\mathcal{F}$ consisting of $k < m$ components. Let $\mathcal{F}'$ consist of m components. The induction hypothesis holds for the graph induced by removing the m-th flower F_m appropriately. We are required to show that $\epsilon_{m-1} < \beta_m \leq \gamma_m < \delta_m \leq \epsilon_m < \omega$ in any c-intervalization.

Clearly $\beta_m \neq \epsilon_{m-1}$ so suppose $\beta_m < \epsilon_{m-1}$. Because $t_1 \in F_m$ is adjacent to all $s \in S_{m-1} = Q_m$ it is easy to see that $\beta_m > \beta_{m-1}$. In fact, the only interval where β_m may be placed is $(\gamma_{m-1} \ldots \delta_{m-1})$. If this were the case, then $I_{t_1 \in F_m} < \delta_{m-1}$. But $t_1 \in F_m$ is adjacent to all $r \in R$ of F_m implying that $\bigcap_{r \in R_{m-1}}$ has non-empty intersection with $\bigcap_{r \in R_m}$. These two sets are colored identically contradicting the proper coloring.

Suppose $\gamma_m < \beta_m$; that is, $max(\{x \mid x \in \bigcap_{y \in t_{1_m} \cup R_m} I_y\}) < min(\{x \mid x \in \bigcap_{y \in t_{1_m} \cup Q_m} I_y\})$. This implies that $max(\{x \mid x \in \bigcap_{r \in R_m} I_r\}) < min(\{x \mid x \in \bigcap_{q \in Q_m = S_{m-1}} I_q\})$. Hence $I_{t_1 \in F_m}$ spans the interval $(\gamma_m \ldots \beta_m)$. But $\epsilon_{m-1} < \beta_m$

and $t_2 \in F_{m-1}$ is adjacent to all vertices in the set S $(= Q)$ of $F_{m-1(=m)}$. Therefore there exists a point x such that $x \in I_{t_1 \in F_m}$ and $x \in I_{t_2 \in F_{m-1}}$ contradicting the proper coloring.

Clearly $\delta_m \neq \gamma_m$ so suppose $\delta_m < \gamma_m$. Now $\delta_m > \delta_{m-1}$ since the colors present at δ_{m-1} are identical to the colors present at δ_m and $t_2 \in F_m$ is adjacent to all vertices of R_m. In fact, the only valid interval for δ_m is $(\epsilon_{m-1} \ldots \beta_m)$. If this were the case, then $\epsilon_{m-1} < I_{t_2 \in F_m} < \beta_m$ implying that $\epsilon_{m-1} < \epsilon_m < \beta_m$. But this contradicts the proper coloring since the vertices of S_m are colored identically to the vertices of $S_{m-1} = Q_m$.

Suppose $\epsilon_m < \delta_m$. Then $max(\{x | x \in \bigcap_{s \in S_m} I_s\}) < min(\{x | x \in \bigcap_{r \in R_m} I_r\})$. Since $S_m \subseteq C_2$, $\omega < min(\{x | x \in \bigcap_{r \in R_m} I_r\})$ which contradicts Claim 1. $\square$

The proofs of the following claims are omitted (see [10] for details). We say that a B component *lands* in the p-th F component if $\gamma_p < I_{b_3} < \delta_p$.

Claim 3 *Consider some B_i in one of the decision components. It must be the case that $\gamma_p < I_{b_3 \in B_i} < \delta_p$ in any c-intervalization of G_{CGI} for $1 \leq p \leq m$.*

Claim 4 *For each decision component, $B_1, B_2 \ldots, B_{m'}$ must land in m' contiguous flowers $F_p, F_{p+1}, \ldots, F_{p+m'-1}$ and furthermore, $p \leq n$.*

Claim 5 *No two distinct elements of QB may land in the same flower F_p.*

Let $F_p, F_{p+1}, \ldots, F_{p+n-1}$ be the n contiguous flowers associated with an edge $(u, v) \in E(G_{IS})$. Let N be the enforcer associated with this edge. Without loss of generality we assume that $u < v$. Let $\zeta = min(\{x | x \in I_{z_2} \cap I_{z_3} \cap I_{z_4}\})$.

Claim 6 *In a c-intervalization, it must be the case that either $\gamma_{p+u} < \zeta < \delta_{p+u}$ or $\gamma_{p+v} < \zeta < \delta_{p+v}$. Furthermore, if a member of QB lands in F_u, then $\gamma_{p+v} < \zeta < \delta_{p+v}$.*

Proof. It is easy to verify that the only intervals where ζ could possibly reside and yet still respect the proper coloring is between γ_i and δ_i for $1 \leq i \leq m$. We first argue that $\beta_p < \zeta < \epsilon_{p+n-1}$ (enforcers stay within their flower pots). But this is clearly the case because of the placement of the $\hat{F}$ vertices where $c(\hat{F}) = slide$. This guarantees that $I_{z_1} > \beta_p$ and $I_{z_5} < \epsilon_{p+n}$.

Now recall that for every F_x where $x \neq u$ or v, a vertex d was created and made adjacent to t_1 and t_2. Therefore, in any c-intervalization, $\exists x \in I_d$ s.t. $x \leq \gamma_x$ and $\exists x \in I_d$ such that $x \geq \delta_x$. Therefore at every point between γ_x and δ_x, $c(d) = block$ is present implying that ζ can not lie in this interval.

To show the second statement in our claim, suppose a member of QB lands in F_u. By a similar argument to that in the proof of Claim 5, it is true that at every point in the interval $(\gamma_u \ldots \delta_u)$ the color *queen* is present. Since $c(z_2) = queen$, ζ can not lie in this interval. $\square$

Claim 7 *G_{IS} has an independent set of size k if and only if G_{CGI} has a c-intervalization using at most $3k + 4$ colors.*

Proof. Suppose G_{CGI} has a c-intervalization $\mathcal{I}$ with $3k + 4$ colors. Let $IS = \{j \mid B_1 \text{ of } \mathcal{B}_i \text{ lands in } F_j, 1 \leq i \leq k\}$. By Claims 4 and 5, $|IS| = k$ and for all $j \in IS$, $1 \leq j \leq n$. We claim that IS constitutes an independent set for G_{IS}. Consider $x \in IS$. By Claim 4, $F_x, F_{x+1}, \ldots, F_{m'+x-1}$ have $B_1, B_2, \ldots, B_{m'}$ land in them respectively and $F_x, F_{n+x}, F_{2n+x}, \ldots, F_{m'+x-1}$ have members of QB land in them. By Claim 6, the enforcer N corresponding to any edge of the form $(x, y) \in E(G_{IS})$ must land in the F component corresponding to y. Therefore $F = F_{pn+y}$, for some $p \in \{0, \ldots, \binom{n}{2} - 1\}$ must not have a member of QB land in it, since if it does N is not able to land anywhere. This implies that F_y does not have a member of QB land in it. Hence $y \notin IS$.

Suppose G_{IS} has a k element independent set and let IS be such a set. For each $x \in IS$, we let B_1 of some unique $\mathcal{B}_i$, land in F_x. By Claim 4, flowers $F_x, F_{x+1}, \ldots, F_{m'+x-1}$ have $B_1, B_2, \ldots, B_{m'}$ land in them respectively. $B_1, B_{n+1}, B_{2n+1}, \ldots, B_{m'}$ are members of QB and therefore the flowers F_x, $F_{n+x}, F_{2n+x}, \ldots, F_{m'+x-1}$ have members of QB land in them. Since IS consitutes an independent set, if $(x, y) \in E(G_{IS})$, then $y \notin IS$ and by a similar chain of reasoning to the above the flowers $F_y, F_{n+y}, F_{2n+y}, \ldots, F_{m'+y-1}$ do not have members of QB land in them. By Claim 6, the enforcer N for edge (x, y) can land in the F component corresponding to y. $\square$

4 Consequences and Connections

From our reduction of *Independent Set* to ICG we can draw two significant conclusions and one tantalizing lead.

Theorem 4 *ICG is hard for* $W[1]$.

The significance of Theorem 4 is that ICG cannot be solved in time $f(k)n^\alpha$, where α is independent of k, unless a similar result can be obtained for the (apparently resistant) problem of determining whether a graph has an independent set of size k. (Similarly for *Clique* and the many other problems in $W[1]$; see [5, 8]. Note that Theorem 4 *conditionally* implies Theorem 3. That is, Theorem 4 implies Theorem 3 under the assumption that $FPT \neq W[1]$. Serendipitously, by the combinatorial reduction of §3 we may also conclude:

Theorem 5 *ICG is NP-complete.*

Proof. Note that in our reduction $f(k)$ is polynomial. $\square$

The NP-completeness of ICG was recently independently proved by Golumbic, Kaplan and Shamir [12] by a reduction that does not satisfy the definition of a parameterized complexity reduction.

The tantalizing lead is the heuristic principle $\star$. The combinatorial mechanism used to prove ICG to be non-finite-state, is functionally the same as that used to prove W-hardness. The first two authors have recently been able to employ this insight to show that *Bandwidth* is hard for $W[1]$. Whether TCG is fixed parameter tractable or hard for $W[1]$ is presently an interesting open problem.

References

1. K. Abrahamson and M. Fellows. *Finite automata, bounded treewidth and well-quasiordering.* Proceedings of the AMS Summer Workshop on Graph Minors (Seattle, 1991), A.M.S. Contemporary Mathematics Series, to appear.

2. S. Arnborg, J. Lagergren, and D. Seese. *Easy problems for tree-decomposable graphs.* J. Algorithms, 12 (1991), 308–340.

3. H. L. Bodlaender. *A linear time algorithm for finding tree-decompositions of small treewidth.* Proceedings of the 25th Annual ACM Symposium on Theory of Computing (1993), 226–234.

4. H. L. Bodlaender, M. R. Fellows and T. Warnow. *Two Strikes Against Perfect Phylogeny.* In: W. Kuich (editor), Proceedings of the 19th International Colloquium on Automata, Languages and Programming (ICALP'92). Springer-Verlag, Berlin, Lecture Notes in Computer Science, volume 623, pp. 273-283.

5. L. Cai, J. Chen, R. Downey and M. Fellows. *The Parameterized Complexity of Short Computation and Factorization.* Technical Report, Department of Computer Science, University of Victoria, June 1993.

6. B. Courcelle. *The monadic second-order logic of graphs I: Recognizable sets of finite graphs.* Information and Computation 85 (1990), 12–75.

7. R. G. Downey and M. R. Fellows. *Fixed parameter intractability.* Proceedings of the Seventh Annual IEEE Conference on Structure in Complexity Theory (1992), 36–49.

8. R. G. Downey and M. R. Fellows. *Fixed parameter tractability and completeness II: on completeness for $W[1]$.* To appear.

9. G. A. Evans. *Combinatoric strategies for genome mapping.* Bioassays 13 (1991), 39–44.

10. M. R. Fellows, M. T. Hallett and H. T. Wareham. *The parameterized complexity of DNA physical mapping.* Technical Report, Computer Science Department, University of Victoria, 1993.

11. M. R. Fellows and M. T. Hallett. *Bandwidth is hard for $W[1]$.* Manuscript, July, 1993.

12. M. Golumbic, H. Kaplan and R. Shamir. *On the complexity of DNA physical mapping.* Technical Report 271/93, Tel Aviv University, January 1993.

13. S. Kannan and T. Warnow. *Triangulating three-colored graphs.* Proceedings Second Annual ACM-SIAM Symposium on Discrete Algorithms (1991), 337–343. To appear, SIAM J. Discr. Math.

14. F. R. McMorris, T. Warnow, and T. Wimer. *Triangulating vertex colored graphs.* To appear, SIAM J. Discr. Math.

168

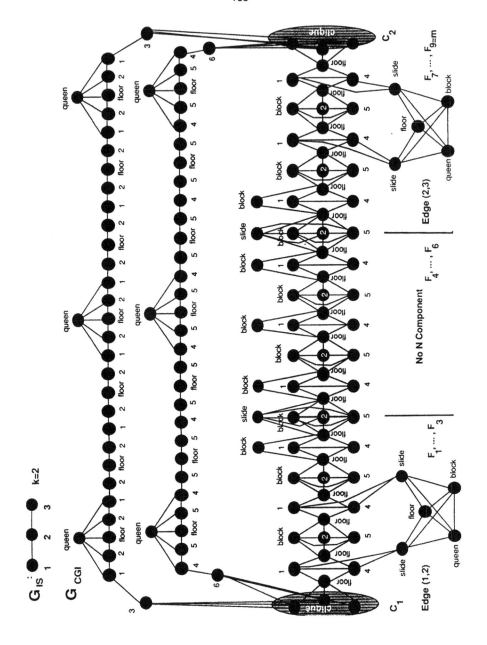

Fig. 2. A Sample Construction of G_{CGI} from G_{IS}.

A Calculus of Random Generation

Philippe FLAJOLET[1], Paul ZIMMERMANN[2], Bernard VAN CUTSEM[3]

[1] Algorithms Project, INRIA Rocquencourt, F-78153 Le Chesnay (France)
<Philippe.Flajolet@inria.fr>
[2] INRIA Lorraine, 615 rue du Jardin Botanique, F-54600 Villers-les-Nancy (France)
<Paul.Zimmermann@loria.fr>
[3] Laboratoire de Modélisation et Calcul, Université Joseph Fourier, BP 53X,
F-38041 Grenoble Cedex (France)
<bvcutsem@imag.fr>

Abstract. A systematic approach to the random generation of labelled combinatorial objects is presented. It applies to structures that are decomposable, *i.e.*, formally specifiable by grammars involving union, product, set, sequence, and cycle constructions. A general strategy is developed for solving the random generation problem with two closely related types of methods: for structures of size n, the boustrophedonic algorithms exhibit a worst–case behaviour of the form $\mathcal{O}(n \log n)$; the sequential algorithms have worst case $\mathcal{O}(n^2)$, while offering good potential for optimizations in the average case. (Both methods appeal to precomputed numerical tables of linear size.)

A companion calculus permits to systematically compute the average case cost of the sequential generation algorithm associated to a given specification. Using optimizations dictated by the cost calculus, several random generation algorithms are developed, based on the sequential principle; most of them have expected complexity $\frac{1}{2}n \log n$, thus being only slightly superlinear. The approach is exemplified by the random generation of a number of classical combinatorial structures including Cayley trees, hierarchies, the cycle decomposition of permutations, binary trees, functional graphs, surjections, and set partitions.

Introduction

This work started with a question arising in statistical classification theory: How can one generate a random "hierarchy"? In combinatorial terms, the generation problem simply amounts to drawing uniformly at random a tree with internal nodes of degree at least 2 and with leaves (external nodes) labelled by distinct integers, the number n of leaves being fixed. The need arises in statistics as one would like to generate at random such hierarchies and compare their characteristics to hierarchical classifications obtained from real–life data.

There are well-known methods for coping with this type of tree generation problems, the general strategy relying on a divide–and–conquer principle: Generate the root with the suitable probability distribution, then recursively generate the root subtrees. Several of the basic principles of this recursive top–down approach have been formalized by Nijenhuis and Wilf in their reference book on combinatorial algorithms [10], by Hickey and Cohen in the case of context–free languages [6], and under a fairly general setting by Greene within the framework of labelled grammars [4]. The present work is in many ways a systematization and a continuation of the pioneering research of these authors.

The class H of all hierarchies can be viewed as a recursively defined type,

$$H = Z + \text{set}(H, \text{card} \geq 2), \tag{1}$$

where $+$ denotes union of types, $\text{set}(H, \text{card} \geq 2)$ builds all unordered combinations of elements of H of cardinality at least 2, and Z designates the initial type of labelled nodes. A class of structures which, like H, admits an equational specification, like (1), is said to be *decomposable*.

The methods which we are going to examine enable us to start from any high level specification of a decomposable class and compile automatically procedures that solve the corresponding random generation problem. Two closely related groups of methods are given: the sequential algorithms are based on a linear search and have worst case time complexity $\mathcal{O}(n^2)$, when applied to objects of size n; the boustrophedonic algorithms are based on a special search technique that proceeds in a bidirectional fashion and they exhibit $\mathcal{O}(n \log n)$ worst case time complexity. The sequential method systematizes some existing technologies set forth by [4, 6, 10]; the boustrophedonic search extends to the realm of random generation an idea of Knuth for finding cycle leaders in permutations [8]. Both methods appeal to precomputed numerical tables of size $\mathcal{O}(n)$ produced by a preprocessing phase of cost $\mathcal{O}(n^2)$ to be effected once only.

Our complexity model is in terms of *arithmetic complexity* where we take unit cost for the manipulation of a large integer. This is justifiable practically on two grounds: (i) optimizations dictated by the arithmetic complexity model are reflected realistically in the global computation times observed; (ii) the computations can be programmed using fixed–precision floating point arithmetic, all our procedures being numerically stable, in which case the arithmetic complexity model directly applies.

In the process of developing such random generation algorithms, several alternative implementation possibilities emerge. It is therefore desirable to have a means of evaluating and comparing the resulting random generation routines provided by the general theory.

One of the main contributions of this work is to introduce in this range of problems a *calculus* that permits to produce automatically generation routines from formal specifications; a companion *cost algebra* of a rather exotic type is developed in order to attain precise average case complexity estimates of the sequential algorithms. The approach thus yields simultaneously a sequential generation algorithm and its associated complexity descriptor in the form of a generating function of average costs.

This paper is an extended abstract. Full proofs can be found in the corresponding journal article [3].

1 Combinatorial structures and constructions

We consider *labelled objects*, which may be viewed as special graphs, where some designated nodes are labelled by distinct integers; the *size* of an object is the number of its labelled nodes, and we further assume that the labelling is canonical in the sense that an object of size n bears labels from the set $[1 .. n]$.

Specifications. We start from the *initial objects* **1** that designates the "empty" structure of size 0 that bears no label, and Z that generically designates a single labelled node of size 1. We operate with the collection of *constructions*,

$$+, \cdot, \texttt{sequence}(\ ), \texttt{set}(\ ), \texttt{cycle}(\ ). \tag{2}$$

There, $(A + B)$ denotes the disjoint union (union of disjoint copies) of A and B; $(A \cdot B)$ consists in forming all pairs with a first component in A and a second component in B; $\texttt{sequence}(A)$ forms sequences of components from A, $\texttt{set}(A)$ forms sets (the order between components does not count), and $\texttt{cycle}(A)$ forms directed cycles. In the product as well as in the composite constructions of sequences, sets, and cycles it is understood that all consistent relabellings are performed. As encountered already with hierarchies, a notation like $\texttt{set}(A, \text{card} \geq k)$ uses a modifier $(\text{card} \geq k)$ to indicate sets that have at least k elements.

The language similar to that of [2] is in the style of programming languages' data types, and is consistent with common practice in combinatorial analysis.

Specification	Objects
$A = Z \cdot \texttt{set}(A)$	Non plane trees
$B = Z + B \cdot B$	Plane binary trees
$C = Z \cdot \texttt{sequence}(C)$	Plane general trees
$D = \texttt{set}(\texttt{cycle}(Z))$	Permutations
$E = \texttt{set}(\texttt{cycle}(A))$	Functional graphs
$F = \texttt{set}(\texttt{set}(Z, \text{card} \geq 1))$	Set partitions
$G = Z + Z \cdot \texttt{set}(G, \text{card} = 3)$	Non plane ternary trees
$H = Z + \texttt{set}(H, \text{card} \geq 2)$	Hierarchies
$K = \texttt{set}(\texttt{cycle}(Z \cdot \texttt{set}(G, \text{card} = 2)))$	3-constrained functional graphs
$L = \texttt{set}(\texttt{set}(\texttt{set}(Z, \text{card} \geq 1), \text{card} \geq 1))$	3–balanced hierarchies
$M = \texttt{sequence}(\texttt{set}(Z, \text{card} \geq 1))$	Surjections

Fig. 1. Eleven basic combinatorial structures and their specifications.

Definition 1. Let $\mathbf{T} = (T_0, T_1, \ldots, T_m)$ be an $(m+1)$–tuple of classes of combinatorial structures. A *specification* of $\mathbf{T}$ is a collection of $m + 1$ equations, with the ith equation being of the form

$$T_i = \Psi_i(T_0, T_1, \ldots, T_m) \tag{3}$$

where Ψ_i is a term built from $\mathbf{1}$, Z, and the T_j, using the standard constructions listed in (2).

We shall also say, for short, that the system (3) is a specification of T_0. A structure that admits a specification is called *decomposable*. The framework of specifications resembles that of context–free grammars for formal languages, but enriched with additional constructions.

Figure 1 displays specifications for eleven basic combinatorial structures that will serve as our driving examples throughout the paper.

Generating functions. We next turn to the enumeration of decomposable structures via generating functions. If C is a class, we let C_n denote the number of objects in C having size n, and introduce the *exponential generating function* (egf)

$$C(z) = \sum_{n=0}^{\infty} C_n \frac{z^n}{n!}. \qquad (4)$$

We also set $c_n = C_n/n!$ and, using the classical notation for coefficients of generating functions, we write $c_n = [z^n]C(z)$. Throughout the paper, we consistently reserve the same groups of symbols for a class, C or T_1, its generating function, $C(z), T_1(z)$, and the enumeration sequence either normalized, $c_n, t_{1,n}$, or not, $C_n, T_{1,n}$.

Theorem 2 (Folk theorem of combinatorial analysis). *(i). Given a specification Σ for a class C, a set of equations for the corresponding generating functions is obtained automatically by the following translation rules:*

$$\begin{cases} C = A + B & \Longrightarrow C(z) = A(z) + B(z) \\ C = A \cdot B & \Longrightarrow C(z) = A(z) \cdot B(z) \\ C = \mathtt{sequence}(A) & \Longrightarrow C(z) = (1 - A(z))^{-1} \\ C = \mathtt{set}(A) & \Longrightarrow C(z) = e^{A(z)} \\ C = \mathtt{cycle}(A) & \Longrightarrow C(z) = \log(1 - A(z))^{-1}. \end{cases}$$

(ii). Given a specification, the corresponding enumerating sequences up to size n are all computable in $\mathcal{O}(n^2)$ arithmetic operations.

2 Standard specifications

In this section, we show how to reduce specifications to standard form. The standard specifications constitute the basis of the random generation procedures to be developed in the paper. The reduction extends the usual Chomsky normal form for context–free grammars. Behind the transformation into standard form, there lies a "quadratization" technique whereby we perform replacements like

$$f = e^g \qquad \Longrightarrow \qquad \frac{d}{dz}f = f \cdot \frac{d}{dz}g, \qquad (5)$$

i.e., we change a highly non–linear construction into a quadratic one. Actually the proper combinatorial equivalent of the analytic operator $\frac{d}{dz}$ is the Θ operator to be introduced below.

The pointing operator plays a vital rôle in the process of random generation as recognized already by Nijenhuis and Wilf [10]. Given a class A of structures, the pointing of A is a class denoted ΘA, $\Theta A = \bigcup_{n=1}^{\infty} (A_n \times [1..n])$, where A_n is the subclass of objects in A having size n and $[1..n]$ is the integer interval $\{1, 2, \ldots, n\}$. In other words, an object in the class ΘA can be viewed as an object of A with the additional property that one of the labels, corresponding to the field in $[1..n]$, is distinguished.

From the definition we have that $C = \Theta A$ implies $C_n = nA_n$. Thus, the introduction of the pointing operation does not affect the conclusions of Theorem 2: the egfs are still computable by the added rule

$$C = \Theta A \qquad \Longrightarrow \qquad C(z) = \Theta A(z), \text{ where } \Theta f(z) = z \cdot \frac{d}{dz}f(z). \qquad (6)$$

Our developments in this section are inspired by Joyal's elegant theory [7] and by Greene's work [4].

Definition 3. Let $\mathbf{T} = (T_0, T_1, \ldots, T_m)$ be a tuple of classes of combinatorial structures. A *standard specification* of $\mathbf{T}$ is a collection of $m + 1$ equations, the ith equation being of one of the forms

$$T_i = 1; \; T_i = Z; \; T_i = U_j + U_k; \; T_i = U_j \cdot U_k; \; \Theta T_i = U_j \cdot U_k, \qquad (7)$$

where each $U_j \in \{1, Z, T_0, \ldots, T_m, \Theta T_0, \ldots, \Theta T_m\}$.

Theorem 4 (Standardization algorithm). *Every decomposable structure admits an equivalent standard specification.*

Proof. The proof is actually a conversion algorithm, that we present by transformation rules. We start with a specification where all composite types (sequences, sets, cycles) have been named.

$\mathbf{S}_0$. *Polynomials.* A polynomial splits up into binary sums and products. For instance, the specification of binary trees $B = Z + B \cdot B$ yields the standard specification: $\{B = Z + B \cdot B\} \implies \{B = Z + U_1; \; U_1 = B \cdot B\}$.

$\mathbf{S}_1$. *Sequences.* The sequence construction is equivalent to a recursive specification,

$$B = \mathsf{sequence}(A) \qquad \implies \qquad B = 1 + A \cdot B. \qquad (8)$$

The equation $B = 1 + A \cdot B$ is to be understood as an isomorphism between structures. What this amounts to is presenting a sequence $s = (s_1, s_2, \ldots, s_k) \in \mathsf{sequence}(A)$ under its equivalent right associative binary form $s \cong (s_1, (s_2, (\cdots)))$.

$\mathbf{S}_2$. *Sets.* The reduction inspires itself of Eq. (5). We claim that

$$B = \mathsf{set}(A) \qquad \implies \qquad \Theta B = B \cdot \Theta A, \qquad (9)$$

this being again understood as a fundamental combinatorial isomorphism: Pointing at a node in a set individuates the component containing the node and the component becomes pointed; this leaves aside a set of components, the non-marked ones.

$\mathbf{S}_3$. *Cycles.* We claim that

$$B = \mathsf{cycle}(A) \qquad \implies \qquad \Theta B = C \cdot \Theta A, \; \; C = \mathsf{sequence}(A), \qquad (10)$$

which reduces cycles to sequences that are already reducible. The meaning is as follows: A pointed cycle of components decomposes into the pointed component and the rest of the cycle; the directed cycle can then be opened at the place designated by the marking and a sequence results.

Similar combinatorial principle apply to the reduction of sequences, sets, and cycles under cardinality constraints. $\qquad \square$

As an illustration, a standard form for hierarchies as defined in (1) is

$$\{H = Z + U_1, \; \Theta U_1 = U_2 \cdot \Theta H, \; \Theta U_2 = U_3 \cdot \Theta H, \; \Theta U_3 = U_3 \cdot \Theta H\}. \qquad (11)$$

3 Basic generation schemes

From the preceding section, it is sufficient to exhibit generation routines for standard specifications. This goal is achieved by means of a set of translation rules or "*templates*", inspired by existing technology of random generation [4, 6, 10]. A *preprocessing stage* furnishes, once and for all in time $\mathcal{O}(n^2)$ and in storage

$\mathcal{O}(n)$ (see Theorem 2), the enumerating sequences, up to size n, of structures intervening in a specification.

Given any class C, recall that $c_n = C_n/n!$ is its normalized counting sequence, from now on assumed to be available. We let gC denote a random generation procedure relative to class C. The general strategy is based on the *divide-and-conquer* priciple.

$\mathbf{T_0}$. *Initial structures.* The generation procedures corresponding to 1 and Z are trivial.

$\mathbf{T_1}$. *Unions.* If $C = A + B$, the probability that a C–structure of size n arises from A is simply a_n/c_n. The random generation procedure uses a uniform variate U drawn uniformly from the real interval $[0, 1]$ to effect the choice.

$\mathbf{T_2}$. *Products.* If $C = A \cdot B$, the probability that a C–structure of size n has an A–component of size k and a B–component of size $n - k$ is

$$\binom{n}{k} \frac{A_k \cdot B_{n-k}}{C_n} \equiv \frac{a_k \cdot b_{n-k}}{c_n}.$$

The random generation procedure results from this equation.

$\mathbf{T_3}$. *Pointing.* Generating A and ΘA are clearly equivalent processes.

Theorem 5 (Sequential random generation). *The templates* $\mathbf{T_0}$, $\mathbf{T_1}$, $\mathbf{T_2}$, *and* $\mathbf{T_3}$ *produce from any standard specification* Σ_0 *a collection of random generation routines* $g\Sigma_0$. *Each routine of* $g\Sigma_0$ *uses precomputed tables consisting of* $\mathcal{O}(n)$ *integers; its worst case time complexity is of* $\mathcal{O}(n^2)$ *arithmetic operations.*

4 Boustrophedonic random generation

It turns out to be possible to combine the ideas underlying standard specifications with others that have also proved useful in detecting cycle leaders in permutations or in transposing rectangular matrices [8], as well as in managing dynamic equivalence relations by means of weighted union–find trees.

The standardization theory implies that all the complexity lies in the random generation of products. More precisely, when measured in the number of while-loops executed, the cost of generating (α, β) by the sequential method is the size of the first component, $|\alpha|$. In fact, a worst–case complexity of $\mathcal{O}(n \log n)$ can be achieved for all decomposable structures. The principle is simply a *boustrophedonic*[4] search.

Theorem 6 (Boustrophedonic random generation). *Any   decomposable structure has a random generation routine that uses precomputed tables of size* $\mathcal{O}(n)$ *and achieves* $\mathcal{O}(n \log n)$ *worst case time complexity.*

Proof. (Sketch) Given a product $C = A \cdot B$, we let K be the random variable denoting the size of the A–component of a C–structure. Amongst C–structures of size n, we have

$$\Pr\{K = k\} = \frac{a_k \cdot b_{n-k}}{c_n},$$

---

[4] Boustrophedonic: turning like oxen in ploughing (Webster).

and we let $\pi_{n,k}$ denote this probability. The idea is to appeal to a special search for the drawing of K with the probability distribution $\{\pi_{n,k}\}_{k=0}^{n}$. Instead of the order of increasing values of k, we explore the possibilities of K in the boustrophedonic order

$$\pi_{n,0}, \pi_{n,n}, \pi_{n,1}, \pi_{n,n-1}, \ldots,$$

that sweeps alternatively from left to right and back. The recurrences translating the costs admit $\mathcal{O}(n \log n)$ solutions (see [5, Sec. 2.2]). □

The purpose of the calculus of rearrangements to be developed in the next sections is precisely to come up with adequate specifications that permit to attain a complexity of $\mathcal{O}(n \log n)$ involving low multiplicative factors by exploiting "natural" regularities present in combinatorial structures. To algorithms designers, the situation resembles that of heapsort —which has guaranteed $\mathcal{O}(n \log n)$ complexity— versus quicksort —which is $\mathcal{O}(n \log n)$ only on average but with small constants—, so that quicksort is often preferred in practice.

5 The cost algebra of sequential generation

We have seen how to compile automatically random generation routines from standard specifications. By the standardization theorem, itself relying on an effective reduction process, the method applies to any decomposable structure. We propose from now on to examine in great detail the cost structure underlying the random generation procedures of the *sequential* group. The cost measure that we adopt counts the number of while-loops executed in procedures corresponding to products. In other words, the cost of generating a product $(\alpha \cdot \beta)$ is simply taken to be the size of the first component, $|\alpha|$. In so doing, we neglect terms that are at worst only $\mathcal{O}(n)$ in terms of the number of integer operations performed. (Our model belongs to the category of *arithmetic complexity* models.)

Consider a procedure gA that generates random elements in a decomposable class A given by a standard specification according to the rules governing Theorem 5; we let γA_n denote its *expected cost*. We define the "total cost" ΓA_n and the *cost generating function* $\Gamma A(z)$ by

$$\Gamma A_n = A_n \times \gamma A_n \qquad \text{and} \qquad \Gamma A(z) = \sum_{n=0}^{\infty} \Gamma A_n \frac{z^n}{n!}.$$

This notion corresponds to that of *complexity descriptor* in [2]. We further abbreviate $\Gamma A(z)$ by ΓA. This notational trick permits us to regard symbolically Γ as an *operator* acting on systems of equations corresponding to specifications.

Theorem 7 (The cost algebra identities). *The cost operator Γ satisfies the identities,*

$$\Gamma Z = \Gamma \mathbf{1} = 0; \quad \Gamma(A + B) = \Gamma A + \Gamma B;$$

$$\Gamma(A \cdot B) = \Gamma A \cdot B + A \cdot \Gamma B + \Theta A \cdot B; \quad \Gamma(\Theta A) = \Theta(\Gamma A). \qquad (12)$$

Binary trees. The rules of the cost algebra allow us to effectively compute complexity descriptors associated with various random generation algorithms. A first illustration, is the cost structure of the generation of binary trees.

Theorem 8 (Binary trees, naïve method). *The generation algorithm for binary plane trees corresponding to the standard specification* $\{B = Z+U_1; \ U_1 = B \cdot B\}$ *has average case complexity*

$$\gamma B_n = \frac{1}{2}\sqrt{\pi}n^{3/2} + \mathcal{O}(n).$$

Schemas. The cost algebra is also powerful enough that we can come up with general results regarding the construction of composite structures. To keep notations simple, we introduce the integral operator $\int f(z) = \int_0^z f(t)\,dt$.

Theorem 9 (Composite schemas). *(i). Let* $C = \mathtt{sequence}(A)$ *be generated according to the standard specification* $\Sigma_0 = \{C = 1 + U_1; \ U_1 = A \cdot C\}$. *Then,*

$$\Gamma C = \frac{\Theta A + \Gamma A}{(1 - A)^2}.$$

(ii). Let $C = \mathtt{set}(A)$ *be generated by the standard specification* $\Sigma_0 = \{\Theta C = C \cdot \Theta A\}$; *then*

$$\Gamma C = e^A \left(\Gamma A + \int \left[\frac{(\Theta A)^2}{z}\right]\right).$$

(iii). Let $C = \mathtt{cycle}(A)$ *be generated by the standard specification* $\Sigma_0 = \{\Theta C = B \cdot \Theta A; \ B = 1 + U_1; \ U_1 = A \cdot B\}$; *then*

$$\Gamma C = \int \left[\frac{1}{z} \frac{2(\Theta A)^2 + \Gamma A \cdot \Theta A + (1 - A) \cdot \Theta \Gamma A}{(1 - A)^2}\right].$$

6 The analysis of cost generating functions

The cost algebra developed in the previous section attains its full dimension when we examine it in the light of asymptotic properties of combinatorial structures. This means that orders of growth of coefficients should be taken into account. The way to do so is to examine the complex analytic structure of intervening generating functions which, as is well known, directly relates to the growth of coefficients (see especially [1], and the systematic use in [2]). More precisely, we can interpret the equations provided by the cost algebra *locally* as analytic relations between singular orders of growth. Consideration of asymptotic properties of structures using the classical arsenal of complex analysis does provide, in all cases of practical interest, valuable guidelines regarding the design of generation algorithms.

Non plane trees. The family of non–plane trees corresponds to the specification $A = Z \cdot \mathtt{set}(A)$. It furnishes a first example where two random generation algorithms derived from combinatorially equivalent specifications lead to rather different complexity behaviours.

We make use of the general principles of the standardization method, our starting point being the pair of combinatorially equivalent specifications

$$\Theta A \cong A + (\Theta A \cdot A) \cong A + (A \cdot \Theta A).$$

Theorem 10 (Non plane trees). *(i). The random generation algorithm for labelled trees corresponding to the standard specification* $\Theta A = A + (\Theta A \cdot A)$

has average cost

$$\gamma A_n = \sqrt{\frac{\pi}{2}} n^{3/2} + \mathcal{O}(n).$$

(ii). The generation algorithm for labelled trees corresponding to the specification $\Theta A = A + (A \cdot \Theta A)$ has average cost

$$\gamma A_n = \frac{1}{2} n \log n + \mathcal{O}(n).$$

Proof. Apply the cost algebra to the specifications, appeal to the variation–of–constant method for corresponding differential equations, and conclude using singularity analysis. □

Optimization transformations. The specification of A, under the form

$$\Theta A - A = (A \cdot \Theta A),$$

in essence generates a family of pointed trees by first generating an unmarked tree, then the rest of the tree containing the mark. The pointed trees are much more numerous than the basic trees, the ratio being $\Theta A_n / A_n = n$. Accordingly, the mark tends to fall on larger portions of the tree. Thus, viewed on the underlying binary parse tree, the random generation has a complexity that, at least in an intuitive probabilistic sense, should behave like a parameter χ of binary trees given by

$$\chi[t_1 \cdot t_2] = \min(|t_1|, |t_2|) + \chi[t_1] + \chi[t_2].$$

This relates to the modified form of path length occurring in boustrophedonic search, whose value on any tree of size n is $\mathcal{O}(n \log n)$. In contrast, the random generation corresponding to the specification

$$\Theta A - A = (\Theta A \cdot A)$$

has a complexity that behaves like standard path length, which is known to be $\mathcal{O}(n^{3/2})$ in such varieties of trees [9]. In order to make this discussion precise, we introduce a formal definition.

Definition 11. Given two generating functions F and G, F *dominates* G, in symbols $F \gg G$, if

$$\frac{f_n}{g_n} \to \infty \qquad \text{as } n \to +\infty.$$

The considerations regarding labelled trees then suggest a simple heuristic:

Big–endian heuristic. Given a standard specification Σ_0, reorganize all comparable pairs in products each time $A \gg B$ using the isomorphism transformation

$$(A \cdot B) \hookrightarrow (B \cdot A).$$

This heuristic applied to the two specifications of non–plane trees leads to the "good choice" with an $\mathcal{O}(n \log n)$ behaviour. A further optimization that this discussion suggests consists in obtaining, as much as possible, specifications where products are *imbalanced* so as to take full advantage of the big–endian heuristic. To that purpose, the Θ operator can be employed. For instance, let us re–examine the binary trees, $B = Z + B \cdot B$. Consider the induced relation obtained by differentiation,

$$\Theta B = Z + \Theta B \cdot B + B \cdot \Theta B.$$

Let K designate the size of the first component in $B \cdot B$, and K' denote the size of the first component in $B \cdot \Theta B$. The expectation of K is $n/2$ while that of K' turns out to be $\mathcal{O}(\sqrt{n})$, so that a global gain of order close to $\mathcal{O}(\sqrt{n})$ is to be anticipated if the big endian heuristic is employed. This dictates a new heuristic:

Differential heuristic. Replace in specifications polynomial relations by differential relations.

Theorem 12 (Binary trees, differential algorithm). *For binary trees, the differential algorithm corresponding to the specification $\Theta B = Z + (B + B) \cdot \Theta B$ has expected complexity*

$$\gamma B_n = \frac{1}{2} n \log n + \mathcal{O}(n).$$

Both non–plane trees and binary trees under the differential algorithm are generated in time asymptotic to $\frac{1}{2} n \log n$. This is in fact a general phenomenon common to many families of trees.

7 Trees, graphs, and iterative structures

All polynomial families of trees as well as functional graphs can be generated in time asymptotic to $\frac{1}{2} n \log n$. Furthermore, the class of iterative structures admits $\mathcal{O}(n)$ random generation algorithms.

Polynomial families of trees. A *polynomial* family of trees is a family defined by allowing only a finite collection Ω of node degrees.

Theorem 13 (Polynomial families). *Consider a polynomial family of non–plane trees defined by*

$$T = Z \cdot \sum_{k \in \Omega} \mathbf{set}(T, \mathrm{card} = k),$$

where each $U_k = \mathbf{set}(T, \mathrm{card} = k)$ is specified by $\Theta U_k = U_{k-1} \cdot \Theta T$. The expected generation time for a random tree of size n, where $[z^n] T \neq 0$, satisfies

$$\gamma T_n = \frac{1}{2} n \log n + \mathcal{O}(n).$$

This theorem applies for instance to non–plane ternary trees (specification G). A similar result holds for hierarchies (specification H) which were our original motivation for considering these questions. The general plane trees defined by $C = Z \cdot \mathbf{sequence}(C)$ are also amenable to a differential algorithm since a relation $U = \mathbf{sequence}(C)$ also implies $\Theta U = U \cdot (U \cdot \Theta C)$. The asymptotic complexity is again of the form $\frac{1}{2} n \log n$.

Functional graphs. Functional graphs, or equivalently finite mappings, are of interest in cryptography and random number generation. They present us with an instance of a structure defined by a specification involving several intermediate classes.

Theorem 14 (Functional graphs). *Functional graphs (E) corresponding to the standard specification*

$$\{\Theta E = \Theta U_1 \cdot E; \ \Theta U_1 = \Theta A \cdot U_2; \ U_2 = 1 + A \cdot U_2; \ \Theta A = A + A \cdot \Theta A\}.$$

are generated in average time

$$\gamma E_n = \frac{1}{2} n \log n + \mathcal{O}(n).$$

A similar result holds for mappings satisfying degree constraints (like specification K).

Iterative structures. We say that a class of structures is *iterative* or *non–recursive* if the dependency graph of the classes entering the unstandardized specification (allowing sequences, sets and cycles) is acyclic. Trees, hierarchies, and functional graphs are typical recursive structures, while permutations, partitions, surjections, and balanced hierarchies of any fixed height are iterative.

Theorem 15 (Iterative structures). *Any iterative class I admits a random generation algorithm of worst case complexity*

$$\gamma I_n = \mathcal{O}(n).$$

The theorem applies to set partitions (specification F), the cycle decomposition of permutations (D), 3–balanced hierarchies (L), and surjections (M). For instance, for surjections, a simple computation based on the cost algebra and Theorem 9 confirms that $\Gamma M = \Theta M$, so that $\gamma M_n = n$, as anticipated. In general the constant in the $\mathcal{O}(n)$ complexity increases with the degree of nesting of the iterative specification.

8 Numerical data

The generation method for decomposable structures has been implemented in the symbolic manipulation system MAPLE by Zimmermann. The complete programme tests specifications for well–foundedness, puts them in standard quadratic form, and compiles two sets of procedures from standard specifications: the counting routines that implement the convolution recurrences, and the random generation routines based on the templates. The whole set, in its current stage, represents some 1500 lines of Maple code. The random generation procedures produced are in the Maple language itself, and they take advantage of the multiprecision arithmetic facilities available in MAPLE.

The version of the Maple programme that was written furthermore compiles random generation routines by automatically implementing a version of the big–endian heuristic based on "probing". As an outcome, all our eleven reference structures are generated in time between 2 and 9 seconds on a machine of 20 Mips for size $n = 400$. Gains involving a factor of about 10 for $n = 400$ result from optimizations dictated by the cost calculus.

9 Unlabelled structures

Many important structures of computer science and combinatorics are *unlabelled*. Work currently under redaction shows that the framework presented here extends to unlabelled combinatorial structures. (The treatment is only made more complex because of the occurrence of Pólya operators.) As a result: (*i*) all unlabelled decomposable structures including context–free languages and term trees of symbolic computation can be generated in worst–case time $\mathcal{O}(n \log n)$; (*ii*). Wilf's **RANRUT** Algorithm [11] has expected case complexity which is $\sim \frac{1}{2} n \log n$.

Acknowledgement. This work was supported in part by the ESPRIT III Basic Research Action Programme of the E.C. under contract ALCOM II (#7141).

References

1. FLAJOLET, P., AND ODLYZKO, A. M. Singularity analysis of generating functions. *SIAM Journal on Discrete Mathematics 3*, 2 (1990), 216–240.
2. FLAJOLET, P., SALVY, B., AND ZIMMERMANN, P. Automatic average–case analysis of algorithms. *Theoretical Computer Science, Series A 79*, 1 (Feb. 1991), 37–109.
3. FLAJOLET, P., ZIMMERMAN, P., AND VAN CUTSEM, B. A calculus for the random generation of labelled combinatorial structures. Research Report 1830, Institut National de Recherche en Informatique et en Automatique, Jan. 1993. 29 pages. Accepted for publication in *Theoretical Computer Science*.
4. GREENE, D. H. *Labelled formal languages and their uses.* PhD thesis, Stanford University, June 1983. Available as Report No. STAN-CS-83-982.
5. GREENE, D. H., AND KNUTH, D. E. *Mathematics for the analysis of algorithms.* Birkhauser, Boston, 1981.
6. HICKEY, T., AND COHEN, J. Uniform random generation of strings in a context–free language. *SIAM Journal on Computing 12*, 4 (1983), 645–655.
7. JOYAL, A. Une théorie combinatoire des séries formelles. *Advances in Mathematics 42*, 1 (1981), 1–82.
8. KNUTH, D. E. Mathematical analysis of algorithms. In *Information Processing 71* (1972), North Holland Publishing Company, pp. 19–27. Proceedings of IFIP Congress, Ljubljana, 1971.
9. MEIR, A., AND MOON, J. W. On the altitude of nodes in random trees. *Canadian Journal of Mathematics 30* (1978), 997–1015.
10. NIJENHUIS, A., AND WILF, H. S. *Combinatorial Algorithms,* second ed. Academic Press, 1978.
11. WILF, H. S. *Combinatorial Algorithms: An Update.* No. 55 in CBMS–NSF Regional Conference Series. Society for Industrial and Applied Mathematics, Philadelphia, 1989.

The bit complexity of distributed sorting

(Extended Abstract)

O. Gerstel [1] and *S. Zaks* [2]

Department of Computer Science
Technion, Haifa, Israel

Abstract

We study the bit complexity of the sorting problem for asynchronous distributed systems. We show that for every network with a tree topology T, every sorting algorithm must send at least $\Omega(\Delta_T \log \frac{L}{N})$ bits in the worst case, where $\{0, 1, .., L\}$ is the set of possible initial values, and Δ_T is the sum of distances from all the vertices to a median of the tree. In addition, we present an algorithm that sends at most $O(\Delta_T \log \frac{L N}{\Delta_T})$ bits for such trees; These bounds are tight if either $L = \Omega(N^{1+\epsilon})$ or $\Delta_T = \Omega(N^2)$. We also present results regarding average distributions. These results suggest that sorting is an inherently non-distributive problem, since it requires an amount of information transfer, that is equal to the concentration of all the data in a single processor, which then distributes the final results to the whole network. The importance of bit complexity - as opposed to message complexity - stems also from the fact that in the lower bound discussion, no assumptions are made as to the nature of the algorithm.

[1] Email address: ORIG@CS.TECHNION.AC.IL
[2] Email address: ZAKS@CS.TECHNION.AC.IL

1 INTRODUCTION

In this paper we study the bit complexity of the sorting problem for asynchronous distributed systems. Such systems are composed of N processors, connected by communication lines, and communicate with one another by exchanging messages through these lines. The processors and the communication lines are modeled by an undirected graph. A protocol (or distributed algorithm) consists of operations of sending or receiving messages and doing local computations at each processor. The bit complexity of a given algorithm for a given network is the maximal number of bits sent during any possible execution of the algorithm on that network.

We study upper and lower bounds for the bit complexity of the sorting problem, in which each processor initially has an identifier and an initial value, and at the end the initial values have to be rearranged according to the initial identifiers. We show that for every network with a tree topology T with N processors, there exists a distribution of initial values and identifiers for which every sorting algorithm must send at least $\Omega(\Delta_T \log \frac{L}{N})$ bits in the worst case, where $\{0, 1, .., L\}$ is the set of possible initial values, and Δ_T is the sum of distances from the median of T to all the other processors in it. We then present an algorithm that sends at most $O(\Delta_T \log \frac{LN}{\Delta_T})$ bits for such trees. These bounds are tight if either $L = \Omega(N^{1+\epsilon})$ or $\Delta_T = \Omega(N^2)$, both of which are realistic assumptions (the first of which requires that the range of possible values is large enough: networks do not have 2^{32} processors, but integers do have this range; the second requires that the tree will not be too flat, e.g. trees with bounded degree). We also discuss the average distribution of initial values, and prove that on the average, every sorting algorithm must send at least $\Omega(\Delta_T \frac{(\frac{L}{N})^N}{\binom{L}{N}} \log \frac{L}{N})$ bits.

These results imply that sorting is an inherently non-distributive problem, since the amount of information that has to be transferred is equal to a non-distributive version of the algorithm, in which one processor collects all the information from all other processors, calculates the final results, and redistributes the values appropriately.

In Section 2 we discuss related works, and in Section 3 we present a formal definition of the model and the problem. The worst case and average case results are discussed in Section 4. We mention a few open problems in Section 5. Some of the proofs in this extended abstract are omitted, while others are only sketched.

2   RELATED WORKS

In [5] the message complexity of the sorting problem for ring networks was studied, where initial values have to be sorted clockwise, starting at any position; it was shown that every sorting algorithm requires $\Omega(N^2 \log \frac{L}{N})$ bits on a ring of size N, and an algorithm was presented that achieves this lower bound; similar results are shown for meshes. While [5] discusses only these two specific topologies, we study the class of all trees, and our results - especially the lower bound - apply for *every* tree in the class.

Our algorithm is a modification of the ones in [6, 3, 5], where the *message complexity* of the sorting problem is studied for tree networks. However, the lower bounds in [6, 3] assume a special class of algorithms, whereas we do not impose any constraints on the design of the algorithm; Moreover, in contrast with [6], in this work we deal with *bit complexity*, and our lower bound applies for *every* tree, and not just for the worst-case tree. A problem closely related to the sorting problem is ranking, and it is discussed in [4, 6].

While all previous works discussed only the worst case distribution of values, this work presents a lower bound on the average distribution as well.

3   PRELIMINARIES

A distributed system is composed of a set of N processors, connected by communication links. A network is viewed as an undirected graph $G = (V, E)$, whose set of vertices is the set of processors $V = \{v_1, ..., v_N\}$, and $(v_i, v_j) \in E$ iff there is a communication line connecting v_i and v_j.

Every processor executes an *algorithm*, that specifies the actions to be taken upon receipt of a message; these actions are sending messages and doing local computations. It is assumed that a message arrives at its destination after an unknown but finite delay. We do not assume that messages on a link arrive at the same order they were sent; nor do we assume that the algorithm in each processor is the same. This is a somewhat relaxed version of the conventional message-passing asynchronous model for networks (see [2]).

An *execution* is a sequence of *events*, that are either sending or receiving a message at a processor according to its algorithm (a local computation can be regarded as part of either of these two types of actions). It is assumed that any non-empty subset of the processors may start the algorithm, and that no two events happen simultaneously. An execution has terminated if all the local algorithms have terminated. Initially each processor has some information concerning the problem it is about to solve, and upon termination of the algorithm each processor has a final value.

The *initial distribution* (or *distribution*) of identifiers and initial values in a distributed system is a function

$$\delta : V \longrightarrow ID \times INIT,$$

where ID is the set of initial identities and $INIT$ is the set of initial values. Given a distribution δ, $ID_\delta(v)$ and $INIT_\delta(v)$ denote the initial identity and the initial value of processor v. Two distributions δ_1 and δ_2 are said to *agree* on a set of processors Q if $\delta_1(v) = \delta_2(v)$ for every $v \in Q$. The set of executions of a given algorithm $\mathcal{A}$ for a given distribution δ is denoted by $\mathcal{EX}(\mathcal{A}, \delta)$.

The *bit complexity* $b(\mathcal{A}, G, \delta)$ of an algorithm $\mathcal{A}$ on a graph G with initial distribution δ, is the maximal total number of bits (in all messages, over all the communication links) sent during any possible execution of $\mathcal{A}$ on G with initial distribution δ. We will write $b(\mathcal{A})$ when G and δ are clear from the text.

We study the *sorting problem*. For this we assume that, for every distribution δ, all initial identifiers $ID_\delta(v_1), ..., ID_\delta(v_N)$ are distinct, while the initial values $INIT_\delta(v_1), ..., INIT_\delta(v_N)$ are not necessarily distinct. An algorithm is solving the sorting problem if it rearranges the initial values according to the initial identifiers; in other words, in every execution every processor v will have a final value $FINAL(v)$, such that the following two conditions are satisfied:

1. The multiset $\{INIT_\delta(v_1), ..., INIT_\delta(v_N)\}$ is equal to the multiset $\{FINAL(v_1), ..., FINAL(v_N)\}$, and
2. $ID_\delta(v) < ID_\delta(w) \Longrightarrow FINAL(v) \leq FINAL(w) \quad \forall v, w \in V$.

Let $T = (V, E)$ be a tree, and for $x, y \in V$ let $d(x, y)$ denote the distance between x and y in the tree. For $v \in V$, define:

$$\Delta_T(v) = \sum_{u \in V} d(u, v),$$

$$\Delta_T = \min_{v \in V} \Delta_T(v).$$

Define the *median* of the tree to be a vertex m satisfying $\Delta_T(m) = \Delta_T$.

Let $\mathcal{F}(n) = \{0, 1\} \times \{1, 2, ..., n\}^*$, i.e. all finite sequences of pairs $< i, j >$ such that $i \in \{0, 1\}$ and $j \in \{1, 2, ..., n\}$. Each such pair is called a *component*.

Lemma 1 ([5]): Let S be a set of σ distinct sequences of $\mathcal{F}(n)$. The total number $\alpha(\sigma, n)$ of components in all the sequences of S satisfies

$$\alpha(\sigma, n) \geq \frac{4\sigma \log(\sigma/10)}{5 \log(2n)} .$$

$\square$

For a given network $G = (V, E)$ with distribution δ, let Q be a subset of V. $CUT(Q)$ contains all edges that have one end in Q and the other end not in Q; namely,

$$CUT(\mathcal{Q}) = \{(v,v') \in E \mid v \in \mathcal{Q}, \ v' \notin \mathcal{Q}\} .$$

For an algorithm $\mathcal{A}$, the *signature* $SIG_\delta(e,c)$ of an execution e of $\mathcal{A}$ on the cut $c = CUT(\mathcal{Q})$ is a sequence in $\mathcal{F}(n)$, where $n = 2|c|$. Each component represents a message arrival event, in the following way: assign the labels $1, ..., n$ to the links of c (two labels per link), each label representing a direction on that link. During the execution e, messages are sent and received on links in c. The i-th component of $SIG_\delta(e,c)$ is $< m, k >$ if the i-th message received is a message with contents m, and it was sent on a link and a direction implied by label k.

Lemma 2 ([5]): Let $D = \{\delta_1, ..., \delta_t\}$ be a set of distributions that agree on a set of processors $\mathcal{Q}$, and let $\mathcal{A}$ be an algorithm. Let $E(D) = \{e_1, ..., e_t\}$ be a set of executions such that $e_i \in \mathcal{EX}(\mathcal{A}, \delta_i)$ for every i. If the executions in $E(D)$ have less than $|D|$ different signatures on $CUT(\mathcal{Q})$, then there exist two executions $e_i, e_j \in E(D)$ for which the final values for processors in $\mathcal{Q}$ are the same.

□

Define a *pairing* P to be a partition of the vertices of the tree into disjoint pairs. This partition leaves at most one vertex unmatched, in case $|V|$ is odd, and this vertex is marked by $free(P)$.

Lemma 3 ([3]): Let T be a tree, and let m be a median of T. Then there exists a pairing P_T, such that all the paths connecting pairs of vertices in P_T pass through m. Furthermore, if $|V|$ is odd then $m = free(P_T)$.

□

4 RESULTS

4.1 Worst case analysis

We first show:

Lemma 4: For every tree T:

1. $\sum_{v \in V} |T_v| = \Delta_T$.
2. $\sum_{v \in V} |T_v| \log |T_v| \geq \Delta_T \log \frac{\Delta_T}{N}$.

□

The next theorem deals with the lower bound:

Theorem 1: Let $\mathcal{A}$ be any sorting algorithm on a tree network T with N nodes, and let $INIT = \{1, ..., L\}$. Then there exists an initial distribution δ, for which the bit complexity $b(\mathcal{A})$ of $\mathcal{A}$ satisfies

$$b(\mathcal{A}) \;=\; \Omega(\Delta_T \log \tfrac{L}{N}).$$

Sketch of proof: Let $T = (V, E)$, and let $P_T = \{(v_1, v_2), (v_3, v_4), ...,$ $(v_{N-1}, v_N)\}$ be the pairing satisfying Lemma 3, where $V = \{v_1, ..., v_N\}$ or $V = \{v_1, ..., v_{N+1}\}$ (in which case $free(P_T) = v_{N+1}$). Let D be a set of initial distributions such that $\delta \in D$ if:

- $ID_\delta(v_i) = i$ for every $v_i \in V$,
- if i is even then $INIT_\delta(v_i) \in \{(i-2)\tfrac{L}{N}, ..., (i-1)\tfrac{L}{N} - 1\}$,
- if i is odd: $INIT_\delta(v_{N+1}) = L$, and $INIT_\delta(v_i) \in \{i\tfrac{L}{N}, ..., (i+1)\tfrac{L}{N} - 1\}$, for $i < N$.

Note that $|D| = \left(\tfrac{L}{N}\right)^N$, since every processor in $\{v_1, ..., v_N\}$ has $\tfrac{L}{N}$ possible values in D.

It is easy to see that in D every pair of vertices $(u, v) \in P_T$ have to exchange values in order to perform sorting.

Direct the tree to be rooted at a median m, and denote by T_v the subtree of this directed tree that is rooted at v. Let $a \in V, a \neq m$ and let b be its father in T. Consider the following partition of V into disjoint sets:

$$V_1 \;=\; V(T_a), \quad V_2 \;=\; \{v \mid \exists w \in V_1 : (v, w) \in P_T\}, \quad V_3 \;=\; V - V_1 - V_2.$$

Consider a subset $D_a \subset D$ of executions such that processors in V_2 have all possible values as in D, while processors in $V_1 \cup V_3$ have some *fixed* value out of the values allowed in D.

From the definition of sorting it follows that for every two distributions in D_a, if the initial values in V_2 are different, then the final values in V_1 will also be different, and that all processors in V_1 agree on all distributions of D_a.

Consider the cut $C = CUT(V_1) = \{(a, b)\}$ that separates V_1 from the rest of the network. Using Lemma 2, it can be shown that there exist at least $|D_a|$ different signatures of executions in $E(D_a)$ on C. Every such signature is in $\mathcal{F}(2)$, as C consists of one link, therefore by Lemma 1, in all these signatures there are at least $\alpha(|D_a|, 2)$ components, each corresponding to sending one bit-message.

Let $K(C, D)$ be the number of bits sent on C among all distributions in D. We have

$$K(C, D_a) \;\geq\; \alpha(|D_a|, 2).$$

Now all the above discussion was true for a specific $D_a \subset D$, and may be applied for every other D_a (there are $\left(\tfrac{L}{N}\right)^{N-|T_a|}$ such $D_a \subset D$). Therefore we have:

$$K(C, D) \;=\; \textstyle\sum_{D_a \subset D} K(C, D_a) \;\geq\; \left(\tfrac{L}{N}\right)^{N-|T_a|} \alpha(|D_a|, 2).$$

Note that all the above discussion may be applied for every $a \neq m$, and not just the fixed a we have chosen. Thus $K(D)$, the total number of bits sent on all the cuts in all distributions of D satisfies

$$K(D) = \sum_C K(C, D) \geq \sum_{a \neq m} \left(\tfrac{L}{N}\right)^{N - |T_a|} \alpha(|D_a|, 2) =$$

$$= \tfrac{4}{5\log 4} \left(\tfrac{L}{N}\right)^N \log\left(\tfrac{L}{N}\right) \sum_{a \neq m} |T_a| = \tfrac{4}{5\log 4} \left(\tfrac{L}{N}\right)^N \log\left(\tfrac{L}{N}\right) \Delta_T,$$

where the last equality follows from Lemma 4. By the pigeon-hole principle there exists a distribution $\delta_0 \in D$ for which at least $\frac{K(D)}{|D|}$ bits are sent on all links (cuts), and therefore the worst case bit complexity satisfies

$$b(\mathcal{A}) \geq \tfrac{K(D)}{|D|} = \tfrac{4}{5\log 4} \Delta_T \log \tfrac{L}{N} = \Omega(\Delta_T \log \tfrac{L}{N}).$$

$\square$

Theorem 2: There exists a sorting algorithm $\mathcal{A}$, such that for every tree network T with N nodes and every initial distribution taken from the set $INIT = \{1, ..., L\}$, the bit complexity $b(\mathcal{A})$ satisfies

$$b(\mathcal{A}) = O(\Delta_T \log \tfrac{LN}{\Delta_T}).$$

Sketch of proof: We use a simple encoding of a set of numbers $\{n_1, n_2, ..., n_k\}$ in the range $\{0, 1, ..., L\}$. This encoding is done by sorting the numbers into ascending order $n_{i_1} \leq n_{i_2} \leq ... \leq n_{i_k}$, and sending the sequence $n_{i_1}, n_{i_2} - n_{i_1}, n_{i_3} - n_{i_2}, \cdots, n_{i_k} - n_{i_{k-1}}$ (it is easy to see how the original sequence can be derived from this sequence). This method requires $O(k \log \tfrac{L}{k})$ bits (rather than $O(k \log L)$ bits; see [5]).

Given the tree T, consider it as rooted at one of its median vertices. The algorithm starts at the leaf processors, which send their identifiers and their initial values to their father. Every internal processor waits to receive an encoding of the lists of identities and initial values accumulated at the sons, decodes them, inserts its own identity and initial value to the corresponding lists and sends an encoding of these new lists to its father. Eventually the root processor receives all the identities and initial values of the processors in the tree, sorts them and sends to each son u the list of the final values that correspond to u and the processors in its subtree. Each internal processor can now determine its own final value and routes to each of its sons the list with the corresponding final values. This process terminates at the leaf nodes. (This algorithm is a modification of the one in [6], to cope with bit complexity.)

Despite the inherently large amount of information transfer, this algorithm does have good local properties, namely low local processing time and low memory utilization.

During the execution of the algorithm two messages are sent on each edge from son to father (containing identities and initial values), and one message is sent from father to son (containing final values). Each message from/to a subtree T_v rooted at v contains at most $3(\,|T_v|\,\log\frac{L}{|T_v|}\,)$ bits, so the total number of bits sent during an execution is:

$$b(\mathcal{A}) \leq 3(\,\textstyle\sum_{v\in V}|T_v|\,\log\frac{L}{|T_v|}\,) \;=\; 3(\,\textstyle\sum_{v\in V}|T_v|\log L \;-\; \sum_{v\in V}|T_v|\log|T_v|\,)$$

and this implies, by Lemma 4.2,

$$b(\mathcal{A}) \;\leq\; 3(\,\Delta_T\log\tfrac{L\,N}{\Delta_T}\,).$$

$\square$

The above two theorems give upper and lower bounds on the bit complexity of sorting on a tree network. These bounds are tight if either the range of initial values in L is significantly larger than the size N of the network (which is usually the case in communication networks), or if the tree is deep enough (e.g., a tree with bounded degrees - paths, binary trees, etc. ; this assumption of bounded degree is common in practice). Formally, we get:

Corollary 1: For every tree network T and set of initial values $INIT = \{1,...,L\}$, if either of the following conditions holds, then the bounds of Theorem 1 and 2 are tight:

1. $L = \Omega(N^{1+\epsilon})$ for some fixed $\epsilon > 0$.
2. $\Delta_T = \Omega(N^2)$.

4.2 Average case analysis

In this sub-section we deal with the expected bit complexity on an average distribution. The proof technique is similar to the one of Theorem 1, but more definitions and lemmata are needed.

We define a *permutation* π as $\pi = \{< u_1, v_1 >, ..., < u_N, v_N >\}$ where $\{u_i\}_{i=1}^{N} = \{v_i\}_{i=1}^{N} = V$. The permutation defines the source and destination (u and v respectively) of the values in the given distribution.

The *distance* $\Gamma_T(\pi)$ of a given permutation is defined by

$$\Gamma_T(\pi) = \textstyle\sum_{<u,v>\in\pi} d(u,v).$$

Every edge e splits the tree into two subtrees T_0^e and T_1^e. Define the *split set*, $\Upsilon_\pi(e)$, as the set of vertices whose source in π is in T_1^e and its destination in T_0^e; namely, $\Upsilon_\pi(e) = \{\,u \mid\, < u, v >\in \pi,\; u \in T_1^e,\; v \in T_0^e\,\}$.

Finally, we define $\bar{b}(\mathcal{A})$ to be the sum of $b(\mathcal{A})$ for all distributions, divided by the total number of distributions.

We show the following two Lemmata:

Lemma 5: $\sum_{e \in E} |\Upsilon_\pi(e)| = \frac{1}{2}\Gamma_T(\pi)$ for every permutation π.

□

Lemma 6: $\sum_\pi \Gamma_T(\pi) \geq \Delta_T N!$ for every tree T.

□

Theorem 3: Let $\mathcal{A}$ be any sorting algorithm on a tree network T, with a set of initial values $INIT = \{1, ..., L\}$. The expected bit complexity $\bar{b}(\mathcal{A})$ of $\mathcal{A}$ satisfies

$$\bar{b}(\mathcal{A} = \Omega(\Delta_T \frac{(\frac{L}{N})^N}{\binom{L}{N}} \log \frac{L}{N}).$$

Sketch of proof: We prove the bound for a fixed distribution of IDs δ_{ID}, and since it does not depend on that distribution, it will also hold for all ID distributions.

Let $\mathcal{D}$ be the set of all distributions of distinct initial values. Define a *simple subset of distributions* $SD \subset \mathcal{D}$ by assigning each processor with ID $i \in \{1, ..., N\}$, the values $\{(i-1)\frac{L}{N}, ..., i\frac{L}{N}\}$.
For a given permutation $\pi = \{< u_1, v_1 >, ..., < u_N, v_N >\}$, and a distribution δ, define δ_π, by

- $ID_{\delta_\pi}(u_i) = ID_\delta(u_i)$ for every i, and
- $INIT_{\delta_\pi}(u_i) = INIT_\delta(v_i)$ for every i,

and let $SD_\pi = \{\delta_\pi | \delta \in SD\}$.
We note the following:

1. Each of the distributions in SD is sorted,
2. $|\mathcal{D}| = \frac{L!}{(L-N)!}$,
3. $|SD| = (\frac{L}{N})^N$, and
4. For every two permutations $\pi \neq \pi'$ $SD_\pi \cap SD_{\pi'} = \emptyset$.

Define $D(\delta, S)$ as the set of distributions in SD_π that have fixed values for the processors outside a given set S, and varying values in S, for a given $\delta \in SD_\pi$; namely:

$$D(\delta, S) = \{\delta' \in SD_\pi | \forall v \in V : v \notin S \implies \delta'(v) = \delta(v)\}.$$

Given an edge e, Lemma 1 may be used for the distributions in $D(\delta, \Upsilon_\pi(e))$, since they all agree on the values in T_0^e, so we have (for some constants C_1, C_2):

$$K(\{e\}, D(\delta, \Upsilon_\pi(e))) \geq \alpha(\,|D(\delta, \Upsilon_\pi(e))|\,, 2\,).$$

It can be shown that for any given distribution $\delta \in SD_\pi$

$$K(\{e\}, SD_\pi) = \sum_{\delta' \in D(\delta_\pi, V \setminus \Upsilon_\pi(e))} K(\{e\}, D(\delta', \Upsilon_\pi(e)))$$

and using Lemma 1 we get

$$K(\{e\}, SD_\pi) \geq C_1 (\frac{L}{N})^{N - |\Upsilon_\pi(e)|} (\frac{L}{N})^{|\Upsilon_\pi(e)|} |\Upsilon_\pi(e)| \log \frac{L}{N}.$$

By

$$K(SD_\pi) \geq \sum_{e \in E} \Upsilon_\pi(e) (\frac{L}{N})^N \log \frac{L}{N}$$

and using Lemma 5, it can be shown that

$$K(SD_\pi) = \sum_{e \in E} K(\{e\}, SD_\pi) \geq C_2 \Gamma_T(\pi)(\frac{L}{N})^N \log \frac{L}{N}.$$

Following our discussion, we now have

$$\bar{b}(\mathcal{A}) = \frac{\sum_{d \in \mathcal{D}} K(d)}{|\mathcal{D}|} \geq \frac{\sum_\pi K(SD_\pi)}{\frac{L!}{(L-N)!}} \geq C_2 \sum_\pi \Gamma_T(\pi) \frac{(\frac{L}{N})^N}{\frac{L!}{(L-N)!}} \log \frac{L}{N}.$$

Finally, using Lemma 6, we get

$$\bar{b}(\mathcal{A}) \geq C_2 \Delta_T N! \frac{(\frac{L}{N})^N}{\frac{L!}{(L-N)!}} \log \frac{L}{N} = \Omega(\Delta_T \frac{(\frac{L}{N})^N}{(\frac{L}{N})} \log \frac{L}{N}).$$

$\square$

Note that if we restrict our discussion to the distributions in $\cup_\pi SD_\pi$ then we get a tight bound of $\bar{b}(\mathcal{A}) = \Theta(\Delta_T \log \frac{L}{N})$.

These results can easily be extended to the case where each processor holds several values (and not just one).

5   OPEN PROBLEMS

We mention the following open problems:

- Our bounds - for the worst case - are tight under the conditions of Corollary 1. Determine all cases where these bounds are tight (for example, the upper bound of Theorem 2 seems to be tight (it is tight for star trees and for simple paths).
- Tighten the gap between the lower and upper bound in the case of average complexity.
- Extend the results to other classes of graphs.

References

1. J.A. Bondy and U.S.R. Murty, *Graph Theory with Applications, North-Holland*, 1976.
2. R.G. Gallager, P.A. Humblet and P.M. Spira, *A distributed algorithm for minimum spanning tree, ACM Trans. on Programming Languages and Systems*, 5, 1, 1983, pp. 66-77.
3. O. Gerstel and S. Zaks, *A new characterization of tree medians with applications to distributed algorithms, 18th International Workshop on Graph-Theoretic Concepts in Computer Science (WG92)*, Frankfurt, Germany, June 1992, Springer-Verlag LNCS 657, pp. 135-144.
4. E. Korach, D. Rotem and N. Santoro, *Distributed algorithms for ranking the vertices of a network, Proceedings of the 13th Southeastern Conference on Combinatorics, Graph Theory and Computing*, 1982, pp. 235-246.
5. M.C. Loui, *The complexity of sorting in distributed systems, Information and Control*, Vol. 60, 1-3, 1984, Vol. 60, 1-3, 1984, pp. 70-85.
6. S. Zaks, *Optimal distributed algorithms for sorting and ranking, IEEE Trans. on Computers*, c-34,4,April 1985, pp. 376-379.

Three-Clustering of Points in the Plane

Johann Hagauer[1] and Günter Rote[2]

[1] Institut für Grundlagen der Informationsverarbeitung, Technische Universität Graz,
Klosterwiesgasse 32/II, A-8010 Graz, Austria;
electronic mail: jhagauer@igi.tu-graz.ac.at
[2] Institut für Mathematik, Technische Universität Graz,
Steyrergasse 30, A-8010 Graz, Austria;
electronic mail: rote@ftug.dnet.tu-graz.ac.at

Abstract. Given n points in the plane, we partition them into three classes such that the maximum distance between two points in the same class is minimized. The algorithm takes $O(n^2 \log^2 n)$ time.

1 Introduction

Overview and Statement of Results. In the classical area of cluster analysis, a given set of items is to be classified into groups (so-called clusters), such that "similar" items belong to the same group and "different" items go into different groups.

A specific case of these problems deals with *geometric* clustering problems, where the items can be represented as points in the plane (or some higher-dimensional space). In these problems the number k of clusters is fixed. Typical clustering criteria are the diameter of a cluster (the maximum distance between two points) or the radius of the smallest enclosing ball. Some previous results include an $O(n \log n)$ algorithm for finding a 2-clustering of a planar point set which minimizes the maximum diameter (Asano, Bhattacharya, Keil, and Yao 1988) and an $O(n \log^2 n / \log \log n)$ time algorithm for finding a 2-clustering which minimizes the sum of the two diameters (Hershberger 1992).

In the present paper we focus on 3-clustering. We present an $O(n^2 \log^2 n)$ algorithm for finding a 3-clustering which minimizes the maximum diameter.

Problem Setting. Capoyleas, Rote, and Woeginger (1991) have shown that any two clusters in an optimal k-clustering are *linearly separable*, for a wide variety of clustering criteria, including the maximum diameter criterion. Since there are only $O(n^2)$ ways to partition an n-point set by a line, this immediately implies polynomial k-clustering algorithms for any fixed value of k. Three clusters can be pairwise separated from each other by three lines. The straightforward application of this fact leads to an algorithm which checks $O(n^6)$ possibilities. A less straightforward approach considers all $O(n^4)$ possibilities how one cluster can be separated from the rest by two rays meeting at a common point, and solving a two-clustering problem for the remaining points by the algorithm of

Asano et al. (1988). This would still lead to a complexity of $O(n^5 \log n)$ for finding a 3-clustering which minimizes the maximum diameter. We will use a more direct approach.

Overview of the algorithm. The maximum diameter can only be one of the $\binom{n}{2}$ distances between the n given points. By a binary search among these values we can therefore reduce the optimization problem to the decision problem of testing the existence of a 3-clustering with a given upper bound δ on the maximum diameter. The cluster which contains the leftmost point P is denoted by A. We test every possible choice for the rightmost point of A separately. This requires some insight into the structure of optimal clusterings and is described in the sequel.

2 Preliminaries

The given set of n points in the plane will be denoted by P. We denote the coordinates of a point u by $u = (u_x, u_y)$. For ease of exposition we assume that no two given points have the same x- or y-coordinate. (This can be achieved by an appropriate rotation.) The line segment joining two points u and v is denoted by $\overline{uv}$, and $d(u,v) = \sqrt{(u_x - v_x)^2 + (u_y - v_y)^2}$ denotes their Euclidean distance. Two sets of points in the plane are said to be *linearly separable*, if they can be strictly separated by a straight line. For a point set S, $\mathrm{diam}(S)$ denotes its diameter, i. e., the maximum distance between two points in S. Our algorithm relies on the the following result of Capoyleas, Rote, and Woeginger (1991).

Proposition 1. *Consider the k-clustering problem of minimizing the maximum diameter. For every point set in the plane, there exists an optimal k-clustering such that each pair of clusters is linearly separable.* ☐

Our approach will be to specify a parameter δ and test for the existence of 3-partitioning such that each cluster has diameter at most δ. We need another easy lemma:

Proposition 2. (Capoyleas, Rote, and Woeginger 1991, proposition 1.) *Let δ be a positive real number, let u and v be two points in the plane at distance less than or equal to δ. Let C_1 and C_2 be the circles with radius δ centered at u and v, let D denote the points in the vertical strip between u and v. Then the part of the region $C_1 \cap C_2 \cap D$ that lies above the line segment $\overline{uv}$ has diameter $\leq \delta$.* ☐

3 Geometric Properties of a Solution

The algorithm to be described searches for all possible linearly separable solutions A, B, C in the following way. The point a with minimum x-coordinate is placed in A. Each point $a' \in P$ satisfying $d(a, a') \leq \delta$ is tested as a candidate for being the rightmost point in A. That is, we will only allow a point

u to be assigned to A if its x-coordinate $u_x < a'_x$. For every point a' we will test in $O(n \log n)$ time whether there is a solution or not. This yields an overall time complexity of $O(n^2 \log n)$ for the decision problem and time complexity of $O(n^2 \log^2 n)$ for the optimization problem.

The pair (a, a') gives rise to the following partition of the point set $P - \{a, a'\}$, see figure 1.

NORTH $:= \{ u \in P \mid u_x < a'_x$ and u is above the segment $\overline{aa'} \}$.
SOUTH $:= \{ u \in P \mid u_x < a'_x$ and u is below the segment $\overline{aa'} \}$.
EAST $:= \{ u \in P \mid u_x > a'_x \}$.
$A_{\text{cand}} := \{ u \in P \mid d(a, u) \leq \delta,\ d(a', u) \leq \delta \}$.

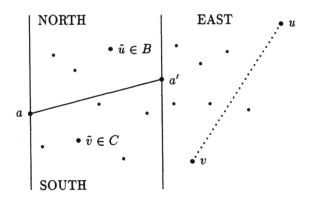

Fig. 1. The partition into NORTH, SOUTH, and EAST.

The following lemma is an immediate consequence of proposition 2.

Lemma 3.

$$\text{diam}(A_{\text{cand}} \cap \text{NORTH}) \leq \delta$$
$$\text{diam}(A_{\text{cand}} \cap \text{SOUTH}) \leq \delta \qquad \qquad \square$$

We call a partition A, B, C of P an (a, a')-*solution* if $a \in A$ is the leftmost point in P, a' the rightmost point in A, $\text{diam}(A) \leq \delta$, $\text{diam}(B) \leq \delta$, and $\text{diam}(C) \leq \delta$, and A, B, and C are linearly separable. In order to find an (a, a')-solution A, B, C we consider three different cases.

Case 1: NORTH $\subseteq A$.
Case 2: SOUTH $\subseteq A$.
Case 3: NORTH $\not\subseteq A$ and SOUTH $\not\subseteq A$.

Case 1 can be treated in the following way. We define $A := \{a, a'\} \cup \text{NORTH} \cup M$, where

$$M := \{ u \in \text{SOUTH} \cap A_{\text{cand}} \mid d(u, v) \leq \delta \text{ for all } v \in \text{NORTH} \} \ .$$

By lemma 3, diam(SOUTH $\cap A_{\text{cand}}) \leq \delta$ which implies diam(M) $\leq \delta$. Therefore placing any point $u \in$ SOUTH $\cap A_{\text{cand}}$ in A does not impose any restriction for other points in SOUTH $\cap A_{\text{cand}}$. Each of these points can be assigned to A independently of the others. Therefore the maximal feasible set A is uniquely defined. We are left with a two-clustering problem for the set of points $P - A$, which can be solved in $O(n \log n)$ time by the algorithm of Asano et al. (1988). Case 2 is treated analogously.

3.1 Case 3: The Initial Classification of Points

Case 3 turns out to be the most difficult case. Since we are looking for separable solutions, it is clear that neither B nor C may contain a point both from NORTH and from SOUTH. Therefore we assume w. l. o. g. $B \cap$ SOUTH $= \emptyset$ and $C \cap$ NORTH $= \emptyset$.

Consider now the set EAST. All points from EAST will be placed either in B or C. The following lemma allows us to make this decision for point pairs which have large distance.

Lemma 4. *Let A, B, C be an (a, a')-solution with $B \cap$ NORTH $\neq \emptyset$ and $C \cap$ SOUTH $\neq \emptyset$. Then for each pair of points $u, v \in$ EAST with distance $d(u, v) > \delta$ the following holds: If $u_y > v_y$ then $u \in B$ and $v \in C$.*

Proof. By assumption there are points $\tilde{u} \in$ NORTH $\cap B$ and $\tilde{v} \in$ SOUTH $\cap C$ (see figure 1). Assume w. l. o. g. $u_x > v_x$. If $\tilde{v}_y < v_y$ then $d(\tilde{v}, u) > d(v, u) > \delta$. Therefore u must be in B. Otherwise, $\tilde{v}_y \geq v_y$, and the segment $\overline{\tilde{u}v}$ crosses either the segment $\overline{\tilde{v}u}$ or the segment $\overline{aa'}$. Either possibility would contradict $u \in C$ and $v \in B$ together with the separability assumption. $\quad\square$

The above lemma and the previous discussion justifies the following initial classification of points:

$A_0 := \{a, a'\}$.
$B_0 := \{ u \in$ NORTH $\mid d(u, a) > \delta$ or $d(u, a') > \delta \} \cup$
$\qquad \{ u \in$ EAST $\mid$ there exists a point $v \in$ EAST, $d(u, v) > \delta$ and $u_y > v_y \}$.
$C_0 := \{ u \in$ SOUTH $\mid d(u, a) > \delta$ or $d(u, a') > \delta \} \cup$
$\qquad \{ u \in$ EAST $\mid$ there exists a point $v \in$ EAST, $d(u, v) > \delta$ and $u_y < v_y \}$.
$AB_{\text{cand}} :=$ NORTH $- B_0$.
$CA_{\text{cand}} :=$ SOUTH $- C_0$.
$BC_{\text{cand}} :=$ EAST $- B_0 - C_0$.

A_0, B_0, and C_0 will be called the *initial sets*, the other three sets the *candidate sets*. We will use the generic terms X_0 and XY_{cand} to refer to any of the respective initial or candidate sets. Note that if B_0 and C_0 are not disjoint, then case 3 cannot lead to a solution, and we may stop immediately.

Lemma 5. *The diameter of all candidate sets is at most δ.*

Proof. The inequalities diam($AB_{\text{cand}}) \leq \delta$ and diam($CA_{\text{cand}}) \leq \delta$ follow from lemma 3. The relation diam($BC_{\text{cand}}) \leq \delta$ follows from the definitions of B_0, C_0, and BC_{cand}. $\quad\square$

3.2 Propagation of Constraints

Every (a, a')-solution A, B, C must satisfy $C \cap AB_{\text{cand}} = \emptyset$, $B \cap CA_{\text{cand}} = \emptyset$, and $A \cap BC_{\text{cand}} = \emptyset$. So far, the placement of points of the initial sets is fixed, whereas the placement of points in candidate sets is not yet fixed. Some point in an initial set may force certain points in a candidate set to be assigned to A, B, or C. Such constraints imposed by the initial assignment may propagate. We describe this situation by two directed graphs $G_1 := (V, E_1)$ and $G_2 := (V, E_2)$ with vertex set $V = P$. There is an arc (u, v) in E_1 if $d(u, v) > \delta$ and one of the following five conditions holds:

1. $u \in B_0$, $v \in AB_{\text{cand}}$
2. $u \in C_0$, $v \in BC_{\text{cand}}$
3. $u \in AB_{\text{cand}}$, $v \in CA_{\text{cand}}$
4. $u \in CA_{\text{cand}}$, $v \in BC_{\text{cand}}$
5. $u \in BC_{\text{cand}}$, $v \in AB_{\text{cand}}$

Similarly, there is an arc (u, v) in E_2 if $d(u, v) > \delta$ and one of the following conditions holds:

1. $u \in B_0$, $v \in BC_{\text{cand}}$
2. $u \in C_0$, $v \in AB_{\text{cand}}$
3. $u \in AB_{\text{cand}}$, $v \in BC_{\text{cand}}$
4. $u \in BC_{\text{cand}}$, $v \in CA_{\text{cand}}$
5. $u \in CA_{\text{cand}}$, $v \in AB_{\text{cand}}$

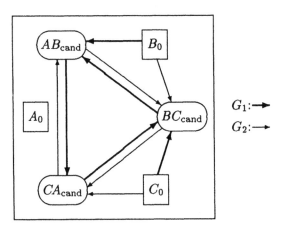

Fig. 2. Schematic representation of the constraint propagation graphs G_1 and G_2. G_1 is shown with thick arcs and G_2 with thin arcs.

Figure 2 gives a schematic representation of these graphs. The two graphs correspond to "counter-clockwise" propagation $(A \to C \to B \to A)$ and to "clockwise" propagation of constraints $(A \to B \to C \to A)$, respectively. (The constraints imposed by the initial set $A_0 = \{a, a'\}$ have already explicitly been taken care of in the definition of B_0 and C_0.) Note that every pair of points $u, v \in AB_{\text{cand}} \cup BC_{\text{cand}} \cup CA_{\text{cand}}$ with $d(u, v) > \delta$ gives rise to two arcs, one in each graph.

Let $Forced_1$ denote the set of points in $AB_{\text{cand}} \cup CA_{\text{cand}} \cup BC_{\text{cand}}$ which are reachable by some path in G_1 starting at a vertex in $B_0 \cup C_0$, and $Forced_2$ the analogous set for G_2.

Lemma 6. *There is an (a, a')-solution A, B, C with $B \cap \text{NORTH} \neq \emptyset$ and $C \cap \text{SOUTH} \neq \emptyset$ (corresponding to case 3) if and only if*

(i) $\text{diam}(B_0) \leq \delta$,
(ii) $\text{diam}(C_0) \leq \delta$, *and*
(iii) the sets $Forced_1$ and $Forced_2$ are disjoint.

Proof. Assume first that the conditions hold. We construct a solution by the following rules. Initially $A = A_0$, $B = B_0$, and $C = C_0$.

Rule 1: If $u \in Forced_1$ then u is assigned to A if $u \in AB_{\text{cand}}$,
$\qquad\qquad\qquad\qquad\quad u$ is assigned to C if $u \in CA_{\text{cand}}$,
$\qquad\qquad\qquad\qquad\quad u$ is assigned to B if $u \in BC_{\text{cand}}$.
Rule 2: If $u \in Forced_2$ then u is assigned to B if $u \in AB_{\text{cand}}$,
$\qquad\qquad\qquad\qquad\quad u$ is assigned to A if $u \in CA_{\text{cand}}$,
$\qquad\qquad\qquad\qquad\quad u$ is assigned to C if $u \in BC_{\text{cand}}$.
Rule 3: If $u \in (AB_{\text{cand}} \cup CA_{\text{cand}} \cup BC_{\text{cand}}) - (Forced_1 \cup Forced_2)$ then classify u as in rule 1.

Rules 1 and 2 express the necessary assignment for the sets A, B, or C that follow from propagating assignments through the graphs, as can be seen by induction on the path length leading to a point $u \in Forced_1 \cup Forced_2$. Rule 3 is an arbitrary decision. The points that fall under this rule could either be handled (consistently) as in rule 1 or as in rule 2. A concise statement of the result of applying the three rules is as follows:

$$A = A_0 \cup (CA_{\text{cand}} \cap Forced_2) \cup (AB_{\text{cand}} - Forced_2)$$
$$B = B_0 \cup (AB_{\text{cand}} \cap Forced_2) \cup (BC_{\text{cand}} - Forced_2)$$
$$C = C_0 \cup (BC_{\text{cand}} \cap Forced_2) \cup (CA_{\text{cand}} - Forced_2)$$

As every point is placed by exactly one rule, the sets A, B and C form a partition of P. To show that we have a valid solution we have to prove that the diameters do not exceed δ. Consider two points u, v with $d(u, v) > \delta$. If both are in initial sets, they cannot be in the same initial set, by conditions (i) and (ii). Therefore they end up in different sets A, B, or C. If at least one of the two points is in a candidate set, lemma 5 implies that they are not in the same candidate set. Therefore one of the arcs (u, v) and (v, u) is either in E_1 or in E_2. Therefore u and v are either both in $Forced_1$ or in $Forced_2$ or in neither set. This means that they are assigned to different sets A, B, or C by rule 1, by rule 2, or by rule 3, respectively.

On the other hand, failure of any of the conditions of the lemma would lead to a contradiction: Conditions (i) and (ii) are necessary by lemma 4. If (iii) does not hold, rules 1 and 2 lead to incompatible conclusions about the assignment of a point $u \in Forced_1 \cup Forced_2$. By the above discussion this makes a solution impossible. $\qquad\square$

4 The Tripartition Algorithm

The algorithm follows the outline in the previous section. We have to test all points $a' \in P$ as possible rightmost points of A. For each a' we determine the sets NORTH, SOUTH, and EAST in linear time, and we test cases 1, 2 and 3 of section 3 (following lemma 3). Let us first look at case 3: After the determination of initial and candidate sets we call the procedures *rule1* and *rule2* below, which in turn rely on the procedure *distribute*. These procedures explore the graphs G_1 and G_2 as they construct them, assigning the points to the sets A, B, or C and propagating constraints as soon as possible.

procedure *rule1*
    $C_{\text{forced}} := C_0$;
    $distribute(C, B)$;
    $B_{\text{forced}} := B_{\text{forced}} \cup B_0$;
    while $B_{\text{forced}} \neq \emptyset$ **do**
        $distribute(B, A)$;
        $distribute(A, C)$;
        $distribute(C, B)$;
end procedure

procedure *rule2*
    $B_{\text{forced}} := B_0$;
    $distribute(B, C)$;
    $C_{\text{forced}} := C_{\text{forced}} \cup C_0$;
    while $C_{\text{forced}} \neq \emptyset$ **do**
        $distribute(C, A)$;
        $distribute(A, B)$;
        $distribute(B, C)$;
end procedure

procedure $distribute(X, Y)$
    $Y_{\text{forced}} := \emptyset$;
    for all points $u \in X_{\text{forced}}$ **do**
    $X := X \cup \{u\}$;
    $New := \{ v \in XY_{\text{cand}} \mid d(u, v) > \delta \}$;
    $Y_{\text{forced}} := Y_{\text{forced}} \cup New$;
    $XY_{\text{cand}} := XY_{\text{cand}} - New$;
end procedure

(A literal interpretation of the procedure *distribute* refers to sets such as BA_{cand}. This are the same as the set AB_{cand}, and in general XY_{cand} and YX_{cand} are identical.)

Rule 3 just assigns the remaining points as follows: $A := A \cup AB_{\text{cand}}$, $B := B \cup BC_{\text{cand}}$, and $C := C \cup CA_{\text{cand}}$. We finally compute the diameters of A, B, and C in $O(n \log n)$ time to check that we really have a solution.

In order to implement the procedure *distribute* efficiently we use the circular hull introduced by Hershberger and Suri (1991). Let δ be fixed. The *circular hull* of a set of points U is the common intersection of all disks of radius δ containing U. (Circular hulls are also known as α-hulls, see Edelsbrunner, Kirkpatrick, and Seidel 1983.) Circular hulls can be constructed in $O(n \log n)$ time. The data structure of Hershberger and Suri (1991) supports the following two main operations:

1. Given a query point u, determine in time $O(\log n)$ a point $v \in U$ such that $d(u, v) > \delta$, if such a point exists.
2. Delete a point from U and update the circular hull in an amortized cost of $O(\log n)$.

The procedures *rule1*, *rule2*, and *distribute* are implemented in the following way: First the circular hulls for the candidate sets XY_{cand} are constructed. For a point $u \in X_{\text{forced}}$ in the procedure *distribute* we can use the circular hull of XY_{cand} to check whether a point $v \in XY_{\text{cand}}$ exists with $d(u, v) > \delta$. Such a point v is deleted from the circular hull of the point set XY_{cand} and inserted in Y_{forced}. Then u is repeatedly used as a query to the circular hull until no more points v are found. The point u is finally removed from XY_{cand} and assigned to X, and it will never again act as query to some circular hull. Therefore the number of queries is bounded by the number of deletions plus the number of points in the initial sets. Since no point is inserted in a circular hull the overall complexity for deletions and queries is $O(n \log n)$. Thus the time bound for the procedures *rule1* and *rule2* is $O(n \log n)$.

Circular hulls can also be used to carry out the remaining operations in $O(n \log n)$ time: the computation of B_0 and C_0 and the treatment of cases 1 and 2. (These tasks can also be solved in $O(n \log n)$ time with furthest-point Voronoi diagrams, a more standard data structure.)

Therefore our tripartition algorithm takes $O(n^2 \log n)$ time, for fixed δ.

Theorem 7. *Given a set of n points in the plane and a real number δ, we can determine in $O(n^2 \log n)$ time whether there is a partition of P into sets A, B, C with diameters at most δ.* $\qquad\qquad\square$

Theorem 8. *Given a set of n points in the plane, we can in $O(n^2 \log^2 n)$ time construct a partition of P into sets A, B, C such that the largest of the three diameters at small as possible.*

Proof. We carry out a binary search on the $\binom{n}{2}$ distances occuring in a point set of cardinality n, using the *tripartition* algorithm. $\qquad\qquad\square$

References

Te. Asano, B. Bhattacharya, M. Keil, and F. Yao: Clustering algorithms based on minimum and maximum spanning trees. In: Proc. 4th Ann. Symp. Comput. Geometry (1988), pp. 252–257

V. Capoyleas, G. Rote, and G. Woeginger: Geometric clusterings. J. Algorithms **12** (1991), 341–356

H. Edelsbrunner, D. G. Kirkpatrick, and R. Seidel: On the shape of a set of points in the plane. IEEE Trans. Inform. Theory **IT-29** (1983), 551–559

J. Hershberger: Minimizing the sum of diameters efficiently. Comput. Geom. Theory Appl. **2** (1992), 111–118

J. Hershberger and S. Suri: Finding Taylored Partitions. J. Algorithms **12** (1991), 431–463

Gossiping in Vertex-Disjoint Paths Mode in d-Dimensional Grids and Planar Graphs *

Extended Abstract

Juraj Hromkovič[†], Ralf Klasing, Elena A. Stöhr, Hubert Wagener[‡]

Department of Mathematics and Computer Science
University of Paderborn,
33095 Paderborn, Germany

Abstract

The communication modes (one-way and two-way mode) used for sending messages to processors of interconnection networks via vertex-disjoint paths in one communication step are investigated. The complexity of communication algorithms is measured by the number of communication steps (rounds). Here, the complexity of gossiping in grids and in planar graphs is investigated. The main results are the following:

1. Effective one-way and two-way gossip algorithms for d-dimensional grids, $d \geq 2$, are designed.

2. The lower bound $2 \log_2 n - \log_2 k - \log_2 \log_2 n - 2$ is established on the number of rounds of every two-way gossip algorithm working on any graph of n nodes and vertex bisection k. This proves that the designed two-way gossip algorithms on d-dimensional grids, $d \geq 3$, are almost optimal, and it also shows that the 2-dimensional grid belongs to the best gossip graphs among all planar graphs.

3. Another lower bound proof is developed to get some tight lower bounds on one-way "well-structured" gossip algorithms on planar graphs (note that all gossip algorithms designed until now in vertex-disjoint paths mode are "well-structured").

---

*This work was partially supported by grants Mo 285/4-1, Mo 285/9-1 and Me 872/6-1 (Leibniz Award) of the German Research Association (DFG), and by the ESPRIT Basic Research Action No. 7141 (ALCOM II).

[†]This author was partially supported by SAV Grant No. 88 and by EC Cooperation Action IC 1000 Algorithms for Future Technologies.

[‡]This author was supported by the Ministerium für Wissenschaft und Forschung des Landes Nordrhein-Westfalen.

1 Introduction and Definitions

This paper is devoted to the problem of information dissemination in prominent interconnection networks. The basic three communication tasks are **broadcast**, **accumulation**, and **gossip** which can be described as follows. Assume that each vertex (processor) in a graph (network) has some piece of information. The **cumulative message** of G is the set of all pieces of information originally distributed in all vertices of G. To solve the **broadcast [accumulation]** problem for a given graph G and a vertex u of G we have to find a communication strategy (using the edges of G as communication links) such that all vertices in G learn the piece of information residing in u [that u learns the cumulative message of G]. To solve the **gossip** problem for a given graph G a communication strategy such that all vertices in G learn the cumulative message of G must be found. Since the above stated communication problems are solvable only in connected graphs, we note that from now on we use the notion "graph" for connected undirected graphs.

The meaning of "a communication strategy" depends on the communication mode. A communication strategy is realized as a **communication algorithm** consisting of a number of **communication steps (rounds)**. The rules describing what can happen in one communication step (round) are defined exactly by the communication mode. Until now, the following two communication modes were considered:

1. **One-way [Two-way] vertex-disjoint paths mode** (1VDP mode [2VDP mode])

   One round can be described as a set $\{P_1, \ldots, P_k\}$ for some $k \in I\!N$, where $P_i = x_{i,1}, \ldots, x_{i,\ell_i}$ is a simple path of length $\ell_i - 1$, $i = 1, \ldots, k$, and the paths are vertex-disjoint. The executed communication of this round in one-way mode consists of the submission of the whole actual knowledge of $x_{i,1}$ to x_{i,ℓ_i} via the path P_i for any $i = 1, \ldots, k$. [The executed communication of this round in two-way mode consists of the complete exchange of the actual knowledge between $x_{i,1}$ and x_{i,ℓ_i} for any $i = 1, \ldots, k$.] The inner nodes of path P_i (nodes different from the end points $x_{i,1}$ and x_{i,ℓ_i}) do not learn the message submitted from $x_{i,1}$ to x_{i,ℓ_i} [exchanged between $x_{i,1}$ and x_{i,ℓ_i}], they are only used to realize the connection between $x_{i,1}$ and x_{i,ℓ_i}.

2. **Listen-in vertex-disjoint paths mode** (LVDP mode)

   As in the previous mode, a round can be described as a set of vertex-disjoint paths. The difference is that here, for each **active path** P (realizing some connection between two vertices), one end-vertex of P broadcasts (submits) its whole knowledge to all other vertices in P. Thus, after the execution of the round all vertices of each path P know the message submitted from the sending endpoint.

The listen-in vertex-disjoint paths mode was introduced in [FHMMM92], where optimal (or almost optimal) broadcast, accumulation, and gossip algorithms for almost all basic interconnection networks were found. This was possible mostly because the LVDP mode is so powerful (it is possible to broadcast in one round in graphs having a Hamiltonian path) that the designed algorithms met the trivial lower bounds. The one-way vertex-disjoint paths mode introduced (but not thoroughly investigated) by Farley [Fa80] is not as powerful as the LDVP mode, and so proving the effectivity of designed algorithms requires to develop new, special lower bound techniques. The 1VPD/2VDP mode has been investigated in [HKS93], where optimal algorithms for paths (approximately $2 \log_2 n$ rounds in both VDP modes), complete graphs ($\log_2 n$ rounds in 2VDP mode and $1.44... \log_2 n$ rounds in 1VDP mode), and flakes are designed, and for several hypercube-like networks of n nodes it is shown that the gossip complexity differs from the gossip complexity of the complete graph K_n of n nodes at most by $O(\log_2 \log_2 n)$ rounds. The optimal broadcast and accumulation algorithms in vertex-disjoint paths modes for most of the fundamental interconnection networks are presented in [FHMMM92, HKS93]. Thus, we concentrate here on the gossip problem in grids (i.e., to search for the gossip complexity of grids in the range given by the complete graph and the path) and in planar networks. Note that the study of VDP modes is of practical importance [BAPRS90, DS87].

Now, let us fix the notation used in this paper. Let for any graph $G = (V, E)$, $V(G) = V$ denote the set of vertices of G, and $E(G) = E$ denote the set of edges of G. For any graph G, let $R(G), R^2(G)$ be the number of rounds (complexity) of the optimal gossip algorithm for G in the 1VDP, 2VDP mode respectively. Also, for any communication algorithm A, let $round(A)$ be the number of rounds of A. Let $b = (1 + \sqrt{5})/2$ throughout the paper.

The paper is organized as follows. Section 2 informally presents the main algorithmic ideas and lower bound proof techniques used in the subsequent sections. Section 3 deals with gossiping in the d-dimensional grid Gr_n^d of $n = m^d$ nodes. We shall establish there

$$(1) \quad R^2(Gr_n^d) \leq (1 + 1/d) \cdot \lfloor \log_2 n \rfloor - \tfrac{1}{2} \cdot \log_2 \log_2 n + O(d),$$

$$(2) \quad R(Gr_n^d) \leq (\log_b 2 + (2 - \log_b 2)/d) \cdot \lfloor \log_2 n \rfloor$$
$$+ (3 - \log_b 2) \cdot (d - 1) + 2d$$
$$= (1.44... + 0.56.../d) \cdot \log_2 n + O(d),$$

for $d \geq 3$. For the 2-dimensional grid Gr_n^2, we get

$$(3) \quad R^2(Gr_n^2) \leq 1.5 \cdot \lfloor \log_2 n \rfloor + 4 = 1.5 \cdot \log_2 n + O(1),$$

$$(4) \quad R(Gr_n^2) \leq (1 + (\log_b 2)/2) \cdot \lfloor \log_2 n \rfloor + 7$$
$$= 1.72... \log_2 n + O(1).$$

Section 4 presents a lower bound method providing

(5) $\quad R^2(G_{n,k}) \geq 2 \cdot \log_2 n - \log_2 k - \log_2 \log_2 n - 2$

for any graph $G_{n,k}$ of n nodes and vertex bisection k.

This yields a tight lower bound to (1):

(6) $\quad R^2(Gr_n^d) \geq (1 + 1/d) \cdot \log_2 n - \log_2 \log_2 n - 2$ for any $d \geq 2$.

A comparison of (1) and (6) shows that $R^2(Gr_n^d) = (1+1/d) \cdot \log_2 n - \Theta(\log_2 \log_2 n)$ for any constant $d \geq 3$. Especially, we get from (5):

(7) $\quad R^2(Pl_n) \geq 1.5 \cdot \log_2 n - \log_2 \log_2 n - 4$

for any planar graph Pl_n of n nodes.

Thus, comparing (3) with (7) we see that the 2-dimensional grid is a very good structure for gossiping among all planar structures.

The lower bound method presented in Section 4 is unable to distinguish between the one-way and the two-way mode, and so the provided lower bounds for the 1VDP mode are not tight to the upper bounds in (2),(4). To overcome this difficulty we develop in Section 5 a new lower bound method enabling to prove at least the lower bound

$$
\begin{aligned}
(8) \quad R(Pl_n) \; &\geq \; (1 + (\log_b 2)/2) \cdot \log_2 n \\
&\quad -(2 - \log_b 2) \cdot \log_2 \log_2 n - O(1) \\
&= \; 1.72... \log_2 n - O(\log_2 \log_2 n).
\end{aligned}
$$

if restricting the class of all gossip algorithms to some subclass of so-called "well-structured" gossip algorithms (as defined in Section 2). Comparing (4) and (8) we see again that Gr_n^2 is one of the best graphs for one-way "well-structured" gossiping among all planar graphs. Note that all algorithms designed here and in [HKS93] are "well-structured".

2 Basic concepts

The aim of this section is to describe the methods used to get upper and lower bounds on the complexity of gossiping in the networks investigated. To get the upper bounds we use a method designing so-called "three-phase algorithms". Let us informally define what a three-phase algorithm is. Let G be any graph. Let $a(G)$ be any subset of nodes of G. We call $a(G)$ the **set of accumulation nodes**, and every node in $a(G)$ is called an **accumulation node**. A **three-phase gossip algorithm** for G according to $a(G)$ works in the following three phases:

1. **Accumulation Phase**

   Divide G into $|a(G)|$ connected components, each component containing exactly one accumulation node of $a(G)$. These components are called **accumulation components**. Each $v \in a(G)$ accumulates the information from the nodes lying in its component.

   {After the first phase, the nodes in $a(G)$ together know the cumulative message of G.}

2. **Gossip Phase**

   Perform a gossip algorithm among the nodes in $a(G)$ in 1VDP (2VDP) mode (i.e. all nodes in $V(G) - a(G)$ are considered to have no information, and they are only used to build disjoint paths between receivers and senders from $a(G)$).

   {After the second phase, every node in $a(G)$ knows the cumulative message of G.}

3. **Broadcast Phase**

   Every node in $a(G)$ broadcasts the cumulative message in its component.

   {After this, all nodes of G know the cumulative message of G.}

In order to construct a really effective gossip algorithm, we shall search for an $a(G)$ in G such that

a) Phase 2 can be performed (almost) as quickly as gossiping in a complete graph of $a(G)$ nodes.

b) The maximal size of a component is as small as possible, which minimizes the time for the first and third phase.

Obviously, every second phase of a three-phase algorithm A corresponds unambiguously to a gossip algorithm C in a complete graph of $|a(G)|$ nodes. We say that C **is implemented in the second phase of** A. All algorithms designed here are three-phase algorithms with second phases implementing an optimal (or almost optimal) gossip algorithm on graphs of $|a(G)|$ nodes.

The lower bound techniques used here are of two distinct kinds. Roughly speaking, the first one (used for graphs with a given vertex bisection in Section 4) is based on the estimations how much information must and can flow through the vertices of a minimal vertex bisection of a given graph in a given number of rounds in order to complete the gossip task. This technique provides good lower bounds for the 2VDP mode, but is too rough for the 1VDP mode. The second technique for the 1VDP mode works only for three-phase gossip algorithms and uses other, special considerations related to bisection width and embeddings into planar graphs.

3 Gossiping in the Grid

In this section we will establish upper bounds on the gossip complexity in grids. First we consider gossiping in two-dimensional grids in detail, and then we sketch a generalization to the d-dimensional case. All presented algorithms follow the 3-phase design.

Let Gr_n^2 be the two-dimensional grid with n vertices and sidelength $m = \sqrt{n}$. By $[i, j]$ we denote the vertex in the i-th row and j-th column. Communication between vertices is established by a set of canonical paths. We say that vertices $v = [i, j]$ and $w = [k, l]$ **communicate via row** t, if the information is passed first vertically from $[i, j]$ to $[t, j]$, then horizontally along the t-th row to $[t, l]$, and then finally vertically from $[t, l]$ to $[k, l]$. With this notation, gossiping in Gr_n^2 can be described as follows. We assume that m is a power of 2. The modifications for removing this assumption are given later. Each column forms an accumulation component. The accumulation vertex for the i-th column is chosen as $[\frac{m}{2}, i]$. Using a recursive doubling technique, accumulation and broadcasting in these components require $\log_2 m$ rounds each. This is optimal for both VDP modes. The second phase implements the optimal gossip algorithm in 2VDP mode and 1VDP mode, respectively, for the complete graph K_m.

The optimal gossiping in two-way mode is easily described in a recursive form:

> **Procedure Gossip($[\frac{m}{2}, 1], \ldots, [\frac{m}{2}, m]$)**
>
> > begin
> > > in parallel do:
> > > > Gossip ($[\frac{m}{2}, 1], \ldots, [\frac{m}{2}, \frac{m}{2}]$)
> > > > Gossip ($[\frac{m}{2}, \frac{m}{2} + 1], \ldots, [\frac{m}{2}, m]$)
> > > end pardo
> > > in parallel for $1 \leq i \leq \frac{m}{2}$ do:
> > > > exchange information between $[\frac{m}{2}, i]$ and $[\frac{m}{2}, m - i + 1]$)
> > > > via row $m - i + 1$
> > > end pardo
> > end

It is now trivial to verify that any pair of communication paths $[\frac{m}{2}, i], \ldots, [\frac{m}{2}, m - i + 1]$ and $[\frac{m}{2}, j], \ldots, [\frac{m}{2}, m - j + 1]$, $i \neq j$, is vertex-disjoint. Since recursive calls that have to be performed in parallel, work on disjoint subgrids, each communication round can be performed in VDP mode. Clearly, the gossip algorithm applied to $a(G)$ requires $\log_2 m$ levels of recursion, i.e. $\log_2 m$ rounds.

Theorem 3.1 $R^2(Gr_n^2) \leq 1.5 \cdot \lfloor \log_2 n \rfloor + 4$.

Proof. If $m = \sqrt{n}$ is no power of 2, we first collect all pieces of the cumulative message into a subgrid not necessarily consisting of consecutive rows of sidelength

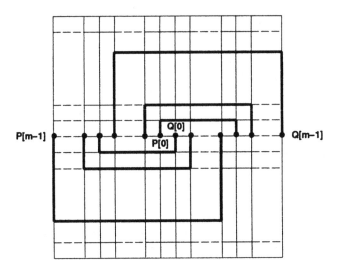

Fig. 1: The communication paths in the one-way mode

$2^{\lfloor \log_2 m \rfloor}$. This requires two communication rounds. Then the above algorithm solves the gossiping in the subgrid. Each of the three phases requires $\lfloor \log_2 m \rfloor$ rounds. Finally, the cumulative message residing in each vertex of the chosen subgrid is distributed to all vertices in two additional rounds. Thus a total of $3 \cdot \lfloor \log_2 m \rfloor + 4 = 3 \cdot \lfloor \log_2 \sqrt{n} \rfloor + 4 \leq 1.5 \cdot \lfloor \log_2 n \rfloor + 4$ rounds suffices. □

Now we turn our attention to the one-way mode of communication.

Theorem 3.2 $R(Gr_n^2) < (1 + (\log_b 2)/2) \cdot \lfloor \log_2 n \rfloor + 7$.

Proof. The proof is analogous to the one of the previous theorem. Only the search for vertex-disjoint paths implementing the optimal one-way gossip algorithm for complete graphs [EM89] is a little more complicated (see Fig. 1 for the implementation of a round of the algorithm given in [EM89], the notation in Fig. 1 is according to [EM89]). The only change in the complexity of the algorithm is the time for the gossiping phase (phase two), which is $\log_b \frac{m}{2} + 3 = \log_b 2 \cdot \log_2 \frac{m}{2} + 3$ now instead of $\log_2 m$. □

We now investigate gossiping in d-dimensional grids Gr_n^d where $n = m^d$ for $d \geq 3$.

Theorem 3.3 *For any* $n = m^d$, $m, d \in I\!N$, $d \geq 3$:

(i) $R^2(Gr_n^d) \leq (1 + 1/d) \cdot \lfloor \log_2 n \rfloor - \frac{1}{2} \log_2 \log_2 n + O(d)$,

(ii) $R(Gr_n^d) \leq (\log_b 2 + (2 - \log_b 2)/d) \cdot \lfloor \log_2 n \rfloor + (3 - \log_b 2) \cdot (d - 1) + 2d$

$\quad = (1.44... + 0.56.../d) \cdot \log_2 n + O(d)$.

Idea of the Proof. Using a generalization of techniques used in the two-dimensional case (the accumulation points form a $(d-1)$-dimensional subgrid, namely $\{(\frac{m}{2}, x_2, \ldots, x_d) \mid 1 \leq x_i \leq m\}$), one can get (ii) and a weaker version of (i) claiming $R^2(Gr_n^d) \leq (1 + 1/d) \cdot \lfloor \log_2 n \rfloor + 2d$. To get (i), a more elaborated technical method is needed requiring too large area to be completely presented in this extended abstract. So, we only give a rough idea: Let $Gr(a, b, c)$ denote the $(a \times b \times c)$-grid. The upper bound (i) is achieved in the following three steps:

1. First, it is shown that, for m being a power of two, gossiping between the vertices of row $\{[\lceil m/\sqrt{\log_2 m}\rceil, i, 1] \mid 1 \leq i \leq m\}$ in $Gr(\lceil 2m/\sqrt{\log_2 m}\rceil, m, 2)$ can be achieved in $\log_2 m$ steps.

2. Then, the previous algorithm is used to gossip in $Gr(r, r, 2)$ in $1.5 \cdot \log_2 n - \frac{1}{2} \cdot \log_2 \log_2 n + 9$ rounds.

3. Finally, 2 is applied to get gossiping in Gr_n^d for $d \geq 3$ in $(1+1/d) \cdot \lfloor \log_2 n \rfloor - \frac{1}{2} \log_2 \log_2 n + O(d)$ rounds.

$\square$

4 A Lower Bound for Two-Way Gossiping

Here, we will derive lower bounds for gossiping in grids. For this purpose, we will prove a general lower bound for graphs with bounded vertex bisection width and then apply this result to grids. Using this approach, we will show the effectivity of our grid algorithms in the 2VDP mode, and also demonstrate that the 2-dimensional grid belongs to the best structures among all planar graphs for gossiping in the 2VDP mode.

For any graph $G = (V, E)$, a **vertex bisector** is a set of vertices $V' \subset V$ such that the removal of the edges incident to the vertices of V' splits G into two equal-sized components G^1 and G^2 (i.e. $||V(G^1)| - |V(G^2)|| \leq 1$). Now, the idea of the lower bound proof for graphs with bounded vertex bisection width is to estimate how much information must and can flow through the vertices of a minimal bisection in a given number of rounds in order to complete the gossip task.

Lemma 4.1 *Let $G_{n,k}$ be a graph of n nodes and vertex bisection k, n even. Then*

$$R^2(G) \geq 2 \cdot \log_2 n - \log_2 k - \log_2 \log_2 n - 2.$$

Proof. Denote the left half of the bisection by A, the right half by B (i.e. $|A|$, $|B| = n/2$) and the vertices of the bisector by V ($|V| = k$). For $i \geq 0$, define

$$I(i) = \sum_{v \in B} I_i(v),$$

where $I_i(v)$ is the number of pieces of information from A known by v after i rounds. We will state an upper bound on $I(i)$. For a complete gossip, all nodes in B must know all the pieces of information from A. Hence, any gossip scheme running in t rounds must satisfy

$$I(t) \geq \left(\frac{n}{2}\right)^2.$$

Combining this condition with the upper bound on $I(i)$ will lead to the desired lower bound on t.

Let us start estimating $I(i)$. Clearly, we have

$$I(0) = 0.$$

Note that after i rounds, $0 \leq i < \log_2 n$, each node can have at most 2^i informations. Hence, in round $i+1$ at most $k \cdot 2^i$ informations can flow from A to B via the bisector V. Also, the number of informations from A already present in B after i rounds, can at most be doubled in round $i+1$. Hence, we have

$$I(i+1) = 2 \cdot I(i) + k \cdot 2^i \quad \text{for all } 0 \leq i < \log_2 n.$$

An easy induction shows that

$$I(i) = i \cdot k \cdot 2^{i-1} \quad \text{for all } 0 \leq i \leq \log_2 n.$$

Thus, after $\log_2 n$ rounds, we have $I(\log_2 n) = (n/2) \cdot k \cdot \log_2 n$.

After i rounds, $i \geq \log_2 n$, each node in A can have at most $n/2$ informations from A. Hence, in round $i+1$ at most $k \cdot n/2$ informations can flow from A to B via the bisector V. Hence, we have

$$I(i+1) = 2 \cdot I(i) + \frac{n}{2} \cdot k \quad \text{for all } i \geq \log_2 n.$$

Again, it can easily be shown by induction that

$$I(i) = \frac{n}{2} \cdot k \cdot \left(2^{i-\log_2 n} \cdot \log_2 n + 2^{i-\log_2 n} - 1\right) \quad \text{for all} \quad i > \log_2 n.$$

As we have seen above, the condition

$$\frac{n}{2} \cdot k \cdot \left(2^{t-\log_2 n} \cdot \log_2 n + 2^{t-\log_2 n} - 1\right) = I(t) \geq \left(\frac{n}{2}\right)^2$$

must hold for any gossip scheme running in t rounds. This can only be true if $t \geq 2 \cdot \log_2 n - \log_2 k - \log_2 \log_2 n - 2$ holds. $\qquad \square$

Now, we are prepared to apply Lemma 4.1 to grids. As the d-dimensional square grid Gr_n^d has vertex bisection $n^{1-1/d}$, Lemma 4.1 yields the following result:

Theorem 4.2

(i) $R^2(Gr_n^d) \geq (1 + 1/d) \cdot \log_2 n - \log_2 n \log_2 n - 2$ *for any* $d \in \mathbb{N}, d \geq 2$,

(ii) $R^2(Gr_n^d) = (1 + 1/d) \cdot \log_2 n - \Theta(\log_2 \log_2 n)$ *for any constant* $d \geq 3$.

Lipton and Tarjan [LT79] gave an upper bound of $\sqrt{8 \cdot n}$ on the vertex bisection of any planar graph Pl_n of n nodes (consider the result in [DDSV93] to see why we use vertex bisection instead of edge bisection in the proof of Lemma 4.1). Applying Lemma 4.1 and recalling Theorem 3.1, we obtain the result that the 2-dimensional square grid Gr_n^2 is one of the best structures among all planar graphs for gossiping in the 2VDP mode.

Theorem 4.3 *For any* $n = m^2$, $m \in \mathbb{N}$, *and for any planar graph* Pl_n *of* n *nodes,*

(i) $R^2(Gr_n^2) \leq 1.5 \cdot \lfloor \log_2 n \rfloor + 4$,

(ii) $R^2(Pl_n) \geq 1.5 \cdot \log_2 n - \log_2 \log_2 n - 4$.

5 A Lower Bound for One-Way Gossiping

The drawback of the lower bound technique successfully used in the previous sections for graphs with a given bisection is that it does not distinguish between the two-way mode and the one-way mode. So, this method is not able to provide tight lower bounds on $R(G)$ for graphs G for which the two-way mode is much more powerful than the one-way mode (which is often the case). We were also not able to modify this technique to provide higher lower bounds by restricting the two-way mode to the one-way mode. (Note that one has the same problem for the standard one-way mode, for which only a few nontrivial lower bounds are known.) Instead of this we have developed another method providing reasonable lower bounds on one-way gossiping $(R(G))$ by the following additional assumptions:

(1') Every gossip algorithm must be a three-phase algorithm.

(2') The only known second phase of any three-phase algorithm must be an implementation of the optimal gossip algorithm in the complete graph K_m given in [EM89] for some $m \in N$.

Note that all gossip algorithms designed for VDP modes until now have the above stated properties (1') and (2'), and their properties lead to simple understanding of the algorithms as well as to an easy implementation. This is also the reason why we are interested in deriving lower bounds for the one-way gossip complexity in this restricted class of gossip algorithms.

Because of the restricted space, we are not able to present this technique in this extended abstract. Thus, we give only the formulation of the lower bound, and the complete proof is given in the full paper.

Theorem 5.1 *Let Pl_n be any planar graph of n nodes, $n \in N$, and let B be a one-way gossip algorithm on Pl_n with the properties (1') and (2'). Then,*

$$round(B) \geq (1 + (\log_b 2)/2) \cdot \log_2 n - (2 - \log_b 2) \cdot \log_2 \log_2 n - d_3$$

for some constant d_3 independent of n.

Since Theorem 3.2 claims that $R(Gr_n^2) \leq (1 + (\log_b 2)/2) \cdot \lfloor \log_2 n \rfloor + 7$, one sees that two-dimensional grids belong to the planar graphs with the quickest gossiping.

6 Conclusion

In this paper we have designed two-way and one-way gossip algorithms for d-dimensional grids, and in some cases we have established also tight lower bounds. The most interesting open problems we see left are the following:

1. We have $1.5 \log_2 n - \log_2 \log_2 n - O(1) \leq R^2(Gr_n^2) \leq 1.5 \log_2 n + O(1)$ for two-dimensional grids and $R^2(Gr_n^d) = (1 + 1/d) \cdot \log_2 n - \Theta(\log_2 \log_2 n)$ for $d \geq 3$. Is it possible to save $O(\log_2 \log_2 n)$ rounds in the 2-dimensional case (by some more elaborated technique), or can the lower bound $1.5 \log_2 n - O(1)$ be achieved?

2. A more powerful lower bound proof method for one-way gossip should be found. The interest is especially in removing the factor $-O(\log_2 \log_2 n)$ from the lower bound of Theorem 5.1 as well as in a generalization of the result of Theorem 5.1, providing the same lower bound by weaker restrictions on the class of one-way gossip algorithms (for instance, by removing the restriction (2') from the assumptions of Theorem 5.1).

References

[BAPRS90] Y. Ben-Asher, D. Peleg, R. Ramaswami, A. Schuster, "The power of reconfiguration", Technical Report, The Hebrew University, Jerusalem, 1990.

[DDSV93] K. Diks, H.N. Djidjev, O. Sykora, I. Vrto, "Edge separators of planar and outerplanar graphs with applications", *Journal of Algorithms* 14 (1993), pp. 258-279.

[DS87] W. Dally, C. Seitz, "Deadlock free message routing in multiprocessor interconnection networks", *IEEE Transactions on Computers* 36 (1987), pp. 547-553.

[EM89] S. Even, B. Monien, "On the number of rounds necessary to disseminate information", *Proc. 1st ACM Symp. on Parallel Algorithms and Architectures* (SPAA'89), 1989, pp. 318-327.

[Fa80] A.M. Farley, "Minimum-Time Line Broadcast Networks", *Networks* 10 (1980), pp. 59-70.

[FHMMM92] R. Feldmann, J. Hromkovič, S. Madhavapeddy, B. Monien, P. Mysliwietz, "Optimal algorithms for dissemination of information in generalized communication modes", *Proc. Parallel Architectures and Languages Europe* (PARLE'92), Lecture Notes in Computer Science 605, Springer Verlag 1992, pp. 115-130.

[HKS93] J. Hromkovič, R. Klasing, E.A. Stöhr, "Gossiping in vertex-disjoint paths mode in interconnection networks", *Proc. 19th Int. Workshop on Graph-Theoretic Concepts in Computer Science* (WG '93), Lecture Notes in Computer Science, Springer Verlag 1993, to appear.

[KL92] M. Klawe, F.T. Leighton, "A tight lower bound on the size of planar permutation networks", *SIAM J. Disc. Math.* 5 (1992), No. 4, pp. 558-563.

[LT79] R.J. Lipton, R.E. Tarjan, "A separator theorem for planar graphs", *SIAM J. Appl. Math.* 36 (1979), No. 2, pp. 177-189.

Fully Dynamic Planarity Testing in Planar Embedded Graphs (Extended Abstract) *

Giuseppe F. Italiano[1], Johannes A. La Poutré[2] and Monika H. Rauch[3]

[1] IBM T.J. Watson Research Center, P.O. Box 704,
Yorktown Heights, NY 10598, USA
[2] Department of Computer Science, Princeton University,
Princeton, NJ 08544, USA
[3] Siemens AG, Corporate Research & Development, ZFE BT SE 14,
D-81730 Munich, Germany

Abstract. We present the first data structure to maintain an embedded planar graph under arbitrary edge insertions, arbitrary edge deletions and queries that test whether the insertion of a new edge would violate the planarity of the embedding. Our data structure supports online updates and queries on an n-vertex embedded planar graph in $O(\log^2 n)$ worst-case time, it can be built in linear time and requires linear storage.

1 Introduction

A standard model for dynamic graph problems involves a sequence of intermixed updates and queries: an update inserts or deletes an edge or isolated vertex and a query asks for certain information about a graph property. The goal is to build a data structure that can support efficiently both updates and queries. If only insertions or only deletions are permitted, then the model is called *semi-dynamic*, and if both insertions and deletions are allowed, the model is called *fully dynamic*.

We call a planar graph that is committed to an embedding *plane*, and use the term *planar* only when the embedding is allowed to change during updates. Let $G = (V, E)$ be a undirected plane graph with n vertices and m edges; planarity implies that $m \leq 3n - 6$. In recent years, there has been a considerable amount of work on dynamic algorithms for plane graphs. Tamassia and Preparata [14] gave dynamic algorithms for transitive closure on st-plane graphs. Eppstein et al. [2] showed how to maintain information about the minimum spanning forest and the connected components of a plane graph in $O(\log n)$ time per operation.

* The first author was on leave from Università di Roma.
  The research of the second author was supported by a NATO Science Fellowship awarded by NWO (the Netherlands Organization for Scientific Research), and partially supported by DIMACS (Center for Discrete Mathematics and Theoretical Computer Science - NSF-STC88-09648).
  This work was done while the third author was at Princeton University, Princeton, NJ 08544.

Hershberger et al. [11] gave an algorithm that maintains the 2-edge-connected components of a plane graph in $O(\log^2 n)$ time per operation. Rauch [12] gave an algorithm that maintains information about 2-vertex connectivity in plane graphs in $O(\sqrt{n \log n})$ time.

In all of these algorithms, insertions expect to be told in which face the edge is to be inserted. If it is not known beforehand that the edge insertion will preserve the given embedding, then the algorithms are not guaranteed to be correct. However, none of these algorithms can test whether this condition is fulfilled within the given time bounds. Indeed, although there is an $O(\log n)$ algorithm by Tamassia [13] to maintain information about the embedding of a plane graph, this algorithm is not able to carry out the full repertoire of updates, since it supports only special cases of deletions. Namely, the algorithm is committed to a precomputed st-numbering [5] of the plane graph, and is not able to support all the deletions that are incompatible with this st-numbering.

This is the motivation to study a fully dynamic algorithm that allows queries of the form "Can an edge be inserted between vertex u and v without destroying the planarity of the embedding?". We call this an algorithm for *fully dynamic planarity testing in plane graphs*. Given a plane graph $G = (V, E)$, it supports the following operations:

Insert(u, v): Insert an edge between the two vertices u and v, if this insertion preserves the planarity of the embedding.

Delete(u, v): Delete the edge (u, v).

Query(u, v): Return *true* if the insertion of an edge between the vertices u and v maintains the planar embedding of the graph and *false* otherwise.

Our data structure supports online updates and queries in $O(\log^2 n)$ time, it can be built in linear time and requires linear space. We mention that if the planar graph is allowed to change its embedding during edge insertions (that preserve planarity), this problem becomes substantially more complex, and the best bound known is $O(\sqrt{n})$ [4]. To obtain our bounds, we develop a novel application of the topology trees of Frederickson [6], augmented with additional data structures. Topology trees represent a graph in a hierarchical way by partitioning the set of vertices into subsets, called *clusters*. Similar to [11], we group neighboring edges that are incident to the same clusters into *edge bundles*. Additionally, we keep bits, called *coverage bits*, at each face that tell us which edges can be embedded inside the face, and show how to maintain these bits dynamically. For any two clusters, we maintain a list of faces into which an edge between the two clusters can be embedded. This is called a *face list*.

2 Topology trees

Our algorithm maintains information about the embedding of an arbitrary plane graph. For the ease of description, we only present our solution to connected graphs. If the graph is not connected, each connected component can be maintained as described below: whenever two components are merged together, their corresponding data structures are joined. Our algorithm uses topology trees [6, 7, 11], which give a recursive decomposition of a graph into vertex clusters. Let $G = (V, E)$ be a connected undirected plane graph. We first perform a standard transformation to convert G into a graph with maximum vertex degree 3 [9]: Suppose $v \in V$ has degree $d > 3$, and is adjacent to vertices $u_1, u_2, \ldots, u_d$ in this cyclic order. In the transformed graph, v is replaced by a *chain* of $d - 1$ dashed edges: namely, we substitute v by d vertices $v_1, v_2, \ldots, v_d$, and add edges (v_i, u_i), $1 \le i \le d$, and *dashed* edges (v_i, v_{i+1}) for $1 \le i \le d - 1$. The order of the edges at v_i is $(v_i, v_{i-1}), (v_i, u_i), (v_i, v_{i+1})$. This transformation creates a plane graph G' which fulfills the following lemma.

Lemma 1. *An edge between vertex u and vertex v can be added to G without destroying the embedding if and only if an edge can be added between a vertex of the chain of u and a vertex of the chain of v without destroying the embedding of G'.*

From now on, we assume that G has maximum vertex degree 3. Throughout the sequence of updates, we maintain a spanning tree T of G that contains all the dashed edges. The *topology tree* is a hierarchical representation of G based on T. Each level of the topology tree partitions the vertices of G into connected subsets called *clusters*. An edge is *incident* to a cluster if exactly one of its endpoints is inside the cluster. Two clusters are *adjacent* if there is a tree edge that is incident to both. The *external degree* of a cluster is the number of tree edges incident to it: any cluster will have maximum external degree 3. A cluster at level 0 is a singleton vertex. A cluster at level $i > 0$ is either (a) the union of two *adjacent* clusters of level $(i - 1)$ such that the external degree of one cluster is 1 or the external degree of both is 2, or (b) one cluster of level $(i - 1)$, if rule (a) does not apply. Notice that a cluster of level $(i - 1)$ belongs to exactly one cluster of level i, and therefore each level of the topology tree defines a partition of the vertices of G (the higher the level, the coarser the partition). There is a node in the topology tree for each cluster formed in this recursive cluster decomposition: each leaf corresponds to a clusters at level 0, and the tree node corresponding to a cluster C at level i is the parent of the cluster(s) of level $(i - 1)$ whose union produced C. Since rules (a) and (b) above reduce the number of clusters at each level by a constant fraction, the height of the topology tree is $O(\log n)$ [7].

Because of the updates on G, the spanning tree T and consequently the topology tree may change. Each topology tree update follows a common routine: we first split the clusters that contain (above others) some specific vertices, and then merge the unbroken clusters, possibly in different ways. We say that we *expand the topology tree* at a vertex u whenever we split all the clusters that contain u (corresponding to the topology tree nodes in the path between u and the tree root). We denote by C_i a cluster at level i and by $C_i(u)$ the cluster at level i that contains the vertex u. To expand the topology tree at a vertex u, we start from the root, and walk down the tree, recursively *expanding the cluster*

$C_i(u)$ at each level. If $C_i(u)$ has only one child $C_{i-1}(u)$, we replace it by $C_{i-1}(u)$. If $C_i(u)$ has two children, we replace $C_i(u)$ by its two children and connect the children appropriately by tree edges with each other and with the other clusters. In either case, we then recursively expand $C_{i-1}(u)$. We stop this expansion when a leaf is reached. A similar procedure expands the topology tree at any constant number of vertices in $O(\log n)$ time. This yields a tree of clusters linked by edges of T, called the *expansion tree*, and having $O(1)$ clusters per level. The following lemma by Hershberger et al. [11] shows that this tree is well structured.

Lemma 2. [11] *If the topology tree is expanded at a constant number of vertices, the sum of the level differences between neighboring clusters in the expansion tree is $O(\log n)$.*

3 Edge Bundles, Coverage Lists and Face Lists

In this section, we present data structures to maintain information about non-tree edges and faces of G. We first describe a data structure, called *edge bundle*, for storing non-tree edges. Our definition of edge bundles is a variation of the data structure defined in [11]. Let C be a cluster, and let e be a non-tree edge incident to C. As described below each edge is assigned a *target*. All edges incident to C are embedded in a circular sequence around C: an *edge bundle* is a maximal subsequence of edges that share the same target. We actually use two different types of targets. Usually, the target of an edge is the least common ancestor (lca) of its endpoints in the topology tree, and an edge bundle is a maximal subsequence of edges at C with the same lca. After the topology tree has been expanded and $O(\log n)$ clusters are created, the target of an edge is changed to be the one cluster (of the $O(\log n)$ created clusters) that contains the other endpoint of the edge. The edge bundles are thus broken up appropriately. The target of a bundle is the target of its edges. The first representation is called *lca-targeting*, the second *precise targeting*. Whenever we expand the topology tree, we use *precise targeting* for edge bundles by defining the targets relatively to a particular expansion of the topology tree: the resulting graph is called *cluster graph*. The data structure we use for an edge bundle is very simple: it is a record containing (1) the number of edges in the bundle, (2) the two extreme (i.e., the first and the last) edges of the bundle, and (3) the target cluster of the bundle.

Whenever we expand a cluster in the topology tree or whenever we merge two clusters together, we compute the list of edge bundles for each cluster involved. This implies that edge bundle lists are subject to splits and merges, and therefore we represent them as balanced binary trees. Recall that each edge bundle has its edge count: we store at each internal node of the balanced tree the sum of the edge counts of the leaves in its subtree. If the topology tree is expanded at a constant number of vertices, no more than $O(\log n)$ edge bundles can be in an edge bundle list, as we will show later. Thus, splitting or merging edge bundle lists can be done in time $O(\log \log n)$.

We next define two additional data structures that maintain information about the faces of G: the coverage lists and the face lists. Let C be a cluster. An edge e of T whose endpoints are both in C is said to be *internal* to C. If (u, v) is a tree edge incident to C with u in C, then u is called a *boundary vertex* of C. We define the *tree path* $p(C)$ *of a cluster* C as the unique tree path of which the end nodes are exactly (all) the boundary vertices of C. Note that if the cluster contains only one boundary vertex, the tree path of the cluster consists of this single vertex only, e.g., if C has external degree 3 or 1.

Let (a, b) be an edge incident to C and let a be contained in C. We define the *projection of a on $p(C)$* to be the vertex closest to a in T lying on $p(C)$. A face is *incident* to C if it is incident to a vertex in C and a vertex not in C. For each face incident to C we define a bit called *coverage bit*. Let (x, y) and (x', y') be the edges incident to F such that x and x' are contained in C. The *coverage bit* of F at C is set to one iff (if and only if) at least one of the vertices on the tree path between the projection of x on $p(C)$ and the projection of x' on $p(C)$ is incident to F. Otherwise it is set to zero. A face that is incident to two edges between cluster C and cluster C' is called a *good* face between C and C' iff the coverage bit at C and the coverage bit at C' is one: an edge between a vertex on $p(C)$ and a vertex on $p(C')$ can be inserted into G iff there exists a good face between C and C'.

We keep a list of the coverage bits of every face that lies between two neighboring edge bundles incident to C or between the first or last edge bundle and a tree edge incident to C. If C has external degree 1, we build one coverage list for C. If it has external degree 2, we create two coverage lists, one for each side of $p(C)$. If C has external degree 3, no coverage list is stored. The same operations as on edge bundle lists are executed on coverage lists. Thus each coverage list is stored in a balanced binary tree with the faces in counterclockwise order of their embedding at C. At each face F, we keep additionally the edge count of the edge bundle that is incident to C and following F in the counterclockwise order of the embedding at C. As in the balanced tree for the edge bundle list, each internal node of the balanced tree records the sum of the edge counts of the leaves in its subtree. Since there are $O(\log n)$ coverage bits in a coverage list (when the topology tree is expanded at a constant number of vertices), each split or merge can be supported in time $O(\log \log n)$.

Additionally, we store at each cluster C a list of good faces between C and some other cluster, called *face list*. Each face in the face list of C is incident to two neighboring edges of the same bundle between C and a cluster C' and it is a good face between C and C'. The face is said to *belong* to this edge bundle. Throughout any sequence of operations, the face list will be subject to splits (whenever an edge bundle is split into two), merges (whenever two edge bundles are joined together), and tests for emptiness (whenever during a query, we check if the edge to be inserted can be embedded inside a bundle). There can be $O(n)$ faces in a face list. To support these primitives in $O(\log n)$ time, we store each face list in a balanced binary tree with the faces in the counterclockwise order of their embedding around C. At each face F, we keep the number of edges incident to C and lying between F and the next good face of C to the left of F or the next tree edge, whichever comes first. At each internal node, we store the sum of the edges at the leaves in its subtree.

4   Recipes

Storing all the edge bundles, coverage lists, and face lists at each cluster yields too much overhead. Hence, we store this information incrementally: we do not store explicitly all these lists, but at each cluster C we store only a "procedure" that, given the lists of C, will tell us how to obtain the lists of C's children. We call this "procedure" the *recipe* of the cluster. More precisely, recipes give pointers to elements of the lists that have to be changed: we call these pointers *location descriptors*. A location descriptor is a number indicating the number of edges incident to C, that are to the left (in the embedding at C) of the edge or face pointed to. We use this definition since the number of edge bundles to the left of either an edge or a face incident to C depends on how edges are bundled together (and therefore can change throughout the algorithm), while the number of edges to their left is independent of the edge bundles. During the expansion or the merging back of the topology tree, each edge bundle list and each coverage list contains $O(\log n)$ elements. Thus it takes time $O(\log \log n)$ to find the edge or face to which a location descriptor points. Computing the location descriptor for an edge or face takes time $O(\log \log n)$ as well. Since there can be $O(n)$ faces in a face list, these operations take time $O(\log n)$ in a face list.

The recipe stored at cluster C depends on the number of children of C and on their external degrees. The constraints on the topology tree yield a total of five cases, viz., C has only one child (*Case 1*), or C has two children of which the external degrees are 3 and 1 (*Case 2*), 1 and 1 (*Case 3*), 2 and 1 (*Case 4*), or 2 and 2 (*Case 5*). We only address here the two cases that are more involved.

Case 4: C has two children with external degree 2 and 1.

Let Y and Z be the children of C, respectively of external degree 2 and 1. We remove each edge bundle between Y and Z from the edge bundle lists and the according coverage bits from the coverage lists (there are at most two such edge bundles, one on each side of the tree edge joining Y and Z). Then we delete the part of the face lists that contains faces belonging to these bundles. Afterwards we concatenate each edge bundle list of Y with one end of the edge bundle list of Z so that edge bundles that are adjacent in the embedding are adjacent in the new edge bundle list and we compute the coverage bits of the faces between them. Each such coverage bit is computed by executing an OR operation on the last coverage bit of the coverage list of one child and on the first coverage bit of the coverage list of the other child. The coverage bits are added to the beginning and to the end of the coverage list of Z. Then the coverage lists of Y and of Z and the face lists are concatenated in the same way as the edge bundle lists. If two newly neighboring edge bundles have the same target, we join them to one edge bundle in the edge bundle list. Then we remove the coverage bit of the face between them from the coverage list and add the face to the face list if it is good (we describe below how we can test for this). If Y has a self–loop, i.e. an edge bundle that is contained in both edge bundle lists, we remove it together with its corresponding bits. Note that such an edge is easy to find. If it exists, Y is the only cluster adjacent to Z. The self–loop is either adjacent to the edges between Y and Z or (if no such edges exist) it is the last edge of one edge bundle list of Y and the first edge of the other edge bundle list.

We describe the recipe for one side of the tree path of Y (the other side being symmetric). The recipe contains the number of edges incident to the left cluster (on this side if it is a degree two cluster). This number is used as a location descriptor to split the edge bundle, coverage, and face list. If there is an edge bundle between Y and Z, we keep its edge count, its extreme edges, the part of the face list which contains faces belonging to the bundle, and the coverage bits (at Y and at Z) of the face defined by the bundle and the tree edge between Y and Z. After removing the edge bundle between Y and Z, if there is no self–loop, we additionally keep the rightmost edge of the left cluster, the leftmost edge of the right cluster, the coverage bit at Y, and the coverage bit at Z of the face between the two edges. It is necessary to store the edges since splitting the cluster might cause splitting an edge bundle and the stored edges are the extreme edges of the two new bundles where the original bundle was split. If there is a self–loop, we store its edge count, its extreme edges, the part of the face list which contains faces belonging to it, and the coverage bits at both sides of Y of the face defined by the self–loop and Y.

Case 5: C has two children both with external degree 2.

Let Y and Z be the children of C. Again, Y and Z can be connected on each side of the tree path by an edge bundle, and these edge bundles have to be removed to build the lists for C, and the coverage and face lists must be updated properly. However, this time it is also possible that an edge bundle starts in Y on one side of the spanning tree path, loops around the whole tree and ends on the other side of the spanning tree in Z: we call this a *loopy* edge bundle. A loopy edge bundle splits the plane into two non-empty parts (each contains at least one cluster). We cannot remove it, since it is the witness that no edge between the inside and the outside clusters can be inserted without violating the planarity of the embedding. Thus, we keep it as long as it separates inside and outside clusters, and remove it only afterwards (as a self–loop in Case 4). Next, we concatenate the bundle lists on each side of the spanning tree path with each other and determine the coverage bits for the faces between them. Afterwards, we add the new coverage bits to the coverage lists and concatenate them and also the face lists appropriately. If two adjacent edge bundles have the same target, they are joined into one edge bundle. Additionally, we remove the according coverage bits from the coverage list and record their faces in the face lists if they are good (we describe below how to test for goodness).

We describe next what we store in the recipe for one side of the tree path of C, the recipe for other side can be found symmetrically. W.l.o.g., let Y be the left cluster on this side. The recipe keeps the number of edges incident to Y. This number is used as a location descriptor to split the edge bundle, the coverage, and the face list when C is expanded. If there is an edge bundle between Y and Z, the recipe also contains this edge bundle (its edge count and extreme edges), the part of the face list which contains faces that belong to the bundle, and the coverage bits (at Y and at Z) that belong to the face between the bundle and the tree edge between Y and Z. After removing this bundle from the edge bundle lists of Y and Z, we store the rightmost edge of Y and the leftmost edge of Z in the recipe. We also keep the coverage bits that are stored at Y and at Z for the face between these two edges. If expanding C splits an edge bundle (on this side of $p(C)$), these edges are extreme edges of the new bundles.

This concludes the case description.

During each merge we might have to test whether up to two faces are good faces. Each face is incident to the two clusters to be merged. To test the goodness of a face we check the coverage bits of all the clusters in the cluster graph that is incident to the face and not contained in the two clusters. Since there are $O(\log n)$ clusters in the cluster graph, this takes time $O(\log n)$.

The creation of the list for C from the lists for C's children requires either the join of two edge bundle, coverage, and face lists (in Case 4 or 5) or the creation of a new coverage list with $O(\log n)$ entries (in Case 2, not described). Concatenating two edge bundle lists or two coverage lists requires to merge two balanced binary trees of size $O(\log n)$ and thus takes time $O(\log \log n)$. Each insertion into a face list and each join of two face lists takes time $O(\log n)$, since $O(n)$ faces can be stored in a face list. Deciding whether a face is good takes time $O(\log n)$ as described below. Thus two clusters can be merged in time $O(\log n)$ to form their parent. A constant number of edge bundles is created during each recipe evaluation, and the target of each edge bundle is set to be the node at which the bundle was created (this node is the lca of the endpoints of the bundle).

Lemma 3. *The expansion of the topology tree at a constant number of vertices takes $O(\log^2 n)$ time and creates $O(\log n)$ edge bundles and coverage bits.*

After the expansion of the topology tree at a constant number of vertices, we have to modify the target of the edge bundles from lca targeting to precise targeting. This process is called *retargeting*. It can be shown that retargeting takes time $O(\log n)$ plus the time for splitting $O(\log n)$ edge bundles. Each split of an edge bundle causes an insertion in an edge bundle list and in a coverage list and a possible deletion in a face list. The insertions take time $O(\log \log n)$, while a deletion takes time $O(\log n)$. Additionally, for each cluster C at which an edge bundle is split and a new face F is created, we have to compute the coverage bit for F at C. Let (x, y) and (x', y') be the two edges of the split bundle which are incident to F. These edges are known since we keep at each edge bundle its extreme edges. Further, let x and x' be in C. To find the coverage bit for F, we run up the path in the topology tree from x and from x' until we find the first node (cluster) which contains both x and x'. The edge bundle containing (x, y) and the edge bundle containing (x', y') cannot be merged together before this node, and they are merged together at C. Thus they are combined either at the lca of x and x' or at one of its ancestors in the topology tree (up to C). At the node C' at which they are merged, the coverage bits of F for both children of

C' are stored. Taking the OR on the two bits gives us the coverage bit for F at C. The node C' can be found by testing whether (x, y) and (x', y') are stored in the recipe starting from the lca of x and x' and going up the topology tree until such a node is found. This takes time $O(\log n)$ per split edge bundle. Since there are $O(\log n)$ edge bundles split altogether, the following lemma is established.

Lemma 4. *The cost of expanding the topology tree at a constant number of vertices to obtain a cluster graph is $O(\log^2 n)$.*

5 Queries

To decide whether a new edge (u, v) between u and v destroys the planar embedding of the graph, we expand the topology tree at the first and last nodes of the dashed chains of u and of v, which results in a cluster graph. By Lemma 3, this graph has $O(\log n)$ clusters, edge bundles, and faces. Then we split the face list(s) at each cluster according to edge bundles. Thus each edge bundle receives a list of good faces that are incident to two edges of the bundle. In the following, a cluster that contains one of the vertices on the dashed chain of x is called a x-*cluster*. Since we expand the extreme nodes of the dashed chain of u, all vertices on the tree path of any u-cluster are vertices of the dashed chain of u. Thus all faces in the face list of an edge bundle incident to a u-cluster are incident to u. This implies that if there is an edge bundle between a u-cluster and a v-cluster and the face list of this bundle is not empty, then there exists a face which is incident to u and to v. Hence, (u, v) can be inserted without destroying the planar embedding. In the other direction, if the face list of such an edge bundle is empty, then (u, v) cannot be embedded into any face belonging to the bundle. If there exists a face F in the cluster graph which is incident to a u-cluster and incident to a v-cluster, then the coverage bit of F at the u-cluster and at the v-cluster is one iff (u, v) can be inserted in F. Thus, (u, v) does not destroy the planar embedding in G iff there exists a face in the cluster graph whose coverage bit at the u-cluster and at the v-cluster is one or there exists an edge bundle between a u-cluster and a v-cluster whose face list is nonempty. Since there are $O(\log n)$ faces and edge bundles in the cluster graph and each can be tested in constant time, we can decide whether (u, v) can be inserted in time $O(\log n)$. Then we merge the topology tree together by just reversing the expansion.

Theorem 5. *We can determine whether an edge between vertices u and v is in accordance with the planar embedding in $O(\log^2 n)$ time.*

6 Updates

In this section we explain the changes to our data structure during graph updates (edge insertions and deletions). Each update consists of three steps. First, the topology tree is expanded at a constant number of vertices to create the cluster graph as in Section 2. Next, we change the targets at the edge bundles to become precise targets, and we insert or delete the edges or vertices in the resulting graph (this part depends on what operation we are executing). Finally, we merge the topology tree back together: the details of the merge will appear later.

To insert an edge (u, v), we consider the following cases: Let x stand for either u or v. If the degree of x before the insertion was 1 or 2, we expand the topology tree at x, create the cluster graph, and give x a new non-tree edge. The new edge is an edge bundle whose face list is empty. Since an edge can only be inserted in a face whose coverage bit is one, the coverage bit of both faces incident the new edge is one. These bits are added appropriately to the coverage list of x. If the degree of x before the insertion was three or more, we must either create or update the dashed chain for x and connect it appropriately. This might cause an edge bundle to be split, and new faces and new edge bundles to be created. With some care, this update might be performed within our time bounds. After making the appropriate modifications at u and v in the cluster graph, we merge the topology tree back together.

Deleting a non-tree edge (u, v) is essentially the inversion of an insertion. For each endpoint x of (u, v) we do the following. If the degree of x before the deletion was at most 3, we expand the topology tree at x, build the cluster graph, and remove the non-tree edge from x. This creates a new face whose coverage bit is one, since both u and v are incident to it. We update the coverage lists of x appropriately. If the degree of x before the insertion was 4 or more, we must either remove or update the dashed path for x. This might cause the merging of edge bundles and require building new face lists, and update the coverage lists, but again it can be handled. After making the appropriate modifications at u and v in the cluster graph, we merge the topology tree back together.

The deletion of a tree edge (u, v) divides the spanning tree T into two parts. We expand the topology tree at both endpoints and create the cluster graph. By processing the edge bundles incident to one of the two faces adjacent to (u, v), we find an edge bundle that connects the two parts of the spanning tree. This can be done in time $O(\log n)$. Now we recursively expand the clusters incident to the edge bundle at the extreme edge (u', v') incident to the face. We expand the topology tree at u' and v' and make (u', v') into a tree edge and (u, v) into a non-tree edge. Then we continue as for the deletion of a non-tree edge.

It remains to show how the clusters of an expanded topology tree are merged back together after an edge insertion or deletion. We do this in three steps. First, we compute the new topology tree for the updated cluster graph which takes $O(\log n)$ time. Second, we switch from precise targeting to lca targeting based on the new topology tree and update the data structure appropriately. This step takes time $O(\log^2 n)$. Third, we compute a recipe for all those clusters that have either an outdated recipe or no recipe at all. We show that in this step we spend amortized time $O(\log n)$ per cluster. Since there are $O(\log n)$ such clusters, the update time will be $O(\log^2 n)$.

In step one we build a new topology tree using the already existing clusters. As shown by Frederickson [7], this requires the expansion of $O(\log n)$ additional clusters, and can be done in $O(\log n)$ time. The new topology tree is used in step two to compute the lca targets. We first compute the precise targets for the newly created edge bundles, which takes time $O(\log^2 n)$. Then we compute the lca of all the edge bundles in the cluster graph based on the new topology tree. This can be done in time $O(\log n \log \log \log n)$ [10]. Adjacent edge bundles with the same target are merged together. This requires removing $O(\log n)$ old edge bundles from the edge bundle list and inserting $O(\log n)$ new bundles. Additionally, $O(\log n)$ coverage bits are deleted from the coverage lists and $O(\log n)$ faces are inserted into the face lists. To determine whether a face should be inserted into the face list we check whether the face is a good face in the same way as above. This takes time $O(\log n)$ per face. Since each operation in a edge bundle or coverage list takes time $O(\log \log n)$ and each operation in a face list takes time $O(\log n)$, the time for updating the data structure because of the lca targeting is $O(\log^2 n)$.

To perform step three we use the notion of *fringe*. The fringe consists of all the nodes in the new topology tree that correspond to clusters in the cluster graph. Note that if a cluster C is below the fringe, it was not expanded, and consequently the edges incident to C did not change. Thus all the location descriptors of C are still valid (recall that a location descriptor is defined independently on how edges are bundled together). The other information in the recipe (e.g., internal edge bundles, extreme edges and coverage bits) is also independent of how the edges leaving C are bundled together, and thus does not depend on changes that happened above C. Consequently, every node below the fringe has a correct recipe, while nodes above the fringe can have either an invalid recipe or no recipe at all (and in both cases a new recipe must be created).

We will create the recipes for the nodes on or above the fringe in a bottom–up fashion, combining for each cluster the lists of its children, and then recording how to reverse this operation in the recipe. At the beginning, each cluster of the partially expanded topology tree has one or two (edge bundle, coverage, and face) lists. Since the edge bundles are lca-targeted, at constant cost per edge bundle we compute for each edge bundle the depths of its targets in the topology tree. We compute the edge bundle, coverage and face lists for the nodes by walking up from the fringe to the root of the topology tree, at each step merging neighboring clusters. To merge two clusters we maintain a data structure at each node on or above the fringe to identify the bundles of a given depth at each cluster along a fringe–to–root path. We store an array of length $O(\log n)$, indexed by depth, of the edge bundles incident to the node (cluster). Each array entry contains a pointer to a list of bundles with the same depth. We also record the total number of edge bundles in the array and maintain a list of the indices of non-empty array fields. We initialize the array at each cluster on the fringe in time $O(\log n)$, then we maintain it as we walk toward the root.

When two clusters C and C' at depth i are merged, we check the edge bundles at depth $(i + 1)$ in the array at either cluster. Since the edge bundles are lca-targeted, the target of all these edge bundles is the parent of C and C'. Note that these are exactly the edge bundles that connect C and C' and that at most three of them are incident to each cluster. By climbing to the root of the balanced tree in the edge bundle list, we can determine for each edge bundle in time $O(\log \log n)$ to which side of $p(C)$ or $p(C')$ it belongs to. With this information we can find a loopy edge between C and C' in constant time if one exists. Now the edge bundle, coverage, and face lists of C and C' are merged according to one of the five cases in Section 4. This takes time $O(\log n)$. Afterwards the edges of the cluster with the fewer edges are copied into the array of the other cluster and the index list is updated appropriately. Since there are $O(\log n)$ edge bundles altogether, each edge bundle is copied at most $O(\log \log n)$ times. Thus the total cost for copying during all merges of two clusters is $O(\log n \log \log n)$. Since the number of edge bundles merged is $O(\log n)$, the cost of merging the topology tree back is $O(\log^2 n)$.

Theorem 6. *There exists a data structure for fully dynamic planarity testing in plane graphs that requires time $O(\log^2 n)$ for any query and for any edge insertion or deletion.*

References

1. A. V. Aho, J. E. Hopcroft, J. D. Ullman, "The Design and Analysis of Computer Algorithms" *Addison-Wesley*, Reading, Massachusetts, 1974.
2. D. Eppstein, G. F. Italiano, R. Tamassia, R. E. Tarjan, J. Westbrook, M. Yung, "Maintenance of a Minimum Spanning Forest in a Dynamic Plane Graph" *J. Algorithms*, 13 (1992), 33–54.
3. D. Eppstein, Z. Galil, G. F. Italiano, A. Nissenzweig, "Sparsification - A Technique for Speeding Up Dynamic Graph Algorithms" *Proc. 33rd FOCS*, 1992, 60–69.
4. D. Eppstein, Z. Galil, G. F. Italiano, T. H. Spencer, "Separator Based Sparsification for Dynamic Planar Graph Algorithms" *Proc. 25th STOC*, 1993, 208–217.
5. S. Even, R. E. Tarjan, "Computing an *st*-numbering" *Theoretical Computer Science*, 2 (1976), 339–344.
6. G. N. Frederickson, "Data Structures for On-line Updating of Minimum Spanning Trees" *SIAM J. Comput.* 14 (1985), 781–798.
7. G. N. Frederickson, "Ambivalent Data Structures for Dynamic 2-edge-connectivity and k smallest spanning trees" *Proc. 32nd FOCS*, 1991, 632–641.
8. Z. Galil, G. F. Italiano, "Maintaining Biconnected Components of Dynamic Planar Graphs" *Proc. 18th ICALP*, LNCS, Springer-Verlag, 1991, 339–350.
9. F. Harary, *Graph Theory*, Addison-Wesley, Reading, MA, 1969.
10. D. Harel, R. E. Tarjan, "Fast Algorithms for Finding Nearest Common Ancestors" *SIAM J. Comput.* 13 (1984), 338–355.
11. J. Hershberger, M. Rauch, S. Suri. "Fully Dynamic 2–Edge–Connectivity in Planar Graphs" *Proc. 3rd Scandinavian Workshop on Algorithm Theory*, LNCS 621, Springer-Verlag, 1992, 233–244.
12. M. Rauch, "Fully Dynamic Biconnectivity in Graphs" *Proc. 33rd FOCS*, 1992, 50–59.
13. R. Tamassia, "A Dynamic Data Structure for Planar Graph Embedding", *Proc. 15th ICALP*, LNCS 317, Springer-Verlag, 1988, 576–590.
14. R. Tamassia, F. P. Preparata, "Dynamic Maintenance of Planar Digraphs, with Applications", *Algorithmica*, 5 (1990), 509–527.

Fully Dynamic Algorithms for Bin Packing: Being (Mostly) Myopic Helps[1]

Zoran Ivković   Errol L. Lloyd

{ivkovich,elloyd}@cis.udel.edu

Department of Computer and Information Sciences, University of Delaware
Newark, DE, 19716

Abstract. The problem of maintaining an approximate solution for *one–dimensional bin packing* when items may arrive and depart dynamically is studied. In accordance with various work on fully dynamic algorithms, and in contrast to prior work on bin packing, it is assumed that the packing may be arbitrarily rearranged to accommodate arriving and departing items. In this context our main result is a *fully dynamic* approximation algorithm for bin packing that is $\frac{5}{4}$–competitive and requires $\Theta(\log n)$ time per operation (i.e. for an *Insert* or a *Delete* of an item). This competitive ratio of $\frac{5}{4}$ is nearly as good as that of the best practical off–line algorithms. Our algorithm utilizes the technique (introduced here) whereby the packing of an item is done with a total disregard for already packed items of a smaller size. This *myopic* packing of an item may then cause several smaller items to be repacked (in a similar fashion). With a bit of additional sophistication to avoid certain "bad" cases, the number of items (either individual items or "bundles" of very small items treated as a whole) that needs to be repacked is bounded by a constant. Further, in the case where there are no departures of items (only dynamic arrivals), we provide a polynomial time approximation scheme such that for any competitive ratio exceeding 1, there is an algorithm having that competitive ratio, and a amortized running time of $\Theta(\log^2 n)$ per *Insert* operation.

1   Introduction

In the (one–dimensional) bin packing problem, a list $L = (a_1, a_2, ..., a_n)$ of items of size $size(a_i)$ in the interval $(0,1]$ is given. The goal is to find the minimum k such that all of the items a_i can be packed into k unit–size bins. That is, for each bin B_i, the sum of the sizes of the items packed into that bin should not exceed 1. Since bin packing is a well known problem, we provide only essential definitions. The reader is referred to [2] for background information and survey of bin packing, together with a number of applications.

For the past quarter century, bin packing has been a central area of research activity in the algorithms and operations research communities. Despite its advanced age (bin packing was one of the original *NP*–complete problems from [14]), bin packing has retained its appeal by being a fertile ground for the study of *approximation algorithms* (more than a decade ago, bin packing was

---

[1] Partially supported by the National Science Foundation under Grant CCR-9120731

labeled "The Problem That Wouldn't Go Away" [2]). While early studies of bin packing focused primarily on off–line and simple on–line algorithms for the one–dimensional problem (and many variants), some recent attention has been devoted to **on–line** and **dynamic** versions [1, 2, 6]. In this paper, we extend these notions to their full generality, by considering **fully dynamic** bin packing, where:

- items may arrive and depart from the packing dynamically, and

- items may be moved from bin to bin as the packing is adjusted to accommodate arriving and departing items.

Each of the earlier works on on–line bin packing differ from the notion of fully dynamic bin packing in one of two ways: either they do not allow an item to be moved from a bin (this has a predictably negative effect on the achievable quality of the packing), or they restrict themselves to dynamic arrivals of items – there are no departures.

In general, *fully dynamic* algorithms are aimed at situations where the problem instance is changing (slowly) over time. This situation occurs often in interactive design processes, and fully dynamic algorithms incorporate these incremental changes without any knowledge of the existence and nature of future changes. The objective of course is to develop fully dynamic algorithms that are "competitive" with existing off–line algorithms [3, 4, 5].

Although the bulk of the existing work on fully dynamic algorithms has been directed toward problems known to be in *P*, some recent attention has been paid to fully dynamic *approximation* algorithms for problems that are *NP*-complete [8, 15]. In this case, being competitive with off–line algorithms means that the quality of the approximation produced by the fully dynamic approximation algorithm should be as good as that produced by the off–line algorithms. Further, the running time per operation (i.e. change) of the fully dynamic algorithm should be as small as possible.

1.1 Bin packing – existing results

The usual measure of the quality of a solution produced by a bin packing algorithm A is its **competitive ratio** $R(A) = \lim_{n \to \infty} max_{OPT(L)=n} \frac{A(L)}{OPT(L)}$, where $A(L)$ and $OPT(L)$ denote the number of bins used for packing of the list L by A and some optimal packing of L. Here, we say that A is **R(A)–competitive**.

In the domain of off–line algorithms, the value of R has been successively improved [2, 12]. Indeed, it has been shown that for any value of $R \geq 1$, there is an $\mathcal{O}(n \log n)$ time algorithm with that competitive ratio [13]. Among algorithms of practical importance, the best result is an $\mathcal{O}(n \log n)$ algorithm for which R is $\frac{71}{60}$ [12].

With respect to on–line bin packing, the problem has been defined strictly in terms of arrivals (*Inserts*) – items never depart from the packing (i.e. there are no *Deletes*). Further, most on–line algorithms have operated under the restriction that each item must be packed in some bin, and it should remain in that bin permanently. In this context, it is shown that for every on–line linear time algorithm A, $R(A) \geq 1.536...$ [2]. Further, the upper bound has been improved over the years to roughly 1.6 [16].

The recent work of [6] focused on a variant of on–line bin packing, again supporting *Inserts* only, in which each item may be moved a constant number of times (from one bin to another). They provide two algorithms: One with a linear running time (linear in n, the number of *Inserts*, which is also the number of items) and a competitive ratio of 1.5, and one with an $\mathcal{O}(n \log n)$ running time and competitive ratio of $\frac{4}{3}$.

Another notion that is related to, but distinct from, fully dynamic bin packing is that of [1], where each item is associated not only with its size, but also with an arrival time and a departure time (interpreted in the natural way). Here, again (and unlike [6]), items cannot be moved once they are assigned to some bin, unless they depart from the system permanently (at their departure time). This variant differs from fully dynamic bin packing in that items are not allowed to be moved once assigned, and through the departure time information. It was shown in [1] that for any such algorithm A, $R(A) \geq 2.5$, and that for their Dynamic *FF* (First Fit), $2.770 \leq R(FF) \leq 2.898$.

1.2 Competitive ratio and running time for fully dynamic approximation algorithms

In this section we discuss the notions of competitiveness and running time in the context of developing fully dynamic approximation algorithms.

We begin by noting that with respect to the definition of *competitive ratio*, there is no need to make a distinction between fully dynamic and off–line algorithms. In each case, these measures reflect the size of the lists produced by the algorithm relative to the size of optimal packings.

With respect to running times, we say that a fully dynamic approximation algorithm B for bin packing has **running time** $\mathcal{O}(f(n))$ if the time taken by B to process a change (an *Insert* or *Delete*) to an instance of n items is $\mathcal{O}(f(n))$. If $\mathcal{O}(f(n))$ is a *worst case* time bound, then B is **uniform**. If $\mathcal{O}(f(n))$ is an *amortized* time bound, then B is **amortized**.

Our general goal in developing fully dynamic approximation algorithms for bin packing is to design algorithms having competitive ratios close to those of the best off–line algorithms. Note however, that this goal, by itself, is not particularly interesting: such a fully dynamic algorithm can always be produced simply by executing an off–line algorithm after every change. In the case of bin packing this would mean that a change to an instance of size n would require time $\Theta(n \log n)$. This is clearly not what is intended nor required in the fully dynamic context.

Thus, in addition to producing algorithms with competitive ratios close to those of the best off–line algorithms, we also require that our algorithms process changes quickly relative to the size of the instance. Of particular interest are algorithms that are, in a sense, the best possible relative to the existing off–

line methods. For bin packing the best known off–line algorithms require time $\Theta(n \log n)$. Thus, a fully dynamic algorithm that runs in time $\Theta(\log n)$ per operation is, in that sense, the best possible. Indeed, the fully dynamic algorithms that we give in sections 2 and 3 run in precisely this time per operation.

1.3   Additional operations

As noted earlier, this paper studies fully dynamic approximation algorithms for bin packing. That is, the algorithms that we present process a sequence of *Inserts* (arrivals) and *Deletes* (departures) of items. Further, all of our algorithms are designed to handle "lookup" queries of the following form. These queries may be interspersed in the *Insert/Delete* sequence:

- *size* – returns in $\mathcal{O}(1)$ time the number of bins in the current packing;

- *packing* – returns a description of the packing in the form of a list of pairs $(x, \text{Bin}(x))$, where $\text{Bin}(x)$ denotes the bin into which an item x is packed, in time linear in the number of items in the current instance.

1.4   What's to come

With the preliminaries concluded, the remainder of the paper is organized as follows: In the next section we review the basic ideas of Johnson's *grouping* [9, 10], and sketch a simple, $\frac{4}{3}$–competitive fully dynamic algorithm derived from the standard off–line algorithm *FFG*. In section 3, we provide the main result. Namely, a fully dynamic algorithm *MMP* that is $\frac{5}{4}$-competitive and requires $\Theta(\log n)$ time per operation. Note that relative to the best off–line algorithms, *MMP* has a running time that is best possible, and it has a competitive ratio that is nearly the equal of the best practical off-line algorithms. This is a surprising result even in terms of off–line bin packing, since it is the first practical bin packing algorithm having a competitive ratio less than $\frac{4}{3}$, that does not rely on packing the items in sorted order (as discussed in section 2, dynamically maintaining a packing based on a sorted list is problematic). That the algorithm is fully dynamic is all the more remarkable.

We note that both the algorithm and the analysis of its competitive ratio are decidedly nontrivial, with the analysis utilizing a specific counting technique to analyze what a "worst–case" situation for the algorithm may constitute. In section 4, we consider the restricted case of "insertions only". Here, the algorithms of sections 2 and 3 are, of course, applicable. Further, we show how to adapt the algorithms of [13] to produce fully dynamic approximation algorithms for any $R > 1$, and having an $\Theta(\log^2 n)$ amortized running time per operation. We note here that the competitive ratio of each algorithm we present is, in fact, tight.

2 Towards full dynamization of bin packing

Motivated by the notion of approximation–competitiveness introduced in the preceding section, a natural approach to the development of fully dynamic bin packing algorithms is to adapt existing methods to work in the dynamic situation. Unfortunately, this is easier said than done. The difficulty is that most of the off–line algorithms perform bin packing in two distinct stages. First, there is a **preprocessing stage** in which the items are organized in some fashion (this reorganization should have a positive effect on the resulting packing). This is followed by a **packing stage** where the actual packing is accomplished. In the off–line situation, this two stage approach is quite natural since the entire list of items is available from the outset. However, in a dynamic environment a two stage process becomes awkward. Consider for example, the algorithm First Fit Decreasing (*FFD*), which is $\frac{11}{9}$-competitive. This algorithm first sorts the items and then packs them in order of decreasing size, using the First Fit packing rule[2]. What about a fully dynamic version of *FFD*? There is, of course, no difficulty in maintaining a sorted list of the elements. *But*, there is a great difficulty in maintaining the packing based on that sorted list, since the insertion (or deletion) of a single item can result in a large number of changes to that packing. It would seem that the packing induced by the sorted list of items is "too specific" to be maintained dynamically, and that perhaps a less specific rule might be of use. Indeed, in this section we consider a weaker notion of *grouping* [9, 10]. In particular, we give a sketch of a fully dynamic bin packing algorithm that is derived from, and is approximation–competitive with the (standard) bin packing algorithm *FFG* (First Fit Grouped).

2.1 Some definitions

Before proceeding, we require a few definitions that will be used throughout the remainder of the paper. In particular, an item a is a B–item (big) if $size(a) \in (\frac{1}{2}, 1]$; a L–item (large) if $size(a) \in (\frac{1}{3}, \frac{1}{2}]$; a S–item (small) if $size(a) \in (\frac{1}{4}, \frac{1}{3}]$; a T–item (tiny) if $size(a) \in (\frac{1}{5}, \frac{1}{4}]$; a M–item (miniscule) if $size(a) \in (0, \frac{1}{5}]$.

Finally, a bin is a B–bin (L,S,T,M) if its largest item is a B–item (L,S,T,M).

2.2 A Grouping Rule Algorithm

Here we provide a brief sketch of the main ideas behind the standard off–line algorithm *FFG*. Details may be found in [9, 10]. There, Johnson considered ways to avoid sorting when preprocessing, and still define a preprocessing rule that would, coupled with a suitable packing rule, have a good competitive ratio. The preprocessing rule proposed, *Grouping–X Rule* runs in time linear in the length of L. Informally, this rule partitions the list into the sublists of B, L, S, T, and M items, and forms a preprocessed list L' that consists of the B–items of L, followed by the L–items of L,..., followed by the M–items of L.

---

[2] Informally, bins are ordered from left to right and an item is packed into the leftmost bin into which it would fit.

The algorithm *FFG* [9, 10] then packs L' via the standard First Fit rule.

Fact 1 FFG *is $\frac{4}{3}$-competitive, and runs in time $\Theta(n \log n)$, where n is the number of items to be packed. [9, 10]*

2.3 A Fully Dynamic Algorithm

The basic ideas involved in designing a fully dynamic algorithm based on *FFG* are outlined here.

We begin by supposing that an item a needs to be inserted using the fully dynamic counterpart of *FFG*. Based on the *Grouping-X rule*, when item a is being packed, that packing should be insensitive to previously packed items of "smaller" types than a. Thus, what would a "see" in the bins? Only the items of its own type or of "larger" type. In this sense, items are myopic in that they can "see" relatively large items, and cannot "see" relatively small ones.

For example, when a B–item needs to be inserted, its vision of the existing packing is such that it "sees" only B–items. This results in inserting that B–item into a bin with no B–items, which possibly "evicts" some of the "smaller" type items. Those items need to be eventually reinserted. Thus, in a similar fashion, the packing these items "see" consists only of items of their own or of "larger" type. Of course, these reinsertions may themselves induce additional reinsertions, and so on. Finally, in order to avoid an excessive number of trivial reinsertions, "little" items are "bundled" together, and are moved from bin to bin as a whole (a more precise description of bundling is postponed until the following section).

Although it appears that this scheme (even with bundling) may allow an unacceptable amount of repacking, this is not the case. In particular, it can be shown that no more than a fixed constant number of items (either individual items or "bundles" of very small items treated as a whole) will need to be repacked as a result of a single *Insert*.

Deletions are implemented in a simple way: the bin in which the item resides is simply emptied, "shut down" (i.e. it never gets touched again, and is not considered to be a part of the packing), and its content, except for the deleted item, are reinserted. Once again, it can be shown that at most a fixed constant number of items will need to be repacked.

Due to space constraints a more detailed description of is omitted, and the following theorem is stated without proof:

Theorem 1 *The fully dynamic version of* FFG *is $\frac{4}{3}$-competitive, and runs in time $\Theta(\log n)$ per* Insert/Delete *operation.*

3 The main result - Mostly Myopic Packing

In this section we describe a fully dynamic algorithm, Mostly Myopic Packing (*MMP*), that is $\frac{5}{4}$-competitive, and that requires time $\Theta(\log n)$ per *Insert/Delete* operation. This algorithm is nearly approximation–competitive with the best practical off-line algorithms. Recall that the algorithm of [12] is $\frac{71}{60}$–competitive, and that *FFD* is $\frac{11}{9}$–competitive. Simple arithmetic shows that our algorithm comes within $\frac{1}{15}$ of the first, and within $\frac{1}{36}$ of the second. This is surprising since both of these off-line algorithms produce packings based on the use of sorting as their preprocessing rule, and, as noted in the previous section, such packings are not conducive to dynamic construction. As an alternative, our method incorporates not only myopic packing, and the notions of grouping and "bundling", but also a careful avoidance of the situations where simpler grouping based algorithms perform poorly. Such situations arise when a sequence of L and S–items are packed into a series of bins containing two L–items, followed by a series of bins that contain three S–items, even though almost all of the items could have been packed into bins containing two L–items and one S–item. Our algorithm will take some pains to be guaranteed of packing a certain portion of these L and S–items into bins containing two L–items and one S–item (we will call these "coalitions").

3.1 Algorithm

Here we describe the main ideas in the *MMP* method. Our goal is to provide an intuitive understanding of those ideas, rather than to provide every detail.

By way of preliminaries, the following ordering of types is assumed: $B \leq L \leq S \leq T \leq M$. Bins containing one B and one L–item, and no more B, L, S, or T–items, will be called bins of type BL[3]. Bins of type BST,BS,BTT,...,LLS,...,TT,T are defined analogously. The relation of **superiority** is defined as the lexicographical ordering over types of bins. For example, a bin of type BL is superior to a bin of type BST.

Finally, bins are assumed to be numbered in such a way that every bin has a unique number with the property that for any two bins the one with the lower number was opened first. For conceptual purposes, we assume that the bins are in increasing order (by number) from left to right.

Now we consider how to *Insert* an item. This is done using three major ideas:

myopic packing – as in the previous section, a K–item a "sees" a packing containing only the items of type K or larger. Based on that view of the packing, a is packed in a First Fit fashion (in a's "K or larger" world). Let B_i be the bin into which a was packed. As in the previous section, this packing of a may result in a forceful eviction of items of smaller types from B_i. Prior to the reinsertion of these items (and unlike the previous section), an attempt is made to pack additional items into B_i so that it is the most *superior* bin type possible. For example, suppose B_i contained one B–item and two T–items, and that a is a S–item. Packing a into B_i would evict both T–items from B_i leaving that bin with a B–item and a (S–item). The construction of the most superior bin type possible would mean, of course, an attempt to locate some T–item that would fit into B_i, thereby constituting a bin of type BST. In order to ensure an $\Theta(\log n)$

---

[3] Note that such bins may contain M–items. However, accounting for M–items will have no substantive effect on the competitive ratio of *MMP*.

running time, a suitable T–item is sought only in the set of currently displaced T–items (at present we have two such items – namely the two evicted from B_i), or in the bins whose type is **inferior** (to be interpreted naturally in the light of the relation of superiority) to the desired type (BST in our example). Here, if an item is taken from some bin, that bin is "shut down", and its contents, together with the content of all the items not currently inserted (initially the two T–items), are reinserted. Intuitively, the discipline of "touching" only the inferior types of bins provides for the desired running time.

LLS–coalitions – The myopic packing method described above behaves rather nicely, **except** when packing a sequence of L–items, followed by a sequence of S–items, if the items arrive in a particularly undesirable order. It may happen that, although almost all the bins constructed could have been of type LLS, the actual packing contains a sequence of bins of type LL, followed by bins of type SSS. This particular situation is in fact the worst possible. To avoid this situation, an amendment is made to the myopic discipline outlined above. Namely, upon arrival, an L–item is authorized (before attempting the usual myopic steps) to try to form a "coalition" with another L–item and an S–item. The latter two can be sought in any bin whose type is inferior to LLS. Similarly, an S–item will, before attempting its regular routine, seek two L–items coming from the bins whose type is inferior to LLS. If two such items are found, a "coalition" is formed. A careful implementation can guarantee that the added complexity of this mostly myopic discipline does not asymptotically add to the running time.

bundling – A key ingredient of our algorithm is that the insertion of an item into a bin may cause the reinsertion of a number of items that were previously packed into that bin. However, since we are interested in performing each operation in time $\Theta(\log n)$, doing these reinsertions one at a time becomes a problem in the case of miniscule items, since an unbounded number of miniscule items may need to be reinserted as the result of a single *Insert* of an item of a larger type. Thus, the idea here is that the M–items in each bin are collected into **bundles** g_i so that: $\forall g_i, size(g_i) \leq \frac{1}{5}$, where $size(g_i) = \sum_{a \in g_i} size(a)$, and for any two bundles in the same bin, their total size exceeds $\frac{1}{5}$.

In the remainder of this paper any reference to M–items implicitly refers to bundles of M–items. This bundling allows for fast and easy manipulation of the M–items, since only the movements of the entire bundles of items are going to be utilized. Note that a bin can contain no more than 9 such bundles, hence no bin can contain more than 9 items.

It is natural to ask whether bundling might be avoided. In section 3.3 we show that it is *not* possible to obtain a competitive ratio of less than $\frac{4}{3}$ unless some bundling (of a non–constant number of items) is utilized.

The details of *Delete* are omitted. However, *Delete* can be appreciated from the standpoint of *Insert*: take the bin in which the item that needs to be deleted is packed; discard that item; take the remaining content of that bin, shut the bin down, and reinsert the remaining content again. The following Lemma is stated without proof:

Lemma 1 MMP *can be implemented to run in time* $\Theta(\log n)$ *per* Insert/Delete *operation.*

3.2 Proof of the competitive ratio

Next, the sketch of the proof of the upper bound on the competitive ratio ($\frac{5}{4}$) is presented. Lower bound examples with a $\frac{5}{4}$ ratio can be given, hence this bound is tight.

The basic idea behind this proof is that we pick an arbitrary optimal packing of a given list L, and perform an extensive analysis of the structure of that packing (in terms of the different bin types and their respective multiplicity). Based on that analysis, we derive lower bounds of quantities L_B, S_B, T_B, and N_{LLS} (see Definition 3). These bounds, coupled with other counting arguments, lead to an analysis by cases. For each of the cases, the competitive ratio of $\frac{5}{4}$ is then proved by algebraic manipulation.

We begin with several preliminary results. The proofs are omitted, as are several additional technical lemmas.

Lemma 2 MMP *will pack* k *L–items,* l *S–items, and* m *T–items into no more than* $\frac{k}{2} + \frac{l}{3} + \frac{m}{4} + 3$ *bins.*

The following Theorem plays a key role:

Theorem 2 *If some optimal packing of* L *contains precisely* α *bins of type* t, t = BL (BS, LLS), *then any* MMP *packing of* L *must contain at least* $\lceil \frac{\alpha}{c} \rceil$ *bins of type* t, *where* $c = 2$ *(2,3).*

From the construction of *MMP* it follows that at any point in time there will be no M–bins, unless all the other bins are "filled", i. e. for each of those other bins B, $gap(B) < \frac{1}{5}$. This suggests that the proof of the following theorem will suffice to establish the competitive ratio:

Theorem 3 *For any list* $L = (a_1, a_2, ..., a_n)$, *where* $size(a_i) > \frac{1}{5}$, $(1 \leq i \leq n)$ *the following holds:*

$$MMP(L) \leq \frac{5}{4}OPT(L) + 4,$$

Proof: Let L denote an arbitrary list whose items are strictly larger than $\frac{1}{5}$. Let OPT denote an arbitrary optimal packing of L. In terms of types of bins, let $f_i, 1 \leq i \leq 30$, denote the number of bins of type i in OPT. Here i corresponds to the i-th type of bin in the lexicographical ordering of bin types. In particular, the first type is BL, the second type is BST,..., the 30-th type is T. In the remainder of the proof it will be shown that *MMP(L)*, the number of bins in the *MMP* packing of L, never exceeds $\frac{5}{4}OPT(L) + 4$, where $OPT(L) = \sum_{i=1}^{30} f_i$ denotes the number bins in OPT. We begin with some definitions.

Definition 1 *The* **front** *types of bins are BL, BST, BS, BTT, BT, B, LLS, and LSTT (types 1,2,...,7, and 11). All the other types are* **back** *types. Let* $\mathcal{L}_{back}$ *(S_{back}, T_{back}) denote the number of L–items (S, T) in the bins of back type. Let* $\mathcal{L}$ *(S, T) denote the total number of L–items (S, T).*

The following equations hold: $\mathcal{L} = f_1 + 2f_7 + f_{11} + \mathcal{L}_{back}$; $S = f_2 + f_3 + f_7 + f_{11} + S_{back}$; $T = f_2 + 2f_4 + f_5 + 2f_{11} + T_{back}$.

Definition 2 *A list of bin indices l, $l = i_1, i_2, ..., i_k$, is a list whose elements are positive integers i_i which correspond to the bin types. $*$ is going to be used as an abbreviation for the list 8,9,10,12,...30. If l denotes a list of bin indices then $f_l = \sum_{i \in l} f_i$*

Definition 3 *Let L_B (S_B, T_B) denote the number of L–items (S, T) in B–bins of the MMP packing of list L. Let N_{LLS} denote the number of bins of type LLS in the MMP packing of list L.*

Using Lemma 2, Theorem 2, and several omitted lemmas, a series of counting arguments can be used to bound $L_B, ..., N_{LLS}$.

It will be assumed that all of the non–B–items that are not *explicitly* accounted for will be packed in the worst possible way (see Lemma 2). By Lemma 2, and the packing properties of *MMP*, the value of *MMP(L)* can be expressed via the quantities defined above as:

$$MMP(L) \leq f_{1,...,6} + N_{LLS} + \frac{\mathcal{L} - L_B - 2N_{LLS}}{2} + \frac{S - S_B - N_{LLS}}{3} + \frac{T - T_B}{4} + 3$$

By substituting from the equations (1) and noting that $\frac{\mathcal{L}_{back}}{2} + \frac{S_{back}}{3} + \frac{T_{back}}{4} \leq \frac{5}{4} f_*$, it follows that the proof may be completed by establishing that:

$$\frac{5}{4} f_{1,...,7,11} + 1 \leq f_{1,...,6} + \frac{1}{2}(f_1 + 2f_7 + f_{11} - L_B) + \frac{1}{3}(f_2 + f_3 + f_7 + f_{11} - S_B - N_{LLS}) +$$
$$\frac{1}{4}(f_2 + 2f_4 + f_5 + 2f_8 - T_B).$$

Based on the bounds on L_B, S_B, T_B, and N_{LLS}, the proof of Theorem 3 is completed by a case analysis. The difficulty of these cases ranges from relatively simple to fairly involved.

□

The results of this section can be summarized in the following Theorem:

Theorem 4 *The fully dynamic bin packing algorithm MMP is $\frac{5}{4}$–competitive and requires $\Theta(\log n)$ per Insert/Delete operation.*

3.3 Is bundling necessary?

The use of bundling is an essential ingredient of our *MMP* algorithm. It allows us to move a number of items from one bin to another in a single operation. Thus, a natural question is whether or not bundling is really necessary. That is, would it be possible to design a fully dynamic algorithm for bin packing that is

$\frac{5}{4}$–competitive (or better!) without using bundling? Such an algorithm would in the processing of a single *Insert/Delete* operation, be able to move at most $\mathcal{O}(1)$ items from one bin to another. The following theorem (proof omitted) provides a negative answer to this question.

Theorem 5 *For any constant c, if A is a fully dynamic algorithm for bin packing that moves no more than c items (worst–case or amortized) per Insert/Delete operation, then the competitive ratio of A is no better than $\frac{4}{3}$.*

4   Inserts only

In this section we give an interesting result on dynamic algorithms when only *Inserts* are allowed. More encompassing than our prior results, the result establishes the existence of a polynomial time *dynamic* approximation scheme[4]. Also in contrast to our earlier results, the running time per operation is *amortized*. In particular, we show:

Theorem 6 *Given a (off–line) bin packing algorithm A that is R_A–competitive, with a running time $T_A(n)$, and given any $\epsilon > 0$, there is a dynamic bin packing algorithm A_ϵ that supports only Inserts. A_ϵ is $(R_A + \epsilon)$–competitive, and requires $\frac{T_A(n)}{n} \log n$ amortized time per Insert operation. Moreover, A_ϵ can be effectively produced.*

Sketch of proof: A_ϵ can be produced as a combination of A and *Next Fit* in the following manner: Initially, *Next Fit* is used in nearly its normal fashion to pack items as they arrive. The only addition to the standard *Next Fit* regime is that a certain level of supervision is built into the algorithm, so that when the desired competitive ratio $R_A + \epsilon$ is about to be exceeded, A is used to repack the entire instance that is present at that point. Intuitively, this (temporarily) improves the packing, since for all reasonable A, their competitive ratio is < 2 (the competitive ratio of *Next Fit*). After that repacking, *Next Fit* is run again, until another repacking is done, and so on. It can be shown in a series of Lemmas that for n *Inserts* there can be no more than $\mathcal{O}(\log n)$ repackings. The theorem follows.
□

Corollary 1 *For every $\epsilon > 0$ there is a dynamic bin packing algorithm A_ϵ that is $(1 + \epsilon)$–competitive, and requires $\mathcal{O}(\log^2 n)$ amortized time per Insert operation.*

Since the above approximation scheme involves amortized running times per operation, it is natural to ask about what can be done using *uniform* running times. Here, the best result remains our algorithm *MMP*, which has a running time of $\Theta(\log n)$ per operation and is $\frac{5}{4}$–competitive (and, of course, it also supports *Deletes*).

---

[4]Defined in the style of [7].

References

[1] E. G. Coffman, M. R. Garey, M. R., and D. S. Johnson. (1983). Dynamic bin packing. *SIAM Journal on Computing* 12, pp. 227–258.

[2] E. G. Coffman, M. R. Garey, and D. S. Johnson. (1984). Approximation algorithms for bin packing: An updated survey. In *Algorithm Design for Computer System Design* (G. Ausiello, M. Lucertini, and P. Serafini, Eds.), pp. 49–106. Springer–Verlag, New York.

[3] D. Eppstein, Z. Galil, G. F. Italiano, and A. Nissenzweig. (1992). Sparsification – A Technique for Speeding up Dynamic Graph Algorithms. *Proceedings of the 33rd IEEE Symposium on Foundations of Computer Science*, pp. 60–69.

[4] G. Frederickson. (1985). Data Structures for On–Line Updating of Minimum Spanning Trees, with Applications. *SIAM Journal on Computing* 14(4), pp. 781–798.

[5] Z. Galil, G. F. Italiano, and N. Sarnak. (1992). Fully Dynamic Planarity Testing. *Proceedings of the 24th ACM Symposium on Theory of Computing*, pp. 495–506.

[6] G. Gambosi, A. Postiglione, and M. Talamo M. (1990). New algorithms for on-line bin packing. In *Algorithms and Complexity, Proceedings of the First Italian Conference*, (G. Aussiello, D. P. Bovet, and R. Petreschi, Eds.), pp. 44–59. World Scientific, Singapore.

[7] M. R. Garey and D. S. Johnson. (1979). *Computers and Intractability: A Guide to the Theory of NP-Completeness.* Freeman, San Francisco.

[8] Z. Ivković and E. L. Lloyd. (1993). Fully Dynamic Maintenance of Vertex Cover. To appear in *Proceedings of the 19th International Workshop on Graph-Theoretic Concepts in Computer Science.*

[9] D. S. Johnson. (1973). *Near-optimal bin packing algorithms.* Doctoral thesis, MIT.

[10] D. S. Johnson. (1974). Fast algorithms for bin packing. *Journal of Computer and System Sciences* 8, pp. 272–314.

[11] D. S. Johnson, A. Demers, J. D. Ullman, M. R. Garey, and R. L. Graham. (1974). Worst-case performance bounds for simple one-dimensional packing algorithms. *SIAM Journal on Computing* 3(4), pp. 299–325.

[12] D. S. Johnson and M. R. Garey. (1985). A 71/60 Theorem for Bin Packing. *Journal of Complexity* 1, pp. 65–106.

[13] N. Karmarkar and R. M. Karp. (1982). An Efficient Approximation Scheme for the One-Dimensional Bin-Packing Problem. *Proceedings of the 23rd IEEE Symposium on Foundations of Computer Science*, pp. 312–320.

[14] R. M. Karp. (1972). Reducibility among Combinatorial Problems. In *Complexity of Computations* (R. E. Miller and J. W. Thatcher, Eds.), pp. 85–103. Plenum, New York.

[15] P. N. Klein and S. Sairam. (1993). Fully Dynamic Approximation Schemes for Shortest Path Problems in Planar Graphs. To appear in *Proceedings of the WADS 1993.*

[16] P. Ramanan, D. J. Brown, C. C. Lee, and D. T. Lee. (1989). On-Line Bin-Packing in Linear Time. *Journal of Algorithms* 3, pp. 305–326.

Increasing the Vertex-Connectivity in Directed Graphs

Tibor Jordán

Department of Computer Science
Eötvös University, Budapest

Abstract. Given a k-vertex-connected directed graph G, what is the minimum number m, such that G can be made k+1-connected by the addition of m new edges? We prove that if a vertex v has in- and out-degree at least k+1, there exists a splittable pair of edges on v. With the help of this statement, we generalize the basic result of Eswaran and Tarjan, and give lower and upper bounds for m which are equal for k=0 and differ from each other by at most k otherwise. Furthermore, a polynomial approximation algorithm is given for finding an almost optimal augmenting set.

1 Introduction

One of the basic graph augmentation problems - with important applications e.g. in the theory of network reliability - is the following: given a k-connected (or k-edge-connected) graph (or digraph) G, what is the minimum number m of edges to be added to G so that the resulting graph (or digraph) is $k + 1$-(edge)-connected?

The answer is completely known only for the edge-connectivity case – see, e.g. [3] –, where several polynomial algorithms were developed for finding an optimal augmenting set, even for the more general case, where the connectivity must be increased by an arbitrary value. The case of vertex-connectivity seems to be more difficult. The complexity of the general problem is still open and the exact value of m (and a polynomial augmenting algorithm) is known only for $k = 0, 1, 2, 3$ in the case of undirected graphs – see [2],[5],[6],[12] – and only for $k = 0$ – see [2] – if the graph is directed. (For directed trees, the more general problem was solved in [11].)

Here we deal with directed graphs and vertex-connectivity, where no bounds or (approximation) algorithms were known so far for $k \geq 1$. One motivation is that in the case of edge-connectivity, the splitting off procedure works very efficiently (the definitions will be given later). With the help of the splitting theorems of Lovász [8] and Mader [9], Frank gave a complete and algorithmic solution for this version in [3]. It is natural to ask whether there exists a splitting theorem for vertex-connectivity, and how can it help in connectivity augmentation.

Simple examples show that the same idea does not work here, it is not possible to split off a vertex completely. See Figure 1. (For $k = 2$.)

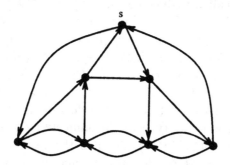

Fig.1. There is no splittable pair on s.

However, as we will see, we can split off two edges from a vertex, if its in-degree and out-degree is at least $k+1$. The second motivation is a paper of Bienstock et al. [1]. They proved an undirected splitting-type theorem and used it for proving a structural statement on the optimal solution of an - undirected - minimal-weight k-connected spanning network. (Our splitting theorem yields a similar claim for the directed case.)

With the help of the splitting off procedure, we can generalize the basic result of Eswaran and Tarjan [2] for $k = 0$, and give lower and upper bounds for m. These bounds give the exact min-max equation for $k = 0$ and the gap between them is at most k for all $k \geq 1$. Our methods yield a polynomial approximation algorithm for every k with the property that the gap between m and the size of its solution set is at most k. (A similar approach for the undirected problem can be found in [7].)

Very recently, using a completely different approach, the exact min-max equality was found for the augmentation number m, even if we want to increase the connectivity by an arbitrary value, see [4]. However, this result does not seem to imply the present one, and in addition, does not yield an efficient combinatorial algorithm for the problem.

In the remaining part of this section we introduce the necessary notations and definitions. A directed graph $G = (V, A)$ is called k-*vertex-connected* (or k-*connected*), if $|V| \geq k + 1$ and the removal of any $k - 1$ or fewer vertices leaves a strongly connected digraph. In our terminology, if we call a digraph k-connected, then it is k-connected but not $k + 1$-connected. Let $\Gamma^-(X)$ denote the set of *in-neighbours* of a set $X \subset V$, that is, $\Gamma^-(X) := \{v \in V - X : (v, u) \in A$ for some $u \in X\}$. Similarly, $\Gamma^+(X)$ denotes the set of *out-neighbours*. $d^-(v)$ $(d^+(v))$ denotes the *in-degree* (*out-degree*) of a vertex $v \in V$. With these notations we get that an at least $k-1$-connected digraph $G = (V, A)$ is k-connected, if and only if $|\Gamma^-(X)| \geq k$ and $|\Gamma^+(X)| \geq k$ for every $X \subset V$ with $|V - X| \geq k$.

A set $X \subset V$ is called *in-tight* (*out-tight*), if $|\Gamma^-(X)| = k$ ($|\Gamma^+(X)| = k$) and $|V - X| \geq k + 1$. Hence increasing the connectivity by one is equivalent to increasing the number of (in- or out-) neighbours for every tight set in G. If a set X is in-tight, $V - X - \Gamma^-(X)$ is out-tight and vica versa. This "complementary" set of a tight set X is denoted by X^*. The maximum number of pairwise disjoint minimal in-tight (out-tight) sets in G is denoted by $b(G)$ ($t(G)$). We do not use indices – b_k or t_k – for denoting the connectivity k of the underlying graph, since it will be clear everywhere. $\beta(G)$ ($\gamma(G)$) denotes the number of the minimal in-tight (out-tight) sets. $M(G)$ denotes the maximum of $\beta(G)$ and $\gamma(G)$. Given a k-connected digraph $G = (V, A)$, a set F of edges on V is called an *optimal augmenting set* if $G' = (V, A \cup F)$ is $k+1$-connected and $|F|$ is minimal subject to this property. We denote this minimum by $m(G)$. An edge e is *critical* in a k-connected digraph G, if $G - e$ is not k-connected. A sequence $C = \{e_0, ..., e_{l-1}\}$ of edges is an *alternating cycle* in a digraph, if there are no directed subpaths of length two in C and we get a cycle after omitting the directions. The edges $e = (s, v)$ and $f = (v, t)$ form a *splittable pair* on the vertex v, if replacing them by a new edge (s, t) does not decrease the connectivity of the graph. This operation is the so-called *splitting off*. We denote the deletion of a set X (consisting of edges or vertices) by $G - X$, the union is sometimes denoted by $G + X$. We denote a directed edge with tail u and head v by uv.

2   The splitting theorem and its consequences

The following statement can be checked easily by observing that every vertex is counted at least as many times on the left-hand side as on the right-hand side.

Lemma 2.1. In a digraph $G = (V, A)$, the functions $|\Gamma^+|$ and $|\Gamma^-|$ are submodular, i.e. the following inequalities hold for any $X, Y \subset V$:

$$|\Gamma^+(X)| + |\Gamma^+(Y)| \geq |\Gamma^+(X \cap Y)| + |\Gamma^+(X \cup Y)|$$
$$|\Gamma^-(X)| + |\Gamma^-(Y)| \geq |\Gamma^-(X \cap Y)| + |\Gamma^-(X \cup Y)|$$

□

From this observation we can derive that – under some conditions – the intersection and the union of two tight sets are tight.

Lemma 2.2. Let X and Y be in-tight (out-tight) sets in a k-connected digraph $G = (V, A)$ with $X \cap Y \neq \emptyset$ and $|V - (X \cup Y)| \geq k$. Then $X \cap Y$

is in-tight (out-tight). Moreover, if $|V - (X \cup Y)| \geq k + 1$, then $X \cup Y$ is also in-tight (out-tight).

Proof. Applying Lemma 2.1 we obtain

$$k + k = |\Gamma^+(X)| + |\Gamma^+(Y)| \geq |\Gamma^+(X \cap Y)| + |\Gamma^+(X \cup Y)| \geq k + k,$$

from which equality holds everywhere. $\square$

Suppose that we try to split off edges from a given vertex s in a k-connected digraph $G = (V, A)$. We have to be sure that after the splitting every set $X \subset V$ with $|V - X| \geq k$ has at least k in- and neighbours again. Therefore, if some in-tight set $S \subset V$, $s \notin S$ satisfies

(*) $\quad |\Gamma^+(s) \cap S| = 1,$

it is not allowed to split off a pair of edges consisting of the edge connecting s and S and an edge connecting $S \cup \Gamma^-(S)$ and s. We say, that an in-tight set S ($s \notin S$) with property (*) is *in-critical* . A set is *out-critical*, if reversing the directions, it is in-critical.

The following theorem will have a key role in the proofs. Since we consider vertex-connectivity, we will always suppose G to be simple.

Theorem 2.3. Suppose that in a k-connected directed graph $G = (V, A)$ the vertex $s \in V$ has $d^-(s) \geq k + 1$ and $d^+(s) \geq k + 1$. Then there is a splittable pair on s.

Proof. By definition, a splitting operation is allowed, if and only if the resulting graph is k-connected again.

Claim I. The edges ts and su can not be split off if and only if there exists an in-critical (or out-critical) set X with $u \in X$ and $t \in X \cup \Gamma^-(X)$ (or $t \in X$ and $u \in X \cup \Gamma^+(X)$).

If there exists a set X of this type, its in- or out-neighbours will form a $k - 1$-element cut-set after splitting off the edges. Conversely, suppose that the resulting graph is not k-connected after splitting off ts and su. Let Y denote a set with – say – $|\Gamma^-(Y)| = k - 1$. Then it is easy to see than Y is in-critical in G.

Let us introduce the following notations. For each in-neighbour t of s let O_t denote the union of the out-critical sets containing t. (Similarly, I_r stands for the union of the in-critical sets containing an out-neighbour r of s.) If there exist no out-critical (or in-critical) sets containing t, the corresponding set is defined by $O_t := \{t\}$. In this case it is said to be a *trivial O-set*.

Since s has at least $k+1$ in- and out-neighbours, and by Lemma 2.2, O_t is out-tight (in addition, out-critical) if it is non-trivial. (Similarly, I_r is in-critical.) Furthermore, applying Lemma 2.2 to $G - s$, we get that $O_t \cap O_u = \emptyset$, if $t \neq u$ are distinct in-neighbours of s, and the same property holds for the I-sets, as well. For a trivial critical set O_p, O_p^* is defined to be $V - p$.

The following statement is clear at this point:

Claim II. Let ts and su be two edges in G. If $u \in O_t^*$ and $t \in I_u^*$, then these edges form a splittable pair.

Indeed, suppose that ts and su are not splittable. Then by Claim I. there exists a – say – in-critical set X containing u with $t \in \Gamma^-(X)$. Thus $X \subseteq I_u$ and $t \notin I_u^*$ follows, contradiction.

We will show that there exists a pair of edges satisfying the conditions of Claim II.

Let $\mathcal{O} := \{O_v : vs \in E\}$, $\mathcal{I} := \{I_w : sw \in E\}$. Suppose – without loss of generality – that $\mathcal{O}$ has at least as many members as $\mathcal{I}$ has. Since the O-sets are pairwise disjoint, there exists a set O_t such that it contains at most one out-neighbour of s – and hence it contains at most one I-set, as well. Since O_t has $k - 1$ out-neighbours in $G - s$, and the I-sets are pairwise disjoint, there exist at least two I-set – let two of them be I_r and I_z – which are disjoint from $\Gamma^+(O_t)$. One of them, say I_r, intersects O_t^*. If I_r – or its in-neighbours – does not contain t, we are done. We claim, that otherwise $O_t \subseteq I_r$ holds. Indeed, $I_r^* \cap O_t$ would be a non-empty set with $k - 1$ neighbours in G, since I_r has no neighbours in $O_t - I_r$. (Observe, that $I_r \cap O_t^*$ – and hence I_r – has k in-neighbours in $V - O_t$.) In this case $I_z \subseteq O_t^*$, hence sz and ts has the desired property. This proves the theorem. □

Remark. Observe, that there exists no trivial O or I set, if every edge with tail or head s is critical. Furthermore, if there exist two trivial sets O_x and I_y, the edges xs and sy can always be split off. (If $x = y$, this means that we can delete them without destroying the k-connectivity.)

Our aim is to prove a good approximation for the optimal augmenting number m. The following lemma gives a sharp lower bound. With the help of Theorem 2.3 we will show that the difference between this lower bound and the optimum is at most k for an arbitrary k-connected digraph.

Lemma 2.4. For a k-connected digraph G, $m(G) \geq \max\{\beta(G), \gamma(G)\}$ holds.

Proof. We claim that any edge in any augmenting set can improve (increase the set of in-neighbours) at most one minimal in-tight set. It is obvious for two disjoint sets. Let us consider two intersecting minimal in-tight sets X and Y. If an edge $e = uv$ improves both of them, then $u \in X^* \cap Y^*$ and $v \in X \cap Y$ holds. Hence $|V - (X \cup Y)| \geq k + 1$, which implies – by Lemma 2.2 – that $X \cap Y$ is in-tight, contradicting the minimality of X. $\qquad\square$

For $k = 0$ equality holds in Lemma 2.4., it was proved by Eswaran and Tarjan in [2]. (Observe, that the minimal in-tight sets are always pairwise disjoint for $k = 0$.) However, for larger values of k this is not the case. Consider the following k-connected digraph C_k. It has $4k + 2$ vertices,

$$V(C_k) = \{v_1, ..., v_k, w_1, ..., w_k, s, t_1, ..., t_k, u_1, ..., u_k, x\}$$

and its edges are given as follows:

$$E(C_k) = \{(t_i, v_j), (w_i, u_j), (u_i, w_j), (v_i, w_j), (u_i, t_j) \mid i, j = 1, ...k\}\cup$$

$$\cup\{(v_i, s), (x, t_i), (s, w_i), (w_i, s), (x, u_i), (u_i, x) \mid i = 1, ..., k\}.$$

It can be checked that $M(C_k) = k+1$, while $m(C_k) \geq 2k+1$, since during the augmentation we need distinct edges for increasing the number of neighbours of the following tight sets: $(v_1, ..., v_k, t_1, ..., t_k)$ as one-element tight sets and the set $\{v_1, ..., v_k, w_1, ..., w_k, s\}$. That is $m(G) - M(G) \geq k$ is possible. See Figure 2. for $k = 1$.

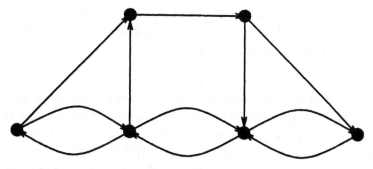

Fig.2. C_1.

This example shows that we can not achieve the minimum by this type of lower bound. However, this example is the worst case, the difference between m and M can not be larger, as we will see later.

The first observation is that if $b(G)$ (or $t(G)$) is large enough, then the minimal in-tight (or out-tight) sets are pairwise disjoint.

Lemma 2.5. Let $b(G) \geq k+1$ $(t(G) \geq k+1)$ in a k-connected digraph G. Then the minimal in-tight (out-tight) sets are pairwise disjoint.

Proof. Let $\mathcal{B}$ be a maximal size ($b(G)$-element) set-system of pairwise disjoint minimal in-tight sets. We claim that $\mathcal{B}$ contains all the minimal in-tight sets in G. Suppose that X is a minimal in-tight set not in $\mathcal{B}$. The maximality of $\mathcal{B}$ implies that X crosses at least one minimal in-tight set W from $\mathcal{B}$, and does not contain any of them, since X is minimal. Thus $|V - (X \cup W)| \geq k$, impossible by Lemma 2.2. $\square$

Corollary. For a k-connected digraph G (i)   if $b(G) \geq k+1$, then $b(G) = \beta(G)$,   (ii)   if $t(G) \geq k+1$, then $t(G) = \gamma(G)$. $\square$

Our next idea is the following: we want to decrease the value $M(G)$ step by step, by adding new edges to the graph. If both of $\beta(G)$ and $\gamma(G)$ are large, the splitting theorem will help. If one of them is already small enough, we can estimate the number of the necessary edges directly. The next result of Mader [10] will be useful for this purpose.

Lemma 2.6. [10] Let F be the set of those critical edges in a k-connected digraph G for which the out-degree of the tail and the in-degree of the head are at least $k+1$. Then F contains no alternating cycles. $\square$

Corollary 2.7. Let F be a minimal set of augmenting edges for a k-connected digraph $G = (V, A)$, that is, $G' = (V, A \cup F)$ is $k+1$-connected and F is minimal – for inclusion – with respect to this property. Then F contains no alternating cycles.

Proof. Observe that every edge (v, w) in an optimal augmenting set is critical (for the $k+1$-connectivity) and has $d^+(v) \geq k+2$ and $d^-(w) \geq k+2$ if it is contained in an alternating cycle in F, since F was added to a k-connected digraph. $\square$

Given a digraph G, construct the following undirected bipartite graph G^D: split every vertex v of G into two parts v_1 and v_2, and connect x_1 to y_2 in G^D if and only if xy is a (directed) edge in G. Observe, that G contains no alternating cycles if and only if G^D is a forest. This straightforward reduction – with Lemma 2.6 – will give good upper bounds for the size of an optimal augmenting set.

Lemma 2.8. Suppose that $X \subseteq V$ covers the minimal in-tight sets and $Y \subseteq V$ covers the minimal out-tight sets in the k-connected digraph $G = (V, A)$. Then $m(G) \leq |X| + |Y| - 1$.

Proof. If we connect all vertices of Y to all vertices of X, the graph becomes $k + 1$-connected. Indeed, suppose that there exists a separating set of size k in the augmented graph. Let T be one of its – in-tight – components. $X \cap T \neq \emptyset$, since X contains a minimal in-tight set. Similarly, $Y \cap X^* \neq \emptyset$, contradicting the existence of the edges between these non-empty intersections.

Now let us choose a minimal augmenting set F from the new edges. Corollary 2.7 implies that F has no alternating cycles. Hence the corresponding bipartite graph (constructed from the subgraph F) has at most $|X| + |Y| - 1$ edges and so has F. (If $v \in X$ but $v \notin Y$, then v_1 is an isolated vertex in G^F.) $\qquad\square$

Lemma 2.9. For a k-connected digraph $G = (V, A)$ let $b(G) \leq k + 1$. Then the minimal in-tight sets can be covered by $k + 1$ vertices.

Proof. If the minimal in-tight sets are pairwise disjoint, we are done.

In the opposite case let X and Y be crossing minimal in-tight sets with $|V - (X \cup Y)|$ maximum. By Lemma 2.3 $|V - (X \cup Y)| \leq k - 1$. Let $x \in X - Y$ and $y \in Y - X$ be arbitrary vertices. We claim that $S = (V - (X \cup Y)) \cup x \cup y$ covers the in-tight sets. If not, there exists a minimal in-tight set W with $S \cap W = \emptyset$. Then $X \cap W \neq \emptyset$, since Y can not contain W. However, by the construction of S, $|V - (X \cup W)| > |V - (X \cup Y)|$, contradiction. The lemma follows. $\qquad\square$

Theorem 2.10. Let $G = (V, E)$ be a k-connected digraph. Then $M(G) \leq m(G) \leq M(G) + k$.

Proof. Lemma 2.4 implies the first inequality. We prove the second one by induction on $q(G) := \min\{b(G); t(G)\}$. Suppose that $q(G) \leq k + 1$. Choose sets $X \subset V$ and $Y \subset V$ such that X covers the minimal in-tight sets and Y covers the minimal out-tight sets. We can assume $|X| \leq \beta(G)$ and $|Y| \leq \gamma(G)$, since they can be choosen to be minimal coverings. Add all the edges yx with $x \in X$ and $y \in Y$ to G. The augmented graph is $k+1$-connected by the choice of X and Y. Let F be a minimal augmenting set contained in this augmentation. By Lemma 2.8 $|F| \leq |X| + |Y| - 1$. Furthermore, $q(G) \leq k + 1$ and Lemma 2.9 give $|F| \leq M(G) + k$.

Suppose that $m(G) > M(G) + k$ holds. Now $q(G) \geq k + 2$ from the previous argument. Let us give an extra vertex s to G and connect one vertex from every – by Lemma 2.5, pairwise disjoint – minimal out-tight set to s and connect s to every minimal in-tight set. Observe, that the augmented graph is $k + 1$-connected. Now $d^+(s) \geq k + 2$ and $d^-(s) \geq k + 2$, hence Theorem 2.3 can be applied. Let ys and sx be a splittable pair. Since the graph remains $k + 1$-connected, $M(G + yx) < M(G)$.

(Observe, that the minimal tight sets of $G + yx$ and G are the same, except the two sets containing x and y, which are lost.) By induction, $m(G + yx) \leq M(G + yx) + k$, and clearly $m(G) \leq m(G + yx) + 1$. Thus $m(G) \leq M(G) + k$ follows, contradiction. $\qquad\square$

We remark that the min-max equation $m = M$ for $k = 0$ follows without using Lemma 2.6. It is not hard to see that the maximum given by somewhat different words in [2] for $k = 0$ equals our value M. Furthermore, observe for $k = 1$ Theorem 2.3 gives the same result (and a bit more) than the more general splitting theorem of Mader in this case.

3 The approximation algorithm

Following the ideas of the proofs, we get an approximation algorithm for the augmentation. In this section we sketch this algorithm and show that its running time is polynomial.

The input is a k-connected directed graph $G = (V, A)$. The algorithm terminates after finding an augmenting set F with $|F| \leq m(G) + k$. During the algorithm, the following property will always hold:

$(\ast\ast)$ $|\Gamma^+(X)| \geq k + 1$ and $|\Gamma^-(X)| \geq k + 1$ for every $X \subset V$ with $|V - X| \geq k + 1$.

The augmented output-graph G'' has property $(\ast\ast)$, hence it is $k+1$-connected.

Algorithm AUG. Computing m of $G = (V, A)$.

1. Add a new vertex s to V. Let $V = \{v_1, \ldots, v_n\}$, and $F := \{sv, vs : v \in V\}$. Let $G' := (V + s, A \cup F)$.

2. For each $i = 1, \ldots, n$ check if $(\ast\ast)$ holds in $G' - sv_i$ and $G' - v_i s$, respectively. If yes, $G' \leftarrow G' - sv_i$ or $G' \leftarrow G' - v_i s$, respectively.

3. If $d^+(s) \geq k + 2$ and $d^-(s) \geq k + 2$, then go to Step 5. Otherwise let $X := \{v \in V : sv \in G'\}$ and $Y := \{v \in V : vs \in G'\}$. Delete s.

4. Let $H := \{yx : y \in Y, x \in X\}$. Let $G' := (V, A \cup H)$. For each edge $h \in H$ check if $(\ast\ast)$ holds in $G' - h$. If yes, let $G' \leftarrow G' - h$, $H \leftarrow H - h$.
   Let $G'' := G'$, $F := H$, stop.

5. For each pair of new edges test if it is possible to split them off without violating $(\ast\ast)$. If yes, split off these edges. Repeat this if $\min\{d^+(s), d^-(s)\} \geq k + 2$. The resulting graph is denoted by

$G' = (V + s, A \cup F')$. Let $X := \{v \in V : sv \in F'\}$ and let $Y := \{v \in V : vs \in F'\}$. Delete s and go to Step 4.

Proof of correctness. We give here a short analysis of the algorithm AUG. A basic subroutine – it is used several times in Step 2.,4. and 5. – is testing whether the condition $(**)$ holds after deleting an edge e or splitting off two edges us and sv. This task can be done by well-known methods, computing a minimum cut between the end-vertices of e, or between u and s – and s and v –, respectively.

The second observation is that after Step 2., $d^+(s) = b(G)$ if $b(G) \geq k + 2$, and $d^+(s) \leq k + 1$, if $b(G) \leq k + 1$. (Similarly for $t(G)$.) This follows from the fact that there must be at least one new edge between s and any of the in-tight sets, and that the set of new edges directed out of s is minimal subject to this property. From this we get that the out-neighbours of s form a minimal covering set of the in-tight sets. Thus the sets X and Y are minimal covers of the in-tight and the out-tight sets, respectively. Hence Step 4. gives a good augmenting set by Lemma 2.8.

Theorem 2.3 guarantees that we can always find a splittable pair of edges in Step 5. if s has at least $k + 2$ in- and out-neighbours. Since there are at most n – as usual, $n := |V|$ – splitting steps and we have to check at most n^2 pairs at every iteration, this step consists of at most $2n^3$ min-cut computations.

Summarizing the above arguments, we get:

Theorem 3.1. The output graph G'' of algorithm AUG is $k+1$-connected and $|F| \leq m(G) + k$. $\qquad\square$

Let us denote by MC the running time of an algorithm which computes a minimum vertex-cut between two vertices in a directed graph. There are well-known max flow - min cut algorithms with running time $O(n^3)$ for this purpose. With this notation the following time-bound for our algorithm can be seen easily. The most time-consuming step is Step 5., where – roughly speaking – sometimes n^3 min-cut computations are necessary. Step 2. and Step 4. require only $2n$ and $(k + 1)^2$ min-cut algorithms, respectively.

Theorem 3.2. The running time of algorithm AUG is $O(n^3 MC)$. $\qquad\square$

Remark. We mention that one can compute the lower bound $\max\{\beta(G), \gamma(G)\}$ in polynomial time. The first to do is determining a tight set separating x and y – if there is any – for every pair of vertices. Then one can get the minimal one subject to this property by decreasing its size step by step, using min-cut computations again with appropriate weights on the vertices. Every minimal tight set arises this way, hence we can choose the distinct sets and compute the values $\beta(G)$ and $\gamma(G)$ easily. (From these observations we get an upper bound n^2 for these numbers.)

Acknowledgement

This work was supported by the Forschungsinstitut für Diskrete Mathematik, Universität Bonn, Sonderforschungsbereich 303 (DFG) and by the OTKA grant no. T4271.

References

1. D.Bienstock, E.F.Brickell, C.L.Monma, On the Structure of Minimum-weight k-connected Spanning Networks, SIAM J. Disc. Math., Vol. **3.**, pp.320-329, 1990

2. K.P.Eswaran, R.E.Tarjan, Augmentation Problems, SIAM J. Comput., **5** , 1976, pp.653-665

3. A.Frank, Augmenting Graphs to Meet Edge-connectivity Requirements, SIAM J. Disc. Math., Vol.5, No.I., pp.25-53, 1992

4. A.Frank, T.Jordán, Minimal Edge-Coverings of Pairs of Sets, submitted

5. T.S.Hsu, On Four-Connecting a Triconnected Graph, In Proc. 33th Annual IEEE Symp. on FOCS, 1992

6. T.S.Hsu, V.Ramachandran, A Linear Time Algorithm for Triconnectivity Augmentation, In Proc. 32th Annual IEEE Symp. on FOCS, 1991, pp.548-559

7. T.Jordán, Optimal and Almost Optimal Algorithms for Connectivity Augmentation Problems, Proc. Third IPCO Conference (G.Rinaldi and L.Wolsey eds.), 1993, pp.75-88

8. L.Lovász, Combinatorial Problems and Exercises, North-Holland, Amsterdam, 1979

9. W.Mader, Konstruktion aller n-fach kantenzusammenhangenden Digraphen, European J. Combin., **3**, 1982 , pp.63-67

10. W.Mader, Minimal n-fach Zusammenhängende Digraphen, Journal of Combinatorial Theory (B),**38**, 1985, pp.102-117

11. T.Masuzawa, K.Hagihara, N.Tokura, An Optimal Time Algorithm for the k-vertex-connectivity Unweighted Augmentation Problem for Rooted Directed Trees, Disc. Applied Math., 1987, pp.67-105

12. T.Watanabe, A.Nakamura, 3-connectivity Augmentation Problems, In Proc. of IEEE Symp. on Circuits and Systems, 1988, pp.1847-1850

On the Recognition of Permuted Bottleneck Monge Matrices [*]

Bettina Klinz[1], Rüdiger Rudolf[1], Gerhard J. Woeginger[2]

[1] Institut für Mathematik B, Technical University Graz, Kopernikusgasse 24, A-8010 Graz, Austria
[2] Institut für Theoretische Informatik, Technical University Graz, Klosterwiesgasse 32/II, A-8010 Graz, Austria

Dedicated to Prof. Rainer E. Burkard on the occasion of his 50th birthday.

Abstract. An $n \times m$ matrix A is called *bottleneck Monge matrix* if $\max \{a_{ij}, a_{rs}\} \leq \max \{a_{is}, a_{rj}\}$ for all $1 \leq i < r \leq n, 1 \leq j < s \leq m$. The matrix A is termed *permuted bottleneck Monge matrix*, if there exist row and column permutations such that the permuted matrix becomes a bottleneck Monge matrix. We first show that the class of permuted 0-1 bottleneck Monge matrices can be recognized in $O(nm)$ time. Then we present an $O(nm(n+m))$ time algorithm for the recognition of permuted bottleneck Monge matrices with arbitrary entries.

1 Introduction

Problem Statement. An $n \times m$ matrix A is called *Monge matrix* if A satisfies the so-called *Monge property*:

$$a_{ij} + a_{rs} \leq a_{is} + a_{rj} \qquad \text{for all } 1 \leq i < r \leq n, 1 \leq j < s \leq m. \qquad (1)$$

This property was introduced by Hoffman [11]. Replacing "+" in (1) by "max" we get the *bottleneck Monge property*:

$$\max \{a_{ij}, a_{rs}\} \leq \max \{a_{is}, a_{rj}\} \qquad \text{for all } 1 \leq i < r \leq n, 1 \leq j < s \leq m. \qquad (2)$$

Matrices A fulfilling property (2) are called *bottleneck Monge matrices* or *max-distribution matrices* (see e.g. Burkard [5], Burkard and Sandholzer [6] or van der Veen [17]).

We call a matrix A *permuted (bottleneck) Monge matrix* iff there are permutations ϕ and ψ of the rows and the columns such that the permuted matrix $A_{\phi,\psi} = (a_{\phi(i)\psi(j)})$ is a (bottleneck) Monge matrix. Formally, three different *recognition* problems for permuted (bottleneck) Monge matrices can be stated as follows:

[*] This research has been supported by the Christian Doppler Laboratorium für Diskrete Optimierung and by the Fonds zur Förderung der wissenschaftlichen Forschung, Project P8971-PHY.

(RP) *Given an $n \times m$ matrix A, decide whether or not there exist two permutations ϕ and ψ such that the permuted matrix $A_{\phi,\psi}$ is a (bottleneck) Monge matrix. In the affirmative case, determine such a pair of permutations (ϕ, ψ).*

($\text{RP}^=$) *Solve* (RP) *under the additional requirement that $\phi = \psi$.*

(RP^r) *Solve* (RP) *when only row permutations, but no column permutations are allowed.*

Amazingly, all results in the literature on recognizing Monge properties are for the sum case only. (See e.g. [7, 9, 13, 14] and the references given in these papers.) In this paper we will present an $O(nm(n+m))$ algorithm for solving the problem (RP) in the bottleneck case. We note that Deineko and Filonenko [9] solved the sum case in $O(nm + m \log m + n \log n)$ time.

Motivation. Many problems in combinatorial optimization become easy or easier, if their input matrices are Monge or bottleneck Monge matrices. For example, assignment and transportation problems with sum or bottleneck objective can be solved by a greedy approach if the cost matrix is (bottleneck) Monge (see Hoffman [11], Burkard [5]). Since these problem classes are not affected by permuting rows and columns of the cost matrix, (RP) plays an important role in the recognition of greedily solvable transportation and assignment problems.

The restricted problem ($\text{RP}^=$) arises in connection with efficiently solvable special cases of the traveling salesman problem (TSP) with sum or bottleneck objective function (see e.g. Gilmore, Lawler and Shmoys [10], Burkard and Sandholzer [6] and van der Veen [17]). Problem (RP^r) occurs for example in flow-shop scheduling problems where we have given n jobs and m fixed ordered machines. All jobs have to pass the machines in the given order and the makespan has to be minimized. Let p_{ij} denote the processing time of job i on machine j. Then an optimal schedule can be found easily if the rows of the matrix $P = (-p_{ij})$ can be reordered such that the resulting matrix is bottleneck Monge (see e.g. the survey paper by Monma and Rinnooy Kan [12]).

Further applications of bottleneck Monge matrices and matrices with related properties can be found in Bein, Brucker and Park [3].

Organization of the paper. In Section 2 we present two characterizations for the class of 0-1 bottleneck Monge matrices. In Sections 3 and 4, we investigate the structure of bottleneck Monge matrices with arbitrary entries. In Section 5 we describe a compact characterization of the set of all pairs of row and column permutations that transform a given 0-1 matrix into a bottleneck Monge matrix. Section 6 derives a recognition algorithm for bottleneck Monge matrices with arbitrary entries. We close the paper with some concluding remarks on problem variations in Section 7.

2   0-1 Bottleneck Monge Matrices

In the sequel we present two characterizations of 0-1 bottleneck Monge matrices. We start with some definitions.

Let A be a 0-1 matrix and denote by $B(A) = (V_1, V_2; E_A)$ its associated *bipartite* graph with a vertex in V_1 for each row of A, a vertex in V_2 for each column of A and an edge $(i, j) \in E_A$ joining vertices $i \in V_1$ and $j \in V_2$ iff $a_{ij} = 1$. A bipartite graph $H = (V_1, V_2; E)$ is said to be *strongly ordered* if for all $(i, j), (i', j') \in E$, where $i, i' \in V_1$ and $j, j' \in V_2$, it follows from $i < i'$, $j' < j$ that $(i, j') \in E$ and $(i', j) \in E$.

A graph $G = (V, E)$ is called *permutation graph*, if there exists a pair (ρ_1, ρ_2) of permutations of the vertex set V such that $(i, j) \in E$ iff vertex i precedes vertex j in ρ_1 and j precedes i in ρ_2.

It can easily be seen that a 0-1 matrix is bottleneck Monge if and only if it does not contain one of the following 2×2 submatrices

$$B_1 = \begin{pmatrix} 1 & 0 \\ 0 & 1 \end{pmatrix}, \qquad B_2 = \begin{pmatrix} 1 & 0 \\ 0 & 0 \end{pmatrix}, \qquad B_3 = \begin{pmatrix} 0 & 0 \\ 0 & 1 \end{pmatrix}.$$

The following observation is now an immediate consequence of Theorem 3 in Chen and Yesha [8].

Observation 1. *A 0-1 matrix A is a bottleneck Monge matrix if and only if its associated bipartite graph $B(A)$ is the complement of a strongly ordered bipartite permutation graph.* □

This result enables us to recognize permuted 0-1 bottleneck Monge matrices in linear time by applying the algorithm of Spinrad, Brandstädt and Stewart [15] which decides whether a given bipartite graph H is a bipartite permutation graph and if so, determines a strong ordering of H.

In the following we present another characterization of 0-1 bottleneck Monge matrices. Let A be an $n \times m$ 0-1 matrix with no rows of all ones. Denote by s_i resp. f_i the position of the first resp. last zero in row i. A 0-1 Matrix A is said to be a *double staircase matrix* if $s_1 \leq s_2 \leq \ldots \leq s_n$, $f_1 \leq f_2 \leq \ldots \leq f_n$ and $a_{ij} = 0$ for all $j \in [s_i, f_i]$. (Matrices with a similar property were introduced independently in Chen and Yesha [8].)

It is easy to see that 0-1 bottleneck Monge matrices with no all ones rows and columns fulfill the following properties: (i) identical rows resp. columns form a contiguous block, (ii) the zeros in each row resp. column are consecutive and finally (iii) all ones rows and columns can be inserted arbitrarily without destroying the bottleneck Monge property.

From now on we assume that A does neither contain identical rows nor identical columns nor all ones rows nor all ones columns. We call such a matrix a *reduced* 0-1 matrix.

Theorem 2. *A reduced 0-1 matrix A is a bottleneck Monge matrix if and only if it is a double staircase matrix.* □

3 Bottleneck Monge Matrices with Arbitrary Entries

3.1 0-1 Bottleneck Monge matrices

From the results in Section 2 we know that *any* 0-1 bottleneck Monge matrix is either a double staircase matrix or can be obtained from a double staircase matrix by inserting an arbitrary number of all ones rows and columns.

3.2 Generalized staircase matrices

Let A be an $n \times m$ matrix and suppose that $\tilde{a}_1 > \tilde{a}_2 > \ldots > \tilde{a}_L$ are the pairwise distinct entries of matrix A. For $1 \leq k \leq L$, we define 0-1 matrices T^k as follows: If entry $a_{ij} \leq \tilde{a}_k$ then the corresponding entry in T^k is 0, otherwise it is 1. The following observation relates the original matrix A to the matrices T^k.

Observation 3. *A is a bottleneck Monge matrix if and only if all matrices T^k, $k = 1, \ldots, L$, are bottleneck Monge matrices.* $\square$

For $k \in \{1, \ldots, L\}$, let R_{2k} contain all rows of A all of whose entries are $> \tilde{a}_k$ and let R_{1k} denote the set of remaining rows. For each row i in R_{1k} we define numbers $s_{ik} := \min\{j : a_{ij} \leq \tilde{a}_k\}$ and $f_{ik} := \max\{j : a_{ij} \leq \tilde{a}_k\}$. We require that a row i from R_{2k} either precedes all rows in R_{1k} or succeeds all rows in R_{1k}. For convenience, we set $s_{ik} = f_{ik} = 1$ in the first case and $s_{ik} = f_{ik} = n$ in the latter case. A matrix A is then termed *generalized staircase matrix* if for all $1 \leq k \leq L$ we have $s_{1k} \leq s_{2k} \leq \ldots \leq s_{nk}$, $f_{1k} \leq f_{2k} \leq \ldots \leq f_{nk}$ and $a_{ij} \leq \tilde{a}_k$ for $j \in [s_{ik}, f_{ik}]$. Obviously, generalized staircase matrices form a proper subclass of bottleneck Monge matrices. This subclass has some nice properties which do not hold for bottleneck Monge matrices in general.

3.3 Two-row bottleneck Monge matrices

Let A be an $2 \times m$ matrix. Then we can show the following results:

(1) There exists always a permutation of the columns such that A becomes bottleneck Monge. Hence *any* $2 \times m$ matrix is a permuted bottleneck Monge matrix. A simple ordering algorithm (see also Burkard [5]) finds an appropriate permutation of the columns in $O(m \log m)$ time.

(2) The set of all column permutations that transform A into a bottleneck Monge matrix can be characterized in a nice way. For example, it can be shown that there is a unique ordering of the columns if all matrix entries are pairwise distinct. Hence, permuted bottleneck $n \times m$ Monge matrices can be recognized in $O(nm + n \log n + m \log m)$ time provided that all nm entries are pairwise distinct.

(3) It can be decided in $O(m)$ time whether a given $2 \times m$ matrix is bottleneck Monge. Hence it can be decided in $O(\min\{n^2 m, m^2 n\})$ time whether a given $n \times m$ matrix is bottleneck Monge. (Simply test all pairs of rows or columns.)

Closely related to Monge matrices is the notion of *Monge sequences*, first introduced by Hoffman. In analogy to the definition of bottleneck Monge matrices, we can define *bottleneck Monge sequences* just by replacing sum by maximum. We mention that the above results on $2 \times m$ bottleneck Monge matrices can be used to devise an algorithm for constructing bottleneck Monge sequences along the lines of the algorithm of Alon, Cosares, Hochbaum and Shamir [2] for Monge sequences.

4 Fundamental Properties of Bottleneck Monge Matrices

Unfortunately, the combinatorial structure of bottleneck Monge matrices is more complicated than that of Monge matrices. The most important difference between the two classes is the following: An $n \times m$ matrix A is a Monge matrix iff

$$a_{i,j} + a_{i+1,j+1} \leq a_{i+1,j} + a_{i,j+1} \qquad \text{for all } 1 \leq i \leq n-1, 1 \leq j \leq m-1.$$

Unfortunately, the condition

$$\max\{a_{i,j}, a_{i+1,j+1}\} \leq \max\{a_{i+1,j}, a_{i,j+1}\} \qquad \text{for all } 1 \leq i \leq n-1, 1 \leq j \leq m-1$$

is however not equivalent to the bottleneck Monge property (2) due to the fact that the *strong cancellation rule* $a \oplus c \leq b \oplus c \implies a \leq b$ is true for $\oplus := +$ but not for $\oplus := \max$.

Another nice property of Monge matrices is that if a matrix A is a Monge matrix, then $-A$ is *totally monotone* (for a definition and further details on totally monotone arrays see Aggarwal, Klawe, Moran, Shor and Wilber [1]). Again, as can easily be seen, this property does not hold for the class of bottleneck Monge matrices. We note, however, that the negative of any generalized staircase matrix (cf. Subsection 3.2) is totally monotone.

Due to these differences between Monge matrices and bottleneck Monge matrices, neither the problem of deciding whether a given matrix is bottleneck Monge nor the problem of recognizing permuted bottleneck Monge matrices can be solved by carrying over the algorithms which are for the sum case. As a consequence our method for recognizing permuted bottleneck Monge matrices is more involved and less efficient than the recognition algorithm of Deineko and Filonenko [9] for permuted Monge matrices.

Despite the negative results mentioned above, several interesting properties of bottleneck Monge matrices can be derived. For lack of space, however, we refrain from giving further details in this abstract. We only introduce some notation which is needed lateron.

Let henceforth A be an $n \times m$ matrix with row set $\{1, 2, \ldots, n\}$ and column set $\{1, 2, \ldots, m\}$. By ε_d we denote the *identity permutation* on the set $\{1, \ldots, d\}$, i.e. $\varepsilon_d(i) := i$ for all $i \in \{1, \ldots, d\}$. For ϕ a permutation on $\{1, \ldots, d\}$, the permutation ϕ^- defined by $\phi^-(i) := \phi(d-i+1)$ is called the *reverse permutation* of ϕ. Accordingly, the *reverse matrix* A^- of matrix A is given by $A^- := A_{\varepsilon_n^-, \varepsilon_m^-}$.

5   Permuted 0-1 Bottleneck Monge Matrices

In this section we present a complete and compact characterization of the set $\mathcal{P}(A) := \big\{ (\phi, \psi) : \ A_{\phi,\psi} \text{ is bottleneck Monge} \big\}$ for 0-1 matrices A.

Let $\text{ZERO}(j)$ denote the set of all rows of matrix A that have a zero entry in column j. Then two columns j_1 and j_2 are said to *intersect* iff they have at least one zero in common, i.e. iff $\text{ZERO}(j_1) \cap \text{ZERO}(j_2) \neq \emptyset$. Accordingly, the undirected graph $I(A) = (C, E)$ with vertex set $C = \{1, \ldots, m\}$ and an edge $(j_1, j_2) \in E$ if and only if the columns j_1 and j_2 intersect, is called *intersection graph* of the matrix A. If the graph $I(A)$ is connected, we also say that the matrix A is *connected*. Likewise, a set of columns of A is said to be connected, if these columns induce a connected matrix.

Let $\tilde{G} = (\tilde{C}, \tilde{E})$ be a connected component of the intersection graph $I(A)$. The matrix A' obtained from A by removing columns which are not contained in $\tilde{C}$ is called a *component* of A. For a component A', define its associated *block* $\bar{A}$ to consist of all rows of A' which contain at least one zero entry. As a consequence of a more general theorem which was omitted here we obtain

Theorem 4. *Let A be an $n \times m$ connected and reduced double staircase matrix. Then $\mathcal{P}(A) = \big\{ (\varepsilon_n, \varepsilon_m), (\varepsilon_n^-, \varepsilon_m^-) \big\}$.* □

Now let A be a connected 0-1 matrix which contains no all ones rows or columns, but which may contain identical rows or columns. It is obvious that such rows and columns must occur contiguously within a double staircase matrix. We refer to a group of contiguous identical rows or columns in matrix A as a *row stripe* resp. *column stripe* of A. In a connected double staircase matrix A the relative order of the row and column stripes is fixed up to the reversal of both orderings, while obviously the rows (columns) within a row (column) stripe may occur in arbitrary order.

Next we deal with 0-1 matrices A which need no longer be connected or reduced, but which still do not contain all ones rows or columns. In this case the set $\mathcal{P}(A)$ can be characterized as follows: The blocks $\bar{A}_1, \ldots, \bar{A}_f$ of A may be arranged according to an arbitrary permutation π on the set $\{1, \ldots, f\}$, but within the block $\bar{A}_q$ the rows and columns must be are arranged according to a pair of permutations $(\sigma_q, \tau_q) \in \mathcal{P}(\bar{A}_q)$ for $q = 1, \ldots, f$.

The remaining case concerns all ones rows and all ones columns. These rows and columns can be placed anywhere within a 0-1 bottleneck Monge matrix. Putting together all these observations, we get a complete description of the set $\mathcal{P}(A)$ for 0-1 matrices A.

We only mention that from the considerations above, a simple $O(nm)$ time algorithm for recognizing permuted 0-1 bottleneck Monge matrices follows. This algorithm is different from the algorithm one obtains from translating the method of Spinrad, Brandstädt and Stewart [15] into matrix-theoretical terms, but does not improve on the complexity of the latter approach. Thus we do not describe any further details of our algorithm.

We close this section with an alternative characterization of permuted 0-1 bottleneck Monge matrices. This result can be seen as extension of Tucker's [16] forbidden submatrix characterization of the set of 0-1 matrices with the consecutive zeros property for both rows and columns. (A 0-1 matrix A is said to have the *consecutive zeros property* for both rows and columns if the rows and columns of A can be permuted such that the zeros in each row and in each column are consecutive.) Let

$$B_4 = \begin{pmatrix} 1\,0\,0\,0 \\ 0\,0\,0\,1 \\ 1\,1\,0\,1 \end{pmatrix}, \qquad B_5 = \begin{pmatrix} 1\,0\,0\,1 \\ 0\,0\,0\,0 \\ 1\,1\,0\,0 \end{pmatrix}.$$

Theorem 5. *Let A be an $n \times m$ 0-1 matrix which has the consecutive zeros property for both rows and columns and let $\mathcal{B}$ be the set of all matrices which can be obtained from matrices in $\{B_4, B_4^T, B_5, B_5^T\}$ by permuting rows and columns. Then A is a permuted 0-1 bottleneck Monge matrix if and only if A does not contain any submatrix from the set $\mathcal{B}$.*

Proof. Omitted. □

6 Recognition of Permuted Bottleneck Monge Matrices

In this section we use the results of the previous section on recognizing permuted 0-1 bottleneck Monge matrices to obtain an efficient algorithm for recognizing permuted bottleneck Monge matrices with arbitrary entries. The main idea of our approach relies on Observation 3. Let again $\tilde{a}_1 > \tilde{a}_2 > \ldots > \tilde{a}_L$ be the sequence of all pairwise distinct entries of matrix A and define the so-called *threshold matrices* T^k, $k = 1, \ldots, L$: If entry $a_{ij} \le \tilde{a}_k$, then the corresponding entry in T^k is 0, otherwise it is 1.

By applying the results of the previous section, we are able to check for each k whether or not the 0-1 matrix T^k is a permuted bottleneck Monge matrix. Moreover, we obtain a characterization of the set $\mathcal{P}_k := \mathcal{P}(T^k) = \{(\phi, \psi) \mid T^k_{\phi,\psi} \text{ is bottleneck Monge}\}$. What remains to be done is to determine the intersection of all sets $\mathcal{P}_k$. Obviously, the original matrix A is permuted bottleneck Monge if and only if this intersection is non-empty.

Define $\mathcal{Q}_k := \bigcap_{q=1}^{k} \mathcal{P}_q$ to be the set of all pairs of permutations (ϕ, ψ) which transform the first k threshold matrices into bottleneck Monge matrices. Our aim is to determine a pair of permutations $(\phi, \psi) \in \mathcal{Q}_L$ or to show that $\mathcal{Q}_L = \emptyset$. However, in general we cannot construct the full set $\mathcal{Q}_L$, since this task is too time-consuming. Instead we will maintain sets $\widetilde{\mathcal{Q}}_k$ with $\widetilde{\mathcal{Q}}_k \subseteq \mathcal{Q}_k$ for all $k = 1, \ldots, L$ such that $\widetilde{\mathcal{Q}}_L = \emptyset$ if and only if $\mathcal{Q}_L = \emptyset$.

(Case 1) We start with the simplest case where all matrices T^k are connected. We furthermore exclude all ones rows and all ones columns for a while from our considerations. Then we are able to compute the sets $\mathcal{Q}_k$ efficiently and have $\widetilde{\mathcal{Q}}_k = \mathcal{Q}_k$. Note that for $k = 1$ these assumptions are trivially true, since all entries of T^1 are equal to 0. We recall that in 0-1 bottleneck Monge matrices with no rows and columns of all ones, identical rows and columns appear contiguously and that for connected 0-1 matrices A with no all ones rows and columns, the set $P(A)$ can be fully described by the order of the row stripes and of the column stripes. Within a stripe the arrangement of the rows or columns is arbitrary. This motivates the following definition.

Let $D_1, \ldots, D_u$ be stripes (sets) of $d_1, \ldots, d_u$ identical elements each. Then the set S of all permutations σ with $\sum_{p=1}^{q-1} d_p < \sigma^{-1}(i) \leq \sum_{p=1}^{q} d_p$ for all $i \in D_q$, $q = 1, \ldots, u$, is called a *stripe permutation*. Furthermore, let S^- denote the *reverse stripe permutation* of S which contains all permutations σ such that $\sigma^- \in S$.

We first show how to intersect two arbitrary stripe permutations S_1 and S_2 and how to construct the corresponding new stripe permutation $S_3 = S_1 \cap S_2$. This intersection process is done recursively. Let $C_1, \ldots, C_u$ and $D_1, \ldots, D_v$ be the different stripes which define S_1 and S_2, respectively. If $S_3 \neq \emptyset$, we must either have $C_1 \subseteq D_1$ or $D_1 \subseteq C_1$; w.l.o.g. suppose $C_1 \subseteq D_1$. Then C_1 is the first stripe of the intersection S_3. The next stripes are obtained recursively as the intersection of the stripe permutations induced by $C_2, \ldots, C_u$ and $D_1 \setminus C_1, D_2, \ldots, D_v$. This algorithm can be implemented to run in linear time (cf. also Rudolf [13]).

It follows from the considerations above that the set P_k and the intersection $Q_{k-1} = \bigcap_{q=1}^{k-1} P_q$ computed so far, can both be represented by a pair of stripe permutations for the rows and the columns, say (R_1, S_1) and (R_2, S_2), respectively. To compute the intersection $P_k \cap Q_{k-1}$ we first determine the following intersections of pairs of stripe permutations: $(R_1 \cap R_2, S_1 \cap S_2)$ and $(R_1^- \cap R_2, S_1^- \cap S_2)$ (note that in the latter case both the ordering of the row and of the column stripes must be reversed). Let (R_3, S_3) be the resulting pair of stripe permutations. Then we have $(\phi, \psi) \in Q_k$ if and only if either $(\phi, \psi) \in (R_3, S_3)$ or $(\phi, \psi) \in (R_3^-, S_3^-)$. Hence the set Q_k can again be represented by a pair of stripe permutations. Summarizing, we are able to construct the full set $Q_k = Q_{k-1} \cap P_k$ in linear time.

(Case 2) Now let us deal with the case when we arrive for the first time at a threshold matrix T^ℓ that is not connected any more, but still does not contain all ones rows or columns. Suppose that T^ℓ induces the blocks $\bar{T}_1, \bar{T}_2, \ldots, \bar{T}_f$. Then the relative order of the blocks is arbitrary, but within each block the sets $P(\bar{T}_q)$ can be completely described by two stripe permutations R_q and S_q for the rows and the columns of $\bar{T}_q$, respectively (instead of R_q and S_q also the reverse permutations R_q^- and S_q^- could be used).

In order to intersect the current set $\tilde{Q}_{\ell-1} = Q_{\ell-1}$ with the set P_ℓ, we thus would have to take into account all possibilities of arranging the blocks of T^ℓ. This task turns out to be too time-consuming. Below we show that it is sufficient to compute a proper subset $\tilde{Q}_\ell$ of the intersection $Q_\ell = Q_{\ell-1} \cap P_\ell$. For that purpose, we determine an appropriate ordering π of the blocks of T^ℓ which is consistent with the current set $Q_{\ell-1}$ and intersect $Q_{\ell-1}$ only with those pairs of permutations within $\mathcal{P}_\ell$ for which the blocks are arranged according to π. The advantage of this approach is that after fixing the ordering π, we are again in the situation that we need to intersect two pairs of stripe permutations.

In the sequel we describe how to find an ordering π of the blocks of T^ℓ which is consistent with the intersection $\mathcal{Q}_{\ell-1}$ determined so far. To that end, suppose that $\mathcal{Q}_{\ell-1}$ is described by the stripe permutations $\mathcal{R}$ and $\mathcal{S}$ which are induced by the row and column stripes $R_1, \ldots, R_u$ and $S_1, \ldots, S_v$, respectively. Let stripe R_q contain r_q rows and stripe C_p contain c_p columns. Then we define numbers $N^c(j)$ resp. $N^r(i)$ for each column j resp. for each row i of matrix T^ℓ by setting

$$N_c(j) := 1 + \sum_{q=1}^{p-1} c_q \text{ for all } j \in C_p, p = 1, \ldots, v \text{ and}$$

$$N_r(i) := 1 + \sum_{q=1}^{t-1} r_q \text{ for all } i \in R_t, t = 1, \ldots, u.$$

Note that $N_c(j)$ denotes the leftmost position of column j within the stripe permutation $\mathcal{S}$, while $N_r(i)$ corresponds to the leftmost position of row i within the stripe permutation $\mathcal{R}$.

Now the numbers $N_c(j)$ and $N_r(i)$ are used to construct an appropriate ordering π of the blocks of T^ℓ. We first compute for each block $\bar{T}_q$ of matrix T^ℓ the following four numbers

$$\alpha_q^c := \min\left\{N_c(j) \mid \text{column } j \in \bar{T}_q\right\}, \qquad \beta_q^c := \max\left\{N_c(j) \mid \text{column } j \in \bar{T}_q\right\}$$

$$\alpha_q^r := \min\left\{N_r(i) \mid \text{row } i \in \bar{T}_q\right\}, \qquad \beta_q^r := \max\left\{N_r(i) \mid \text{row } i \in \bar{T}_q\right\}.$$

The ordering π is obtained by sorting the blocks such that $\alpha_{\pi(1)}^c \leq \ldots \leq \alpha_{\pi(f)}^c$, $\beta_{\pi(1)}^c \leq \ldots \leq \beta_{\pi(f)}^c$, $\alpha_{\pi(1)}^r \leq \ldots \leq \alpha_{\pi(f)}^r$ and $\beta_{\pi(1)}^r \leq \ldots \leq \beta_{\pi(f)}^r$. If there is no such ordering, the intersection $\tilde{\mathcal{Q}}_{\ell-1} \cap \mathcal{P}_\ell$ is empty implying that also $\mathcal{Q}_L = \emptyset$. Ties are broken arbitrarily.

By merging the stripe permutations representing the blocks of T^ℓ according to the ordering π constructed above, we obtain a pair $(\mathcal{R}', \mathcal{S}')$ of stripe permutations. There is only one difficulty in the merging process: How do we choose the right orientation for each block of T^ℓ. To that end, we have to ensure that in each block the rows and columns are ordered according to increasing values of N_r and N_c, respectively. (If this is impossible, the input matrix A cannot be permuted into a bottleneck Monge matrix.)

(Case 3) The only remaining case concerns threshold matrices T^k with all ones rows or columns. Such rows and columns are treated as follows: If an all ones row or column is created in T^k that was not present in T^{k-1}, the position of this row or column is arbitrary with respect to T^k and any matrix T^ℓ with $\ell > k$. Its position, however, is determined by the position of that stripe of the set $\tilde{\mathcal{Q}}_{k-1}$ in which this row or column was located in step $k-1$. Hence we can choose an arbitrary position within this stripe, fix the new all ones row or column at this position and disregard it in all further steps.

Theorem 6. *For a permuted bottleneck Monge matrix A, the algorithm described above detects a pair (ϕ, ψ) such that $A_{\phi,\psi}$ is bottleneck Monge.*

Proof. Omitted. □

To analyse the time complexity of our algorithm we observe that the algorithm goes through L rounds with $O(nm)$ time. By this we get a complexity of $O(Lnm)$ that is $O(n^2 m^2)$ in the worst case. In the remaining part of this section we show how to improve this worst case bound for the general case to $O(nm(n+m))$.

The key observation is that not every threshold value $\tilde{a}_k$ contributes a new information about the set $\widetilde{Q}_L$. Thus, we introduce *critical thresholds*, i.e. thresholds $\tilde{a}_k$ such that at least one stripe in the pair of stripe permutations representing the set $\widetilde{Q}_{k-1}$ is partitioned into two stripes. This only happens if there are two rows or columns within this stripe which become different in the k-th threshold matrix T^k, whereas they were identical in T^q for $q < k$. Hence, for a critical threshold $\tilde{a}_k$ we always have $\widetilde{Q}_k \neq \widetilde{Q}_{k-1}$ and for non-critical thresholds $\widetilde{Q}_k = \widetilde{Q}_{k-1}$ (with the trivial exception of $\widetilde{Q}_k = \emptyset$). Therefore, skipping thresholds which are not critical can do no harm, provided that we check in the end whether the constructed permutations indeed yield a bottleneck Monge matrix.

Lemma 7. *Let A be an $n \times m$ matrix. Then there are at most $m - 1$ thresholds which are critical with respect to the columns of A, and at most $n - 1$ thresholds are critical with respect to the rows. Furthermore, starting from the critical threshold $\tilde{a}_k$, the next critical threshold $\tilde{a}_\ell$ can be determined in $O(nm)$ overall time.* □

Combining all previous considerations and results we obtain an algorithm, which either finds a pair (ϕ, ψ) which permutes the given input matrix A into a bottleneck Monge matrix $A_{\phi,\psi}$ or proves that this is not possible. It is easy to see that this algorithm can be implemented in $O(nm(n+m))$ time. Note that there are at most $O(n+m)$ critical thresholds and hence at most $O(n+m)$ rounds. Each round takes $O(nm)$ time. The final check of property (2) can be performed in $O(nm^2)$ time (see Section 3).

Finally we mention that the following two-phase approach yields a slight improvement over the above algorithm. We start with applying a modified version which only constructs an ordering of the columns and disregards the rows. This clearly takes $O(nm^2)$ time since there are at most $O(m)$ critical thresholds. A corresponding ordering of the rows as explained in Section 7 can then be determined in $O(\min\{n^2 m, nm^2 \log n\})$ time. Combining these considerations with the algorithm above, we get an $O(\min\{nm(n+m), nm^2 \log n\})$ time algorithm for recognizing permuted bottleneck Monge matrices with arbitrary entries.

7 Summary and Concluding Remarks

We studied the following recognition problem: Given an $n \times m$ matrix A, either construct a pair of permutations (ϕ, ψ) for the rows and columns of A such that the permuted matrix $A_{\phi,\psi}$ becomes a bottleneck Monge matrix or determine that no such permutations exist. We presented an algorithm which solves this problem in $O(nm(n+m))$ for matrices with arbitrary entries.

Finally, we mention some variations of the recognition problem (RP) which can be solved by extending the techniques of Section 6.

For example, the problem (RP^r) in which the ordering of the columns has to remain fixed can be solved in $O(\min\{n^2m, nm^2\log n\})$ time. The first time bound is achieved by a straightforward modification of the algorithm above. (Note that in this case there are at most $O(n)$ critical thresholds.) The second bound is based on the results on two-row bottleneck Monge matrices in Section 3. For each $n \times 2$ submatrix A' of the given matrix A we compute the set of all row permutations transforming A' into a bottleneck Monge matrix. Due to the simple structure of these sets, it is easy to compute a permutation ϕ which is feasible for all submatrices and hence also for A. The overall complexity of this approach is $O(nm^2\log n)$. (We need $O(n\log n)$ time per submatrix.)

The problem (RP$^=$) where we demand $\phi = \psi$ can be solved in $O(n^3)$ time (details will be described in a forthcoming paper).

For certain problem classes the bottleneck Monge property is however too weak, and it must therefore be replaced by a stronger property. For example, Bein, Brucker and Park [3] introduced the following property ($*$): For all $1 \leq i < r \leq n$ and $1 \leq j < s \leq m$ we either have $\max\{a_{ij}, a_{rs}\} < \max\{a_{is}, a_{rj}\}$ or $\max\{a_{ij}, a_{rs}\} = \max\{a_{is}, a_{rj}\}$ and $\min\{a_{ij}, a_{rs}\} \leq \min\{a_{is}, a_{rj}\}$. Thanks to the fact that a 0-1 matrix fulfills property ($*$) if and only if it is a Monge matrix, we can recognize the class of matrices that can be permuted to fulfill ($*$) in $O(nm(n+m))$ by essentially the same approach as used for the class of permuted bottleneck Monge matrices. (We use the algorithm of Deineko and Filonenko [9] as a subroutine to decide whether a given 0-1 matrix is Monge.)

References

1. A. Aggarwal, M.M. Klawe, S. Moran, P. Shor and R. Wilber, Geometric applications of a matrix-searching algorithm, *Algorithmica* **2**, 195–208, 1987.

2. N. Alon, S. Cosares, D.S. Hochbaum and R. Shamir, An algorithm for the detection and construction of Monge sequences, *Linear Algebra and its Applications* **114/115**, 669–680, 1989.

3. W.W. Bein, P. Brucker and J.K. Park, Application of an algebraic Monge property, unpublished manuscript, presented at the 3rd Twente Workshop on Graphs and Combinatorial Optimization, Enschede, The Netherlands, June 1993.

4. K.S. Booth and G.S. Lueker, Testing for the consecutive ones property, interval graphs, and graph planarity using PQ-tree algorithms, *Journal of Computer and System Sciences* **13**, 1976, 335-379.

5. R.E. Burkard, On the role of bottleneck Monge matrices in combinatorial optimization, manuscript, Institute of Mathematics, University of Technology, Graz, Austria, June 1993, submitted to *Operations Research Letters*.

6. R.E. Burkard and W. Sandholzer, Efficiently solvable special cases of bottleneck travelling salesman problems, *Discrete Applied Mathematics* **32**, 1991, 61–76.

7. K. Cechlárová and P. Szabó, On the Monge property of matrices, *Discrete Mathematics* **81**, 1989, 123–128.

8. L. Chen and Y. Yesha, Efficient parallel algorithms for bipartite permutation graphs, *Networks* **22**, 1993, 29–39.

9. V.G. Deineko and V.L. Filonenko, On the reconstruction of specially structured matrices, *Aktualnyje Problemy EVM: programmirovanije*, Dnepropetrovsk, DGU, 1979, 43–45, (in Russian).

10. P.C. Gilmore, E.L. Lawler and D.B. Shmoys, Well-Solved Special Cases, in: E.L. Lawler at al. (eds.), *The Traveling Salesman Problem: A Guided Tour of Combinatorial Optimization*, John Wiley, Chichester, 1985, 87–143.

11. A.J. Hoffman, On simple linear programming problems, in: *Convexity, Proc. Symposia in Pure Mathematics*, Vol. 7, (ed. V. Klee), American Mathematical Society, Providence, RI, 1961, 317–327.

12. C.L. Monma, A.H.G. Rinnooy Kan, A concise survey of efficiently solvable special cases of the permutation flow shop-problem, *RAIRO Recherche opérationelle* **17**, 1983, 105–119.

13. R. Rudolf, Recognition of d-dimensional Monge arrays, Technical Report 230-92, Institute of Mathematics, University of Technology, Graz, Austria, 1992, to appear in *Discrete Applied Mathematics*.

14. R. Shamir, A fast algorithm for constructing Monge sequences in transportation problems with forbidden arcs, Report 136/89, Tel Aviv University, 1989, to appear in *Discrete Mathematics*.

15. J. Spinrad, A. Brandstädt and L. Stewart, Bipartite permutation graphs, *Discrete Applied Mathematics* **18**, 1987, 279–292.

16. A. Tucker, A structure theorem for the consecutive ones property, *Journal of Combinatorial Theory* **12(B)**, 1972, 153–162.

17. J.A.A. van der Veen, Solvable cases of the traveling salesman problem with various objective functions, Ph.D. thesis, University Groningen, The Netherlands, 1992.

Computing Treewidth and Minimum Fill-In: All You Need are the Minimal Separators*

T. Kloks[1]** , H. Bodlaender[1], H. Müller[2] and D. Kratsch[2]

[1] Department of Computer Science, Utrecht University, P.O. Box 80.089,
3508 TB Utrecht, The Netherlands
[2] Fakultät für Mathematik und Informatik, Friedrich-Schiller-Universität,
Universitätshochhaus, 07740 Jena, Germany

Abstract. Consider a class of graphs $\mathcal{G}$ having a polynomial time algorithm computing the set of all minimal separators for every graph in $\mathcal{G}$. We show that there is a polynomial time algorithm for TREEWIDTH and MINIMUM FILL-IN, respectively, when restricted to the class $\mathcal{G}$. Many interesting classes of intersection graphs have a polynomial time algorithm computing all minimal separators, like permutation graphs, circle graphs, circular arc graphs, distance hereditary graphs, chordal bipartite graphs etc. Our result generalizes earlier results for the TREEWIDTH and MINIMUM FILL-IN for several of these classes. We also consider the related problems PATHWIDTH and INTERVAL COMPLETION when restricted to some special graph classes.

1 Introduction

In many recent investigations in computer science, the notions of treewidth and path-width play an increasingly important role. One reason for this is that many problems, including many well studied NP-complete graph problems, become solvable in polynomial and usually even linear time, when restricted to the class of graphs with bounded tree- or pathwidth [2, 4, 7]. Of crucial importance for these algorithms is, that a tree-decomposition or path-decomposition of the graph is given in advance. Much research has been done in finding a tree-decomposition with a reasonable small treewidth. Recently, it was shown in [8] that a linear time algorithm exists to find an optimal tree-decomposition for a graph with bounded treewidth. However, the constant hidden in the 'big oh', is exponential in the treewidth, limiting the practicality of this algorithm.

The problem 'Given a graph $G = (V, E)$ and an integer k, is the treewidth of G at most k' is NP-complete, even when only complements of bipartite graphs G are allowed as input graphs [3] and it also remains NP-complete on bipartite graphs [20]. The problem 'Given a graph $G = (V, E)$ and an integer k, is the pathwidth of G at most k' is NP-complete on cobipartite graphs [3], bipartite graphs [20] and triangulated graphs [17].

The treewidth can be computed in polynomial time on triangulated graphs (trivially), cographs [10], circular arc graphs [28], chordal bipartite graphs [23], permutation graphs [9], circle graphs [22] and distance hereditary graphs [1]. Since many NP-complete problems remain NP-complete when restricted to some of these classes, it is

---

* The work of the first and second author has been supported partially by the ESPRIT Basic Research Action of the EC under contract No. 7141 (project ALCOM II.)
** This author is supported by the Foundation for Computer Science (S.I.O.N.) of the Netherlands Organization for Scientific Research (N.W.O.). Current address: Department of Mathematics and Computing Sciences, Eindhoven University of Technology, P.O.Box 513, 5600 MB Eindhoven, The Netherlands, Email: ton@win.tue.nl.

of great importance to be able to use the algorithms for graphs of small treewidth for these problems.

The TREEWIDTH problem is the problem of finding a triangulated graph H with smallest maximum clique size having the given graph G as spanning subgraph. The PATHWIDTH problem is the problem of finding an interval graph H with smallest maximum clique size having the given graph G as spanning subgraph.

The following two problems have similar goals. The MINIMUM FILL-IN problem is the problem of finding a triangulated graph H having the given graph G as spanning subgraph such that the number of added edges is minimum. The INTERVAL COMPLETION problem is the problem of finding an interval graph H having the given graph G as spanning subgraph such that the number of added edges is minimum. The MINIMUM FILL-IN problem is also called CHORDAL GRAPH COMPLETION and is of great importance in relation with the performance of Gaussian elimination on sparse matrices, see [26].

The problem 'Given a graph $G = (V, E)$ and an integer k, is it possible to add at most k edges to G such that the new graph is triangulated', is NP-complete on cobipartite graphs [30] and on bipartite graphs [29]. On the other hand, in [27] an $O(n^5)$ algorithm for bipartite permutation graphs, a proper subclass of the chordal bipartite graphs, is given. Furthermore, a minimum fill-in can be computed in linear time on cographs [12], a proper subclass of the permutation graphs. The problem 'Given a graph $G = (V, E)$ and an integer k, is it possible to add at most k edges to G such that the new graph is an interval graph', is known to be NP-complete on line graphs [14].

It was shown in [9] that for every cocomparability graph G, i.e. G is a complement of a transitive orientable graph, the triangulated graph H having the graph G as a spanning subgraph and smallest maximum clique size is indeed an interval graph. Thus, treewidth and pathwidth coincide on cocomparability graph, a class of the perfect graphs containing permutation graphs, interval graphs and trapezoid graphs. R. Möhring [18] showed that this result is extendable to AT-free graphs (asteroidal triple-free graphs). Thus, on AT-free graphs, which contain cocomparability graphs as a proper subclass while they are no longer a subclass of perfect graphs, treewidth and pathwidth still coincide [24].

Moreover, the 'phase method', described in [18] yields a very general result which also applies to finding a triangulated graph H having the graph G as a spanning subgraph such that the number of added edges is minimum. Hence, if the graph G is AT-free, then such a graph H is always an interval graph. Consequently, the problems MINIMUM FILL-IN and INTERVAL COMPLETION coincide on AT-free graphs as did TREEWIDTH and PATHWIDTH [24].

The main subject of the paper is to extend significantly the knowledge about the algorithmic complexity of the four mentioned problems when restricted to some special class of graphs by the follwing general approach:

Given a polynomial $P(n)$. Let $\mathcal{G}_P$ be a class of graphs having an algorithm computing for every $G \in \mathcal{G}_P$ the set of all minimal separators in $P(n)$ time, where n is the number of vertices of G. In this paper we show that for all graphs in such a class $\mathcal{G}_P$, the TREEWIDTH and the MINIMUM FILL-IN can be computed in polynomial time. Notice that many interesting classes of graphs have a polynomial time algorithm computing the set of all minimal separators on all graphs of the class. Such classes are permutation graphs, circular permutation graphs, trapezoid graphs, circle graphs, circular arc graphs, distance hereditary graphs, chordal bipartite graphs, cocomparability graphs of bounded dimension, weakly triangulated graphs, etc. Our results generalize earlier results for TREEWIDTH and MINIMUM FILL-IN for some of these classes, see for example [1, 9, 21, 22, 28].

Furthermore, it is shown that MINIMUM FILL-IN can be solved in linear time on k-almost trees and on partial 2-trees and in time $O(n^2)$ on cotriangulated graphs. Finally, it is shown that the problem 'Given a chordal bipartite graph $G = (V, E)$ and an integer k, is the pathwidth of G at most k', remains NP-complete on bipartite distance hereditary graphs, a proper subclass of the chordal bipartite graphs.

2 Preliminaries

In this section we start with some necessary definitions and results. We consider only finite, undirected and simple graphs $G = (V, E)$. We always denote the number of vertices of G by n and the number of its edges by e. For definitions and properties of graph classes not given here we refer to [11, 15, 19, 20].

Definition 1. A graph H is *triangulated* if it does not contain a chordless cycle of length at least four as an induced subgraph. A *triangulation* of a graph G is a graph H with the same vertex set as G such that H is triangulated and G is a subgraph of H.

In this paper we show that the two problems TREEEWIDTH and MINIMUM FILL-IN are solvable for those graphs for which the set of minimal separators can be computed in polynomial time.

If $G = (V, E)$ is a graph and $W \subseteq V$ a subset of vertices then we use $G[W]$ as a notation for the subgraph of G *induced* by the vertices of W.

Definition 2. Given a graph $G = (V, E)$ and two non adjacent vertices a and b, a subset $S \subset V$ is an a, b-*separator* if the removal of S separates a and b in distinct connected components. If no proper subset of S is an a, b-separator then S is a *minimal a, b-separator*. A *minimal separator* is a set of vertices S for which there exist non adjacent vertices a and b such that S is a minimal a, b-separator.

The following lemma, which must have been rediscovered many times, appears for example as an exercise in [15].

Lemma 3. *Let S be a separator of the graph $G = (V, E)$. Then S is a minimal separator if and only if there are two different connected components of $G[V - S]$ such that every vertex of S has a neighbor in both of these components.*

We use Dirac's characterization of triangulated graphs [13].

Lemma 4. *A graph G is triangulated if and only if every minimal separator is a clique.*

Definition 5. A *minimal triangulation* H of a graph $G = (V, E)$ is a triangulation such that the following two conditions are satisfied.

1. If a and b are non adjacent in H then every minimal a, b-separator in H is also a minimal a, b-separator in G.
2. If S is a minimal separator in H and C is the vertex set of a connected component of $H[V \setminus S]$ then $G[C]$ is also connected.

In [9] the following theorem is shown.

Theorem 6. *Let H be a triangulation of a graph G. There exists a minimal triangulation H' of G such that H' is a subgraph of H.*

We now show that we can restrict the triangulations to be considered somewhat more.

Definition 7. Let Δ be the set of all minimal separators of a graph $G = (V, E)$. For a subset $\mathcal{C} \subseteq \Delta$ let $G_{\mathcal{C}}$ be the graph obtained from G by adding edges between vertices contained in the same set $C \in \mathcal{C}$. If the graph $G_{\mathcal{C}}$ is a minimal triangulation of G such that $\mathcal{C}$ is exactly the set of all minimal separators of $G_{\mathcal{C}}$, then $G_{\mathcal{C}}$ is called an *efficient* triangulation.

Notice that for each $C \in \mathcal{C}$, the induced subgraph $G_{\mathcal{C}}[C]$ is a clique.

Theorem 8. *Let H be a triangulation of a graph G. There exists an efficient triangulation $G_{\mathcal{C}}$ of G which is a subgraph of H.*

Proof. Take a minimal triangulation H' which is a subgraph of H such that the number of edges of H' is minimal (theorem 6). We claim that H' is efficient. Let $\mathcal{C}$ be the set of minimal vertex separators of H'. We prove that $G_{\mathcal{C}} = H'$.

Since every minimal separator in a triangulated graph is a clique, it follows that $G_{\mathcal{C}}$ is a subgraph of H'. Consider a pair of vertices a and b which are adjacent in H' but not adjacent in G. Remove the edge from the graph H'. Call the resulting graph H^*. Since the number of edges of H' is minimal, it follows that H^* has a chordless cycle. Clearly this cycle must have length four. Let $\{x, y, a, b\}$ be the vertices of this square. Then x and y are non adjacent in H'. But then a and b are contained in every minimal x, y-separator in H'. It follows that a and b are also adjacent in $G_{\mathcal{C}}$. □

Corollary 9. *There exists an efficient triangulation such that the maximum clique has a number of vertices equal to the treewidth of the graph plus one, and there exists an efficient triangulation with a minimum number of edges (hence realizing minimum fill-in).*

3 Treewidth and minimum fill-in on graph classes having a polynomial time algorithm computing all minimal separators

3.1 Blocks

In this subsection, let $G = (V, E)$ be a graph and let Δ be the set of minimal separators of G.

Definition 10. A *block* is a pair (S, C) where $S \in \Delta$ is a minimal separator in $G = (V, E)$ and C is the vertex set of a connected component of $G[V \setminus S]$.

Definition 11. A block (S, C) is *feasible* if there exists a set $\mathcal{T} \subseteq \Delta$ such that the following conditions are satisfied:

1. Each $T \in \mathcal{T}$ is contained in $S \cup C$,
2. $S \in \mathcal{T}$ and
3. $G_{\mathcal{T}}[S \cup C]$ is triangulated.

The set $\mathcal{T}$ is called a *realizer* for (S, C).

Lemma 12. *Let $T \subseteq \Delta$ be a set of minimal separators such that G_T is an efficient triangulation. Let $S \in T$ and let $C_1, \ldots, C_t$ be the vertex sets of the connected components of $G[V \setminus S]$. Then each block (S, C_i) is feasible. Moreover, the set $T_i \subseteq T$ consisting of those separators that are fully contained in $S \cup C_i$, is a realizer for the block (S, C_i) and $G_{T_i}[S \cup C_i] = G_T[S \cup C_i]$.*

Proof. Consider a block (S, C_i). Let $T_i \subseteq T$ be the subset of separators that are contained in $S \cup C_i$. Then clearly $S \in T_i$. Let $G_i = G_{T_i}[S \cup C_i]$. We show that $G_i = G_T[S \cup C_i]$. This proves the theorem since $G_T[S \cup C_i]$ is triangulated.

First notice that G_i is a subgraph of $G_T[S \cup C_i]$. Assume that there are non adjacent vertices a and b in G_i which are adjacent in G_T. Then at least one of a and b is an element of C, otherwise they would be adjacent in G_i since $S \in T_i$. There is a minimal vertex separator $T \in T$ which is not contained in $S \cup C_i$ such that $a, b \in T$. Since T is a clique in G_T, T can not contain vertices in different connected components of $G_T[V \setminus S]$. ☐

Corollary 13. *There exists a minimal separator S and realizers T_i, one for each block (S, C_i), such that G_T, with $T = \cup T_i$, is a triangulation with the minimum number of edges over all triangulations of G.*

Corollary 14. *Let the treewidth of G be k. There exists a minimal separator S and realizers T_i, one for each block (S, C_i), such that G_T with $T = \cup T_i$, is a triangulation of G such that the number of vertices in each clique of G_T is at most $k + 1$.*

3.2 Nice blocks

Let $P(n)$ be a polynomial and let $\mathcal{G}_P$ be a class of graphs for which the set of minimal vertex separators can be computed by a $P(n)$ time algorithm for every graph of the class. As before, let $G = (V, E)$ be a graph in $\mathcal{G}_P$ and let Δ be the set of minimal separators in G.

Definition 15. *Let $B = (S, C)$ be a block. Let $N \subseteq S$ be the set of vertices of S which have, in the graph G, a neighbor in C. Let U be a realizer for B. The realizer U is small, if in $G_U[S \cup C]$, the set of vertices in S which have a neighbor in C is N.*

The following lemma states that we can restrict ourselves to small realizers.

Lemma 16. *Let $B = (S, C)$ be a block. Assume there is an efficient triangulation G_T such that $S \in T$. Let $U \subseteq T$ be the set of separators that are contained in $S \cup C$. Then U is a small realizer.*

Proof. Let N be the set of vertices in S which have a neighbor in C in the graph G. Let $H = G_U[S \cup C]$. Let N' be the set of vertices in S which have a neighbor in C in the graph H. Since H is a triangulation of $G[S \cup C]$, $N \subseteq N'$. Assume there exists a vertex $a \in N' \setminus N$.

Hence, there exists a vertex $z \in C$ such that z and a are adjacent in H but not in G. Since they are not adjacent in G, there must exist a minimal separator $W \in U$ such that $a, z \in W$. Since G_T is efficient, W is also a minimal separator in G_T. Let x and y be vertices such that W is a minimal x, y-separator in G_T. Then W is also a minimal x, y-separator in G since G_T is a minimal triangulation of G.

First assume that x and y are both not in C. We claim that $W \setminus \{z\}$ is also an x, y-separator in G_T, which is a contradiction. Indeed, any path from x to y passing through z and no other vertex of W must visit S at least twice. Since S is a clique in G_T there is a shortcut in the path which avoids z. But this is not possible since such a path must pass through some vertex of $W \setminus \{z\}$.

Now assume $x \in C$. We claim that $W \setminus \{a\}$ is also an x, y-separator, which is again a contradiction. Since W is also a minimal x, y-separator in G there is a path from x to y in G passing through a but using no other vertex of W. But this is clearly not possible since a has no neighbors in C in the graph G. $\qquad\square$

Definition 17. Let $B = (S, C)$ be a block. Let N be the set of vertices of S which have a neighbor in C in the graph G. For a vertex $z \in C$ let $D_1^z, \ldots, D_t^z$ be the vertex sets of the connected components of $G[C \setminus \{z\}]$. For each D_i^z, let N_i^z be the set of vertices of $S \cup \{z\}$ which have a neighbor in D_i^z in the graph G. The block B is called *nice* if one of the following conditions hold:

1. $G_{\{S\}}[S \cup C]$ (i.e., the graph induced by $S \cup C$ with edges between vertices of S added) is triangulated, or
2. There exists a vertex $z \in C$ such that each N_i^z is a minimal separator and all blocks (N_i^z, D_i^z) are nice.

Lemma 18. *Let $B = (S, C)$ be a block. Assume there is an efficient triangulation G_T such that $S \in T$. Then B is nice.*

Proof. The proof is by induction on the number of vertices in C. If $|C| = 1$ then $G_{\{S\}}[S \cup C]$ is triangulated, and hence B is nice. Assume the lemma holds for all blocks $B' = (S', C')$ with $|C'| < |C|$. Let G_T be an efficient triangulation with $S \in T$. Let N be the set of vertices in S that have a neighbor in C in the graph G. Let $\mathcal{U} \subseteq T$ be the set of separators that are contained in $S \cup C$. By lemma 16, $\mathcal{U}$ is a small realizer for B. Since $G_{\mathcal{U}}[S \cup C]$ is triangulated and S is a clique in $G_{\mathcal{U}}[S \cup C]$, there must exist a vertex $z \in C$ which is adjacent to N in S in the graph $G_{\mathcal{U}}[S \cup U]$.

Let D_i and N_i be sets defined as follows. $D_1, \ldots, D_t$ are the connected components of $G_T[C \setminus \{z\}]$. For each i let N_i be the set of vertices in $\{z\} \cup N$ which have a neighbor in D_i in the graph G_T. Then each $N_i \in T$. This shows that (N_i, D_i) is a block, and since G_T is a minimal triangulation it follows that $D_i = D_i^z$.

By Lemma 16 N_i^z is exactly the set of vertices in N_i which have a neighbor in D_i^z. This implies that $N_i = N_i^z$. Hence all sets N_i^z are minimal separators. By induction the blocks (N_i^z, D_i^z) are nice. This proves the theorem. $\qquad\square$

3.3 Algorithm

In this subsection we show that TREEWIDTH and MINIMUM FILL-IN can be solved in polynomial time for a class of graphs $\mathcal{G}_P$, having a $P(n)$ time algorithm computing the set of all minimal separators for every $G \in \mathcal{G}_P$, for some polynomial $P(n)$. We concentrate on TREEWIDTH but it is easy to adapt the algorithm for MINIMUM FILL-IN. Readers familiar with the algorithm given in [3] for recognition of partial k-trees, will note strong similarities.

Definition 19. Let B be a nice block. The *weight* of B, $w(B)$, is the minimum, over all efficient triangulations G_T with $S \in T$, of the maximum number of vertices in a clique of $G_T[S \cup C]$.

The following lemma shows how the weight of a nice block can be determined. We use the same notation as in Definition 17.

Lemma 20. *Let $B = (S, C)$ be a nice block. If $G_{\{S\}}[S \cup C]$ is triangulated then $w(B)$ is the number of vertices in the maximum clique of $G_{\{S\}}[S \cup C]$. Otherwise, $w(B)$ is the minimum over all vertices $z \in C$ for which all blocks $B_i = (N_i^z, D_i^z)$ are nice, of $\max(|S|, |N| + 1, \max_i w(B_i))$.*

Proof. The claim is clearly true if $G_{\{S\}}[S \cup C]$ is triangulated. Consider an efficient triangulation G_T with $S \in \mathcal{T}$. Let N be the set of vertices that have a neighbor in C in the graph G. There must exist a vertex z in C that is adjacent to all vertices of N in G_T. The lemma follows by induction if we assume that for the smaller blocks (N_i^z, D_i^z) the lemma is true. □

The algorithm for the treewidth can be implemented as follows.

step 1 First compute a complete list of all minimal separators.
step 2 Make a list of all blocks (S, C).
step 3 Sort the blocks (S, C) according to increasing number of vertices in the component C.
step 4 Consider the blocks one by one in this order. For each block $B = (S, C)$ determine if B is nice and, if so, determine its weight as follows. If $G_{\{S\}}[S \cup C]$ is triangulated then B is nice and the weight $w(B)$ is equal to the maximum number of vertices in a clique of $G_{\{S\}}[S \cup C]$. Otherwise, if $G_{\{S\}}[S \cup C]$ is not triangulated, then for each choice of $z \in C$ determine the components D_i^z and the sets N_i^z. Check if all N_i^z are minimal separators. If so, check if all blocks (N_i^z, D_i^z) are nice. If this holds, compute the weight using Lemma 20. Minimize this over all choices of z.
step 5 For all separators S such that all incident blocks (S, C) are nice, compute the maximum weight of the blocks. Give the separator this weight. Determine the minimum of the weights of the separators.

Notice that this algorithm can be implemented to have polynomial running time. By assumption, the first step can be performed in $P(n)$ time. Suppose, the number of minimal separators of a graph in $\mathcal{G}_P$ with n vertices is at most $Q(n)$. Clearly, $Q(n) \leq P(n)$. The bottlenecks of the algorithm can be step 1 and step 4. To analyse the costs of step 4, notice that the weight of each block can be computed in $O(n^3)$ time, if we do some preprocessing on the separators (it can be checked in $O(n)$ time if a set N_i^z is a minimal separator). This shows that the algorithm can be implemented to run in $O(P(n) + n^3 Q(n))$ time.

Theorem 21. *Let $P(n), Q(n)$ be some polynomials. Let $\mathcal{G}_P$ be a class of graphs G, having a $P(n)$ time algorithm computing the set of all minimal separators for every $G \in \mathcal{G}_P$, such that each graph with n vertices in $\mathcal{G}_P$ has at most $Q(n)$ minimal separators. Then the TREEWIDTH problem as well as the MINIMUM FILL-IN problem can be solved by a $O(P(n) + n^3 Q(n))$ time algorithm for all graphs of $\mathcal{G}_P$.*

As a remark we give the main ideas for adapting the algorithm to MINIMUM FILL-IN. Analogously to Definition 19 the weight $\hat{w}(B)$ of a nice block $B = (S, C)$ is the minimum, over all efficient triangulations G_T with $S \in \mathcal{T}$, of the number of edges in $G_T[S \cup C]$. Similar to Lemma 20 $\hat{w}(B)$ of a nice block B fullfills: If $G_{\{S\}}[S \cup C]$ is triangulated then $\hat{w}(B)$ is the number of edges in $G_{\{S\}}[S \cup C]$. Otherwise, $\hat{w}(B)$ is the minimum over all vertices $z \in C$ for which all blocks $B_i = (N_i^z, D_i^z)$ are nice, of $(\sum_{i=1}^{r}(\hat{w}(B_i) - \binom{|N_i^z|}{2}))) + |N| + \binom{|S|}{2}$, if exactly r blocks appear.

4 Some special graph classes

Many well-known classes of graphs have a polynomial time algorithm computing the set of all minimal separators for any graph of the class, particularly classes of intersection graphs:

Permutation graphs can be defined as the intersection graphs of line segments between two parallel lines. They have $O(n^2)$ minimal separators which is shown by a scanline argument in [9]. A generalization of the permutation graphs are the **circular permutation graphs**, which can be defined as the intersection graphs of paths between two concentric circles in the plane, such that no two paths cross each other more than once. A similar scanline argument as in [9] shows that there are $O(n^4)$ minimal separators, which can be computed in $O(n^5)$ time (= the time needed to output the separators). (The main difference is that a minimal separator is now characterized by *two* scanlines.) Another generalization of permutation graphs are the **cocomparability graphs of bounded dimension**. Using the characterization from [16], it can be shown that the number of minimal separators is polynomial (but exponential in the dimension), using techniques, resembling the scanline argument in [9].

Trapezoid graphs are the intersection graphs of trapezoids between two parallel lines where two parallel sides of any trapezoid are segments of the parallel lines. Here a subset $S \subseteq V$ is a minimal separator of $G = (V, E)$ iff it consists of all vertices corresponding to the trapezoids crossing a scanline, i.e. a line segment with both endpoints on the parallel lines, and every $s \in S$ has a neighbour in every component of $G - S$. There are $O(n^2)$ non-equivalent scanlines, hence trapezoid graphs have at most $O(n^2)$ minimal separators.

Circle graphs are the intersection graphs of chords of a circle. Here a scanline is a chord and in a similar way we get that circle graphs have at most $O(n^2)$ minimal separators. **Circular arc graphs** are the intersection graphs of arcs of a circle. They have at most $O(n^2)$ minimal separators which is again shown by a scanline argument, whereby scanlines are chords of the circle.

Chordal bipartite graphs are the bipartite graphs not having a chordless cycle of length greater than or equal to six. Their minimal separators are either a neighbourhood of a vertex or the intersection of two maximal complete bipartite subgraphs. Thus, chordal bipartite graphs have at most $O(e^2)$ minimal separators since they have at most $O(e)$ maximal complete bipartite subgraphs. (It was already shown in [23] that the treewidth of a chordal bipartite graph can be computed in time $O(e^3)$).

Cotriangulated graphs (graphs that are the complement of a triangulated graph) have a linear number of minimal separators, since these are exactly the neighbourhood sets $N(v)$ for all vertices $v \in V$.

Distance hereditary graphs, a subclass of the circle graphs, have at most $O(n)$ minimal separators which can be seen by analyzing the increase of this number when doing one-vertex-extensions (see definition 28 and also [1, 5]).

Weakly triangulated graphs are graphs G such that G and $\overline{G}$ do not contain an induced cycle of length at least five. The class contains many well-known classes like chordal bipartite graphs, distance hereditary graphs, triangulated graphs and cotriangulated graphs. Weakly triangulated graphs can have at most $O((n+e)^2)$ minimal separators. The proof of this non-trivial fact will be given in the full version.

All the above mentioned graph classes have the property that for every graph G of the class there is a set of candidates containing all minimal separators of G such that the size of this candidate set is bounded by a polynomial in n and e and there is a polynomial time algorithm computing this candidate set. Every graph class with these properties has a polynomial time algorithm for computing the set of all minimal separators for any graph of the class, since it can be checked in linear time whether a certain candidate $S \subseteq V$ is indeed a minimal separator by using Lemma 3. The problems TREEWIDTH and MINIMUM FILL-IN are solvable in polynomial time when restricted to any of these graph classes by theorem 21. Nevertheless, we consider a few of them in more detail.

Without proof, we mention the following results, which will be shown in the full paper: $O(nk)$, and $O(n^2)$ algorithms for treewidth and minimum fill-in, respectively, for permutation graphs, and for trapezoid graphs, provided permutation or trapezoid diagrams are given, and $O(nk^{d-1}d)$ and $O(n^d d)$ algorithms for treewidth and minimum fill-in for cocomparability graphs of dimension d, provided an f-diagram which is the concatenation of d permutation diagrams is given (see [16]).

4.1 k-almost trees and graphs with treewidth at most 2

The following lemma is not hard to prove.

Lemma 22. *Suppose vertex v has exactly two neighbors w, x in graph $G = (V, E)$. There exists a triangulation $H = (V, F)$ of G with a minimum number of edges (hence minimum fill-in), such that v has still degree 2 in H, and $\{v, w\} \in F$.*

A graph $G = (V, E)$ is an k-almost tree, if for every every spanning tree T of G and biconnected component of G, there are at most k edges in the biconnected component that do not belong to T. An almost tree with parameter 1 is also called a *cactus*.

Let k be a fixed constant. One can show that every minimal separator in an k-almost tree has size $O(k)$. Hence, minimum fill-in can be solved in polynomial time for k-almost trees. A linear time algorithm can be obtained, using lemma 22. (As the treewidth of an almost tree with parameter k is at most $k + 1$ [6], the treewidth of these graphs can be computed in linear time [8]).

Theorem 23. *Let k be a fixed constant. The minimum fill-in problem can be solved in linear time for k-almost trees.*

Proof. It is sufficient to solve the problem for biconnected graphs. By applying lemma 22 and removing simplicial vertices, we reduce in linear time the problem to finding the minimum fill-in on a biconnected k-almost tree without vertices of degree 2. The latter has size $O(k)$, hence allows the problem to be solved in constant time. □

A similar technique works on the graphs with treewidth 2. It should be noted, that graphs with treewidth 2 can have an exponential number of minimal separators. An interesting and surprisingly still open problem is to compute the minimum fill-in for graphs with bounded treewidth.

Theorem 24. *The minimum fill-in problem can be solved in linear time for graphs with treewidth at most 2.*

4.2 Cotriangulated graphs

Cotriangulated graphs are the class of graphs containing exactly the complements of triangulated graphs. The following three theorems, showing that TREEWIDTH and PATHWIDTH are solvable in polynomial time on cotriangulated graphs, are given in [20]. Theorem 25 describes the structure of triangulations of cotriangulated graphs.

Theorem 25. *Let $G = (V, E)$ be triangulated. Let H be a triangulation of $\overline{G}$. There is a maximal clique with vertex set C in G such that $H[V \setminus C]$ is a clique.*

Theorem 26. *Let $\overline{G}$ be a cotriangulated graph. There is an $O(n^2)$ algorithm to find a tree-decomposition with width equal to the treewidth of G.*

Theorem 27. *There exists a polynomial time algorithm which, given a cotriangulated graph $\overline{G}$, computes a path-decomposition of width equal to the pathwidth of $\overline{G}$.*

Theorem 25 is also the key for designing an $O(n^2)$ algorithm solving the MINIMUM FILL-IN problem on cotriangulated graphs.

Clearly, for any triangulated graph $G = (V, E)$ the minimal triangulations of $\overline{G}$ are split graphs with independent sets forming the maximal cliques of G. This leads to the following algorithm. First we construct the complement G of $\overline{G}$. Next for each vertex $v \in V$ compute the degree $d(v)$ in G. Then compute the list of maximal cliques of G. That is a list of at most n cliques. For each clique C the corresponding fill-in is $\binom{V \setminus C}{2} \cap E$. The cardinality of this set is $|E| + \binom{|C|}{2} - \sum_{v \in C} d(v)$. The set with smallest cardinality is a minimum fill-in. It is easy to see that each step of the algorithm takes $O(n^2)$ time.

4.3 Chordal bipartite graphs

In this subsection we deal with chordal bipartite graphs and bipartite distance hereditary graphs. A bipartite graph is chordal bipartite if each of its cycles of length at least six has one chord, and it is bipartite distance hereditary if each of these cycles has two chords [5]. As mentioned above, TREEWIDTH and MINIMUM FILL-IN are polynomial time solvable both on chordal bipartite graphs and distance hereditary graphs. In contrast, we show the intractability of determining the pathwidth of bipartite distance hereditary graphs.

Let d_G be the distance metric on $G = (V, E)$. A (connected) graph $G = (V, E)$ is *distance hereditary* iff for each induced connected subgraph H of G holds: $\forall x, y \in V(H)$ $d_H(x, y) = d_G(x, y)$. The following is given in [5].

Definition 28. Let $G' = (V', E')$ be a graph, $x' \in V'$, $x \notin V'$. We define the graph $G := \text{OVE}(G', x', x)$ as follows: $G = (V, E)$ with $V := V' \cup \{x\}$ and
1. OVE = PV — the *pendant vertex* operation: $E := E' \cup \{xx'\}$,
2. OVE = FT — the *false twin* operation: $\quad E := E' \cup \{xy : y \in N(x')\}$,
3. OVE = TT — the *true twin* operation: $\quad E := E' \cup \{xy : y \in N[x']\}$.

In [5] is shown that a graph is bipartite distance hereditary iff it can be generated from the K_1 by a sequence of one vertex extensions PV and FT, and distance hereditary, iff additionally the true twin operation in allowed.

Theorem 29. *The problem 'Given a bipartite distance hereditary graph $G = (V, E)$ and an integer k, is the pathwidth of G at most k', is NP-complete.*

Proof. The reduction here is from a version of MINIMUM CUT LINEAR ARRANGEMENT, which is NP-complete for trees with poynomially bounded edge weights [25]. We consider a tree $T = (V, E)$ with $V = \{x_1, \ldots x_n\}$, polynomially bounded weights $w(e)$ on the edges $e \in E$, and a bound k. We define $W = 1 + \sum_{e \in E} w(e)$. W.l.o.g. we may assume $k < W - 1$.

The construction of our reduction graph $G = (X, Y, F)$ is as follows. We start with a copy of T, subdivide each edge $\{x_i, x_j\}$ by a new vertex $y_{i,j}$ and add a pendant vertex y_i to x_i for all i with $1 \le i \le n$. We obtain G from this enlarged tree by multiplying (in sense of false twins) the vertices x_i by W (these copies form the set X_i), the vertices y_i by $2W$ (these copies form the set Y_i), and the vertices $y_{i,j}$ by $w(\{x_i, x_j\})$ (these copies form the set $Y_{i,j}$). The color classes of G are $X = \bigcup_{i=1}^{n} X_i$ and $Y = \bigcup_{i=1}^{n} Y_i \cup \bigcup_{\{x_i, x_j\} \in E} Y_{i,j}$.

Clearly G is a bipartite distance hereditary graph. The maximal complete bipartite subgraphs of G are of the form $G[X_i \cup X_j \cup Y_{i,j}]$ for an edge $\{x_i, x_j\} \in E$ or $G[X_i \cup Y_i \cup \bigcup_{x_j \in N(x_i)} Y_{i,j}]$ for $1 \le i \le n$.

The remainder of the NP-completeness proof is a matter of routine now. For each maximal complete bipartite subgraph we have to complete one color class to obtain a triangulation of G. Hence a path decomposition of G uses for each i, $1 \le i \le n$, a set containing X_i and this leads to a linear layout of V. □

5 Conclusions

A direct consequence of theorem 21 and the result of [24] (cf. section 1) is that for any class $\mathcal{G}_P$ which has polynomial time algorithm computing the set of all minimal separators for every $G \in \mathcal{G}_P$, and which is a subclass of the class of the AT-free graphs, PATHWIDTH and INTERVAL COMPLETION, are solvable by polynomial time algorithms.

We end with some open questions. It would be interesting to find more well known classes of graphs, for which theorem 21 can be applied. What is the complexity of INTERVAL COMPLETION on split graphs? What is the algorithmic complexity of MINIMUM FILL-IN, INTERVAL COMPLETION and of PATHWIDTH, when restricted to graphs of bounded treewidth? For these latter problems, the usual techniques for graphs with bounded treewidth seem not to work, or at least it will be difficult to see how they could be applied.

References

1. Anand, R., H. Balakrishnan and C. Pandu Rangan, Treewidth of distance hereditary graphs, To appear.
2. Arnborg, S., Efficient algorithms for combinatorial problems on graphs with bounded decomposability — A survey. *BIT* **25**, (1985), pp. 2–23.
3. Arnborg, S., D. G. Corneil and A. Proskurowski, Complexity of finding embeddings in a k-tree, *SIAM J. Alg. Disc. Meth.* **8**, (1987), pp. 277–284.
4. Arnborg, S. and A. Proskurowski, Linear time algorithms for NP-hard problems restricted to partial k-trees. *Disc. Appl. Math.* **23**, (1989), pp. 11–24.
5. Bandelt, H. J. and H. M. Mulder, Distance-Hereditary Graphs, *Journal of Combinatorial Theory*, Series **B 41**, (1986), pp. 182–208.
6. Bodlaender, H., Classes of graphs with bounded treewidth, Technical Report RUU-CS-86-22, Department of Computer Science, Utrecht University, 1986.
7. Bodlaender, H., A tourist guide through treewidth, Technical report RUU-CS-92-12, Department of Computer Science, Utrecht University, Utrecht, The Netherlands, 1992. To appear in: *Acta Cybernetica*.
8. Bodlaender, H., A linear time algorithm for finding tree-decompositions of small treewidth, In *Proceedings of the 25th Annual ACM Symposium on Theory of Computing*, 1993, pp. 226–234.

9. Bodlaender, H., T. Kloks and D. Kratsch, Treewidth and pathwidth of permutation graphs, Technical report RUU-CS-92-30, Department of Computer Science, Utrecht University, Utrecht, The Netherlands, (1992). To appear in: *Proceedings of the 20th International Colloquium on Automata, Languages and Programming* (1993).

10. Bodlaender, H. and R. H. Möhring, The pathwidth and treewidth of cographs, In *Proceedings 2nd Scandinavian Workshop on Algorithm Theory*, Springer Verlag, Lecture Notes in Computer Science 447, (1990), pp. 301–309.

11. Brandstädt, A., Special graph classes — a survey, Schriftenreihe des Fachbereichs Mathematik, SM-DU-199 (1991), Universität-Gesamthochschule Duisburg.

12. Corneil, D. G., Y. Perl, L. K. Stewart, Cographs: recognition, applications and algorithms, *Congressus Numerantium*, 43 (1984), pp. 249–258.

13. Dirac, G. A., On rigid circuit graphs, *Abh. Math. Sem. Univ. Hamburg* 25, (1961), pp. 71–76.

14. Garey, M. R. and D. S. Johnson, *Computers and Intractability: A Guide to the Theory of NP-completeness*, San Francisco, 1979.

15. Golumbic, M. C., *Algorithmic Graph Theory and Perfect Graphs*, Academic Press, New York, 1980.

16. Golumbic, M. C., D. Rotem, J. Urrutia, Comparability graphs and intersection graphs, *Discrete Mathematics* 43, (1983), pp. 37–46.

17. Gustedt, J., On the pathwidth of chordal graphs, To appear in *Discrete Applied Mathematics*.

18. Habib, M. and R. H. Möhring, Treewidth of cocomparability graphs and a new order-theoretic parameter, Technical Report 336/1992, Technische Universität Berlin, 1992.

19. Johnson, D. S., The NP-completeness column: An ongoing guide, *J. Algorithms* 6, (1985), pp. 434–451.

20. Kloks, T., *Treewidth*, Ph.D. Thesis, Utrecht University, The Netherlands, 1993.

21. Kloks, T., Minimum fill-in for chordal bipartite graphs, Technical Report RUU-CS-93-11, Department of Computer Science, Utrecht University, 1993.

22. Kloks, T., Treewidth of circle graphs, Technical Report RUU-CS-93-12, Department of Computer Science, Utrecht University, 1993.

23. Kloks, T. and D. Kratsch, Treewidth of chordal bipartite graphs, *10th Annual Symposium on Theoretical Aspects of Computer Science*, Springer-Verlag, Lecture Notes in Computer Science 665, (1993), pp. 80–89.

24. Möhring, R. H., private communication.

25. Monien, B. and I. H. Sudborough, Min Cut is NP-complete for Edge Weighted Trees, *Theoretical Computer Science* 58 (1988), pp. 209–229.

26. Rose, D. J., Triangulated graphs and the elimination process, *J. Math. Anal. Appl.*, 32 (1970), pp. 597–609.

27. Spinrad, J., A. Brandstädt, L. Stewart, Bipartite permutation graphs, *Discrete Applied Mathematics*, 18 (1987), pp. 279–292.

28. Sundaram, R., K. Sher Singh and C. Pandu Rangan, Treewidth of circular arc graphs, To appear in *SIAM Journal on Discrete Mathematics*.

29. Tarjan, R. E., Decomposition by clique separators, *Discrete Mathematics* 55 (1985), pp. 221–232.

30. Yannakakis, M., Computing the minimum fill-in is NP-complete, *SIAM J. Alg. Disc. Meth.* 2, (1981), pp. 77–79.

Block Gossiping on Grids and Tori: Deterministic Sorting and Routing Match the Bisection Bound

Manfred Kunde

Fakultät für Informatik, Technische Universität München
Arcisstr. 21, 80290 München, Fed. Rep. of Germany

Abstract. Deterministic sorting and routing on r-dimensional $n \times ... \times n$ grids of processors is studied. For $h - h$ problems, $h \geq 4r$, where each processor initially and finally contains at most h elements, we show that the general $h - h$ sorting as well as $h - h$ routing problem can be solved within $hn/2 + o(hr^2n)$. That is, the bisection bound is asymptotically tight for deterministic $h - h$ sorting and $h - h$ routing. On an r-dimensional torus, a grid with wrap-arounds, the number of transfer steps is $hn/4 + o(hrn)$, again matching the corresponding bisection bound. This shows that inspite of the fact that routing problems contain more information at the beginning than the sorting problems there is no substantial difference between them on grids and tori. The results are possible by a new method where subsets of packets and information are uniformly distributed to the whole grid.

1 Introduction

Getting the right data to the right place within a reasonable amount of time is one of the most challenging and important tasks facing the designer of any large-scale general purpose parallel machine [14]. In this paper we present deterministic routing and sorting algorithms on mesh-connected arrays where the number of parallel data transfers asymptotically matches the trivial bisection bound.

An $n_1 \times ... \times n_r$ mesh-connected array or grid is a set $mesh(n_1, ..., n_r)$ of $N = n_1 n_2 ... n_r$ identical processors where each processor $P = (p_1, ..., p_r)$, $0 \leq p_i \leq n_i - 1$, is directly connected to its nearest neighbours only. A processor $Q = (q_1, ..., q_r)$ is called nearest neighbour of P if and only if the distance between them is exactly 1. For a grid without wrap-around connections the distance is given by $d(P, Q) = |p_1 - q_1| + \cdots + |p_r - q_r|$. For grids with wrap-around connections (tori) we define $d_{wrap}(P, Q) = \min(|p_1 - q_1|, n_1 - |p_1 - q_1|) + \cdots + \min(|p_r - q_r|, n_r - |p_r - q_r|)$.

For partial $h - h$ *routing problem* each processor initially receives at most h elements. Each packet has a destination address specifying the processor to which it has to be sent. Each processor is destination of at most h packets. The routing problem is to transport each packet to its address. We will also study *full $h - h$* problems where the grid contains exactly hN elements. Full $h - h$ routing problems are closely related to the $h - h$ sorting problem where the packets are considered as elements from a linearily ordered set. Each packet or element in a processor P is assumed to lie in a (memory) place (P, j), $0 \leq j < h$. For a given j the set of places $\{(P, j) | P \in mesh\}$ is called the j-th layer. There are at most h disjoint layers, numbered from 0 to $h - 1$. The places are indexed by an index function g, which is a bijective mapping from $mesh \times layer$ onto $\{0, ..., hN - 1\}$. The sorting problem (with respect to g) is to transport i-th smallest element to the place indexed with $i - 1$.

For a full $h - h$ routing problem one can supply each packet with one of the indices (for a given indexing g) of the places concerning to the destination processor of that packet. In this manner the full $h - h$ routing problem can be viewed as an $h - h$ sorting problem and therefore be solved by a sorting algorithm.

The model of computation is the conventional one (see [13, 14]) where exchange of data can happen only between nearest neighbours. The main restriction is that during one step interval at most one packet (as an atomic unit) can be transported on each directed channel between neighbouring processors. Packets may be stored in a processor until a limited buffer is filled. For complexity considerations only external transport steps are counted, i.e. operations within a processor, especially between different layers, are ignored.

For 2-dimensional $n \times n$ meshes without wrap-arounds several $1 - 1$ sorting algorithms have been proposed for a buffersize of 1 which all need about $3n + o(n)$ steps [16, 18, 19]. Park and Balasubramanian [15] showed that the same number of $3n + o(n)$ steps is also sufficient for $2 - 2$ sorting, i.e. for a doubled loading. In [9] we showed that the $1 - 1$ sorting problem can be solved deterministically in $2.5n + o(n)$ transport steps. A randomized algorithm for $1 - 1$ sorting was recently presented by Kaklamanis and Krizanc [3] sorting in only $2n + o(n)$ steps with high probability. For $1 - 1$ routing Leighton et. al. presented optimal deterministic algorithms (with constant buffer size) which are exactly matching the distance bound of $2n - 2$ steps [12, 17].

For $h - h$ problems on an $n \times n$ mesh it was shown in [9, 10] that sorting can be done by $hn + O(hn^{2/3})$ transport steps for $h \geq 4$, which is approximately twice the bisection bound. In the following improvements of this result were possible mainly by randomized algorithms as shown by Kaufmann, Rajasekaran and Sibeyn [4, 5, 6]. These randomized algorithms were mainly based on randomized routing alogrithms of Valiant and Brebner [20]. By the randomized approach the first time the simple bisection bound could be matched with high probability, for $h - h$ routing [5] as well as for $h - h$ sorting [6], provided $h \geq 8$. In this paper we show that it is possible to solve $h - h$ routing and sorting problems within $\max\{4n, hn/2\} + o(hn)$ steps deterministically. That means that for $h \geq 8$ the bisection bound can also asymptotically be matched by deterministic sorting and routing algorithms.

For wrap-around meshes Kaufmann and Sibeyn [6] gave a modified randomized $h - h$ sorting algorithm with $\max\{2n, hn/4\} + o(hn)$ steps. We show that without any modification our deterministic sorting and routing algorithms on wrap-around meshes also need only $\max\{2n, hn/4\} + o(hn)$ steps.

The method presented in this paper is different from those in [2, 3, 5, 6] where splitters are sent to a center block of the mesh and the obtained information is then broadcasted from this center block to all other blocks. In the new approach of this paper each block sends different information to all blocks. So it is a kind of all-to-all scattering between blocks. That means that this operation is more powerful than gossiping between blocks where each block broadcasts the same information to all other blocks, which is also called an all-to-all broadcasting. However, at a central point of our approach the informations sent to different blocks do not differ too much: they are approximately the same. Thus one can say that we have a kind of *pseudo all-to-all broadcasting* or *approximate block gossiping*.

The methods can straightforwardly be extended to r-dimensional meshes. Several algorithms for $h-h$ routing and sorting on r-dimensional grids have been otained in the past five years (see [2, 3, 4, 5, 6, 7, 8, 11]). The so far best deterministic $h-h$ sorting algorithm needed only $hn + o(hr^2n)$ steps [9]. In this paper we show that it is possible to solve $h-h$ routing and sorting problems within $\max\{2rn, hn/2\} + o(hr^2n)$ steps deterministically, i.e. only half the number as before. For $h \geq 4r$ this number of steps cannot be improved substantially, because it asymptotically matches the trivial bisection bound. On wrap-around meshes the number of steps is even halved to $\max\{rn, hn/4\}+o(hr^2n)$. Both numbers of steps are recently reported for randomized algorithms which solve the problems only with high probability (by Kaufmann and Sibeyn [6]).

Two further points are to be mentioned. The new deterministic sorting algorithms work for a broad variety of indexing schemes. And in contrast to former sorting algorithms they are optimally working as well for layer last indexings as for layer first indexings. This high flexibility with respect to indexing schemes is accompanied by a demand for a small buffersize. In many of the cases for each processor a buffer for only h packets is sufficient.

Finally, the results can be used for the so-called off-line routing (see for example [1]) where it is allowed to determine a route, as good as possible, for all packets in advance and outside the given grid. We show that there is no real improvement for off-line routing over on-line $h-h$ routing on an r-dimensional grid. At our present knowledge this shows a contrast to other types of nets (as hypercubes etc.).

2 Sorting on two-dimensional arrays

On an r-dimensional cubes with sidelength n for $h-h$ sorting and $h-h$ routing a lower bound of $hn/2$ transport steps is valid for all $r \geq 1$, the so-called *bisection bound*. (See for example [11]). For an r-dimensional torus the bisection bound is $hn/4$ transport step.

In this section we will show that the bisection bound can be matched asymptotically for multi-packet sorting on two-dimensional $n \times n$ grids. We start with some basic definitions.

2.1 Basic notations and definitions

As special submeshes so-called *blocks* and subsets of blocks are frequently used. Let $[a_i, b_i]$ denote an interval of integers with $0 \leq a_i \leq b_i < n$. An interval of processors is given by $([a_1, b_1], [a_2, b_2]) = \{(p_1, p_2) \mid a_i \leq p_i \leq b_i, i = 1, 2\}$. If $a_i = b_i = p_i$ then $p_i = [a_i, b_i]$ is used. The interval $[0, n-1]$ is abbreviated by $*$. An *interval of processors along the i-th axis* is called an i-*tower*, e.g. $(p_1, *)$ is a 2-tower.

1. Let a *block parameter* $k \geq 2$ be an integer such that k^3 is a divisor of n. For this k an $n \times n$ grid is divided into k^2 blocks. For arbitrary $k_i, 0 \leq k_i \leq k-1, i = 1, 2$, let $b_i = k_i n/k$. Then a *block* is defined by $B(k_1, k_2) = ([b_1, b_1 + n/k - 1], [b_2, b_2 + n/k - 1])$.

2. An i-*brick* is a subinterval of places of length n/k in a given layer j of a block B and an i-tower T. More formally, a brick is described by a set $(B \cap T) \times \{j\}$, where $B \cap T \neq \emptyset$ and j denotes a layer, $0 \leq j < h$.

That is, each block has sidelength $b = n/k$ and contains $(n/k)^2$ processors. Note that a processor $P = (p_1, p_2)$ lies in block $B(\lfloor kp_1/n \rfloor, \lfloor kp_2/n \rfloor)$. Obviously, there are exactly k^2 blocks in the whole mesh. The bricks within each block will be basic tools for the algorithms presented in this paper. Note that in each block there are $h(n/k)^2/(n/k) = h(n/k)$ i-bricks for an arbitrary axis i.

For the sorting problem several index functions have been used in the literature. We will introduce indexings for all kinds of 2-dimensional tuples.

1. The *lexicographical indexing* is defined by $lex_{2,n}(p_1, p_2) := p_1 n + p_2$, and the *reversed lexicographical indexing* by $rev_{2,n}(p_1, p_2) := p_2 n + p_1$. The *snake-like lexicographical indexing* g is given by $lex\_snake_{2,n}(p_1, p_2) = p_1 n + p_2$ for even p_1, and equals $(p_1 + 1)n - 1 - p_2$ for odd p_1.

2. For $P = (p_1, p_2)$ let $k_i = \lfloor k p_i/n \rfloor$, $i = 1, 2$. Then the *blockwise indexing* $g\_block_{2,n,k}$ with respect to an block indexing $g_{2,k}$ and an internal indexing $g'_{2,n/k}$ is given by

$$g\_block_{2,n,k}(p_1, p_2) = (n/k)^2 g_{2,k}(k_1, k_2) + g'_{2,n/k}(p_1 \bmod (n/k), p_2 \bmod (n/k)).$$

If $g_{2,k} = lex\_snake_{2,k}$ and $g'_{2,n/k} = lex_{2,n/k}$ we speak of the *blockwise snake-like indexing*.

3. For h layers the *layer last indexing of places* g_L with respect to an indexing of *processors* g is defined by $g_L(P, j) = hg(P) + j$, and the *layer first indexing* g_F is given by $g_F(P, j) = n^2 j + g(P)$, $P \in mesh$, $0 \le j \le h - 1$.

4. An indexing g is said to be *continuous*, if for all P, Q with $g(P) = g(Q) + 1$ we have $d(P, Q) = 1$, i.e. P and Q are neighbouring r-tuples.

The subscripts $2, n, k$ are omitted if it is obvious which actual values are meant. For submeshes the indexings are used in the corresponding manner. Normally we assume that blocks are numbered from 0 to $k^2 - 1$ by any continuous indexing g like the snake-like lexicographical indexing. (In this case the distance between two blocks B and C is given by $d(B, C) = \min\{d(P, Q) | P \in B, Q \in C\}$.) Furthermore, we assume that the i-bricks within the blocks are internally indexed by a layer-first indexing scheme.

We will first present uniaxial algorithms, which have turned out to be helpful in designing more complex algorithms. An algorithm is said to be *uniaxial*, if at each time step all processors use either rows only or columns only for sending packets.

2.2 The uniaxial sorting algorithm

The sorting procedures will use sorting of $n/k \times n/k$ blocks where the block parameter k is given in such a way that n/k^3 is an integer. Choose for example $k = n^{1/3}$. The indexing used for the whole sorting algorithm is layer last blockwise indexing. The indexing g for the blocks has to be continuous. For instance, the blockwise snake-like indexing is of this type.

Algorithm 1 (h-layer-uniaxial sort)

1. Sort uniaxially all $n/k \times n/k$ blocks with respect to rev_F (layer first reversed lexicographical indexing).
2. Perform a uniaxial all-to-all transportation, with brick i going to block $i \bmod k^2$ (with respect to the lexicographical indexing of the blocks).
3. Sort uniaxially all blocks with respect to layer first lexicographical indexing lex_F (for $n/k \times n/k$ meshes).
4. Perform a uniaxial all-to-all transportation with respect to the continuous block indexing g, where all those bricks with internal index i are going to the block with index $\lfloor i k^3/(nh) \rfloor$ with respect to g.

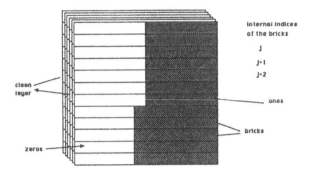

Fig. 1. Bricks in the dirty layer

5. Sort uniaxially with respect to layer last blockwise continuous indexing g all neighbouring pairs of blocks
a) with block indices $2i$ and $2i+1$, $i = 0, \ldots, k^2/2$,
b) with block indices $2i - 1$ and $2i$, $i = 1, \ldots, k^2/2$.

We will first show that the above algorithm really sorts. The complexity analysis will be done later. That the algorithm is uniaxial will become clear by the construction in the folling subsections.

Theorem 1. On a two-dimensional $n \times n$ grid with a buffer of h elements the algorithm 1 uniaxially solves the $h - h$ sorting problem with respect to arbitrarily given continuous block indexing.

Proof. The proof will be done by the help of the zero-one principle which says that it is sufficient to show that all loading of zeros and ones can be sorted by the algorithm. (See e.g. [13].)

Let in the beginning the two-dimensional grid contain an arbitrary loading of zeros and ones, with z zeros, $0 \leq z < hn^2$. After the all-to-all transport each block B_j contains exactly n/k^3 bricks from all blocks, each single brick from a successive region of k^2 bricks in a layer of that blocks. Let z_j be the number of zeros in block B_j in the beginning, i.e. $z = \sum_{j=0}^{k^2-1} z_j$. Then block B_i receives either $\lfloor z_j/k^2 \rfloor$ or $\lceil z_j/k^2 \rceil$ zeros from block B_j, because in step 1 all blocks are sorted with respect to layer-first reversed lexicographical indexing. Note that there is at most one region of k^2 successive bricks where the number of zeros may differ by at most 1. In all other regions the number of zeros in the corresponding bricks are the same. (Figure 1 .) Hence the number a_i of zeros in each block B_i is bounded by

$$\sum_{j=0}^{k^2-1} \lfloor z_j/k^2 \rfloor \leq a_i \leq \sum_{j=0}^{k^2-1} \lceil z_j/k^2 \rceil .$$

Therefore, $|a_i - a_j| \leq k^2$.

After sorting the blocks we have $\lfloor a_j/(n/k) \rfloor$ clean bricks of zeros in each block B_j. The only dirty brick maybe that with number $x_j = \lfloor a_j/(n/k) \rfloor + 1$. Since

$|a_i - a_j| \leq k^2$ and $k^2 \leq n/k$ we get $|x_j - x_i| \leq 1$. That means, in all blocks the only possibly dirty brick are those with internal index $x = \min x_i$ or $x + 1 = \min x_i + 1$.

By the all-to-all transportation successive bricks go into the same block or into successive blocks. Thus all dirty bricks go into at most two successive blocks. Therefore at most two neighbouring blocks may be dirty, because the block indexing g is continuous. The indexing of the processors within the blocks is of minor importance. After the final sorting of pairs of blocks all blocks and the total mesh is sorted.

Note that the continuous block indexings form a large class of index schemes which contains all kinds of snake-like indexings, spirals and so on. Moreover, it is worthwhile mentioning that in step 4. of the above algorithm layer first indexings can also be chosen. In this case, however, we have to guarantee that in each layer j we have continuous block indexings g_j. For these block indexings we have to demand that $g_j^{-1}(k^2 - 1) = g_{j+1}^{-1}(0)$ for $j = 0, \ldots, h - 2$. Then all those bricks with internal index i are going to the block with index $\lfloor (i \bmod (n/k))k^3/n \rfloor$ with respect to g_j, where $j = \lfloor ik/n \rfloor$. Thus for sorting we can choose suitable index functions from a broad variety of indexing schemes. This high flexibility with respect to indexings is accompanied by a demand for only a small buffersize as shown in the following sections.

The method used in the above algorithm is different from those in [2, 3, 5, 6] where splitters are sent to a center block of the mesh and the obtained information is then broadcasted from this center block to all other blocks. By the all-to-all transportations in algorithm 1 each block sends different packets (and information) to all blocks. So it is a kind of all-to-all scattering between blocks. That means that this operation is more powerful than gossiping between blocks where each block broadcasts the same information to all other blocks, which is also called an all-to-all broadcasting.

However, in step 2. the informations sent to different blocks do not differ too much: they are approximately the same. This fact is reflected in the nearly the same number of ones (or zeros), which is sent by a block to all other blocks in the case of a loading of zeros and ones. For arbitrary integers that means that representatives or splitters from each sorted interval of length n/k are sent to all blocks. By this way all the blocks receive not exactly, but approximately the same information. Thus one can say that we have a kind of *pseudo all-to-all broadcasting* or *approximate block gossiping*.

2.3 All-to-all transportation for one layer

Now have a look at the all-to-all transportation. The basic operations will take place on a $k \times k$ array of $1 \times n/k$ bricks or, which is the same, a $k \times n$ array of processors. Such a two-dimensional subgrid also will be refered as a k-stripe or just stripe (of bricks). We will map the bricks from the same column j of bricks to all other columns of bricks. This can be done in several ways. The basic routing procedure on each row of processors are shifts. A shift of i positions describes a routing problem on a one-dimensional array of processors where each packet from processor j, $0 \leq j < n$, has to be routed to processor $(j + i) \bmod n$. A shift of $n - i$ positions is the same as one of $-i$ positions. In this case we sometimes speak of a negative shift of i positions. For a $1 - 1$ problem a shift of i positions, $i \geq 1$, costs $\max(i, n - i)$ steps on a linear array. On ring it only needs $\min(i, n - i)$ steps.

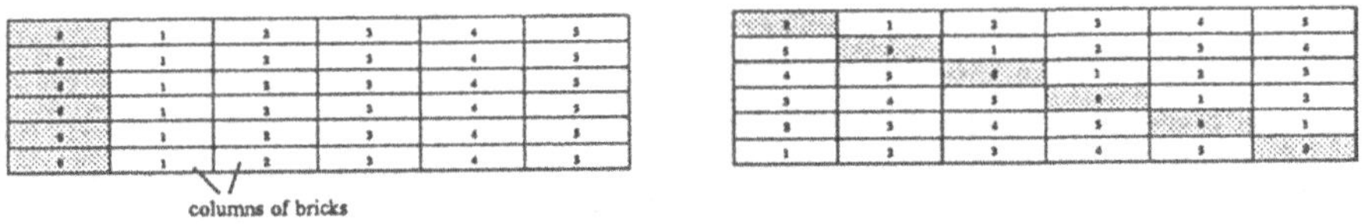

columns of bricks

Fig. 2. Uniform Brick Transport

Let π be any permutation on $0, \ldots, k-1$.

Algorithm 2 (Uniform-Brick-Transport)

$ubt(k, n/k, \pi)$

For all $i, j = 0, \ldots, k-1$ do in parallel: shift brick with coordinates (i, j) onto brick $(i, (j + \pi(i)) \bmod k)$

i.e. for all processors in brick (i, j) shift contents of processor (p_1, p_2) to processor $(p_1, ((p_2 + \pi(i)n/k) \bmod n))$.

By the above algorithm the j-th column of bricks receives exactly one brick from all other columns of bricks. The case where π is the identity is illustrated in figure 2. The transport cost for at most one packet at each processor is depending on the maximal distance. For a torus connection the shift costs at most $n/2$ steps. If there are no wrap-arounds at most $n - n/k$ steps are sufficient. Note that within a column of bricks we can easily permute the bricks in such a way that a brick (i, j) with brick destination (i', j') is first mapped onto that brick (u, j) that is destined to (u, j') under any standard brick transportation with respect to a π. After that transportation the brick (u, j') is then mapped on brick (i', j'). This can be done with all bricks in parallel. The additional cost is only $2k$. That means that on an $k \times k$ array of bricks of length n/k an arbitrary brick transposition of the above type can be performed within $2k + n$ transport steps on grid without wrap-around and within $2k + n/2$ steps on a torus.

In a similiar way we can distribute bricks in rows of blocks, i.e. an $n/k \times k$ array of bricks. A *uniform all-to-all mapping of bricks*, *unif*, is a permutation on the set of all bricks, which sends from each block exactly n/k^2 bricks (from the n/k bricks) to each of the k blocks. That is, $|unif(B_i) \cap B_j| = n/k^2$ for all blocks $B_i, B_j, 0 \le i, j < k$. A special uniform all-to-all mapping, the standard transportation, is given as follows. Divide the row of blocks into n/k^2 sucessive k-stripes. In each of these stripes the uniform brick transport ubt with respect to a permutation π has to be performed. Under this mapping in each block there are n/k^2 bricks destined for an arbitrarily given block j. In each block i transport those n/k^2 bricks destined to block j under the mapping *unif* to those n/k^2 bricks for block j under the standard mapping utb. This operation costs at most n/k steps. Then perform the standard transportation $ubt(k, n/k, \pi)$. After this operation within each block a further permutation of bricks can be performed within n/k steps. By the above discussion we get the following lemma.

Lemma 2. An arbitrary uniform all-to-all mapping of bricks on an $n/k \times k$ array of bricks of length n/k can be performed within $n + 2n/k$ transport steps. On a torus the number of transport steps is only $n/2 + 2n/k$.

We will now use the uniform brick transportation to formulate the all-to-all mapping on the whole grid in one layer. Obviously, an analogous brick transportation can be used for bricks transported along columns.

Algorithm 3 (Basic All-to-all mapping)

$ata(k, n/k, \pi)$

For all $j, l = 0, \ldots, k - 1$ and all $i = 0, \ldots, n/k - 1$ do in parallel:

1. In all rows j of blocks perform a brick transportation such that a 2-brick with internal index i goes into block $(j, i \bmod k)$.
2. In all blocks $B(j, l)$ make an orthogonal transposition, such that the i-th 2-brick along the rows becomes the i-th 1-brick along the columns.
3. In all columns j of blocks perform a brick transportation such that brick with original brick number i goes into block $(\lfloor i/k \rfloor \bmod k, j)$, i.e. the block with index $i \bmod k^2$ with respect to the lexicographical indexing of the blocks.

By lemma 2 we immediately get that the above algorithm needs $2n + O(n/k)$ transport steps on a grid without wrap-arounds and $n + O(n/k)$ steps on a torus. Assume that we have an arbitrarily given block indexing g. Let $g^{-1}(i)$ denote that block with index i, $0 \le i < k^2$, with respect to g. Let $a_i = lex(g^{-1}(i))$ be the index of block $g^{-1}(i)$ under the lexicographical indexing lex. Within each block perform now a permutation of bricks such that all bricks with internal index from $\{j | j \equiv i \bmod k^2\}$ are mapped onto those bricks with internal index from $\{j' | j' \equiv a_i \bmod k^2\}$. This costs $O(n/k)$ steps. Then perform algorithm 3 which now maps those bricks with original internal index i to the block with index $i \bmod k^2$ with respect to g.

Lemma 3. On an $n \times n$ grid an all-to-all mapping with respect to an arbitrarily given block indexing g, where all 2-bricks (1-bricks) with internal index i are mapped into block with index $i \bmod k^2$ with respect g, can uniaxially be performed within $2n + O(n/k)$ transport steps. On a torus the number of steps reduces to $n + O(n/k)$.

2.4 All-to-all transportation for h layers

The above algorithm gives some freedom in selecting π. We will use this to handle h layers simultanously instead of one after another. For doing so we introduce the concept of dense bundles of shifts which was presented first in [9].

Let us start the discussion with $h - h$ routing problems on linear arrays. Assume that the packets of layer j, $0 \le j < n$, have to be shifted i_{j+1} positions, $0 \le i_{j+1} < n$. We then say that the bundle of h shifts $bundle(i_1, \ldots, i_h)$ has to be performed. A bundle of h shifts $bundle(i_1, \ldots, i_h)$, $h \ge 4$, is said to be a *dense bundle* if and only if there is an $m \in \{\lfloor h/2 \rfloor, \lceil h/2 \rceil\}$ with $0 \le i_1 \le \cdots \le i_m \le n/2 \le i_{m+1} \le \cdots \le i_h \le n$, and for some small deviations ϵ_1, ϵ_2, with $|\epsilon_1|, |\epsilon_2|$ both in $o(n)$, we have

$$\sum_{j=1}^{m} i_j = hn/8 + \epsilon_1, \text{ and } \sum_{j=m+1}^{h} (n - i_j) = hn/8 + \epsilon_2.$$

As a first illustration of dense bundles observe that if a bundle of h shifts $bundle(i_1, \ldots, i_h)$ is dense, then $bundle(n - i_h, \ldots, n - i_1)$ is dense, too. The term "dense" is motivated by the next fundamental lemma, which shows that dense bundles of shifts can optimally be performed on linear arrays, which means that dense streams of packets have to pass all cuts.

Lemma 4. For $h \geq 4$ a dense bundle of h shifts can be performed in $hn/8 + max(|\epsilon_1|, |\epsilon_2|)$ transport steps on a ring, and in $hn/4 + |\epsilon_1| + |\epsilon_2|$ steps on a linear array.

Proof. The technical proof is omitted here and can be found in [10].

A dense bundle of shifts is defined on a linear array or ring. In [9, 10] it has been shown that for all $h \geq 4$ there are well-behaving permutations $\pi_1, \ldots, \pi_h$ over $\{0, \ldots, k-1\}$ such that $bundle(\pi_1(i) \cdot n/k, \ldots, \pi_h(i) \cdot n/k)$ form dense bundles of shifts for all $i, 0 \leq i < k$. For example, in the case $h = 4$ we can choose the four permutations $\pi_1(i) = i$, $\pi_2(i) = k/2 - i$, $\pi_3(i) = k/2 + i$, and $\pi_2(i) = k - i$. The corresponding deviations fulfill in all the cases $|\epsilon_1|, |\epsilon_2| \leq n/k$.

Thus we obtain the following lemma.

Lemma 5. On an $n \times n$ grid the all-to-all mapping with respect to a block indexing g in h layers, $h \geq 4$, maps all 2-bricks (1-bricks) with internal index i into the block with index $i \bmod k^2$ with respect to g. The mapping can uniaxially be performed within

$hn/2 + O(hn/k)$ transport steps on a array without wrap-arounds, and

$hn/4 + O(hn/k)$ steps on a torus.

Proof. For layer $j, 0 \leq j < h$, take permutation π_{j+1} from the set of well-behaving permutations $\pi_1, \ldots, \pi_h$. Assume that the basic all-to-all mapping for one layer is to be performed in layer j with permutation π_{j+1}. In each block (a, b) transport the respective hn/k^3 bricks to the hn/k^3 bricks in the h layers which will be transported to block (c, d) under the basic all-to-all mappings with respect to indexing g and permutation π_{j+1} in layer j. This costs at most hn/k steps. Then perform the basic all-to-all mapping with the dense bundles of shifts first along the rows which costs $hn/4 + O(hn/k)$ steps. Hereafter perform the same operation along the columns, which consumes the same number of steps. Hence in total $hn/2 + O(hn/k)$ steps are needed. Of course, the all-to-all mapping first along the columns, then along the rows, takes the same number of steps.

Obviously, the number of transport steps on the ring is only $hn/4 + O(hn/k)$, by lemma 4.

For algorithm 1 we can now give the number of transport steps needed. Note that the sorting of blocks in the beginning, in the middle, and at the end only costs $O(hn/k)$ steps. Two times an all-to-all mapping has to be performed, each one costs $hn/2 + O(hn/k)$ transport steps on a array without wrap-around, and $hn/4 + O(hn/k)$ transport steps on a torus (by lemma 5). Thus we can state the following theorem.

Theorem 6. On a two-dimensional $n \times n$ grid with a buffer of h elements, $h \geq 4$, the $h - h$ sorting problem (with respect to any continuous blockwise indexing) can uniaxially be solved with

a) $hn + O(hn/k)$ steps for on a grid without wrap-around

b) $hn/2 + O(hn/k)$ steps on a torus.

For all $h \geq 4$ the number of steps asymptotically matches the multi-section bound for uniaxial algorithms.

2.5 Overlapped sorting

The so far presented algorithms were uniaxial algorithms. As a problem with sorting on grids it was mentioned in [9] that the indexings seem to be not well-suited for overlapping. For probabilistic sorting this problem was recently solved [6]. We now show that optimal deterministic sorting can be done for a broad variety of blockwise indexing schemes.

Algorithm 4 (overlapped h-layer-sort)
1. Sort all $n/k \times n/k$ blocks with respect to rev_F (layer last reversed lexicographical indexing).
2. Colour all places (p_1, p_2, l) in processor (p_1, p_2) with colour $(p_1 + p_2 + l)$ mod 2. Within each processor put the contents of places with colour 0 into layers $0, \ldots, h/2 - 1$ and those of places with colour 1 into layers $h/2, \ldots, h - 1$.
3. Perform the uniaxial sorting algorithm 1 in colour 0 layers with row-column-row-column orientation and in colour 1 layers with column-row-column-row orientation, for both the colours with respect to the given continuous block indexing g.
4. Sort (in all layers) with respect to layer last blockwise indexing g all neighbouring pairs of blocks, first with block indices $2i$ and $2i + 1$, $i = 0, \ldots, k^2/2 - 1$, then with block indices $2i - 1$ and $2i$, $i = 1, \ldots, k^2/2 - 1$.

Theorem 7. On a two-dimensional $n \times n$ grid with a buffer of h elements, $h \geq 8$, the $h - h$ sorting problem (with respect to any continuous blockwise indexing) can be solved with
a) $hn/2 + O(hn^{2/3})$ steps on a grid
b) $hn/4 + O(hn^{2/3})$ steps on a torus.
For all $h \geq 8$ the number of steps asymptotically matches the bisection bound.

Proof. Assume there has been an arbitrary, but fixed loading of zeros and ones in the beginning. Let z be the total number of zeros in the mesh and z_j be the number of zeros in block B_j in the beginning. After the first sorting of blocks in each block there is at most one processor which may contain zeros and ones. All the other processors are clean. i.e. contain either zeros or ones. Hence after the second step in each block the number of zeros $z_j^{(0)}$ in layers with colour 0 and the number of zeros $z_j^{(1)}$ in layers with colour 1 differ at most by 1. Therefore, in the whole grid the difference between the number of zeros $z^{(0)}$ in layers with colour 0 and the number of zeros $z^{(1)}$ in layers with colour 1 is at most k^2.

After the overlapped sorting the only dirty block in colour 0 is that with index $\lceil z^{(0)}/(k^2 h/2) \rceil$ and in colour 1 is that block with index $\lceil z^{(1)}/(k^2 h/2) \rceil$. Therefore, the index of dirty blocks differ at most by 1. Since the chosen indexing of blocks is the same for both the colours, the respective dirty blocks are neighbouring or even the same. Thus by the final sorting of pairs of blocks the total grid can be sorted.

The number of steps is the sum of the number of steps for the uniaxial sorting algorithm 1 for $h/2$ layers, which is $hn/2 + O(hn/k)$, and the number of steps for the sorting of blocks in the beginning and at the end, which is $O(hn/k)$. Since k is about $n^{1/3}$ the theorem is proven.

3 Routing on two-dimensional grids

The above sorting algorithm can directly be applied on full $h - h$ routing problems. For doing so, assume an order of the packets according to the snake-like blockwise lexicographical indexing of their addresses. Applied on a partial $h-h$ routing problem with a total loading of 75 percent, for example, the sorting algorithm would route the packets to wrong destinations. However, for packet routing problems we can use the fact that packets know their final destination. This fact is used in third step of the following algorithm.

Algorithm 5 (h-layer routing)
Route(n, n, h)
1. Sort packets in all $n/k \times n/k$ blocks with respect to rev_F (layer first reversed lexicographical indexing).
2. Perform an all-to-all transportation along rows and columns, with brick i going to block $i \bmod k^2$ (with respect to the lexicographical indexing of the blocks).
3. Sort all blocks with respect to layer first lexicographical indexing lex_F. Within the blocks send packets with block address j, $0 \leq j < k^2$, to bricks with internal indices $\{jhn/k^3, \ldots, jhn/k^3 + hn/k^3 - 1\}$.
4. Perform an all-to-all transportation along rows and columns, where the brick with internal index i is going to the block with index $\lfloor ik^3/(hn) \rfloor$, where the blocks are indexed with respect to the snake-like lexicographical block indexing.
5. Within pairs of neigbouring blocks send packets to their final destination.

In step 4 it may happen that too many packets want to go to their transport bricks. However, by a similar argument as used in the proof of theorem 1 one can show that after step 2. in each block the number of packets destined for an arbitrary block j is at most $hn/k^3 + e_j$, where $\sum_j e_j \leq k^2 \leq n/k$. In this case at most e_j packets have to be routed either to brick with internal index $jhn/k^3 - 1$ or to that with index $jhn/k^3 + hn/k^3$, bricks to be transported to block $j - 1$ or $j + 1$ respectively. It can be shown that this routing within the blocks takes only $O(n/k)$ additional steps.

Theorem 8. On a two-dimensional $n \times n$ grid with a buffer of h elements, $h \geq 8$, the partial $h - h$ routing problem can be solved within
a) $hn/2 + O(hn^{2/3})$ transport steps, and
b) $hn/4 + O(hn^{2/3})$ steps on a torus.
For all $h \geq 8$ the number of steps asymptotically matches the bisection bound.

4 Higher-dimensional grids

The algorithms of the previous sections can easily be extended to grids with dimension $r \geq 3$. A basic tool in this context is a generalization of the fundamental all-to-all mapping which maps bricks from each r-dimensional $n/k \times \ldots \times n/k$ block uniformly onto all other r-dimensional blocks. For the lack of space the details are omitted here and will be presesented in a full paper. The result is that on an r-dimensional $n \times n \times \ldots \times n$ grid with a buffer of h elements, $h \geq 4r$, the $h - h$ sorting problem and the $h - h$ routing problem can be solved within
$hn/2 + O(hr^2 n^{r/(r+1)})$ transport steps, and

$hn/4 + O(hr^2 n^{r/(r+1)})$ steps on a torus.
The number of transport steps asymptotically matches the bisection bound.

References

[1] Annexstein, F.and Baumslag, M. A unified approach to off-line permutation routing on parallel networks. Proc. of SPAA 90. Iland of Crete, 1990, pp. 398–406.

[2] Kaklamanis, C., Krizanc, D., Narayanan, L. and Tsantilas, T. Randomized Sorting and Selection on Mesh-Connected Processor Arrays. Proceedings of SPAA 91. Hilton Head, South Carolina, 1991, pp. 17–28.

[3] Kaklamanis, C., Krizanc Optimal Sorting on Mesh-Connected Processor Arrays. Proceedings of SPAA 92. San Diego, California, 1992, pp. 50–59.

[4] Kaufmann, M., Rajasekaran, S., Sibeyn, J. Matching the Bisection Bound for Routing and Sorting on the Mesh. Proc. of SPAA 92. San Diego, California, 1992, pp. 31–40

[5] Kaufmann, M., S., Sibeyn, J. Optimal Multi-packet Routing on the Torus. Proc. Third Scandinavian Workshop on Algorithm Theory, 1992, pp. 118–129.

[6] Kaufmann, M., S., Sibeyn, J. Optimal $k - k$ Sorting on Meshes and Tori. Submitted.

[7] Kunde, M. Routing and Sorting on Mesh-Connected Arrays. In J.H.Reif (ed.). Proceedings of the 3rd AWOC 88, Lect. Notes Comp. Sci. 319, Springer, Berlin, 1988, pp. 423–433.

[8] Kunde, M. Balanced Routing: Towards the Distance Bound on Grids. Proceedings of SPAA 91. Hilton Head, South Carolina, 1991, pp. 260–271.

[9] Kunde, M. Concentrated Regular Data Streams on Grids: Sorting and Routing Near to the Bisection Bound. Proc. of FOCS 91. San Juan, Puerto Rico, 1991, pp. 141–150.

[10] Kunde, M. Routing and Sorting on Grids. Habilitationsschrift. TU Munich, Munich, 1991, 177 pp..

[11] Kunde, M. and Tensi, T. k-k routing on multidimensional mesh-connected arrays. J. of Parallel and Distributed Computing, 11 (1991), pp. 146-155

[12] Leighton, T., Makedon, F. and Tollis, I. A 2n-2 Step Algorithm for Routing in an n x n Array with Constant Size Queues. Proc. of SPAA 89. Santa Fe, 1989, pp. 328–335.

[13] Leighton, T. Introduction to Parallel Algorithms and Architectures: Arrays - Trees - Hypercubes. Morgan Kaufmann, San Mateo, California, 1992.

[14] Leighton, T. Methods for Message Routing in Parallel Machines. Proceedings of the 1992 ACM Symposium on Theory of Computation, STOC 92. 1992, pp. 77–96.

[15] Park, A. and Balasubramanian, K. Reducing Communication Costs for Sorting on Mesh-Connected and Linearly Connected Parallel Computers. J. of Parallel and Distributed Computing, 9 (1990), pp. 318–322

[16] Ma, Y., Sen, S. and Scherson, I.D. The distance bound for sorting on mesh-connected processor arrays is tight. Proceedings FOCS 86, pp. 255–263.

[17] Rajasekaran, S. and Overholt, R. Constant Queue Routing on a Mesh. In Choffrut, C. and Jantzen, M. (eds.), Proceedings of STACS 91,Lect. Notes Comp. Sci., vol. 480, Springer, Berlin, 1991, pp. 444–455.

[18] Schnorr, C.P. and Shamir, A. An optimal sorting algorithm for mesh-connected computers. Proceedings STOC 1986. Berkley, 1986, pp. 255–263.

[19] Thompson, C.D. and Kung, H.T. Sorting on a mesh-connected parallel computer. CACM 20 (1977), pp. 263–270.

[20] Valiant, L.G. and Brebner, G.J. Universal schemes for parallel communication, Proceedings STOC 81, pp. 263–277.

The Complexity of Scheduling Trees with Communication Delays

(extended abstract)

Jan Karel Lenstra[1], Marinus Veldhorst[2*], Bart Veltman[1]

[1] Department of Mathematics and Computing Science, Eindhoven University of Technology, P.O. Box 513, 5600 MB Eindhoven, the Netherlands
[2] Department of Computer Science, Utrecht University, P.O. Box 80.089, 3508 TB Utrecht, the Netherlands. E-mail: marinus@cs.ruu.nl

Abstract. We consider the problem of finding a minimum-length schedule on m machines for a set of n unit-length tasks with a forest of intrees as precedence relation, and with unit interprocessor communication delays. First we prove that this problem is NP-complete; second we derive a linear time algorithm for the special case that m equals 2.

1 Introduction

We consider the allocation and scheduling problem that occurs when a parallel program is implemented on a parallel computer. More precisely, when one views the program as a collection of processes and the computer as a collection of processors, the problem is to determine for each process on which processor and at what time it must be executed so as to minimize the overall completion time of the program.

In addition to the execution times required by the processes, there may be communication delays, which occur when processes need intermediate results computed by other processes on other processors. During a communication delay, processors may execute processes that have obtained all necessary data. The communication delay depends both on the parallel program with its input and on the parallel computer. The former determines the number of data items to be communicated, the latter determines the amount of time needed for communicating these data items.

In this paper we consider a number of special cases of this problem. Switching to terminology commonly used in scheduling theory, we will speak about tasks and machines rather than processes and processors, and we will use the shorthand notation of [14] to denote specific types of scheduling problems.

---

* The research of this author was done while he was on sabbatical leave at the Department of Mathematics and Computing Science, Eindhoven University of Technology, Eindhoven, the Netherlands. Moreover it was partially supported by ESPRIT Basic Research Action No. 7141 (project ALCOM II: *Algorithms and Complexity*) and by EC Cooperative Action IC-1000 (project ALTEC: *Algorithms for Future Technologies*).

The basic optimization problem is denoted by $P|prec, c_{jk}|C_{\max}$. Here P indicates that there are m identical parallel machines, where m is part of the problem instance. Alternatively, Pm indicates that m is fixed and specified as part of the problem type, and $P\infty$ indicates that the number of machines is unrestrictively large, i.e., $m \geq n$. $prec$ denotes that a precedence relation is imposed on the n tasks $v_1, \ldots, v_n$. c_{jk} denotes that if task v_j precedes v_k and these two tasks are scheduled on different machines, then there is an interprocessor communication delay of c_{jk} time units; if v_j and v_k are scheduled on the same machine, there is no communication delay. The fact that execution lengths of tasks are not mentioned, means that they are part of the problem instance. $C_{\max}$ refers to the optimality criterion: in any schedule each task v_j has a completion time C_j, and we want to minimize the maximum completion time $C_{\max} \equiv \max_j C_j$. To this optimization problem corresponds a decision problem, denoted by $P|prec, c_{jk}|C_{\max} \leq D$ in which we only want to know whether there is a schedule such that the maximum completion time does not exceed D.

We are interested in the special case that the precedence graph is a forest of intrees, and the execution lengths and interprocessor communication delays are all equal to 1. We present two results: $P|tree, c_{jk} = 1, p_j = 1|C_{\max} \leq D$ is NP-complete, and $P2|tree, c_{jk} = 1, p_j = 1|C_{\max}$ can be solved in linear time. The first result is somewhat surprising because the problem can be solved in linear time in the absence of communication delays (cf. [8]).

Well-known results on related problems are:

- $P|\emptyset, c_{jk} = 0|C_{\max} \leq D$ is NP-complete (cf. [6]);
- $P|prec, c_{jk} = 0, p_j = 1|C_{\max} \leq D$ is NP-complete (cf. [11]), where $P2|prec, c_{jk} = 0, p_j = 1|C_{\max}$ can be solved in polynomial time (cf. [5]);
- $P|bipartite, c_{jk} = 1, p_j = 1|C_{\max} \leq 4$ is NP-complete, while $P|prec, c_{jk} = 1, p_j = 1|C_{\max} \leq 3$ can be solved in polynomial time (cf. [7, 9]);
- $P\infty|prec, c_{jk} = 1, p_j = 1|C_{\max} \leq 6$ is NP-complete, while $P\infty|prec, c_{jk} = 1, p_j = 1|C_{\max} \leq 5$ can be solved in polynomial time (cf. [7]);
- $P\infty|tree\ of\ depth\ \leq 2, c_{jk}|C_{\max} \leq D$ is NP-complete (cf. [4]), while $P\infty|tree, c_{jk} \leq \min_j p_j|C_{\max} \leq D$ can be solved in $O(n)$ time (cf. [3]);
- $Pm|tree, c_{jk} = 1, p_j = 1|C_{\max}$ can be solved in $O(n^{2m})$ time (cf. [12]). If $m = 2$, the problem can even be solved in $O(n^2)$ time (cf. [9]).

For a more extensive overview we refer to [14], which is revised and extended in [13].

Our linear-time algorithm differs fundamentally from the approach in [9] that leads to a quadratic time algorithm. While in the latter, subtrees of tasks are scheduled based on their size, we use list scheduling to get an $O(n)$ time bound. Unfortunately our algorithm cannot easily be adjusted to give optimal schedules in case 3 or more machines are available.

Afrati et. al. (cf. [1]) and Prastein (cf. [10]) study also the relation between communication and scheduling. However, in their model the communication does not lead to a delay in time, but it is a separate criterion. These researchers consider the problem of simultaneously minimizing the maximum completion time and the total amount of communication.

The remainder of this paper is organized as follows. First we give some preliminaries. In Sect. 2 we give a sketch of the NP-completeness proof, and in Sect. 3 the linear-time algorithm for $P2|tree, c_{jk} = 1, p_j = 1|C_{\max}$ is presented. We conclude with some open problems.

Preliminaries In the remainder of this paper we assume unit execution lengths of tasks and unit communication delays. V is the set of n tasks, and the precedence relation is given as a directed graph $F = (V, A)$, which is a forest of intrees. A schedule S consists for each task $v \in V$ of a 2-tuple $(t(v), m(v))$ of integers satisfying the following two conditions:

C.1 $0 \leq t(v)$ and $1 \leq m(v) \leq m$ for each $v \in V$,
C.2 for all v and w in V with $v \neq w$ and $m(v) = m(w)$, $t(v) \neq t(w)$.

We say that v is scheduled at time $t(v)$ on machine $m(v)$. S is a feasible schedule if in addition condition **C.3** is satisfied.

C.3 if $(u, v) \in A$ then:

$$t(v) \geq \begin{cases} t(u) + 1 \text{ if } m(u) = m(v) \\ t(u) + 2 \text{ otherwise} \end{cases}$$

A task v is ready at time t, if given the schedule until and including time $t - 1$, v can be scheduled without violating conditions **C.1**, **C.2**, and **C.3**.
Given two lists $H = (h_1, h_2, \ldots, h_k)$ and $E = (e_1, \ldots, e_p)$. h_1 is at the front of H, h_i occurs before (after) h_j in H if $i < j$ ($i > j$). E is a sublist of H if there are indices $i_1 < i_2 < \cdots < i_p \leq k$ such that $h_{i_j} = e_j$ for each j ($1 \leq j \leq p$).

2 The NP-completeness Result for Arbitrary m

In this section we show that the problem $P|tree, c_{jk} = 1, p_j = 1|C_{\max} \leq D$ is NP-complete.

Theorem 1. *The $P|tree, c_{jk} = 1, p_j = 1|C_{\max} \leq D$ problem is NP-complete.*

Proof. Obviously the problem is in NP. To prove that it is NP-hard, we use a reduction from the NP-complete problem Satisfiability.

Satisfiability Given a set U of variables and a collection C of clauses over U, does there exist a truth assignment for C?

Given an instance (U, C) of Satisfiability, define a threshold value b as $b = 2|C| + 4$ and let the number of machines be given by $m = 2|U| + \sum_{c \in C} |c| + 1$.
For each variable x we introduce a task $\hat{x}$ and two *variable chains* consisting of $b - 2$ tasks each; one of these chains corresponds to the literal x and the other corresponds to the literal $\bar{x}$. Both chains precede $\hat{x}$; as illustrated in Fig. 1. Let $c_1, \ldots, c_{|C|}$ be an arbitrary ordering of the clauses in C. For each clause c_i ($1 \leq i \leq |C|$) we introduce $|c_i|$ *clause chains* consisting of $2i - 1$ tasks each;

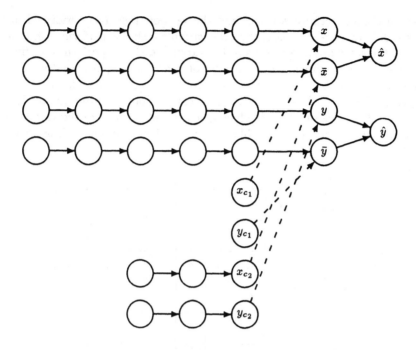

Fig. 1. Variable tasks and clause tasks corresponding to a Satisfiability instance with $c_1 = (x, \bar{y})$ and $c_2 = (\bar{x}, y)$.

there is a one-to-one correspondence between these chains and the literals that constitute c_i. We introduce precedence constraints between the variable tasks and the clause tasks, as follows. If the occurrence of variable x in c_i is unnegated, then the last task (x_{c_i}) of the clause chain corresponding to this occurrence has to precede the last task of the variable chain corresponding to the literal x. If the occurrence of x in c_i is negated, then x_{c_i} has to precede the last task of the variable chain corresponding to $\bar{x}$. For an illustration see the dashed lines of Fig. 1.

Finally, we introduce a total of $b + |U| + \sum_{i=1}^{|C|} \{|c_i|(b - 2i - 3) + 1\}$ dummy tasks, which form a number of chains. First, there is a single chain of length b. Second, there are $|U|$ unit-length chains, each consisting of a single task. Finally, for each clause c_i there are $|c_i|$ chains of dummy tasks; one is of length $b - 2i - 2$ and the other chains are of length $b - 2i - 3$. For each dummy chain of length l, $l < b$, its last task has to precede the $(l+2)$nd task of the dummy chain of length b. Hence, in a feasible schedule of length b each dummy chain is scheduled on a single processor, the first task of such a chain is executed in time slot 1, and the execution of the chain is without interruption.

Suppose that a truth assignment for the Satisfiability instance (U, C) exists. Given such a truth assignment, one can construct a schedule of length b as follows. If variable x is true, then the variable task x is performed in time slot

$b-1$ on the same processor as $\hat{x}$ and the variable task $\bar{x}$ is performed in time slot $b-2$. If variable x is false, then the variable task $\bar{x}$ is performed in time slot $b-1$ on the same processor as $\hat{x}$ and the variable task x is performed in time slot $b-2$. Let x_{c_i} be an unnegated occurrence of x in c_i and suppose that variable x is true, then x_{c_i} can be performed in time slot $b-3$. Now, the tasks corresponding to clause c_i are executed by the same machines that execute dummy chains of length $b-2i-2$ and $b-2i-3$. Thus, given the assignment of the variable tasks one can easily construct a feasible schedule of length b; for an illustration see Fig. 2.

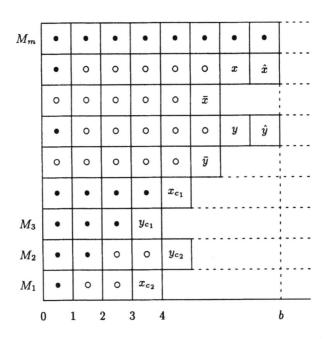

Fig. 2. A schedule of length b.

Conversely, suppose that there exists a feasible schedule of length at most b. The dummy tasks impose a structure on such a schedule, as illustrated in Fig. 2 by the black dots. At most $|U|$ non-dummy tasks can be processed in time slot 1. Given a variable x, at least one of the two chains corresponding to x has to begin its execution at 0. Thus, for each variable x exactly one of its chains is processed in time period $[0, b-2]$, and exactly one of its chains is processed in time period $[1, b-1]$. The latter is executed by the same machine as $\hat{x}$; the literal corresponding to this chain is regarded to be true. Given a clause c_i, the corresponding clause chains can only be executed by machines that execute dummy chains of length at most $b-2i-2$. It follows that the clause chains of c_i are executed by the machine that executes a dummy chain of length $b-2i-2$

and the $|c_i| - 1$ machines that each execute a dummy chain of length $b - 2i - 3$. At least one of these clause chains completes in time slot $b - 3$; it has to precede a true literal, since otherwise the schedule would not be feasible. Hence, from the schedule we can derive a truth assignment for the corresponding Satisfiability instance. □

3 A Linear Time Algorithm for $m = 2$

In this section we consider n tasks in a forest $F = (V, A)$ of intrees, with unit execution times and unit communication delays. Though we develop a linear (i.e., $O(n)$) time algorithm to find an optimal schedule for F on $m = 2$ machines, we try to formulate results for arbitrary m.

Our algorithm is based on list scheduling. The list scheduling approach consists of two stages. In the first stage one creates a list L of tasks. In the second stage one generally scans for increasing time slots t the list L (starting at the front) to find as many as possible ready tasks, but not more than m, and tries to schedule them on the available machines at time t. We modify stage 2 slightly in order to deal with the communication delays: two or more siblings can only be scheduled at the same time if another sibling remains unscheduled and hence will be scheduled at a later time. In the full version of the paper a precise formulation of the algorithm $Tassign$ will be given for the time assignment of the modified stage 2. In Fig. 3 an example is given which justifies the modification.

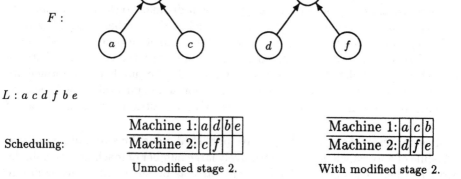

$L : a\ c\ d\ f\ b\ e$

Scheduling:

Machine 1:	a	d	b	e
Machine 2:	c	f		

Unmodified stage 2.

Machine 1:	a	c	b
Machine 2:	d	f	e

With modified stage 2.

Fig. 3. Example of modified stage 2 of the list scheduling algorithm.

We first prove that this stage 2 runs in $O(mn)$ time if L is a topological order of F. Then we will show how to create in linear time a specific topological order that leads to an optimal schedule in case of two machines.

Definition 2. If during the execution of algorithm $Tassign$ a task v is the ith task found for some i $(1 \leq i \leq m)$ at some time slot t, then v is an *ith-choice task*.

Actually algorithm *Tassign* assigns only time units, but not machines, to tasks.

Proposition 3. *Suppose each vertex v has been assigned a time $t(v)$ by algorithm Tassign. Then in $O(n)$ time each task v can be assigned a machine $m(v)$ such that the set of 2-tuples $(t(v), m(v))$ constitutes a feasible schedule.*

With $L^{(t)}$ we denote the list L as it is in algorithm *Tassign* at the moment that the algorithm is applied to time slot t. Hence $L^{(0)} = L$.

Lemma 4. *Let L be a topological order of F. Then the sequence of ith-choice tasks is a sublist of L.*

As a consequence we have that the number of tasks scheduled at time t is non-increasing in t.

Theorem 5. *If L is a topological order of the tasks in the forest of intrees, then algorithm Tassign can be implemented to run in $O(nm)$ time.*

Now we take a closer look at stage 1 of the list scheduling algorithm. We want to design an efficient algorithm that creates a topological order L such that the application of algorithm *Tassign* to L yields an optimal schedule. Unfortunately we were only able to prove optimality for $m = 2$, and we found a counterexample for the case of three machines.

Like Hu (cf. [8]) did for the case without communication delays, we create a list in which tasks are ordered according to their distance to the root. In the case without communication delays, this distance is just the number of arcs. It gives a tight lower bound on the difference between the completion times of a task and of its root successor. Incorporating communication delays, the concept of distance should be adjusted to take into account not only the number of intermediate tasks, but also the amount of time required for communication; and it should give precise schedule times of tasks in case of an unrestrictively large number of machines.

Several researchers (cf. [2], [3]) have published algorithms to find an optimal schedule of a forest when an unrestrictively large number of machines is available. Most of them schedule each task also as early as possible, and hence compute for each task v the value $est(v)$ defined as:

Definition 6. Let F be a forest of intrees, and v a task of F. Then the *earliest starting time $est(v)$* of v is the time that v is scheduled in an optimal schedule of $F(Pred(v) \cup \{v\})$ in case of an unrestrictively large number of machines, where $F(Pred(v) \cup \{v\})$ is the subgraph of F induced by $Pred(v) \cup \{v\}$ in which $Pred(v)$ denotes the set of predecessors of v.

The values $est(v)$ can be computed in $O(n)$ time according to the following recurrence relation:

$$est(v) = \begin{cases} 0 & \text{if } v \text{ is a leaf} \\ 1 + est(u_0) & \text{if } v \text{ has exactly one child } u_0 \text{ such that } est(u) \le est(u_0) \\ & \text{for all children } u \text{ of } v \\ 2 + est(u_0) & \text{if } v \text{ has at least two different children } u_0 \text{ and } u_1 \text{ such} \\ & \text{that } est(u) \le est(u_0) = est(u_1) \text{ for all children } u \text{ of } v. \end{cases}$$

Next we assign to each task a so-called *scheddist*-label. It comprises a distance concept between a task and an imaginary root that has each root of F as child. Obviously this imaginary root has *scheddist*-label 0. If a non-leaf task v has *scheddist*-label d then exactly one child u_0 of v has $scheddist(u_0) = 1 + d$ and moreover $est(u_0) \ge est(u)$ for all children u of v. All other children u of v have label $scheddist(u) = 2 + d$. Observe that the *scheddist*-labeling is not unique.

With an unrestrictively large number of machines and v having label d, there is an optimal schedule in which the imaginary root is scheduled d time slots later than v. While *est* deals with earliest starting time, *scheddist* is closely related to a latest starting time in an optimal schedule.

Definition 7. Let v be a non-leaf in F. Task u is the *eldest child* of v if u is a child of v and $scheddist(u) = 1 + scheddist(v)$. u is the *eldest sibling* of u' if u is the eldest child of the parent of u'. (Roots are considered to have a common imaginary parent.)

The basic idea is more or less to schedule tasks in decreasing order of *scheddist*-label, and in this way an eldest child is scheduled later than its siblings. For a proof that the schedule is optimal, we use a list L with an additional property:

Let v and v' be tasks. Let w be the eldest sibling of v if v is not the eldest child of its parent; in the other case let w be the parent of v. w' is defined similarly with respect to v'. If v occurs before v' in L, then w occurs before w' in L.

In the full version of the paper we will give a precise formulation of the algorithm *CreateList* that creates the list L.

The equivalent of the longest path (in the case without communication delays) is formed by a longest sequence of tasks that starts with a leaf and then proceeds with the eldest sibling or the parent.

Definition 8. A sequence $(x_k, x_{k-1}, \ldots, x_0)$ of different tasks is a *sibling path* of length k from x_k to x_0 if for each i $(1 \le i \le k)$ x_{i-1} is the parent or the eldest sibling of x_i, and moreover $scheddist(x_i) = 1 + scheddist(x_{i-1})$.

Lemma 9. *Let x be a task such that $est(x) = k$. Then there is a sibling path P to x with length k. If $k > 1$, then the next to last task in P is the eldest child of x.*

Proposition 10. *A sibling path of F is a sublist of L.*

Now we come to a fundamental lemma, necessary for the proof that the obtained schedule is optimal in case of two machines.

Lemma 11. *Let S be a schedule of a forest F, obtained by applying the algorithms CreateList and Tassign to F. Suppose there is a time interval $[t - k, t]$ of k time units $(k > 1)$ in which $a_k, a_{k-1}, \ldots, a_1$ and $b_k, \ldots, b_1$ are the first-choice and second-choice tasks, respectively. Suppose that for each i $(2 \le i \le k)$ b_i occurs in L after a_{i-1}. Then $(a_k, a_{k-1}, \ldots, a_1)$ is a sibling path and $(a_{k-1}, a_{k-2}, \ldots, a_1)$ is a path in F.*

Lemma 12. *Suppose the schedule S satisfies the conditions of lemma 11; and let k be such that the second-choice task at time $t - k - 1$ occurs in L before a_k. Then all tasks scheduled before time $t - k$ are predecessors of the parent of a_k.*

With these two lemmas it appears that under the conditions of lemma 11 the subgraph of F induced by $\{a_k, \ldots, a_1\}$ and all predecessors of the parent of a_k is scheduled optimally.

Theorem 13. *Application of the algorithms CreateList and Tassign together gives an optimal schedule for each forest F of intrees on two machines.*

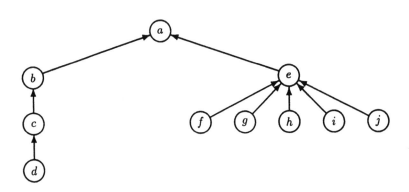

Fig. 4. A counterexample for three machines.

In case of three machines the combination of *CreateList* and *Tassign* does not necessarily yield an optimal schedule. Consider the graph of Fig. 4. $est(b) = est(e) = 2$. If b is chosen to have $scheddist(b) = 2$, then the created list becomes $L = (j, i, h, g, d, f, d, e, b, a)$ and *Tassign* will find a schedule with completion time 6. If on the other hand e is chosen to have $scheddist(g) = 2$, then $L = (d, j, i, h, g, c, f, b, d, a)$ and *Tassign* will find a schedule with completion time 5.

4   Open Problems and Conclusion

One open problem is the complexity of $P2|prec, c_{jk} = 1, p_j = 1|C_{\max}$. The same problem but without communication delays is solvable in polynomial time (cf. [5]), but is NP-hard in the model of Prastein (cf. [10]). Another question is whether there exist good and efficient approximation algorithms for the problem $P|tree, c_{jk} = 1, p_j = 1|C_{\max}$.

After the research for this paper was finished, E.L. Lawler notified us that he recently proved that each critical path schedule on a transformed graph (cf. [12]) differs by no more than $m - 2$ units of time from the optimal schedule; this implies optimality in case $m = 2$. It is easy to prove that our algorithm yields such a critical path schedule. This gives an answer on the question on the existence of approximation algorithms.

References

1. F. Afrati, C. H. Papadimitriou, and G. Papageorgiou. Scheduling dags to minimize time and communication. In *Proc. 3rd Conf. VLSI Algorithms and Architectures AWOC*, LNCS, vol. 319, pages 134–138, Springer-Verlag, Berlin, 1988.
2. F. D. Anger, J.-J. Hwang, and Y.-C. Chow. Scheduling with sufficient loosely coupled processors. *J. of Parallel and Distributed Computing*, 9:87–92, 1990.
3. P. Chrétienne. A polynomial algorithm to optimally schedule tasks on a virtual distributed system under tree-like precedence constraints. *Europ. Jrnl. Operational Res.*, 43:225–230, 1989.
4. P. Chrétienne and C. Picouleau. The basic scheduling problem with interprocessor communication delays. In *Proc. Summerschool on Scheduling Theory and its Applications, Bonas, France*, pages 81–100. INRIA, 1992.
5. M. Fujii, T. Kasami, and K. Ninomiya. Optimal sequencing of two equivalent processors. *SIAM J. Appl. Math.*, 17:784–789, 1969. Erratum. SIAM J. Appl. Math. 20:141, 1971.
6. M. R. Garey and D. S. Johnson. *Computers and Intractability, A guide to the theory of NP-completeness.* W. H. Freeman and Co., San Francisco, CA, 1979.
7. J. A. Hoogeveen, B. Veltman, and J. K. Lenstra. Three, four, five, six, or the complexity of scheduling with communication delays. Technical Report BS-R9229, CWI, Amsterdam, The Netherlands, 1992.
8. T. C. Hu. Parallel sequencing and assembly line problems. *Operations Res.*, 9:841–848, 1961.
9. C. Picouleau. *Etude de problèmes d'optimisation dans les systèmes distribués.* PhD thesis, Univers. Pierre et Marie Curie, Paris, France, 1992.
10. M. L. Prastein. Precedence-constrained scheduling with minimum time and communication. Technical Report UILU-ENG-87-2207, ACT-75, Coordinated Science Laboratory, Dpt of CS, Univ of Illinois at Urbana-Champaign, Urbana-Champaign, IL, USA, 1987.
11. J. D. Ullmann. NP-complete scheduling problems. *J. Comput. Syst. Sci.*, 10:384–393, 1975.
12. T. A. Varvarigou, V. P. Roychowdhury, and T. Kailath. Scheduling in and out forests in the presence of communication delays, 1992. Unpublished manuscript.

13. B. Veltman. *Multiprocessor scheduling with communication delays*. PhD thesis, Department of Mathematics and Computing Science, Eindhoven University of Technology, Eindhoven, The Netherlands, 1993.

14. B. Veltman, B. J. Lageweg, and J. K. Lenstra. Multiprocessor scheduling with communication delays. *Parallel Computing*, 16:173–182, 1990.

Optimal Tree Contraction on the Hypercube and Related Networks

Ernst W. Mayr[1] and Ralph Werchner[2]

[1] Institut für Informatik, Technische Universität, München,
mayr@informatik.tu-muenchen.de
[2] Fachbereich Informatik, J.W. Goethe-Universität, Frankfurt am Main,
werchner@informatik.uni-frankfurt.de

Abstract. An optimal tree contraction algorithm for the boolean hypercube and the constant degree hypercubic networks, such as the shuffle exchange or the butterfly network, is presented. The algorithm is based on novel routing techniques and, for certain small subtrees, simulates optimal PRAM algorithms. For trees of size n, stored on a p processor hypercube in in-order, the running time of the algorithm is $O(\lceil \frac{n}{p} \rceil \log p)$. The resulting speed-up of $O(p/\log p)$ is optimal due to logarithmic communication overhead, as shown by a corresponding lower bound.

1 Introduction

Tree contraction is a fundamental technique for solving problems on trees. A given tree is reduced to a single node by repeatedly contracting edges, *resp.*, merging adjacent nodes. This operation can be used for problems like *top-down* or *bottom-up algebraic tree computations* [ADKP89], which themselves can be applied to solve the membership problem for certain subclasses of languages in DCFL [GR86] or to evaluate expressions consisting of rational operands and the operators $+, -, \cdot, /$.

In the context of parallel processing the objective is to contract the tree using a small number of stages, each consisting of independent edge contractions. Brent [B74] was the first to show that a logarithmic number of such stages is sufficient, and he applied this result to the restructuring of algebraic expression trees, producing logarithmic depth. Subsequent work [MR85] [GR86] [GMT88] [KD88] [ADKP89] concentrated on the efficient parallel computation of contraction sequences, and eventually resulted in work optimal logarithmic time tree contraction algorithms on the EREW PRAM.

We consider the tree contraction problem on the binary hypercube and on similar networks. In this model, computing a suitable contraction sequence is more difficult, and the additional problem of routing pairs of nodes to be contracted to common processors arises. For the contraction of paths, both problems become trivial and can be solved in logarithmic time by a parallel prefix operation [S80]. For arbitrary trees, we combine two known (PRAM) contraction techniques with a recursive approach. To achieve a logarithmic running time we separate the local communication operations from the few long distance communication steps and perform them in appropriately sized subcubes. The necessary routing steps are performed by new logarithmic time algorithms designed for special classes of routings. Our algorithm contracts a tree of size n on a p processor binary hypercube in $O(\left\lceil \frac{n}{p} \right\rceil \log p)$ steps, which we also show is asymptotically optimal by a matching lower bound.

2 Fundamental Concepts and Notation

We first give a short description of our model of computation. A *network of processors* is a set of processors, interconnected by bidirectional communication links. Each processor has the capabilities of a RAM, a unique processor-id and additional instructions to send or receive one machine word to respectively from a direct neighbor. We assume the word length of the processors to be $\Theta(\log p)$ bits where p is the number of processors. The topological structure of the communication links can be described by an undirected graph (V, E).

A *d-dimensional hypercube* is a network of processors represented by the graph $G = (V, E)$ with

$$V = \{0, 1\}^d \,,$$
$$E = \left\{ (u, v) \,\middle|\, u \text{ and } v \text{ differ in exactly one bit} \right\} \,.$$

A network with bounded degree and a structure very similar to the hypercube is the *d-dimensional shuffle-exchange*. The processors are numbered as they are in the hypercube. Two processors are connected by a link if their ids differ only in the last bit (*exchange* edges) or if one id is a cyclic shift by one position of the other id (*shuffle* edges). A compendium of results concerning hypercubes and shuffle-exchange networks can be found in [L92].

A tree contraction is the reduction of a given binary tree to a single node, proceeding in stages. In a stage, a set of disjoint pairs of adjacent nodes in the current tree is selected, and the edges connecting these pairs are contracted, *i.e.*, each pair of nodes is merged into a single node. The contractions are not allowed to produce a node with more than two children.

For the tree contraction problem on networks of processors we assume that the data comprising each node can be stored in a constant number of processor words and that two adjacent nodes stored by the same processor can be merged in constant time.

On networks the complexity of the tree contraction problem depends on the representation of the given tree. We assume the in-order sequence of the nodes to be evenly distributed over the sequence of processors ordered by processor-id. To uniquely describe the structure of the tree in this way, each nontrivial subtree is assumed to be enclosed by a pair of parentheses. To contract a tree given in pre- or post-order, it can be transformed to in-order without asymptotically increasing the time complexity using the routing algorithm of [MW92].

If the tree is given by an arbitrary, but balanced distribution of the nodes linked by pointers, a list ranking on the Euler tour of the tree and a corresponding routing operation can be used to transform the tree into in-order representation. On a p processor hypercube, this transformation can be carried out in $O(\left\lceil \frac{n}{p} \right\rceil \log^2 p \log\log\log p \log^* p)$ steps for trees of size n. The dominating part is the time required for the list-ranking.

The algorithm given in the next section uses the following basic operations known to be executable in logarithmic time on a hypercube or shuffle-exchange network:

- parallel-prefix-operation and segmented parallel-prefix-operation [S80];
- monotone routing (the order of the data items is unchanged) [NS81];
- sparse enumeration sort (sorting p^α data items on a p processor hypercube, for fixed $\alpha < 1$) [NS82];
- parentheses-structured routing (routing between matching pairs in a well-formed word of parentheses) [MW92].

3 Tree Contraction on the Hypercube

In this section we show how trees of size n can be contracted on a p processor hypercube or hypercubic network in $O(\lceil \frac{n}{p} \rceil \log p)$ steps. We concentrate on the contraction of trees with at most p nodes in logarithmic time on the hypercube. For inputs of a size $n > p$, a hypercube of size $\Theta(n)$ can be simulated with a slowdown of $\Theta(\frac{n}{p})$. Because of the simple communication structure, the hypercube algorithm can be simulated on a shuffle exchange network of the same size with just constant slowdown. Furthermore, due to a general simulation result [S90], any algorithm on the shuffle exchange network can be simulated with constant slowdown on each of the other so called hypercubic networks, such as the de Bruijn network, the cube connected cycles network, or the butterfly network.

We give an outline of the contraction algorithm. It proceeds in three phases:

1. We identify and contract small subtrees in a recursive call of the algorithm.
2. The remaining tree may still be quite large, but, if so, has a structure that allows a significant further reduction by eliminating the leaves and contracting chains of nodes with a single child each to one edge, an operation similar to the "compact" operation in [MR85].
3. To contract the remaining tree using the PRAM algorithm of [KD88], we first compute the communication structure of this algorithm when executed on the remaining tree. After rearranging the nodes of the tree in the hypercube accordingly, each step of the PRAM tree contraction algorithm can be simulated in constant time.

3.1 Contraction of Subtrees

In the first phase, we divide the processors of the d-dimensional hypercube into contiguous blocks of size $2^{\lfloor \frac{3}{4}d \rfloor}$. These blocks are $\lfloor \frac{3}{4}d \rfloor$-dimensional subcubes, say $C_1, C_2, \ldots, C_{2^{\lceil d/4 \rceil}}$. We determine those subtrees entirely contained in a single C_i. For this, we first compute the *height*, i.e., the level of nesting, of the parentheses enclosing the subtrees. An opening parenthesis with height h is part of a local subtree iff there is a parenthesis to its right, but within its block, with a height $\leq h$. Performing a parallel-prefix-computation within each block C_i, we can compute for each parenthesis the minimum height of the parentheses to its right within the block. Comparing this value with its own height, each opening parenthesis can determine whether it is matched within its block. An analogous computation is performed for the closing parentheses.

In a recursive call to this algorithm, for all subcubes C_i in parallel, the identified subtrees are contracted. In general, a subcube C_i may contain a sequence of several maximal local subtrees. This causes the need to make the algorithm capable of contracting a sequence of trees. We only describe how to contract a single tree, the modifications to the algorithm for dealing with a sequence of trees, however, are quite straightforward.

3.2 Compaction

Denote by T, T', and T'' the trees at the beginning of the algorithm, after the first phase, and after the second phase, respectively. Let L be the set of pairs (u, v), where v is a leaf of T', u is its parent, and v is not the left sibling of another leaf. We reduce T' to T'' by eliminating all pairs of nodes in L (see Figure 1).

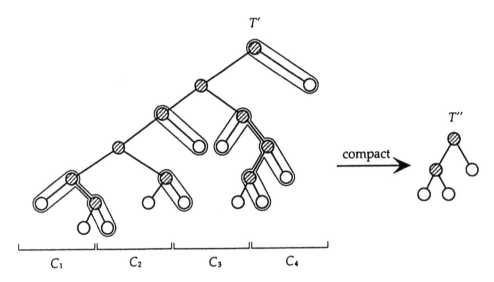

Fig. 1. The compact operation applied to T'

This reduction is equivalent to one "rake" operation (eliminating the leaves) and a repeated "compress" operation (contracting the maximal chains of nodes with a single child). Together, they were called a "compact" operation in [MR85]. The result is again an extended binary tree, *i.e.*, a tree with each node having zero or two children.

We first prove that the resulting tree is small. Then we show how the compact operation can be carried out on the hypercube in logarithmic time.

Lemma 1. *Starting with an expression tree of size at most 2^d, applying the first and second phase of our algorithm, the resulting tree T'' is an extended binary tree with at most $2^{\lceil d/4 \rceil} - 1$ leaves.*

Proof. Each leaf of T' corresponds to a maximal subtree of T initially stored totally within one of the blocks $C_1, C_2, \ldots, C_{2^{\lceil d/4 \rceil}}$. Assign to each leaf the index of its subcube. Then the sequence of numbers assigned to the leaves in T' is sorted and two leaves that are siblings are assigned different numbers. As the only leaves of T' surviving in T'' are those whose right sibling is also a leaf, they are all assigned distinct numbers. Since these numbers are in the range from 1 to $2^{\lceil d/4 \rceil} - 1$, T'' has at most $2^{\lceil d/4 \rceil} - 1$ leaves. $\qquad\qquad\qquad\qquad\square$

To perform the compact operation efficiently on the hypercube we first switch some siblings in the tree such that for all $(u, v) \in L$, the leaf v is a right child of u. For our in-order representation of the tree this means that we have to make a modification according to the following pattern:

$$\ldots (v\, u(\ldots)) \ldots \quad \longrightarrow \quad \ldots ((\ldots) u'\, v) \ldots , \qquad (1)$$

where u is modified to u' to denote that its children have been interchanged.

The required routing belongs to a special class of partial permutations called *parentheses structured* routings [MW92]. To define this class consider the following situation: Some of the processors are storing a parenthesis. Together with some of the opening parentheses data items are stored. The sequence of parentheses is well-formed, and each data item has to be routed from its opening parenthesis to the processor storing the matching closing parenthesis. A partial permutation that can be described in this way is called a parentheses structured routing. The algorithm proposed in [MW92] performs the routing in logarithmic time on the hypercube, the shuffle exchange network or any other hypercubic network. In (1), the routing of the data item (u, v) can be guided by the pair of parentheses enclosing the subtree rooted at u.

Let $(u_1, v_1), (u_2, v_2), \ldots, (u_r, v_r)$ be a maximal chain of tupels from L in T', *i.e.*, u_i is a child of u_{i+1} for $1 \leq i < r$, and u_r is either the root or the sibling of an internal node, and v_1's sibling w is either a leaf or the parent of two internal nodes. After the above routing step, this chain is, in in-order notation, a sequence

$$\ldots u_1 v_1) u_2 v_2) \ldots) u_r v_r \ldots .$$

In each such chain, the edges (u_i, v_i) are contracted by routing the leaves v_i one position to the left. The remaining path $u_1, \ldots, u_r$ is contracted to a single node by a segmented parallel-prefix-operation. Finally, the resulting node is routed to w by a sparse-enumeration-sort, and the two nodes are merged.

3.3 Simulation of PRAM Contraction Algorithm

In the third and last phase, we simulate the logarithmic time PRAM tree contraction algorithm proposed in [KD88] or [ADKP89]. Although Lemma 1 guarantees that there are only very few nodes left, a step-by-step simulation of the PRAM algorithm, performing a routing phase for each parallel random access step of the PRAM, would result in a suboptimal algorithm.

In our approach, we first compute the communication structure of the complete execution of the PRAM algorithm (without a complete simulation). Then, we route the nodes that will have to communicate to adjacent hypercube processors. After rearranging the data, we are able to simulate the PRAM tree contraction algorithm without slowdown.

In [KD88] the tree is reduced using the operation "rake". This operation, applied to a leaf l of the tree, involves three nodes: the leaf l, its parent p and its sibling s. The two edges connecting these nodes are contracted and the nodes l and p are eliminated while node s survives.

If the given tree has m leaves, the contraction algorithm proceeds in $2 \cdot \lfloor \log m \rfloor$ stages called $[1,1], [1,r], [2,1], [2,r], \ldots, [\lfloor \log m \rfloor, 1], [\lfloor \log m \rfloor, r]$. The leaves are numbered left to right from 1 to m. In stage $[i,1]$ (resp., $[i,r]$) the rake operation is applied to all leaves numbered by odd multiples of 2^{i-1} that are *left* (resp. *right*) children. In this fashion, no node is involved in more than one rake operation in each stage (see Figure 2 for an example).

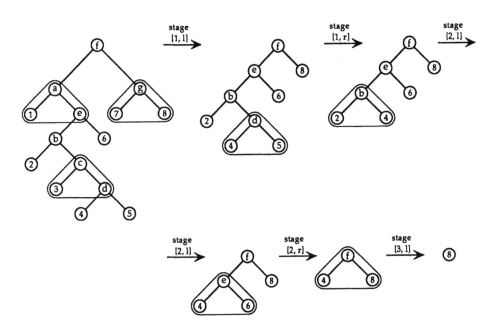

Fig. 2. Contraction of T''

We apply this contraction algorithm to the tree T''. For a stage x, let T_x be the reduced tree just before stage x. The contraction of T'' can be represented by a tree $\tilde{T}$ (see Figure 3, continuing the example). For each stage x of the contraction procedure there is one level of $\tilde{T}$ containing exactly the nodes of T_x. In $\tilde{T}$, a node v with children v_1, v_2, v_3 corresponds to the contraction of v_1, v_2, v_3 to v. The depth of $\tilde{T}$ is bounded by $\lfloor d/2 \rfloor$ since T'' has $m < 2^{\lceil d/4 \rceil}$ leaves. Thus $\tilde{T}$ can easily be embedded in a d-dimensional cube with dilation 2. To contract T'', we compute this embedding of $\tilde{T}$, route the nodes of T'' to the corresponding leaves of $\tilde{T}$, and perform the contraction along the (embedded) edges of $\tilde{T}$. Note that the nodes of T'' can be routed in logarithmic time by a sparse enumeration sort followed by a

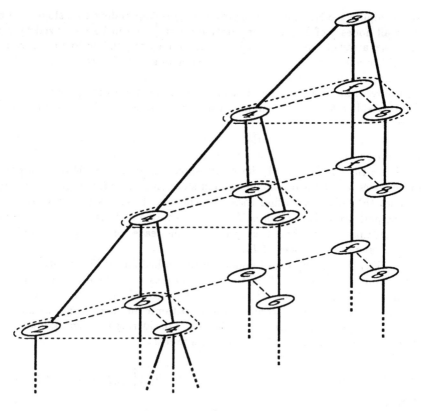

Fig. 3. The top 4 levels of the corresponding $\tilde{T}$

monotone routing. The major problem is to compute their destinations according to the embedding of $\tilde{T}$.

For each node v of $\tilde{T}$, we define $p(v) \in \{0, 1, 2, 3\}^{\lfloor d/2 \rfloor}$ as the description of the path from the root of $\tilde{T}$ to v. The i-th element of $p(v)$ is $j \in \{1, 2, 3\}$ if the i-th edge on this path points to a j-th child, and an appropriate number of 0's is appended to pad $p(v)$ to length $\lfloor d/2 \rfloor$. Substituting each number of $p(v)$ by its two-bit-representation, the hypercube address of v in the embedding of $\tilde{T}$ can easily be obtained. It remains to show how to compute $p(v)$ efficiently for the leaves of $\tilde{T}$, i.e. the nodes of T''.

Investigating the tree contraction algorithm, the following facts can easily be derived:

(a) The nodes of $T_{[i,1]}$ are the leaves of T'' numbered by multiples of 2^{i-1} and their lowest common ancestors in T''.

(b) Let v be a node of $\tilde{T}$ on a level corresponding to some stage x of the contraction. The descendants of v in $\tilde{T}$ are those nodes of T'' that have been contracted into v. These are v itself and those nodes u connected in T'' (considered as an undirected tree) to the parent of v by a path containing (including the endpoints) no node of T_x.

We compute the elements of the strings $p(v)$ corresponding to a stage x in parallel for all nodes v of $\tilde{T}$. The computation for stage x can be visualized by a two dimensional array of processors, with each column corresponding to a path from the root of T''' to a leaf and with each row corresponding to an internal node of T'''. The operations performed on the array are prefix computations and broadcast operations along the rows and the columns. Placing rows and columns in appropriately sized subcubes yields a logarithmic running time while using no more than $\lfloor d/2 \rfloor 2^{d/2+2}$ processors.

For stage x, we first compute the nodes of T_x using Fact (a) above. Simulating one stage of the contraction, the behavior of each node v of T_x determines $j \in \{1, 2, 3\}$ so that v is a j-th child in $\tilde{T}$ on the level corresponding to stage x. This j is broadcast to those nodes u of T''' which are descendants of v in $\tilde{T}$ since it is part of their string $p(u)$. Fact (b) tells how to find these descendants. First, v's parent broadcasts j along the path towards the root of T''' until reaching a node of T_x. Then, the nodes reached on this segment broadcast j towards the leaves of T''', again stopping the broadcast operation at nodes of T_x.

From $p(u)$ the destination of node u in the embedding of $\tilde{T}$ is easily computed. After routing the nodes of T''' to their destinations each stage of the contraction is simulated in constant time. Thus, the three phases of our hypercube contraction algorithm for trees of size p consist of a logarithmic number of parallel steps and a recursive call in $\lfloor \frac{3}{4}d \rfloor$-dimensional subcubes. Considering the remark made at the beginning of this section we obtain our main result:

Theorem 2. *A tree of size n can be contracted in $O(\lceil \frac{n}{p} \rceil \log p)$ steps on a p processor hypercube.*

The running time of this algorithm can be improved by a constant factor using the fact that no s-node of some stage $[i, 1]$ can be an l-node in the following stage $[i, r]$. Hence, the 4 address bits corresponding to two consecutive stages $[i, 1]$ and $[i, r]$ can never be 0111 or 1111. The remaining 7 possibilities can be encoded into 3 bits, and the algorithm can be modified making the recursive call for $\lceil d/3 \rceil$-dimensional subcubes.

The primitive operations used in the algorithm and the final simulation of the PRAM tree contraction algorithm can all be performed on the shuffle exchange network in logarithmic time. Combined with a general simulation result in [S90], we have:

Corollary 3. *A tree of size n can be contracted on any p processor hypercubic network in $O(\lceil \frac{n}{p} \rceil \log p)$ steps.*

The results of Theorem 2 and Corollary 3 can be shown to be asymptotically optimal for any input size if we assume that the only operations allowed on the nodes of the tree is copying, routing and contracting two nodes.

In this case, the number of messages required to contract a tree T, with each node u of T stored in processor $p(u)$ of a network G, is at least $\frac{1}{2} \sum_{1 \leq i < k} \text{dist}_G(u_i, u_{i+1})$, for any path $u_1, \ldots, u_k$ in T.

It is easy to construct a tree with n nodes where this lower bound is $\Omega(n \log p)$ assuming a balanced distribution of the in-order sequence of nodes in a p processor hypercube.

4 Dynamic Expression Evaluation

As an application, the contraction algorithm given in the previous section can be used to evaluate algebraic expressions of size n on a p processor hypercube in $O(\left\lceil \frac{n}{p} \right\rceil \log p)$ steps, provided that the operators in the expression satisfy certain closure properties [ADKP89]. Furthermore, the value of each subexpression can be computed within the same time bound: after contracting the tree, we reverse the contraction steps, starting with a single node, ending up with the original tree, and maintaining an expression tree with the values of all subexpressions already computed.

Tree contraction can be applied to the evaluation of algebraic expressions with rational numbers and the operators $+, -, \cdot, /$, but the model of computation has to be adjusted appropriately. As the numbers computed in the evaluation may grow rapidly we have to supply the network with the ability to communicate rational numbers and to perform the basic arithmetic operations on rational numbers in constant time. We call this model of computation a *rational number network*.

It can be shown that our expression evaluation algorithm on the rational number hypercube is asymptotically optimal for each input size. Thus, the maximum speed-up achievable for the expression evaluation problem on the hypercube is $O(p/\log p)$. Note that, on the EREW-PRAM, the optimal speed-up of $\Theta(p)$ can be achieved for input sizes in $\Omega(p \log p)$.

Intuitively, the lower bound for the rational number hypercube derives from the fact that the initial distribution of the operands in the hypercube can be chosen in such a way that nodes adjacent in the expression tree are stored in distant processors of the hypercube. Consider, for example, expressions of the type

$$x_1 \cdot ((x_2 \cdot \ldots ((x_{p-1} \cdot ((x_p \cdot y_p) + y_{p-1})) \ldots + y_2)) + y_1)$$

where the operands x_i and y_i are stored at maximal distance from each other. In our proof, Abelson's argument [A80] bounding the communication complexity from below by the rank of a certain matrix is applied simultaneously to the d partitions dividing the d-dimensional hypercube into two $d-1$-dimensional subcubes. For the constant degree hypercubic networks, the lower bound can be obtained considering a small bisector of the network.

Theorem 4. *Expressions containing n rational numbers and the operators $+, -, \cdot, /$ can be evaluated, including all subexpressions, on a p processor rational number hypercube or hypercubic network in $O(\left\lceil \frac{n}{p} \right\rceil \log p)$ steps, and this bound is optimal.*

5 Conclusion and Open Problems

Using a recursive approach, we have shown how to contract trees efficiently on hypercubes and related networks. Our technique of precomputing the communication structure of a PRAM contraction algorithm, rearranging the data and simulating the algorithm without any further overhead can be used for the contraction of small trees in networks. It can also be applied to the construction of NC^1 circuits for the expression evaluation problem which is simpler than the one proposed in [BCGR92].

Due to the recursive call and the usage of the parentheses structured routing algorithm, the constant factor in the worst case running time of our contraction algorithm is quite large. But the algorithm performs much better for random trees. As noted in [PS91], with high probability most of the nodes are contained in local subtrees and will be eliminated in the recursive call or even in the reduction within a single processor.

For rational number hypercubes and hypercubic networks, the maximal speed-up achievable for the expression evaluation problem is $O(p/\log p)$. We conjecture that a similar result holds for all constant degree networks, but we have so far been unable to apply the known lower bound results for communication complexity to this case.

References

[A80]   H. Abelson. Lower bounds on information transfer in distributed computations. *Journal of the ACM*, 27:384–392, 1980.

[ADKP89]   K. Abrahamson, N. Dadoun, D.G. Kirkpatrick and T. Przytycka. A simple parallel tree contraction algorithm. *Journal of Algorithms*, 10:287–302, 1989.

[B74]   R. Brent. The parallel evaluation of general arithmetical expressions. *Journal of the ACM*, 21:201–206, 1974.

[BCGR92]   S. Buss, S. Cook, A. Gupta and V. Ramachandran. An optimal parallel algorithm for formula evaluation. *SIAM Journal on Computing*, 21:755–780, 1992.

[GMT88]   H. Gazit and G.L. Miller and S.-H. Teng. Optimal tree contraction in the EREW model. In: Tewksbury, Stuart K. and Bradley W. Dickinson and Stuart C. Schwartz (eds.): *Concurrent Computations: Algorithms, Architecture, and Technology*. Plenum Press: New York-London (1988), 139–156.

[GR86]   A. Gibbons and W. Rytter. An optimal parallel algorithms for dynamic expression evaluation and its applications. *Proceedings of the 6th Conference on Foundations of Software Technology and Theoretical Computer Science*, Springer Verlag, LNCS-241:453–469, 1986.

[KD88]   S.R. Kosaraju and A.L. Delcher. Optimal parallel evaluation of tree-structured computations by raking. *Proceedings of the 3rd Aegean Workshop on Computing: VLSI Algorithms and Architectures*, 101-110, 1988.

[L92]   F.T. Leighton. *Introduction to Parallel Algorithms and Architectures*. Morgan Kaufmann Publishers, 1992.

[MW92]   E. W. Mayr and R. Werchner. Optimal routing of parentheses on the hypercube. *Proceedings of the 4th Annual ACM Symposium on Parallel Algorithms and Architectures*, 109-117, 1992.

[MR85]   G. L. Miller and J. H. Reif. Parallel tree contraction and its applications. *Proceedings of the 31st Annual Symposium on Foundations of Computer Science*, 478–489, 1985.

[NS81]   D. Nassimi and S. Sahni. Data broadcasting in SIMD computers. *IEEE Transactions on Computers*, C-30:101–107, 1981.

[NS82]   D. Nassimi and S. Sahni. Parallel permutation and sorting algorithms and a new generalized connection network. *JACM*, 29:642–667, 1982.

[PS91]   G. Pietsch and E. Schömer. Optimal parallel recognition of bracket languages on hypercubes. *8th Annual Symposium on Theoretical Aspects of Computer Science*, Springer Verlag, LNCS-480:434–443.

[S90]   E. J. Schwabe. On the computational equivalence of hypercube-derived networks. *Proceedings of the 2nd Annual ACM Symposium on Parallel Algorithms and Architectures*, 388-397, 1990.

[S80]   J.T. Schwartz. Ultracomputers. *ACM Transactions on Programming Languages and Systems*, 2:484–521, 1980.

Evolution of an algorithm [*]

Mike Paterson

Department of Computer Science
University of Warwick
Coventry CV4 7AL, UK

1 Extended Abstract

In this talk I shall trace the origins and subsequent development of a novel priority queue algorithm. The story spans twenty years of intermittent collaboration between Mike Fischer and myself.

The source of this saga still holds some mystery. In 1973 while working at MIT, we came across an interesting algorithm of unknown origin for computing the transitive closure of symmetric Boolean matrices. The algorithm involves two passes through the matrix, row-by-row, in the course of which some rows are OR'ed into later rows. We were first interested in the non-trivial proof of correctness which seemed to be needed. Our investigations in this direction led to the notion of 'skew closure', and an algebraic treatment which clarified the structure of the algorithm somewhat. This work was largely completed by 1977, and was presented at a 1980 symposium in Zürich in honour of Ernst Specker [1].

The algorithm was also posing further problems concerning its running time. On a RAM an $n \times n$ Boolean matrix obviously requires only $O(n^2)$ elementary operations, or, with some care, $O(n \log n)$ *row* operations. However, since transitive closure is an operation of fundamental importance in complexity (and because this was the mid-1970's) it seemed natural to explore the time required for a Turing Machine implementation. A naive approach uses $\Theta(n^3)$ or even $\Theta(n^4)$ steps. The hardest task in the algorithm is to send $\Theta(n)$ rows of the matrix forward to be OR'ed with later rows. Our more efficient implementation used a form of 'postal service', designed to transport quantities of 'mail' efficiently to arbitrary forward addresses. In our application a mail item is a row and the addresses are the row numbers $\{1, \ldots, n\}$. Our solution to this 'linear postman problem' operates recursively, splitting in two the range of addresses at each stage. For a range of n addresses, the depth of recursion is $O(\log n)$ and $O(n \log n)$ operations on mail items are made. For the transitive closure application, each operation on a mail item (row) takes $O(n)$ time, and so a time of $O(n(\log n)^2)$ is achieved.

Fischer gave a seminar on the 'mailcarrier problem' at IBM Yorktown in 1975. Other applications of this algorithm soon occurred to us, including search-and-update problems for large sequentially-accessed databases, and a Turing machine implementation of some topological sorting problems. In 1977, Nick Pippenger wrote a paper about fast simulation of logical networks by machines without random-access

---

[*] This work was partially supported by the ESPRIT II BRA Programme of the EC under contract 7141 (ALCOM II).

storage [2]. This seemed such a good application for our mailcarrier that we promptly invited Pippenger to be a coauthor. In 1978 and 1979, we were happily giving seminars on three continents (I recall MIT, Aachen, INRIA, Urgench, ...) on the mailcarrier/postman problem and its several applications. The triply-authored paper focussing on these applications was planned and drafted, but, 15 years on, is still 'in preparation'.

As the 1980's began, Fischer and I realised that, with only minor alterations to the algorithm, we could have a general priority queue algorithm, able to process n operations in $O(n \log n)$ steps. This algorithm, which we called 'Quickprior', worked by recursively splitting the sequence of queue requests. What irked us for some time was that the operation of the recursion was too rigid, merely counting the number of requests, and failing to take advantage of any favourable arrangements of insertions and deletions. During a sequence of mutual visits between the Universities of Washington and Warwick, we worked through Quickprior and Quickerprior, rigid versus flexible splitting, and various metrics of progress, refining the algorithm by degrees. At the same time we were developing the techniques we needed to prove good upper bounds on the running time. We tried in-tolls, out-tolls, birth and death taxes, even supertax, in our efforts to make sense of the behaviour of our algorithms.

After many tries and many setbacks we achieved the 'Fishspear Algorithm' which *is* able to take advantage of 'easy' sequences of queue requests. We make this precise in our paper [3], in which asymptotically matching upper and lower bounds are given. Though Quickprior does maintain the queue elements in a partial order which we can now describe as a 'fishspear', it does not have the more sensitive control structure which gives the improved bound.

With Fishspear we have an algorithm which, in the worst case, uses fewer comparisons than the usual heap implementation, yet, unlike the heap, it can take considerable advantage from the more favourable cases. Its key feature, however, arises from its Turing machine ancestry: it uses only sequential storage, e.g., a fixed number of stacks or storage tapes. The importance of this property is that on a typical computer system with paged virtual memory the locality of addresses accessed ensures that the number of 'page faults' (when a block of data has to be brought from some background storage into the primary storage area) is greatly reduced. This becomes increasingly significant as the size of the priority queue to be stored increases.

Fishspear received its name and was exposed to the world at FOCS'84. We did some further polishing, experimental implementation and testing before submitting a paper to JACM in 1987. It was accepted for publication in 1990, but it was 1992 before we were finally satisfied. Publication is now imminent.

Although this is a report on twenty years of gradual progress, there is still much to do. We have an upper bound for the number of comparisons used but we do not know how closely the algorithm actually approaches this bound. We understand the algorithm well enough to prove this performance bound, but we still lack the insight needed to make the proof transparent. Thus we have no easy way to predict the effect on performance of small changes to the algorithm.

The most pressing practical problem is to provide an implementation to translate the superior performance in terms of numbers of comparisons into a faster overall algorithm. Although the data comparisons are performed entirely by the merging of (parts of) pairs of sorted lists, the overheads concerned with the movement of data and keeping track of list sizes make the present implementations uneconomic in many typical applications [4],[5].

I look forward to continuing work with old and new friends and colleagues on an intriguing problem which has given me pleasure and frustration for most of my professional career.

References

1. M.J. Fischer and M.S. Paterson, "The fast skew-closure algorithm," *L'Enseignement Mathématique XXVI* 3-4, 1980, 345-360.
2. N. Pippenger, "Fast simulation of combinational logic networks by machines without random-access storage," *Proc. 15th Allerton Conf. on Communication, Control and Computing*, 1977, 25-33.
3. M.J. Fischer and M.S. Paterson, "Fishspear: a priority queue algorithm," *Proc. 25th Ann. IEEE Symp. Foundations of Computer Science*, 1984, 375-386. An improved version is to be published in JACM in 1993.
4. E.M. Fischer, "Priority queues and tape based machines," Yale University Computer Science Senior Project, 1989.
5. T. Watts, "Implementation and testing of a new priority queue algorithm," University of Warwick Final Year Computer Science Project, 1993, 101pp.

Mesh Connected Computers with Fixed and Reconfigurable Buses: Packet Routing, Sorting, and Selection[1]

Sanguthevar Rajasekaran
Department of CIS, University of Pennsylvania

Abstract Mesh connected computers have become attractive models of computing because of their varied special features. In this paper we consider two variations of the mesh model: 1) a mesh with fixed buses, and 2) a mesh with reconfigurable buses. Both these models have been the subject matter of extensive previous research. We solve numerous important problems related to packet routing, sorting, and selection on these models. In particular, we provide lower bounds and very nearly matching upper bounds for the following problems on both these models: 1) Routing on a linear array; and 2) $k - k$ routing, $k - k$ sorting, and cut through routing on a 2D mesh for any $k \geq 12$. In addition we present greedy algorithms for $1 - 1$ routing, $k - k$ routing, cut through routing, and $k - k$ sorting that are better on average and supply matching lower bounds. We also show that sorting can be performed in logarithmic time on a mesh with fixed buses. As a consequence we present an optimal randomized selection algorithm. Further, we provide a selection algorithm for the mesh with reconfigurable buses whose time bound is significantly better than the existing ones. Most of our algorithms have considerably better time bounds than known algorithms for the same problems.

1 Introduction

1.1 Definition of Problems

Packet routing problem in any fixed connection network is to send packets from their origins to destinations such that at most one packet goes through any wire (or edge) at any time. This problem is vital to parallel computing and has attracted a vast amount of research.

A special case of the routing problem is *partial permutation routing*, where ≤ 1 packet originates from any node and ≤ 1 packet is destined for any node. A packet routing algorithm is judged by 1) its *run time*, i.e., the time taken by the last packet to reach its destination, and 2) its *queue length*, i.e., the maximum number of packets any node will have to store during routing. Contentions for edges can be resolved using a *priority scheme*. Furthest destination first, furthest origin first, etc. are examples of priority schemes.

The problem of $k - k$ routing is one where at most k packets originate from any node and at most k packets are destined for any node. In cut through routing, we think of each packet as consisting of k pieces (also called *flits*). At any time, the flits of a packet can not be broken, i.e., all the flits should be in contiguous nodes. The problem of $k - k$ sorting on any fixed connection machine is the problem of sorting where exactly k packets are input at any node. Given a set of N numbers and an $i \leq N$, the problem of selection is to find the ith smallest element from out of the N given numbers.

1.2 Definition of Models

The fixed connection machines assumed in this paper are: 1) the mesh connected computer with fixed buses (denoted as M_f), and 2) the mesh with reconfigurable buses (denoted as M_r). The basic topology of a two dimensional mesh is an $n \times n$ square grid with one processor per grid point. When necessary, the MIMD model is assumed where every processor can communicate with all its neighbors in one unit of time. In M_f we assume that each row and each column has been augmented with a broadcast bus. Only one message can be broadcast along any bus at any time, and this message can be read by all the processors connected to this bus in the same time unit.

In the model M_r, processors are connected to a reconfigurable broadcast bus. At any given time, the broadcast bus can be partitioned (i.e., reconfigured) dynamically into subbuses with the help of

---

[1]This research was supported in part by an NSF Research Initiation Award CCR-92-09260 and an ARO grant DAAL03-89-C-0031.

locally controllable switches. For instance, in an $n \times n$ mesh, the different columns (or rows) can form subbuses. Even within a column (or row) there could be many subbuses, and so on. There are related models such as PARBUS (see e.g., [5]). Our algorithms are applicable to these models as well, in which case the stated time bounds will change retaining optimality. The model we assume is essentially the same as the PARBUS, except that we assume the edges of the mesh are bidirectional. This in particular means that if a bus is of length one, then it is bidirectional otherwise it is unidirectional (i.e., only one message can be broadcast).

Both M_r and M_f are becoming popular because of the absence of diameter consideration and commercial implementations [1]. Even as theoretical models both M_r and M_f are very interesting. For instance, sorting of n keys can be performed in $O(1)$ time on an $n \times n$ mesh M_r, whereas we know that sorting needs $\Omega(\frac{\log n}{\log \log n})$ time even on the CRCW PRAM (given only a polynomial number of processors).

1.3   Previous and New Results

Meshes with fixed buses have been studied by various researchers in the past (see e.g., [16, 7, 2, 9]). An equally impressive amount of work has been done on the mesh with reconfigurable buses as well (see e.g., [1, 10]).

Our results are summarized in Tables I and II. In these Tables, q stands for queue size and $[\sqrt{}]$ denotes new results presented in this paper. Define $\log^{(1)} x$ as $\log x$, and $\log^{(i)} x$ as $\log(\log^{(i-1)} x)$ (for any integer $i \geq 2$). Then, $\log^* x$ is nothing but the smallest i such that $\log^{(i)} x \leq 1$. Due to space limit, we omit the proofs of many Lemmas and Theorems. These proofs can be found in [13].

We present algorithms for $k - k$ routing, $k - k$ sorting, and cut through routing on both M_f and M_r that are very nearly optimal. For instance, our $k - k$ sorting algorithm on M_r runs in time $\frac{kn}{2} + o(kn)$ for the worst case input (with high probability), and we also prove a lower bound of $\frac{kn}{2}$. We also present algorithms that are better on average for $k - k$ routing, $k - k$ sorting, and cut through routing on M_f and M_r. These algorithms are also very nearly optimal upto small lower order additive terms. All the above algorithms need a queue size of only $k + o(k)$ with high probability and extend to higher dimensional meshes as well. The stated time bounds hold with high probability.

In addition, we show that random routing on a linear array M_f can be performed in $0.382n + o(n)$ steps and the same can be performed in a 2D mesh in time $0.536n + o(n)$ steps. In contrast, $\frac{2}{3}n$ is a lower bound for worst case permutation routing on a linear array as well as a 2D mesh [9]. A logarithmic time sorting algorithm is also presented for M_f.

For the problem of selection on M_f, we present an optimal $O(n^{1/3})$ time randomized algorithm. (This algorithm selects from out of n^2 elements on an $n \times n$ mesh M_f.) The best known previous algorithm is due to Kumar and Raghavendra and has a run time of $O(n^{1/3}(\log n)^{2/3})$ [7]. In [7] a lower bound of $\Omega(n^{1/3})$ is proven for selection and related problems. Our selection algorithm also runs in $O(n^{1/4})$ time on an $n^{5/4} \times n^{3/4}$ mesh, which is optimal and is an improvement over the run time of $O(n^{1/4} \log n)$ that the algorithm of Chen, Chen, and Chen [2] has.

A number of results are known for selection on M_r. The best known previous randomized algorithm is due to Doctor and Krizanc [3]. Their algorithms can select in 1) $O(\log n)$ expected time assuming that each input permutation is equally likely; or 2) randomized $O(\log^2 n)$ time (with no assumptions). The best known previous deterministic algorithm is due to Hao, MacKenzie, and Stout [4]. They show that selection can be done in 1) $O(\log^* n)$ time assuming that each input permutation is equally likely; or in 2) $O(\log n)$ time without making any assumptions. In this paper we show that selection can be done using a randomized algorithm that runs in 1) $O(\log^* n)$ expected time assuming that each input permutation is equally likely (This is an independent work); or 2) $O(\log^* n \log \log n)$ time (with no assumptions). These algorithms have the same asymptotic run times on M_r and PARBUS. Finally, we show that given n elements, the maximum of these elements can be found in $O(1)$ time (deterministically) on an $n \times n^\epsilon$ mesh M_r or PARBUS, where ϵ is a constant > 0.

Note: The proof technique we introduce in this paper for analyzing $k - k$ routing on both M_r and M_f is very simple and and might find independent applications. Also, we believe that the algorithms we present for concentration on M_r (given in connection with selection) might be applicable elsewhere.

PROBLEM	M_f (Worst Case)	M_f (Average)
Permutation Routing	$n + O(\frac{n}{e}) + o(n)$ [9]	$0.536n + o(n)$ [√]
$k - k$ Routing	$\frac{kn}{3} + o(kn)$ [√]	$\frac{kn}{6} + o(kn)$ [√]
$k - k$ Sorting	$\frac{kn}{3} + o(kn)$ [√]	$\frac{kn}{6} + o(kn)$ [√]
Cut Thro' Routing	$\frac{kn}{3} + 1.5n + o(kn)$ [√]	$\frac{kn}{6} + n + o(kn)$ [√]
Selection	$O(n^{1/3})$ [√]	$O(n^{1/3})$ [√]

Table I: Algorithms for M_f

PROBLEM	M_r (Worst Case)	M_r (Average)
Permutation Routing	$n + O(\frac{n}{r}) + o(n)$ [14]	$n + o(n)$ [√]
$k - k$ Routing	$\frac{kn}{2} + o(kn)$ [√]	$\frac{kn}{4} + o(kn)$ [√]
$k - k$ Sorting	$\frac{kn}{2} + o(kn)$ [√]	$\frac{kn}{4} + o(kn)$ [√]
Cut Thro' Routing	$\frac{kn}{2} + 1.5n + o(kn)$ [√]	$\frac{kn}{4} + n + o(kn)$ [√]
Selection	$O(\log^* n \log\log n)$ [√]	$O(\log^* n)$ [4] [√]

Table II: Algorithms for M_r

1.4 Some Definitions

We say a randomized algorithm uses $\widetilde{O}(g(n))$ amount of any resource (like time, space, etc.) if there exists a constant c such that the amount of resource used is no more than $c\alpha g(n)$ with probability $\geq 1 - n^{-\alpha}$ on any input of length n and for any α. Similar definitions apply to $\widetilde{o}(g(n))$ and other such 'asymptotic' functions.

By *high probability* we mean a probability of $\geq 1 - n^{-\alpha}$ for any fixed $\alpha \geq 1$ (n being the input size of the problem at hand). Let $B(n,p)$ denote a binomial random variable with parameters n and p, and let 'w.h.p.' stand for 'with high probability'. In the analysis of our randomized algorithms we make use of Chernoff bounds.

2 Locally Optimal Routing on a Linear Array M_f

Leung and Shende [9] have shown that routing on an n-node linear array with a fixed bus needs at least $\frac{2n}{3}$ steps in the worst case. They also matched this lower bound with an algorithm that runs in $\frac{2n}{3}$ steps on any input. Thus as far as the worst case input is concerned, the permutation routing problem on a linear array has been solved.

An interesting question is: 'Can we perform optimal routing on a linear array for *any* input?' In the case of a conventional array, the maximum distance any packet will have to travel is clearly a lower bound for routing on a given input and this lower bound can be matched with an algorithm as well [15]. In the case of M_f it is not even clear what a lower bound will be for any input. In this section we prove such a lower bound and present an algorithm that matches this lower bound (up to a $o(n)$ lower order term).

A Lower Bound: Let $\mathcal{L}$ be an n-node linear array with a fixed bus and π be a permutation to be routed. If the number of packets that have to travel a distance of d or more is n_d, then $\min\{d, n_d\}$ is a lower bound for the routing time of π (for each $1 \leq d \leq n$).

Proof: From among the packets that have to travel a distance of d or more, if there exists at least one packet that never uses the bus, a lower bound for routing is d. On the other hand, if all of these packets use the bus one time or the other, clearly, n_d steps will be needed. □

The above observation yields the following

Lemma 2.1 *Routing a permutation π needs at least* $\max_d \{\min\{d, n_d\}\}$ *steps on a linear array with a fixed bus.*

Our algorithm for routing uses a subroutine for calculating prefix sums:

Lemma 2.2 *[16] Prefix sums of n numbers can be computed on an n-node linear array M_f in $O(\sqrt{n})$ time steps. $\Omega(\sqrt{n})$ time is needed for computing prefix sums.*

Locally Optimal Routing: The idea of this algorithm is to exploit Lemmas 2.1 and 2.2. We first compute $d' = \max_d\{\min\{d, n_d\}\}$. We route all the packets that have to travel a distance of d' or more using the bus, and the other packets are routed using the edge connections under the furthest destination first priority scheme.

We claim that d' can be computed in $O(\sqrt{n}\log n)$ time. Observe that 1) For a given d, we can compute n_d in $O(\sqrt{n})$ time using Lemma 2.2; and 2) As d increases n_d remains nonincreasing (it might decrease more often). Thus we could determine d' using a binary search on the values of d.

Once we determine d', we can perform routing in time $\max\{d', n_{d'}\}$. Thus we arrive at

Theorem 2.1 *There exists an algorithm for routing on a linear array that is optimal for every input up to a $o(n)$ lower order term.*

Observation. In the case of a conventional linear array, if L is the maximum distance any packet has to travel, we can perform routing in L steps, provided we know the value of L. But for the above algorithm, no such information about the input is needed.

Routing on a linear array M_r: Rajasekaran and McKendall [14] have presented an algorithm for routing a permutation that runs in $\frac{3}{4}n$ steps. The obvious lower bound is $\frac{n}{2}$. The problem of optimal routing still remains open.

3 $k - k$ Routing and Cut Through Routing on M_f

In this section we present algorithms for $k - k$ routing and cut through routing on a two dimensional mesh with fixed buses whose run times are $\frac{kn}{3} + \tilde{o}(kn)$, and $\frac{kn}{3} + 1.5n + \tilde{o}(kn)$, respectively, whenever $k \geq 12$. The algorithm we use resembles the one given by Rajasekaran in [12] but there are many crucial differences. A lower bound of $\frac{kn}{3}$ applies, proof of which can be found in [13].

3.1 Routing on a Linear Array

The algorithm for $k - k$ routing on a 2D mesh consists of 3 phases where each phase corresponds to routing along a linear array. Here we state and prove a Lemma that will prove useful in analyzing all the three phases of the mesh algorithm.

Problem 1. There are a total of $\epsilon kn + o(kn)$ packets in an n-node linear array (for some constant $\epsilon \leq 1$), such that the number of packets originating from or the number of packets destined for any successive i nodes is $\leq \epsilon ki + o(kn)$ (for any $1 \leq i \leq n$). Route the packets.

Lemma 3.1 *Problem 1 can be solved in time $\frac{\epsilon kn}{3} + \tilde{o}(kn)$ under the model M_f if $k \geq \frac{3}{\epsilon}$.*

Proof. Details of the proof can be found in [13]. We provide a brief summary. Let the nodes of the array be named $1, 2, \ldots, n$. Certain packets (call them *special packets*) will be routed using the bus, whereas the other packets will be routed using edge connections under the furthest destination first priority scheme. Whether or not a packet is special is decided by a coin flip. A packet can become special with probability $\frac{1}{3}$.

Using Chernoff bounds, the number of special packets is $\frac{\epsilon kn}{3} + \tilde{o}(kn)$. We could perform a prefix sums computation in $o(n)$ time and arrive at a schedule for these special packets. Thus the special packets can be routed within the stated time bound.

Observe that the number of non special packets originating from or destined for any successive i nodes is $\frac{2}{3}\epsilon ki + \tilde{o}(kn)$ (for any $1 \leq i \leq n$). Let $\beta = \frac{2}{3}\epsilon$.

Consider only non special packets whose destinations are to the right of their origins. Let i be an arbitrary node. It suffices to show that any packet that ever wants to cross node i from left to right will do so within $\frac{\beta kn}{2} + \tilde{o}(kn)$ steps. Ignore the presence of packets that do not have to cross node i.

The number of packets that want to cross node i from left to right is $\min\{\beta ki, \ \beta k(n-i)\} + \tilde{o}(kn)$. The maximum of this number over all i's is $\frac{\beta kn}{2} + \tilde{o}(kn)$. It immediately follows that the non special packets will be done in $\frac{\beta kn}{2} + \frac{n}{2} + \tilde{o}(kn)$ steps. But our claim is slightly stronger. If node i is busy transmitting a packet at every time unit, the result follows. There may be instances when i may be idle. We show that even after accounting for the idle time of i, transmission will be complete within the stated time bound. $\square$

Corollary 3.1 *$k - k$ routing (for any $k \geq 2$) on a (conventional) linear array can be performed in $\frac{kn}{2}$ steps using the furthest destination first priority scheme.*

3.2 Routing on an $n \times n$ Mesh

Next we show that $k - k$ routing can be completed using a randomized algorithm in time $\frac{kn}{3} + \tilde{o}(kn)$. This algorithm has three phases. This three phase algorithm will be employed later on in many other contexts as well. Call this algorithm *Algorithm B*.

Algorithm B

To start with each packet is colored red or blue using an unbiased coin. The algorithm used by red packets is described next. Blue packets execute a symmetric algorithm using, at any given time, the dimension unused by the red packets. Let q be any red packet whose origin is (i, j) and whose destination is (r, s).

Phase I: q chooses a random node in the column of its origin (each such node being equally likely). If (i', j) was the node chosen, it traverses along column j up to this node.

Phase II: q travels along row i' up to column s.

Phase III: The packet q reaches its destination traversing along column s.

A blue packet in phase I chooses a random node in the row of its origin and goes there along the row. In phase II it traverses along the current column to the row of its destination and in phase III it travels along the current row to its destination. Because of the MIMD model assumed in this paper, and because all the three phases are disjoint there will not be any conflict between blue and red packets.

Our algorithm for $k - k$ routing is Algorithm B with some crucial modifications. We only describe the algorithm for red packets. Blue packets execute a symmetric algorithm. Routing of packets in phase II is done using the algorithm of section 3.1. The algorithm for routing in phases I and III is slightly different. We describe the algorithm for phase I and the same is used in phase III as well.

Algorithm for phase I: Consider the task of routing along an arbitrary column. Let $\mathcal{A}$ and $\mathcal{C}$ be the regions of the first $\frac{n}{\sqrt{12}}$ nodes and the last $\frac{n}{\sqrt{12}}$ nodes of this n-node linear array, respectively. Let $\mathcal{B}$ be the region of the rest of the nodes in the array. Any packet that originates from $\mathcal{A}$ whose destination is in $\mathcal{C}$ and any packet that originates from $\mathcal{C}$ with a destination in $\mathcal{A}$ will be routed using the bus. Scheduling for the bus is done using a prefix sums operation. The rest of the packets are routed using the edge connections employing the furthest destination first priority scheme.

Theorem 3.1 *The above algorithm for $k - k$ routing runs in time $\frac{kn}{3} + \tilde{o}(kn)$, the queue length being $k + \tilde{o}(k)$, for any $k \geq 12$.*

Proof. Both the number of blue packets and the number of red packets is $B(kn^2, 1/2)$. Thus w.h.p. these two numbers will be nearly the same. Further, the number of blue (red) packets that will participate in row routing of phases I and III (phase II) is $B(kn, 1/2)$ each. Also, the number of red (blue) packets that participate in column routing of phases I and III (phase II) is $B(kn, 1/2)$ each. Thus using Lemma 3.1 (with $\epsilon = 1/2$) we can show that phase II can be completed in $\frac{kn}{6} + \tilde{o}(kn)$ steps, for any $k \geq 6$.

We claim that phase I and phase III can be completed within $\frac{kn}{12} + \tilde{o}(kn)$ steps each, for any $k \geq 12$. We only provide the analysis for phase I, since phase III is just the reverse of phase I.

Analysis of phase I: Consider only blue packets and an arbitrary row. If i is any node in this row, it suffices to show that all the packets that ever want to cross node i from left to right will do so within $\frac{kn}{12} + \tilde{o}(kn)$ steps. In the following case analysis we only obtain an upper bound for the number of packets that want to cross i. But i may not be busy transmitting a packet at every time unit. We could employ the proof technique of section 3.1 to show that the given time is enough even after accounting for possible idle times of i. Observe that the number of packets that originate from region $\mathcal{A}$ with a destination in region $\mathcal{C}$ is $\frac{kn}{24} + \tilde{o}(kn)$.

Case 1. $i \leq \frac{n}{\sqrt{12}}$: The number of packets that have to cross node i from left to right is $\leq \frac{ki}{2}\left(\frac{n-i-n/\sqrt{12}}{n}\right) + \tilde{o}(kn)$. This number is $\leq \frac{kn}{24}(\sqrt{12} - 2) + \tilde{o}(kn) = \frac{(\sqrt{3}-1)kn}{12} + \tilde{o}(kn)$, for any i in region $\mathcal{A}$.

Case 2. $\frac{n}{\sqrt{12}} < i < n - \frac{n}{\sqrt{12}}$: In this case the number of packets that want to cross i from left to right is $\leq \frac{ki}{2}\left(\frac{n-i}{n}\right) - \frac{kn}{24} + \tilde{o}(kn)$. This number is no more than $\frac{kn}{12} + \tilde{o}(kn)$ for any i in region $\mathcal{B}$.

Case 3. $i > n - \frac{n}{\sqrt{12}}$: Number of packets that have to cross i is $\leq \frac{k(i-n/\sqrt{12})}{2}\left(\frac{n-i}{n}\right) + \widetilde{o}(kn)$, which is $\leq \frac{(\sqrt{3}-1)kn}{12} + \widetilde{o}(kn)$ for any i in region $\mathcal{C}$.

Thus phase I (and phase III) can be completed within $\frac{kn}{12} + \widetilde{o}(kn)$ steps.

Queue size analysis. The total queue length of any successive $\log n$ nodes is $\widetilde{O}(k \log n)$ (because the expected queue length at any single node is k implying that the expected queue length in $\log n$ successive nodes is $k\log n$; now apply Chernoff bounds). One could employ the technique of Rajasekaran and Tsantilas [15] to distribute packets locally such that the number of packets stored in any node is $\widetilde{O}(k)$. The queue length can further be shown to be $k + \widetilde{o}(k)$ using the same trick. $\square$

Corollary 3.2 *If $k = O(n^\nu)$ for some constant $\nu < 1$, the queue length of the above algorithm is only $k + \widetilde{O}(1)$.*

The following Theorem pertains to $k - k$ routing on r-dimensional meshes.

Theorem 3.2 *$k-k$ routing on an $r-$ dimensional mesh M_f can be performed within $\frac{kn}{3}+\widetilde{O}(krn^{(r-1)/r})$ steps, the queue length being $k+\widetilde{o}(k)$, as long as $k \geq 12$. If $k = O(n^\nu)$, the queue length is only $k+\widetilde{O}(1)$.*

Similar Theorems can be proven for cut through routing as well. The proofs of the following Theorems are omitted:

Theorem 3.3 *Cut through routing can be completed in time $\frac{kn}{3} + \frac{3}{2}n + \widetilde{o}(kn)$ on M_f, the queue length being $k + \widetilde{o}(k)$ for any $k \geq 12$.*

Theorem 3.4 *Cut through routing on an r-dimensional mesh M_f can be performed in $\frac{kn}{3} + (r+1)\frac{n}{2} + \widetilde{O}(krn^{(r-1)/r})$ steps, the queue length being $k + \widetilde{o}(k)$, for any $k \geq 12$. If $k = O(n^\nu)$ for some constant $\nu < 1$, the queue length is only $k + \widetilde{O}(1)$.*

The following theorems pertain to M_r proofs of which have been omitted:

Theorem 3.5 *$\frac{kn}{2}$ is a lower bound for $k-k$ routing, $k-k$ sorting, and cut through routing on the mesh M_r. We can perform these in time $\frac{kn}{2} + \widetilde{o}(kn)$ and $\frac{kn}{2} + 1.5n + \widetilde{o}(kn)$, respectively, the queue length being $k + \widetilde{o}(k)$, for any $k \geq 8$.*

Theorem 3.6 *$k-k$ routing, $k-k$ sorting, and cut through routing on an $r-$ dimensional mesh M_r can be performed within time $\frac{kn}{2} + \widetilde{O}(krn^{(r-1)/r})$, $\frac{kn}{2} + \widetilde{O}(krn^{(r-1)/r})$, and $\frac{kn}{2} + (r+1)\frac{n}{2} + \widetilde{O}(krn^{(r-1)/r})$, respectively, the queue length being $k + \widetilde{o}(k)$, for any $k \geq 8$. If $k = O(n^\nu)$, the queue length is only $k + \widetilde{O}(1)$.*

4 $k - k$ Sorting

Many optimal algorithms have been proposed in the literature for $1 - 1$ sorting on the conventional mesh. A $2n + o(n)$ step randomized algorithm has been discovered for sorting by Kaklamanis, Krizanc, Narayanan, and Tsantilas [6]. But $2n - 2$ is a lower bound for sorting on the conventional mesh. Recently Rajasekaran and McKendall [14] have presented an $n + o(n)$ randomized algorithm for routing on a reconfigurable mesh, where it was shown that sorting can be reduced to routing easily if there exists a mechanism for broadcasting.

We obtain the following

Theorem 4.1 *$k - k$ sorting on an $n \times n$ mesh M_f can be performed in $\frac{kn}{3} + \widetilde{o}(kn)$ steps, the queue length being $k + \widetilde{o}(k)$, for any $k \geq 12$.*

Theorem 4.2 *$k - k$ sorting on M_r can be completed in $\frac{kn}{2} + \widetilde{o}(kn)$ time, the queue size being $k + \widetilde{o}(k)$, for any $k \geq 8$.*

On the conventional mesh, there exists a randomized algorithm for $k - k$ sorting that runs in $\frac{kn}{2} + 2n + \widetilde{o}(kn)$ time [12].

5 Algorithms with Better Average Performance

In this section we present algorithms for routing and sorting that perform better on average than the worst case behavior of algorithms presented in previous sections. The average case behavior assumed here is that each packet is destined for a random location (this notion being the same as the one assumed in [8]). Leighton [8] has shown that the greedy algorithm for $1-1$ routing on the conventional mesh indeed runs in time $2n - \tilde{o}(n)$, the queue size at any node being no more than 4 plus the number of packets destined for this node. (The greedy algorithm referred to here is: a packet originating from (i, j) with a destination at (r, s) is sent along row i up to column s, and then along column s up to row r. Also, the high probability involved in the definition of $\tilde{o}()$ here is over the space of all possible inputs.)

In a conventional mesh, it is easy to see that if a single packet originates from each node and if this packet is destined for a random node, then there will be at least one packet that has to travel a distance of $\geq 2n - o(n)$ with high probability. Thus $2n - o(n)$ is a lower bound even on average (compared with the $2n - 2$ lower bound for the worst case $1 - 1$ permutation routing time).

However, on a mesh with fixed buses, there seems to be a clear separation of the average case and the worst case. For instance, on a linear array $1-1$ routing needs $\frac{2n}{3}$ steps in the worst case, whereas in this section we show that on average it only takes $\frac{(3-\sqrt{5})n}{2} \approx .382n$ steps. We also prove similar results for routing on a $2D$ mesh, $k-k$ routing, $k-k$ sorting, and cut through routing.

The following Lemmas are folklore and will prove helpful in our analysis: Let $z_1, z_2, \ldots, z_m$ be $0, 1$ valued independent random variables such that Prob.$[z_j = 1] = p_j$ for $1 \leq j \leq m$. Let $S^m = \sum_{j=1}^m z_j$ and the expectation of S^m be $\mu = E[S^m] = \sum_{j=1}^m p_j$. We are interested in the probability that S^m is above or below its expectation. The following Lemma bounds the probability that S^m is below its mean.

Lemma 5.1 For $0 \leq T < \mu$, Prob.$[S^m < T] \leq e^{-(\mu-T)^2/(2\mu)}$.

The next Lemma bounds the probability that S^m is above its mean.

Lemma 5.2 For $\mu < T \leq 2\mu$, Prob.$[S^m \geq T] \leq e^{-(T-\mu)^2/(3\mu)}$.

5.1 The Case of a Linear Array M_f

Problem 2. Let $\mathcal{L}$ be a linear array with n nodes numbered $1, 2, \ldots, n$. There is a packet at each node to start with. The destination of each packet could be any of the n nodes all with equal probability. Route the packets.

Lemma 5.3 Problem 2 can be solved in time $\frac{(3-\sqrt{5})}{2}n + \tilde{o}(n)$ steps.

Proof. Make use of the optimal algorithm given in section 2. The claim is that the algorithm will terminate within the specified time.

For some fixed d, $1 \leq d \leq n$, let $\mathcal{A}$ stand for the first d nodes, $\mathcal{B}$ stand for the next $(n - 2d)$ nodes, and $\mathcal{C}$ stand for the last d nodes of $\mathcal{L}$.

From among the packets originating from region $\mathcal{A}$, the expected number of packets that have to travel a distance of d or more is $\frac{(n-d)+(n-d-1)+\ldots+(n-2d+1)}{n} = d - \frac{1.5d^2}{n} + \frac{d}{2n}$. From among the packets originating from region $\mathcal{B}$, the expected number of packets that will travel a distance of d or more is $\frac{(n-2d)(n-2d+1)}{n}$ which simplifies to $n - 4d + 4\frac{d^2}{n} + 1 - \frac{2d}{n}$. Also, the expected number of packets that have to travel d or more distance from region $\mathcal{C}$ is $\frac{(n-d)+(n-d-1)+\ldots+(n-2d+1)}{n} = d - \frac{1.5d^2}{n} + \frac{d}{n}$.

Summing, the total expected number of packets that will travel a distance of d or more is $= E_d = \frac{d^2+n^2-2dn+n-d}{n} = \frac{(n-d)(n-d+1)}{n}$. Using Lemma 5.2, the actual number of packets is only $\tilde{o}(n)$ more than the expectation. The algorithm will run for $d' + \tilde{o}(n)$ time where d' is such that $E_{d'} \approx d'$. d' can be seen to be no more than $\frac{(3-\sqrt{5})n}{2} + O(1)$. $\square$

5.2 The Case of a 2D Mesh M_f

Problem 3. There is a packet to start with at each node of an $n \times n$ mesh M_f. The destination of each packet is a random node in the mesh, each such node being equally likely. Route the packets.

Lemma 5.4 Problem 3 can be solved in time $2(2 - \sqrt{3})n + \tilde{o}(n) \approx 0.536n + \tilde{o}(n)$ using the greedy algorithm. The queue size at each node is no more than 2 plus the number of packets destined for that node.

Proof. The greedy algorithm referred to is the following: Initially each packet is colored red or blue each color being equally likely. A red packet travels along the row of its origin up to its destination column in phase I. In phase II this red packet travels along its destination column to its actual destination. A blue packet executes a symmetric algorithm, i.e., it travels along the column of its origin in phase I and in phase II it travels along its destination row.

Assume that the two phases of the greedy algorithm are disjoint. Then, each phase is nothing but routing along a linear array. The number of packets originating from any i successive nodes can be seen to be $\frac{i}{2} + \tilde{o}(n)$. This fact, together with a computation similar to that given in the proof of Lemma 5.3, implies that the expected number of packets that have to travel a distance of d or more in any of the phases is $= E_d = \frac{(n-d)(n-d+1)}{2n} + \tilde{o}(n)$. Thus, a single phase will terminate in time $d' + \tilde{o}(n)$ where d' is such that $E_{d'} \approx d'$. One can see that d' is nearly $= (2 - \sqrt{3})n + \tilde{o}(n)$.

The proof of the queue size is cumbersome. However, we could modify the greedy algorithm using the trick described in [15]. The idea is based on the fact that the expected queue size at the end of phase I at any node is 1. This in particular means that the total queue size in any successive n^ϵ nodes (for some constant $\epsilon < 1$) is $n^\epsilon + \tilde{O}(\sqrt{n^\epsilon \log n})$. Thus we could group the processors in each row into groups of n^ϵ processors each, and locally distribute packets destined for each group. The extra queue size then will not exceed 2 w.h.p. $\square$

5.3 $k - k$ Routing, $k - k$ Sorting, and Cut Through Routing

We analyze the expected behavior of greedy algorithms for $k - k$ routing, $k - k$ sorting and cut through routing. The greedy algorithm referred to here (for M_f) is also the last two phases of Algorithm B. Scheduling for the bus in both the phases is done exactly as in phase I of algorithm in section 3.2.

Theorem 5.1 *The greedy algorithms for random $k - k$ routing, $k - k$ sorting, and cut through routing on M_f terminate in an expected time of $\frac{kn}{6} + \tilde{o}(kn)$, $\frac{kn}{6} + \tilde{o}(kn)$, and $\frac{kn}{6} + n + \tilde{o}(kn)$, respectively, if $k \geq 12$. The queue length can be adjusted to be $k + \tilde{o}(k)$. $\frac{kn}{6} - \tilde{o}(kn)$ is a lower bound on the expected run time for all these three problems.*

Theorem 5.2 *Random $k - k$ routing, $k - k$ sorting, and cut through routing on M_r can be realized in time $\frac{kn}{4} + \tilde{o}(kn)$, $\frac{kn}{4} + \tilde{o}(kn)$, and $\frac{kn}{4} + n + \tilde{o}(kn)$, respectively, for any $k \geq 8$. $\frac{kn}{4} - \tilde{o}(kn)$ is an expected time lower bound for these problems.*

6 Selection on M_r

The problem of selection on M_r has been studied by many researchers [3, 4]. Doctor and Krizanc's randomized algorithms can select in 1) $O(\log n)$ expected time assuming that each input permutation is equally likely; or 2) $O(\log^2 n)$ time with no assumptions. Hao, MacKenzie, and Stout [4] show that selection can be done in: 1) $O(\log^* n)$ time assuming a uniform distribution on all possible inputs; or 2) $O(\log n)$ time with no assumptions. In this section we show that selection on M_r can be performed in: 1) $O(\log^* n)$ expected time assuming that each input permutation is equally likely (Though the same result is mentioned in [4], our work is independent); and 2) Randomized $O(\log^* n \log\log n)$ time, with no assumptions. Our selection algorithms are based on Rajasekaran's hypercube selection algorithm [11].

6.1 Some Basics

The following Lemmas will be employed in our selection algorithm:

Lemma 6.1 *Jang, Park, and Prasanna [5]: If each node in an $n \times n$ mesh M_r has a bit, the number of 1's can be computed in $O(\log^* n)$ time.*

Lemma 6.2 *For any $1 \leq r \leq n$, elements in the first r rows of an $n \times n$ mesh M_r can be sorted in $O(r)$ time. (This Lemma has been discovered independently by [Jang and Prasanna], [Jenq and Sahni], and [Nakano, Peleg, Schuster]).*

Problem 4. Consider an $n \times n$ mesh M_r. Say there are ℓ_i elements arbitrarily distributed in row i, for $1 \leq i \leq n$. Let $\ell = \max\{\ell_1, \ell_2, \ldots, \ell_n\}$. For each i, concentrate the elements of row i in the first ℓ_i columns of row i.

Lemma 6.3 *Problem 4 can be solved in time $O(\ell)$ if ℓ is given.*

Proof. It is easy to solve Problem 4 in $O(\log n)$ time using the prefix sums algorithm of Miller, Prasanna, Reisis, and Stout [10]. In order to solve this problem in $O(\ell)$ time we use the following algorithm: There are ℓ rounds in the algorithm (for $t = 1, 2, \ldots, \ell$).

for $t := 1$ to ℓ do
(* Computation is local to each row i, $1 \le i \le n$ *)

> **Step 1.** If node j in row i has an element, then it sends a 1 to its right; at the same time it opens its W–E switch so that any message from left is blocked. If node j has no element, it simply closes its W–E switch so that any message from left is simply forwarded to the right.

> **Step 2.** If a node has an element and if it receives a 1 from left it simply accepts failure in this round. The node with an element which does not receive a 1 from left broadcasts its packet so that it can be concentrated in column t of row i. Realize that there can be only one such node that gets to broadcast in any round. The node that gets to broadcast will not participate in any future rounds, whereas every other node with an element will participate in the next round.

Clearly, the above algorithm runs correctly in $O(\ell)$ time. The above algorithm, though very simple, brings out the power of reconfiguration. $\square$

Along the same lines we can prove the following

Lemma 6.4 *If each node of an $n \times n$ mesh M_r has a bit, we can compute the boolean AND and boolean OR of these n^2 bits in $O(1)$ time.*

Lemma 6.5 *Problem 4 can be solved in $O(\ell)$ time even if ℓ is unknown. Within the same time, we will also be able to estimate ℓ.*

Proof. The idea is to check after each round (of algorithm given for Lemma 6.3), whether all the elements have been concentrated or not. Since this checking can be done in $O(1)$ time (c.f. Lemma 6.4), the algorithm will still run in $O(\ell)$ time. The second part of the Lemma is obvious. Call this algorithm as Algorithm C. $\square$

Corollary 6.1 *Say there are only ℓ elements (arbitrarily distributed) in an $n \times n$ mesh M_r. Then we could compute the prefix sums of these elements in time $O(\ell)$ even if ℓ is not given.*

Lemma 6.6 *If i is any row and j is any column of an $n \times n$ mesh M_r, and if each node of row i has an element, then, we can copy the elements of row i into column j in constant time.*

Proof. Broadcast the elements of row i along the columns so that elements of row i appear along the diagonal. Now perform another broadcast of the diagonal elements along the rows. $\square$

6.2 The Algorithm for the Uniform Case

The selection algorithm to be described assumes that each input permutation is equally likely. We are given n^2 elements and we have to find the ith smallest element. Assume that each element (or key) is *alive* to start with.

Algorithm Select1

repeat forever
Step 1. Count the number of *alive* keys using the algorithm of Lemma 6.1. Let N be this number. If N is $\le n^{1/3}$ then *quit* and go to Step 7;
Step 2. Each *alive* element includes itself in a sample S with probability $\frac{n^{1/3}}{N}$. The total number of keys in the sample will be $\tilde{\Theta}(n^{1/3})$;
Step 3. Concentrate the sample keys and sort them. Let ℓ_1 be $\text{select}(S, i\frac{s}{N} - \delta)$ and let ℓ_2 be $\text{select}(S, i\frac{s}{N} + \delta)$, where $\delta = d\sqrt{s \log N}$ for some constant d $(> c\alpha)$ to be fixed;
Step 4. Broadcast ℓ_1 and ℓ_2 to the whole mesh;
Step 5. Count the number of *alive* keys $< \ell_1$ (call this number N_1); Count the number of *alive* keys $> \ell_2$ (call this number N_2); If i is not in the interval $(N_1, N - N_2]$, go to Step 2 else let $i := i - N_1$;
Step 6. Any *alive* key whose value does not fall in the interval $[\ell_1, \ell_2]$ *dies*;
end repeat
Step 7 Concentrate the *alive* keys and sort them. Output the ith smallest key from this set.

Theorem 6.1 *The above algorithm selects in $O(\log^* n)$ expected time assuming that each input permutation is equally likely on M_r as well as the PARBUS model.*

Proof. We show that the *repeat* loop is executed no more than 11 times w.h.p. and that each of the above seven steps can be performed in $\widetilde{O}(\log^* n)$ time.

In Step 3 and Step 7 we make use of Lemma 6.5 to concentrate the keys along the rows. If ℓ is the maximum number of sample keys in any row, we sort the first ℓ columns using Lemma 6.2. The crucial fact is that the value of ℓ in any iteration is $\widetilde{O}(1)$. Notice that in any iteration, there are only $\widetilde{\Theta}(n^{1/3})$ sample keys and these sample keys will be uniformly distributed among all the n rows. Expected number of packets in any row will be $\Theta(\frac{n^{1/3}}{n})$, immediately implying that the number of sample keys in any row is $\widetilde{O}(1)$.

Steps 2, 4, and 6 take $O(1)$ time each. Counting in steps 1 and 5 takes $O(\log^* n)$ time (c.f. Lemma 6.1). Given that ℓ is $\widetilde{O}(1)$, steps 3 and 7 take $\widetilde{O}(1)$ time each. Thus the theorem follows. $\square$

6.3 A Selection Algorithm for the General Case

The algorithm to be used for the general case is the same as the algorithm Select1 with some crucial modifications. In the uniform case, in any iteration, the alive and sample keys will be uniformly distributed among the nodes of the mesh. Thus an expected $O(1)$ number of iterations (of the *repeat* loop) sufficed. The same need not hold in general. For instance, the alive keys after the first iteration may appear concentrated in a small region of the mesh (e.g., in a $\sqrt{N} \times \sqrt{N}$ submesh). The same might be the case after every iteration. Thus it seems $\Omega(\log \log n)$ iterations will be needed. Also, concentrating the sample keys (in order to identify ℓ_1 and ℓ_2) now becomes more complicated, for the same reason namely, these sample keys may not appear uniformly distributed among the nodes. Next we present the algorithm:

Each element (or key) is *alive* to start with.

Algorithm Select2

repeat forever

Step 1. Count the number of *alive* keys using the algorithm of Lemma 6.1. Let N be this number. If N is $\leq \log \log n$ then *quit* and go to Step 7;

Step 2. Each *alive* element includes itself in a sample S with probability $\frac{N^{1/6}}{N}$. The total number of keys in the sample will be $\widetilde{\Theta}(N^{1/6})$;

Step 3.1. Concentrate the sample keys as follows: Some of the keys will be concentrated along the rows and the others will be concentrated along the columns. This is done by simultaneously concentrating the keys along the rows as well as along the columns. I.e., perform one round of Algorithm C concentrating along the rows, followed by one round of Algorithm C concentrating along the columns, and so on. If a node succeeds in concentrating its element along the row (say), this element will be eliminated from future consideration. In particular, in the next round, this node will behave as though it does not have any element. Concentration stops when each sample key has been concentrated either along the row or along the column. Let r and c stand for the maximum number of rows and columns, respectively, used for concentrating the sample keys. Let $\ell = \max\{r, c\}$. (We'll show that ℓ is $\widetilde{O}(1)$.)

Step 3.2. Now copy the ℓ or less rows of sample keys into columns (using the algorithm of Lemma 6.6). This copying is done one row at a time.

Step 3.3. Sort the $\leq 2\ell$ columns of sample keys using Lemma 6.2. Let ℓ_1 be select$(S, i\frac{s}{N} - \delta)$ and let ℓ_2 be select$(S, i\frac{s}{N} + \delta)$, where $\delta = d\sqrt{s \log N}$ for some constant d ($> c\alpha$) to be fixed;

Steps 4,5,6. Execute Step 4, Step 5, and Step 6 of the algorithm Select1;

end repeat

Step 7. Concentrate the *alive* keys in the first row. This can be done for instance by broadcasting one key at a time. Realize that if there are only ℓ elements in the mesh, we can perform a prefix computation (c.f. Corollary 6.1) to arrive at a schedule for the broadcasts in $O(\ell)$ time. Output the ith smallest key from this set.

Theorem 6.2 *The above algorithm runs in time $\widetilde{O}(\log^* n \log \log n)$ on M_r as well as on PARBUS.*

Proof. It suffices to show that the value of ℓ in Step 3.1 is $\widetilde{O}(1)$. The rest of the steps can be analyzed as before.

If there are N alive keys at the beginning of some iteration of the *repeat* loop, then each alive key will be included in the sample with probability $\frac{N^{1/6}}{N}$.

How many rows will have more than 12 sample keys? Realize that if there are p_i alive keys in some row i, the expected number of sample keys from this row will be $p_i N^{-5/6}$. Classify a row as either *dense* or *sparse*, depending on whether it has $> N^{4/6}$ alive keys or $\leq N^{4/6}$ alive keys, respectively. For any sparse row, applying Chernoff bounds, the number of sample keys in this row can not be more than 6β with probability $\geq (1 - N^{-\beta})$, for any $\beta \geq 1$. Since there are at most N rows in the mesh, the expected number of sparse rows that have > 12 sample keys is $\leq N^{-1}$. This in turn means that the number of sparse rows with > 12 sample keys is $\tilde{O}(1)$. That is, every row (with the exception of $\tilde{O}(1)$ of them) with > 12 elements has to be dense.

Sample keys in the sparse rows (except for $\tilde{O}(1)$ of them) will potentially get concentrated along the rows. Also notice that there can be at most $N^{2/6}$ dense rows. Even if these dense rows and the sparse rows with > 12 sample keys are such that each column (when restricted to these rows) is completely filled with alive keys, the number of sample keys in each column can only be $\tilde{O}(1)$ and hence these sample keys will get concentrated along the columns.

In the above algorithm, if N is the number of alive keys at the beginning of any iteration, then at the end of this iteration the number of alive keys is no more than $N^{11/12}$ w.h.p. Here, the high probability is with respect to the current size of the problem, i.e., N. Therefore we conclude that the expected number of iterations of the *repeat* loop is $O(\log \log n)$. We can also show that the number of iterations is $\tilde{O}(\log \log n)$. Moreover, given that ℓ is $= \tilde{O}(1)$, each iteration of the *repeat* loop takes $\tilde{O}(\log^* n)$ time. Step 7 runs in $O(\log \log n)$ time. $\square$

A Note on Optimality: It is easy to see that a single step of computation on the mesh M_r can be simulated in $O(1)$ time on the Parallel Comparison Tree (PCT) model of Valiant [17]. The effect of reconfiguration can be achieved for free on the PCT, since the later charges only for the comparisons performed. Thus it will follow that selection needs $\Omega(\log \log n)$ time on the mesh M_r using any deterministic comparison based algorithm. (The same fact is mentioned in [4] as well). We believe that $\Omega(\log \log n)$ is a lower bound for selection on M_r even using a randomized comparison algorithm. This is an interesting open problem.

Along the same lines we can show the following

Theorem 6.3 *Maximal selection from out of n elements can be deterministically performed on an $n \times n^\epsilon$ mesh M_r or PARBUS in $O(1)$ time. Here ϵ is any constant > 0.*

7 Selection on M_f

In this section we show that selection on an $n \times n$ mesh M_f can be performed within $\tilde{O}(n^{1/3})$ steps. The best known previous algorithm is due to Kumar and Raghavendra [7] and it runs in $O(n^{1/3}(\log n)^{2/3})$ time. In [7], a lower bound of $\Omega(n^{1/3})$ is proven for selection and related problems and hence our selection algorithm is optimal. Our algorithm also runs in an optimal $\tilde{O}(n^{1/4})$ time on an $n^{5/4} \times n^{3/4}$ mesh M_f. In contrast, the best known previous selection algorithm on an $n^{5/4} \times n^{3/4}$ mesh had a run time of $O(n^{1/4} \log n)$ [2].

The algorithm used is the same as Select1. We only mention the modifications to be made: In Step 1 and Step 5, counting is done using the $O(n^{1/3})$ time algorithm for prefix sums due to Kumar and Raghavendra [7]. In Step 3, concentration is done by broadcasting one key at a time. The concentrated keys can be sorted using our logarithmic time sorting algorithm for M_f. Step 7 is identical to Step 3. The rest of the steps are easy and the run time of the whole algorithm can be analyzed using the fact that the *repeat* loop is executed no more than 11 times w.h.p. We get:

Theorem 7.1 *The above selection algorithm runs in $\tilde{O}(n^{1/3})$ time. Also, selection on an $n^{5/4} \times n^{3/4}$ mesh M_f can be performed in an optimal $\tilde{O}(n^{1/4})$ time.*

8 Conclusions

In this paper we have addressed numerous important problems related to packet routing, sorting, and selection on a mesh with fixed buses and on a mesh with reconfigurable buses. Many existing best known results have been improved. Some remaining open problems are: 1) Can the randomized algorithms given in this paper be matched with deterministic algorithms?; 2) Can sorting be performed in time asymptotically less than $\log n$ on a fixed dimensional mesh M_f with a polynomial number of processors?

Acknowledgements

I am grateful to Sunil Shende for introducing me to this area, and to Dipak P. Doctor and Danny Krizanc for supplying me with a copy of their manuscript [3]. I would also like to thank Viktor K. Prsanna and Assaf Schuster for providing me with a number of relevant articles.

References

[1] Y. Ben-Asher, D. Peleg, R. Ramaswami, and A. Schuster, The Power of Reconfiguration, Journal of Parallel and Distributed Computing, 1991, pp. 139-153.

[2] Y-C. Chen, W-T. Chen, and G-H. Chen, Efficient Median Finding and Its Application to Two-Variable Linear Programming on Mesh-Connected Computers with Multiple Broadcasting, Journal of Parallel and Distributed Computing 15, 1992, pp. 79-84.

[3] D.P. Doctor and D. Krizanc, Three Algorithms for Selection on the Reconfigurable Mesh, Technical Report TR-219, School of Computer Science, Carleton University, February 1993.

[4] E. Hao, P.D. McKenzie and Q.F. Stout, Selection on the Reconfigurable Mesh, Proc. Frontiers of Massively Parallel Computation, 1992.

[5] J. Jang, H. Park, and V.K. Prasanna, A Fast Algorithm for Computing Histograms on a Reconfigurable Mesh, Proc. Frontiers of Massively Parallel Computing, 1992, pp. 244-251.

[6] C. Kaklamanis, D. Krizanc, L. Narayanan, and Th. Tsantilas, Randomized Sorting and Selection on Mesh Connected Processor Arrays, Proc. ACM Symposium on Parallel Algorithms and Architectures, 1991.

[7] V.K.P. Kumar and C.S. Raghavendra, Array Processor with Multiple Broadcasting, Journal of Parallel and Distributed Computing 4, 1987, pp. 173-190.

[8] T. Leighton, Average Case Analysis of Greedy Routing Algorithms on Arrays, in Proc. ACM Symposium on Parallel Algorithms and Architectures, pp. 2-10, July 1990.

[9] J. Y-T. Leung and S. M. Shende, Packet Routing on Square Meshes with Row and Column Buses, in Proc. IEEE Symposium on Parallel and Distributed Processing, Dallas, Texas, Dec. 1991, pp. 834-837.

[10] R. Miller, V.K. Prasanna-Kumar, D. Reisis, and Q.F. Stout, Meshes with Reconfigurable Buses, in Proc. 5th MIT Conference on Advanced Research in VLSI, 1988, pp. 163-178.

[11] S. Rajasekaran, Randomized Parallel Selection, Proc. Tenth International Conference on Foundations of Software Technology and Theoretical Computer Science, 1990. Springer-Verlag Lecture Notes in Computer Science 472, pp. 215-224.

[12] S. Rajasekaran, $k - k$ Routing, $k - k$ Sorting, and Cut Through Routing on the Mesh, Technical Report, Department of CIS, University of Pennsylvania, Philadelphia, PA 19104, October 1991.

[13] S. Rajasekaran, Mesh Connected Computers with Fixed and Reconfigurable Buses: Packet Routing, Sorting, and Selection, Technical Report MS-CIS-92-56, Department of CIS, Univ. of Pennsylvania, July 1992.

[14] S. Rajasekaran and T. McKendall, Permutation Routing and Sorting on the Reconfigurable Mesh, Technical Report MS-CIS-92-36, Department of Computer and Information Science, University of Pennsylvania, May 1992.

[15] S. Rajasekaran and Th. Tsantilas, Optimal Routing Algorithms for Mesh-Connected Processor Arrays, Algorithmica, vol. 8, 1992, pp. 21-38.

[16] Q.F. Stout, Mesh-Connected Computers with Broadcasting, IEEE Trans. Computers 32, 1983, pp. 826-830.

[17] L.G. Valiant, Parallelism in Comparison Problems, SIAM Journal on Computing, vol. 14, 1985, pp. 348-355.

An Efficient Parallel Algorithm for the Layered Planar Monotone Circuit Value Problem*

Vijaya Ramachandran and Honghua Yang

Dept. of Computer Sciences, Univ. of Texas at Austin, Austin Texas 78712

Abstract. A planar monotone circuit (PMC) is a Boolean circuit that can be embedded in the plane and that contains only AND and OR gates. A layered PMC is a PMC in which all input nodes are in the external face, and the gates can be assigned to layers in such a way that every wire goes between gates in successive layers. Goldschlager, Cook & Dymond and others have developed NC^2 algorithms to evaluate a layered PMC when the output node is in the same face as the input nodes. These algorithms require a large number of processors $(\Omega(n^6)$, where n is the size of the input circuit). In this paper, we give an efficient parallel algorithm that evaluates a layered PMC of size n in $O(\log^2 n)$ time using only a linear number of processors on an EREW PRAM. Our parallel algorithm is the best possible to within a polylog factor, and is a substantial improvement over the earlier algorithms for the problem.

1 Introduction

A *Boolean circuit* is a circuit whose wires do not form directed cycles. The problem of evaluating a Boolean circuit, given the values of all its inputs, is called the *circuit value problem* (CVP). This is a central problem in the area of algorithms and complexity. Ladner [11] has shown that the CVP is *P*-complete under log space reductions. Some special cases of the CVP have been studied, among which the *monotone circuit value problem*, where the circuit has only AND and OR gates, and the *planar circuit value problem*, where the circuit has a plane embedding, have been shown to be *P*-complete by Goldschlager [7].

One interesting special case of CVP is the *planar monotone circuit value problem* (PMCVP). A *planar monotone circuit* (PMC) is a Boolean circuit that has a plane embedding and that contains only AND and OR gates. A *layered* PMC (LPMC) is a PMC in which all inputs are in the external face, and the gates can be assigned to layers in such a way that every wire goes between gates in successive layers. We shall use LPMCVP to denote the problem of evaluating an LPMC. Goldschlager [5, 6], Dymond & Cook [2], and Mayr [12] have shown that a special case of the LPMCVP, where the output node is in the same face as the input nodes of the LPMC, is in NC^2. An NC^2 algorithm for the LPMCVP, without the restriction that the output node be in the same face as the input nodes, is given in Yang [19]; this algorithm uses the straight-line code parallel evaluation technique of Miller, Ramachandran & Kaltofen [13]. In recent work, NC^3 algorithms for the general PMCVP are given in [19] and Delcher & Kosaraju [1]. All of these algorithms use a large number of processors.

---

* This work was supported by Texas Advanced Research Program Grant 003658480 and NSF Grant CCR 90-23059. Email addresses: {vlr, yanghh}@cs.utexas.edu.

A straightforward approach for evaluating an LPMC is to evaluate gates layer by layer. This method is inherently sequential. In [5] Goldschlager introduced a technique that is based on the planarity and monotonicity of the problem, to solve the special case of the LPMCVP cited above by working on *valid segments* instead of individual gates. Here a valid segment denotes a sequence of adjacent gates at a layer that are evaluated to be 1. This technique has also been used in [12, 19]. A common drawback in these results is that a large number of processors $(\Omega(n^6))$ is required. Therefore, the problem of reducing the processor requirement becomes an interesting and challenging question. This is the problem we address in this paper. We present an efficient parallel algorithm for evaluating an LPMC using a new approach, namely *segment propagation with circuit transformation*. Our algorithm does not require the output node to lie in the same face as the input nodes and it runs in $O(\log^2 n)$ time using n processors on an EREW PRAM, where n is the size of the circuit. This is within a polylog factor of the best possible. Our algorithm is an improvement over the previous algorithms [5, 6, 2, 12, 19] for LPMCVP in that earlier algorithms used indirect methods such as the relationship between sequential space and parallel time or the parallel evaluation of straight-line code to place the problem in NC. By using direct techniques, we are able to obtain an efficient parallel algorithm for its solution. Very recently, we have developed an efficient parallel algorithm [16] for the general PMCVP that runs in polylog time using only a linear number of processors on an EREW PRAM. The algorithm in [16] builds on the algorithm presented in this paper, but it is substantially different in that it works on the dual graph of an embedded PMC instead of exploiting a layered structure.

The parallel computation model we use here is the EREW PRAM model. For PRAM models and techniques for designing efficient algorithms on a PRAM, see Karp & Ramachandran [8]. The algorithmic notation in this paper is from Tarjan [17] and Ramachandran [14]. Due to page limitation, some proofs and algorithms have been shortened or omitted. For details, see [15].

2 Preliminaries

Definition 2.1 A *contour* is a simple closed curve in the plane. A contour divides the plane into two parts: the part that is unbounded is called the *outer part* of the contour, the part that is bounded is called the *inner part* of the contour. We say contour c_1 *encloses* contour c_2 if c_2 is in the inner part of c_1.

Definition 2.2 A *layered PMC* (LPMC) C is defined to be a sequence of layers with layer numbers $1, 2, \ldots, d$ that has the following properties:

(1) Layer 1 consists of all input nodes of C and layer d consists of the output gate of C. A layer between 1 and d consists of AND and OR gates with one or two inputs.

(2) The gates at layer l are located on a contour that encloses the contour of layer $l + 1$, for each l, $1 \le l < d$.

(3) At each layer l, gates are numbered increasingly from 0 to $n_l - 1$ in the clockwise direction on the contour, where n_l is the number of gates at layer l.

(4) Each wire goes from layer l to layer $l + 1$ without crossing another wire. For each wire in C, there is a directed path in C containing that wire that goes from

an input node to the output gate.

An example of an LPMC is shown in Figure 1. We assume that the computation on the gate numbers at layer l follows a modular arithmetic with modulus n_l, where n_l is the number of gates at layer l. The following definitions are with respect to an embedded LPMC C with d layers and an output gate t.

Definition 2.3 The *left (right) neighbor* of a gate i at layer l is the gate $i - 1$ $(i + 1)$ at layer l. Two gates are *adjacent* if one is the left neighbor of the other.

If a layer has only one gate g, then the left (right) neighbor of g is g itself.

Definition 2.4 A *segment* S_{lij} is a sequence of adjacent gates $i, i + 1, \ldots, j$ at layer l. The two end gates i and j are called the *starting gate* and the *ending gate* of the segment, denoted by $start(S_{lij})$ and $end(S_{lij})$, respectively. A segment is *empty* if it contains no gate. A nonempty segment S_{lij} is *circular* if it contains all gates at layer l, i.e., either layer l has only one gate or $j = i - 1$. Two segments S_{lij} and $S_{li'j'}$ are *adjacent* if $i' = j + 1$ or $i = j' + 1$. Two segments are *overlapped* if they have at least one gate in common.

When there is no confusion, we will omit the subscripts of a segment, and a segment means a nonempty segment unless otherwise specified. Given an input assignment Φ to C, we can define the value of a segment in C.

Definition 2.5 A segment is *valid* if every gate in the segment evaluates to 1. A valid segment S_{1ij} at layer 1 is called a *maximal valid input segment* (MVIS), if either S_{1ij} is circular or both input nodes $i - 1$ and $j + 1$ have value 0.

Note that due to the planarity and the monotonicity of C, an MVIS uniquely propagates a largest valid segment to the next layer, which in turn uniquely propagates another largest valid segment to the next layer. This propagation terminates when an empty segment is encountered or the last layer of C is reached. Since a segment is described by its layer number and its starting and ending gates, we can characterize the segment propagation in terms of starting and ending gates of segments at adjacent layers and the relations among them.

For each gate g in the embedding of C, there is a clockwise cyclic ordering of the wires adjacent to g. By our definition of an LPMC, the output wires of g are consecutive in the cyclic ordering. We define the output wire of g that appears immediately before the left input wire of g in the cyclic ordering to be the *leftmost output* of g, and the output wire of g that appears immediately after the right input wire of g to be the *rightmost output* of g. The following definition is based on the fact that a two-input AND gate evaluates to 1 iff both inputs of the AND gate are 1, while an OR gate evaluates to 1 if one input is 1.

Definition 2.6 The *left leg (right leg)* of a gate g at layer l, $1 \leq l \leq d - 1$, denoted by $leftleg(g)$, is a gate at layer $l + 1$ defined as follows: Let g' be the gate at layer $l + 1$ receiving the leftmost (rightmost) output l_{out} (r_{out}) of g.

(1) If g' is a one-input gate or an OR gate, or g' is a two-input AND gate and l_{out} (r_{out}) is the left (right) input of g', then g' is the left (right) leg of g.

(2) If g' is a two-input AND gate and l_{out} (r_{out}) is the right (left) input of g', then the right (left) neighbor of g' is the left (right) leg of g.

Definition 2.7 The *left leg* and the *right leg* of a segment S_{lij}, denoted by $leftleg(S_{lij})$ and $rightleg(S_{lij})$, are defined to be $leftleg(i)$ and $rightleg(j)$, respectively. A segment S is a *twisting segment* if its validity cannot guarantee any nonempty segment at the next layer to be valid.

Lemma 2.1 *A segment S is a twisting segment iff it satisfies one of the following three cases which are shown in Figure 2.*
Case 1: S consists of a single gate with exactly one output which is connected to an input of a two-input AND gate at the next layer.
Case 2: S consists of a single gate with exactly two outputs connected to two neighboring two-input AND gates at the next layer. The left output is connected to the right input of the left AND gate; the right output is connected to the left input of the right AND gate.
Case 3: S consists of two adjacent gates, each with exactly one output. The two outputs of S are connected to two neighboring two-input AND gates at the next layer. The left output is connected to the right input of the left AND gate; the right output is connected to the left input of the right AND gate.

Proof Sketch: If S contains at least three gates, then since each gate has at most two inputs and at least one output, $leftleg(S)$ is to the left of $rightleg(S)$ and the validity of S guarantees at least two gates $leftleg(S)$ and $rightleg(S)$ at the next layer to be valid. The cases where S contains one or two gates are limited and a careful case analysis shows that the above three cases are the only three cases of a twisting segment. Here we only show that a segment S as described in Case 1 is a twisting segment. Let S contain only one gate g_1, which has only one output wire o. If o goes to a one-input gate g' or an OR gate g', then the validity of g_1 guarantees the validity of g'; if o goes to a two-input AND gate g' (i.e., Case 1), then the validity of g_1 cannot guarantee the validity of g'. Hence S is a twisting segment. □

Definition 2.8 Let S be a segment at a layer l. Then S *derives* an empty segment iff S is a twisting segment; S *derives* a segment S' at layer $l+1$, iff S is not a twisting segment and $start(S') = leftleg(S)$, $end(S') = rightleg(S)$.

Definition 2.9 A segment S at a layer l *propagates* a segment S' at a layer l', if 1) $l = l'$ and $S = S'$, or 2) S derives a segment S'' and S'' propagates S'. A segment S' is the *terminating segment* of a segment S if S propagates S' and S' derives an empty segment.

Clearly, a terminating segment of S must be a twisting segment, but the reverse need not be true. See Figures 1 & 3.

Lemma 2.2 *If a segment S in an LPMC C is valid, then the segments propagated by S are also valid.*

Proof: It suffices to prove that if a segment S_{lij} is valid and S_{lij} derives a segment $S_{(l+1)i'j'}$, then $S_{(l+1)i'j'}$ is valid. Since $i' = leftleg(i)$ and $j' = rightleg(j)$, by the planarity and the monotonicity of C, every gate in $S_{(l+1)i'j'}$ excluding i' and j' evaluates to 1. By Definition 2.6, if i' is a one-input gate or an OR gate, then i is an input to i' and i' evaluates to 1; if i' is a two-input AND gate and i is the left input of i', then the right input of i' is also from S_{lij} and i' evaluates to 1; otherwise, $i'-1$ is a two-input AND gate and i is the right input of $i'-1$, then i' receives its inputs from S_{lij} and i' evaluates to 1. Similarly, we can show that j' evaluates to 1. Hence $S_{(l+1)i'j'}$ is valid. □

Note that the derive and propagate relations may not give us the biggest possible segment at the next layer that is guaranteed to be valid for the special case when a valid segment S is circular, since the segment propagated by S at the next layer may omit an AND gate. But this case will be handled easily in our algorithm since once we detect a valid circular segment in the circuit, the output of the circuit must be 1 and our job is done.

Definition 2.10 We call a valid segment propagated by an MVIS in an LPMC a *valid propagated segment* (VPS) of the MVIS.

Our algorithm works by propagating all MVIS's in an LPMC. Figure 1 shows the MVIS's and their propagation. We now give a high level description of our algorithm. Given an LPMC C, we repeatedly perform the following two steps.

1. For each gate g, determine if it is in a VPS as follows:
   (a) Find the number of VPS's starting and ending at g, respectively.
   (b) Use the above numbers to determine if g is in at least one VPS.
2. Transform C to a smaller LPMC with the same output value.

In the following sections we will fill in the details of the algorithm. In Section 3, we present our method for the computation needed in step 1. In Section 4, we describe our approach for step 2. In Section 5, we give the overall algorithm for the LPMCVP and its correctness and complexity analysis. We show that each iteration of steps 1&2 can be performed efficiently in parallel, and that the number of MVIS's in C goes down by a factor of 2 with each iteration.

3 Segment Propagation in Parallel

Our goal is to find out, for each gate in an LPMC C with an input assignment Φ, whether the gate is in a VPS (see Definition 2.10) propagated by some MVIS. A naive approach is to propagate each MVIS individually, which requires $\Theta(n^2)$ processors in the worst case. Our approach is to propagate all the MVIS's in C together. We assume that the VPS's propagated by different MVIS's are different from each other even though they could be the same segment. The intuition is that it is good enough to obtain, for each gate, how many VPS's start and end at this gate, since we can then find out whether a gate g is in a VPS by simply checking whether there is a VPS that starts at g or at a gate to the left of g, and ends at g or at a gate to the right of g (i.e., whether some VPS contains g).

3.1 Tracing the Starting and the Ending Gates of the VPS's

Let C be an embedded LPMC with an input assignment Φ.

Definition 3.1 The *left leg tree (right leg tree)* of C is a directed tree $T_L = (V, E_L)$ $(T_R = (V, E_R))$ defined as follows: V is the set of input nodes and gates in C; a directed edge $(g_i, g_j) \in E_L$ iff $g_j = leftleg(g_i)$ $((g_i, g_j) \in E_R$ iff $g_j = rightleg(g_i))$. Both trees are rooted at the output gate of C.

Note that an edge in either tree does not necessarily correspond to a wire of C. We are actually interested in a subtree of T_L and a subtree of T_R that depend on the input assignment Φ to C as described in the following definition.

Definition 3.2 The *left propagation tree T_{L_Φ} (right propagation tree T_{R_Φ})* of C under Φ is the subtree of T_L (T_R) that is reachable by the starting (ending) input nodes of the MVIS's in C. (The T_{L_Φ} and the T_{R_Φ} of the circuit in Figure

1 are shown in Figures 4 & 5.) For two gates g and g' in $T_{L_{\circ}}$ $(T_{R_{\circ}})$, we say g is a *predecessor* of g' and g' is a *successor* of g in $T_{L_{\circ}}$ $(T_{R_{\circ}})$ if there is a directed path from g to g' in $T_{L_{\circ}}$ $(T_{R_{\circ}})$. For a gate g at layer l in $T_{L_{\circ}}$ $(T_{R_{\circ}})$, $PRED_L(g)$ $(PRED_R(g))$ is the set of the proper predecessors of g in $T_{L_{\circ}}$ $(T_{R_{\circ}})$, and $M_L(g)$ $(M_R(g))$ is the set of MVIS's in C whose starting (ending) input nodes are either g or predecessors of g in $T_{L_{\circ}}$ $(T_{R_{\circ}})$.

Informally, $M_L(g)$ $(M_R(g))$ is the set of MVIS's whose VPS's at layer l start (end) at g provided that these MVIS's can propagate to layer l. However, it is possible that some of the MVIS's may have terminated propagation before reaching layer l. The next few lemmas address this issue.

Definition 3.3 For each gate g at layer l in $T_{L_{\circ}}$ $(T_{R_{\circ}})$, $TERM_L(g)$ $(TERM_R(g))$ is the set of MVIS's whose terminating segments (see Definition 2.9) start (end) at g, and $PROP_L(g)$ $(PROP_R(g))$ is the set of MVIS's whose propagated segments at layer l indeed start (end) at g. (So, $TERM_L(g) \subseteq PROP_L(g) \subseteq M_L(g)$.) For a gate not in $T_{L_{\circ}}$ or $T_{R_{\circ}}$, its corresponding sets are empty.

We focus on computations on $T_{L_{\circ}}$. The computations on $T_{R_{\circ}}$ are symmetric.

Lemma 3.1 *For a gate g in $T_{L_{\circ}}$, $PROP_L(g) = M_L(g) \setminus \cup_{g' \in PRED_L(g)} TERM_L(g')$.*

Proof: By the definitions of $M_L(g)$, $TERM_L(g)$ and $PROP_L(g)$. □

Corollary 3.1 $|PROP_L(g)| = |M_L(g)| - \sum_{g' \in PRED_L(g)} |TERM_L(g')|.$

Proof: The proof is based on the following two facts: 1. $TERM_L(g) \subseteq PROP_L(g) \subseteq M_L(g)$. 2. $TERM_L(g) \cap TERM_L(g') = \phi$ for any $g' \neq g$. □

We are interested in obtaining $|PROP_L(g)|$ $(|PROP_R(g)|)$ for all g in $T_{L_{\circ}}$ $(T_{R_{\circ}})$, since that would take care of step 1.(a) of our high level algorithm given in Section 2. Since $|M_L(g)|$ can be easily computed using Euler-tour technique [18] on $T_{L_{\circ}}$, the remaining problem is to compute $|TERM_L(g)|$.

Definition 3.4 An MVIS M is *left-twisted* at a gate g at layer l if g is a successor of $start(M)$ in $T_{L_{\circ}}$ and the segment between g and the successor of $end(M)$ in $T_{R_{\circ}}$ at layer l is a twisting segment (see Definition 2.7). $TWIST_L(g)$ is the set of MVIS's that are left-twisted at g.

Note that M may be left-twisted more than once along the directed path in $T_{L_{\circ}}$ from $start(M)$. However, the very first gate on the path at which M is left-twisted is the starting gate of its terminating segment as shown below.

Lemma 3.2 *For a gate g in $T_{L_{\circ}}$, $TERM_L(g) = TWIST_L(g) \setminus \cup_{g' \in PRED_L(g)} TWIST_L(g')$.*

Proof: $TWIST_L(g) \setminus \cup_{g' \in PRED_L(g)} TWIST_L(g')$ contains the MVIS's that are left-twisted for the first time at gate g. By Definition 3.3, The lemma holds. □

For each gate g in $T_{L_{\circ}}$, $TWIST_L(g)$ and $TWIST_R(g)$ can be computed directly from the $M_L(g)$ and the $M_R(g)$ as follows: Let g_1 be the right neighbor of g and g_2 the left neighbor of g.

$$TWIST_L(g) = \begin{cases} M_L(g) \cap M_R(g) & \text{if } g \text{ forms a twisting segment of Case 1 or 2} \\ M_L(g) \cap (M_R(g) \cup M_R(g_1)) & \text{if } g \text{ and } g_1 \text{ form a twisting segment of Case 3} \\ \phi & \text{otherwise} \end{cases}$$

$$TWIST_R(g) = \begin{cases} M_L(g) \cap M_R(g) & \text{if } g \text{ forms a twisting segment of Case 1 or 2} \\ M_R(g) \cap (M_L(g) \cup M_L(g_2)) & \text{if } g_2 \text{ and } g \text{ form a twisting segment of Case 3} \\ \phi & \text{otherwise} \end{cases}$$

Since the $TWIST_L(g)$ need not be disjoint for different g with predecessor-successor relation, to compute $|TERM_L(g)|$, we need to study the relation among the $TWIST_L(g)$. Before proving the following lemma, we first observe that if g' is a successor of g in T_{L_Φ}, then $M_L(g') \supseteq M_L(g)$; if g and g' do not have predecessor-successor relation in T_{L_Φ}, then $PROP_L(g) \cap PROP_L(g') = M_L(g) \cap M_L(g') = \phi$. This observation also holds if we replace the subscripts L with R.

Lemma 3.3 *Let g be a gate in T_{L_Φ} and let g' be a successor of g in T_{L_Φ}. Then either $TWIST_L(g) \cap TWIST_L(g') = \phi$, or $TWIST_L(g) \subseteq TWIST_L(g')$.*

Proof: We consider the case where g forms a single gate twisting segment (Cases 1 & 2) and the case where g and its right neighbor g_2 form a twisting segment of two gates (Case 3). Note that in the latter case g and g_2 share the same immediate successor in T_{L_Φ} and the same immediate successor in T_{R_Φ}. Let g_l and g_r be the immediate successors of g in T_{L_Φ} and T_{R_Φ}, respectively. Then $TWIST_L(g) \subseteq M_L(g_l) \cap M_R(g_r)$. Since g' is a successor of g and hence a successor of g_l in T_{L_Φ}, $M_L(g_l) \subseteq M_L(g')$. Assume $TWIST_L(g) \cap TWIST_L(g') \neq \phi$. Let M be an MVIS such that $M \in TWIST_L(g)$ and $M \in TWIST_L(g')$. Then $M \in M_R(g_r)$ and either $M \in M_R(g')$ (g' forms a twisting segment of Cases 1 or 2) or $M \in M_R(g'')$, where g'' is the right neighbor of g' in C (g' and g'' form a twisting segment of Case 3). If $M \in M_R(g')$, then since $M_R(g_r) \cap M_R(g') \neq \phi$, g' is a successor of g_r in T_{R_Φ} and $M_R(g_r) \subseteq M_R(g')$. Hence $TWIST_L(g) \subseteq M_L(g_l) \cap M_R(g_r) \subseteq M_L(g') \cap M_R(g') \subseteq TWIST_L(g')$. If $M \in M_R(g'')$, then since $M_R(g_r) \cap M_R(g'') \neq \phi$, g'' is a successor of g_r in T_{R_Φ} and $M_R(g_r) \subseteq M_R(g'')$. Hence $TWIST_L(g) \subseteq M_L(g_l) \cap M_R(g_r) \subseteq M_L(g') \cap M_R(g'') \subseteq TWIST_L(g')$. □

Based on the two preceding lemmas, we give the following definition.

Definition 3.5 For a gate g with a nonempty $TWIST_L(g)$ and a successor g' of g in T_{L_Φ}, $TWIST_L(g)$ is *immediately enclosed* by $TWIST_L(g')$, denoted by $TWIST_L(g) \subseteq_I TWIST_L(g')$, iff $TWIST_L(g) \subseteq TWIST_L(g')$ and no g'' on the directed path from g to g' in T_{L_Φ} satisfies $TWIST_L(g'') \subseteq TWIST_L(g')$.

Lemma 3.4 *For a gate g in T_{L_Φ},*

(1) $TERM_L(g) = TWIST_L(g) \setminus \bigcup_{g' \in PRED_L(g) \wedge TWIST_L(g') \subseteq_I TWIST_L(g)} TWIST_L(g')$;

(2) $|TERM_L(g)| = |TWIST_L(g)| - \sum_{g' \in PRED_L(g) \wedge TWIST_L(g') \subseteq_I TWIST_L(g)} |TWIST_L(g')|$.

Proof: By Lemma 3.2 and Lemma 3.3, (1) holds. Note that no g' in the summation in (2) have predecessor-successor relation. Let g'_1 and g'_2 be two gates that satisfy (2). Then $TWIST_L(g'_1) \cap TWIST_L(g'_2) = \phi$. Hence (2) holds. □

Procedure 1 contains the steps to compute $|PROP_L(g)|$ for each gate g in T_{L_Φ} and $|PROP_R(g)|$ for each gate g in T_{R_Φ}.

Procedure 1: Computing $|PROP_L(g)|$ and $|PROP_R(g)|$

1.1. Find the MVIS's in C and assign to them a total order from left to right;

1.2. Construct T_{L_Φ} and T_{R_Φ} of C;

1.3. Compute $M_L(g)$ for each gate g in T_{L_Φ} and $M_R(g)$ for each g in T_{R_Φ};

1.4. Compute $TWIST_L(g)$ and $TWIST_R(g)$ from the $M_L(g)$ and the $M_R(g)$;

1.5. Find the enclosure relation $\subseteq_I$ among the $TWIST_L(g)$ and the $TWIST_R(g)$;

1.6. Compute $|TERM_L(g)|$ and $|TERM_R(g)|$ for each g using Lemma 3.4;

1.7. Compute $|PROP_L(g)|$ and $|PROP_R(g)|$ for each g using Corollary 3.1;

Up to this point, we have shown how to calculate $|PROP_L(g)|$ for each gate g in T_{L_Φ} and $|PROP_R(g)|$ for each gate g in T_{R_Φ}. We will show later in Section 5 that all the steps described above can be implemented in $O(\log n)$ time using a linear number of processors on an EREW PRAM.

3.2 Verifying if a Gate is in a Valid Propagated Segment

Suppose that layer l of an LPMC C consists of a sequence of gates $g_1, g_2, \ldots g_{n_l}$. We first construct an array $A_l = (|PROP_L(g_1)|, -|PROP_R(g_1)|, |PROP_L(g_2)|, -|PROP_R(g_2)|, \ldots, |PROP_L(g_{n_l})|, -|PROP_R(g_{n_l})|)$. We then compute the prefix sums in array A_l. Let S_{g_i} denote the prefix sum value at the position of $|PROP_L(g_i)|$ in A_l and let S'_{g_i} denote the prefix sum value at the position of $-|PROP_R(g_i)|$ in A_l. Recall that we treat the VPS's propagated by different MVIS's as different even though some could be the same segment. Intuitively, S_{g_i} is the number of the VPS's at layer l starting at g_i or at a gate to the left of g_i minus the number of the VPS's at layer l *ending at a gate to the left of* g_i; $S'(g_i)$ is the number of the VPS's at layer l starting at g_i or at a gate to the left of g_i minus the number of the VPS's *ending at* g_i *or at a gate to the left of* g_i.

Definition 3.6 A gate g at layer l in C is a *rear gate* if g is in at least one VPS and g is the ending gate of all VPS's at layer l that contain g. The VPS's at layer l *form a ring* if there is at least a VPS at layer l and there is no rear gate at layer l. Let min_l be the smallest prefix sum value in array A_l of C at layer l.

Lemma 3.5 *If there is a rear gate at layer l, then a gate g_i is in a valid propagated segment iff $S_{g_i} > min_l$.*

Proof: Let g_r be a rear gate at layer l. Let k be the number of the VPS's with their starting gates to the right of their ending gates in the sequence of gates $g_1, g_2, \ldots g_{n_l}$ at layer l (i.e., the wrap-around segments). Note that the starting gates of such segments must be to the right of g_r in the sequence and the ending gates of such segments must be to the left of or equal to g_r in the sequence. By Definition 3.6, all VPS's whose starting gates are g_r or gates to the left of g_r must end at g_r or at gates to the left of g_r in the sequence of gates $g_1, g_2, \ldots g_{n_l}$ at layer l. Let k' be the number of the VPS's whose starting gates are g_r or gates to the left of g_r. Then, $S'_{g_r} = k' - k' - k = -k$. Similarly for a gate g_t not in any VPS at layer l, we have $S'_{g_t} = -k$, and $S_{g_t} = -k$ since $|PROP_L(g_t)| = |PROP_R(g_t)| = 0$. For a gate g_i in a VPS at layer l, we have the following three cases: Case 1: g_i is a rear gate. Then $S_{g_i} = S'_{g_i} + |PROP_R(g_i)| > -k$ since $S'_{g_i} = -k$ and $|PROP_R(g_i)| > 0$. Case 2: g_i is to the right of a rear gate g_j but g_i is not a rear gate. Then $S_{g_i} \geq S'_{g_i} = S'_{g_j} + \sum_{j<k\leq i} |PROP_L(g_k)| - \sum_{j<k\leq i} |PROP_R(g_k)|$. Since g_j is a rear gate but g_i is not, there is a VPS starting from a gate between g_{j+1} and g_i (inclusive) and ending at a gate to the right of g_i. Hence $S_{g_i} \geq S'_{g_i} > S'_{g_j} = -k$, Case 3: g_i is to the left of a rear gate g_j but g_i is not a rear gate. Similarly $S_{g_i} \geq S'_{g_i} > -k$. Hence $min_l = -k$ and the lemma holds. $\qquad\square$

The iff condition in Lemma 3.5 no longer holds when the VPS's at layer l form a ring. The following lemma deals with this situation.

Lemma 3.6 *If the valid propagated segments at layer l form a ring, then there exists a layer l' ($l' < l$) that has a rear gate and every gate at layer l' is in valid propagated segments.*

After performing segment propagation on an LPMC C, if the output t of C is in a VPS or there is a layer l' where all gates are in VPS's, then we are done. Otherwise we proceed to perform circuit transformation.

4 Circuit Transformation

We transform an LPMC C with an input assignment Φ to another LPMC C' with an input assignment Φ' such that C and C' have the same output value but the number of the MVIS's in C' is reduced by at least half.

Definition 4.1 An MVIS M *meets* another MVIS M' in C if there exist two segments S and S' at a layer in C such that M propagates S, M' propagates S', and S and S' are either adjacent or overlapped. Further, we define an equivalence relation among all MVIS's in C using the transitive closure of *meet*. Let $[M_1], [M_2], ..., [M_m]$ be the equivalence classes of the MVIS's in C induced by this equivalence relation. We transform C to C' using the following two rules. Initially, let C' be the same as C and let Φ' be the same as Φ.

Replacement Rule: For a non-singleton equivalence class $[M_i]$ in C, set the value of the input nodes between the MVIS's in $[M_i]$ to 1 in Φ', i.e., replace $[M_i]$ by a single MVIS in C' starting at the starting input node of the leftmost MVIS in $[M_i]$ and ending at the ending input node of the rightmost MVIS in $[M_i]$.

Simplification Rule: For a singleton equivalence class $[M_i]$ in C, remove the MVIS in $[M_i]$ and the valid segments propagated by the MVIS from C'.

Lemma 4.1 *The number of the MVIS's in C' is at most half of that in C.*

Proof: This lemma follows immediately from the previous two rules. □

Lemma 4.2 *C' outputs 1 iff C outputs 1; after removing all gates unreachable to the output t and reassigning gate numbers at each layer in C', C' is an LPMC.*

Proof: In the Replacement Rule, by the definition of *meet*, the value of the input nodes between the MVIS's in an equivalence class $[M_i]$ in C does not affect the output t of C. For the Simplification Rule, we consider a singleton equivalence class that consists of an MVIS M. If an OR gate or a one-input AND gate receives an input from a VPS propagated by M, then the gate would also be in a VPS propagated by M. If the two inputs of a two-input AND gate are from a single VPS propagated by M, then the gate would also be in a VPS propagated by M. The two inputs of a two-input AND gate cannot come from two different VPS's, one propagated by M and the other propagated by another MVIS M', since M and M' would be adjacent and would be in the same equivalence class and M would not be in a singleton equivalence class. Therefore the only gates whose inputs are removed as a result of applying the Simplification Rule are the two-input AND gates with exactly one input removed since it is from a VPS propagated by M. But the value of such AND gates depend only on the other input since we know that the removed input has value 1. Hence the value of the output t of C is not changed in C'.

Since only one input wire of a two-input AND gate can possibly be removed from C, each gate in C' has fan-in at least 1. Removing all gates unreachable to the output gate t from C' will not change the fan-in of the remaining gates in C', and the fan-out of each remaining gate and input nodes in C' (except t) is non-zero. Hence the lemma holds. □

5 The Algorithm for the LPMCVP

We now present our parallel algorithm for evaluating an LPMC of size n. The correctness of this algorithm is guaranteed by the previous lemmas.

Algorithm 1: Evaluating an LPMC
Input: An LPMC C of size n, an input assignment Φ and an output t.
Output: The value of the output t of C.
1. **do** {{Repeat the steps between **do** and **od**.}}
{{Steps 2-4 check the loop exit conditions.}}
2.   **if** $t = 1$ or $t = 0 \longrightarrow$ **stop**
3. | all gates at some layer are valid $\longrightarrow$ $t:=1$, **stop**
4. | all input nodes have value 0 $\longrightarrow$ $t:=0$, **stop**
    fi;
{{Steps 5-8 propagate valid segments from the MVIS's in C.}}
5.   Apply Procedure 1 to compute $|PROP_L(g)|$ and $|PROP_R(g)|$ for each g;
6.   Construct the array A_l as described in Section 3.2 and calculate its prefix sums for each layer l, let min_l be the minimum prefix sum value in A_l;
7.   **pfor** each gate g at layer $l \longrightarrow$
8.   **if** $S_g > min_l \longrightarrow g := 1$ **fl**
    rofp;
{{Steps 9-12 transform C to C'.}}
9.   Find the equivalence classes induced by the transitive closure of *meet* in C;
10.   Transform C to C' with an input assignment Φ' using the Replacement Rule and the Simplification Rule;
11.   Remove all gates from C' that are unreachable to the output t of C' and all wires adjacent to the removed gates;
12. $C := C'$, $\Phi := \Phi'$, reassign gate numbers to the gates in C at each layer
    od

Theorem 1 *Algorithm 1 takes $O(\log^2 n)$ time using n processors on an EREW.*

Proof Sketch: Since each iteration of the **do** loop either transforms circuit C to an equivalent circuit C' with at most half of the number of the MVIS's (by Lemma 4.1) or stops, there are at most $\log n$ iterations.

For the complexity of segment propagation, all steps except step 1.5 in Procedure 1 can be implemented in $O(\log n)$ time using n processors on an EREW PRAM by applying standard techniques such as prefix sums [10, 8], Euler-tour [18, 8], and tree evaluation [4, 9]. The solution for step 1.5 in Procedure 1 is not obvious. We have given a procedure in [15] that finds the enclosure relation in $O(\log n)$ time using a linear number of processors. We represent the enclosure relation among the $TWIST_L(g)$ as an *enclosure forest EFL* (see Figure 4) such that a gate g' is the parent of a gate g in EFL iff $TWIST_L(g) \subseteq_I TWIST_L(g')$. For details, see [15]. Steps 7-8 can be solved optimally in $O(\log n)$ time using a prefix sums algorithm.

For the complexity of circuit transformation, in step 9-10 the equivalence classes among the MVIS's in C can be found by applying an optimal EREW algorithm [3] for finding connected components of a planar undirected graph. We first construct an undirected graph $G = < V, E >$ from C (after performing steps 2-8) as follows: $V = \{g : g$ is a valid gate in $C\}$, $E = \{e : e$ is a wire connecting two valid gates in $C\} \cup \{e = (g_1, g_2) : g_1$ and g_2 are two adjacent valid gates at a layer in $C\}$. Then the MVIS's in C that are in the same equivalence class (and only those MVIS's) are in the same connected component of G, and an MVIS that is in a singleton equivalence class together with its VPS's are in a connected component in G. We have given a procedure in [15] that implements step 11 in $O(\log n)$ time using n processors on an EREW PRAM. Step 12 can be performed by computing prefix sums at each layer. □

6 Conclusion

We have presented an efficient EREW PRAM algorithm for the LPMCVP that runs in $O(\log^2 n)$ time using n processors where n is the size of the input circuit. Our result is a substantial improvement over earlier results [2, 5, 6, 12, 19] for this problem, in terms of processor efficiency and the solution of the problem using direct algorithmic techniques. The parallel complexity of our algorithm is within a polylog factor of the best possible.

References

1. Delcher, Arthur L. and Kosaraju, S. Rao *"An NC Algorithm for Evaluating Monotone Planar Circuits"* Manuscript, submitted to SICOMP, 1992.
2. Dymond, Patrick W. and Cook, Stephen A. *"Hardware Complexity and Parallel Computation"* IEEE Symp. on Foundations of Comp. Sci., 1980, p360-372.
3. Gazit, Hillel *"An Optimal Deterministic EREW Parallel Algorithm for Finding Connected Components in a Low Genus Graph"* Proc. 5th Int. Parallel Processing Symp., April 1991, page 84.
4. Gibbons, A. M. and Rytter, W. *"An Optimal Parallel Algorithm for Dynamic Expression Evaluation and its Applications"* Symp. on Foundations of Software Technology and Theoretical Comp. Sci., Springer-Verlag, 1986, p453-469.
5. Goldschlager, Leslie M. *"A Space Efficient Algorithm for the Monotone Planar Circuit Value Problem"* Info. Proc. Letters, Vol. 10, No. 1, February 1980, p25-27.
6. Goldschlager, Leslie M. *"A Unified Approach to Models of Synchronous Parallel Machines"* Proc. 10th Ann. ACM Symp. on Theory of Comp., May 1978, p89-94.
7. Goldschlager, Leslie M. *"The Monotone and Planar Circuit Value Problems Are* log *Space Complete for P"* SIGACT News, Vol. 9, 1977, p25-29.
8. Karp, Richard M. and Ramachandran, Vijaya *"Parallel Algorithms for Shared Memory Machines"* Handbook of Theo. Comp. Sci., J. Van Leeuwen, ed., North Holland, 1990, p869-941.
9. Kosaraju, S. R. and Delcher, A. L. *"Optimal Parallel Evaluation of Tree-Structured Computations by Ranking"* Proc. 3rd Aegean Workshop on Computing, Springer-Verlag LNCS 319, 1988, p101-110.
10. Ladner, R. E. and Fischer, M. J. *"Parallel Prefix Computation"* JACM, vol. 27, 1980, p831-838.
11. Ladner, R. E. *"The Circuit Value Problem is* log *Space Complete for P"* SIGACT News, 1975, p18-20.
12. Mayr, Ernst W. *"The Dynamic Tree Expression Problem"* Proc. 1987 Princeton Workshop on Algorithms, Architecture and Technology Issues for Models of Concurrent Computation, Chap. 10, 1987, p157-179.
13. Miller, Gary L., Ramachandran, Vijaya and Kaltofen, Erich *"Efficient Parallel Evaluation of Straight-Line Code and Arithmetic Circuits"* SIAM J. Comput., Vol. 17, No. 4, August 1988, p687-695.
14. Ramachandran, Vijaya *"Parallel Open Ear Decomposition with Applications to Graph Biconnectivity and Triconnectivity"* Invited chapter in Synthesis of Parallel Algorithms, J. H. Reif, editor, Morgan-Kaufmann, 1993 p275-340.
15. Ramachandran, Vijaya and Yang, Honghua *"An Efficient Parallel Algorithm for the Layered Planar Monotone Circuit Value Problem"* Tech. Rep., TR 93-10, CS Dept, UT Austin, 1993.
16. Ramachandran, Vijaya and Yang, Honghua *"An Efficient Parallel Algorithm for the General Planar Monotone Circuit Value Problem"* Manuscript, 1993.
17. Tarjan, R. E. *"Data Structures and Network Algorithms"* SIAM, PA, 1983.
18. Tarjan, R. E. and Vishkin, U. *"An Efficient Parallel Biconnectivity Algorithm"* SIAM J. Comput., vol 14, 1985, p862-874.
19. Yang, Honghua *"An NC Algorithm for the General Planar Monotone Circuit Value Problem"* Proc. 3rd IEEE Symposium on Parallel and Distributed Processing, Dec. 1991, p196-203.

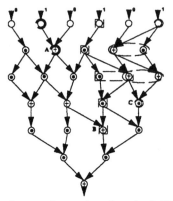

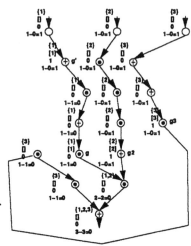

The MVIS and its propagated segments that terminate at A are in thick dotted circles.
The MVIS and its propagated segments that terminate at B are in dotted squares.
The MVIS and its propagated segments that terminate at C are in dotted circles.

Figure 1. An LPMC C and the propagation of its MVIS's.

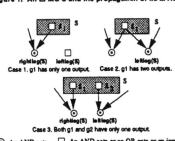

rightleg(S)   leftleg(S)
Case 1. g1 has only one output.

leftleg(S)
Case 2. g1 has two outputs.

rightleg(S)   leftleg(S)
Case 3. Both g1 and g2 have only one output.

⊙ An AND gate   □ An AND gate or an OR gate or an input node

▨ A segment   ⟶ A wire in circuit C

Figure 2. Three cases of a twisting segment.

The left propagation tree of C, T_L

{ } — M_L [ ] — $TWIST_L$ n — $[TERM_L]$ = — $|PROP_L|$

<1,–5,g> <1,–2,g'> <1,2,g'> <1,5,g> <2,–5,g2> <2,5,g2> <3,–4,g3> <2,4,g3>

EFL   g' ○ 1   g3 ○ 1
      g ○ 1   g2 ○ 1

Figure 4. The left propagation tree and the EFL of C.

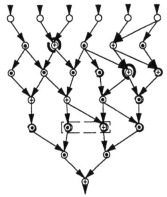

The segments in dotted circles are twisting segments of Case 1.
The segments in thick dotted circles are twisting segments of Case 2.
The segment in the dotted square is a twisting segment of Case 3.

Figure 3. The twisting segments in C.

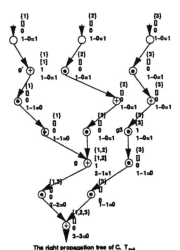

The right propagation tree of C, T_R

{ } — M_R [ ] — $TWIST_R$ n — $[TERM_R]$ = — $|PROP_R|$

<1,–5,g> <1,–2,g'> <1,2,g'> <2,5,g> <3,–4,g3> <3,4,g3>

EFR   1 ○ g'   1 ○ g3
      2 ○ g

Figure 5. The right propagation tree and the EFR of C.

Randomized Routing on Meshes with Buses

Jop F. Sibeyn[1], Michael Kaufmann[2] and Rajeev Raman[3]

[1] Max-Planck-Institut für Informatik, Im Stadtwald, 6600 Saarbrücken, Germany. E-mail: jopsi@mpi-sb.mpg.de
[2] Wilhelm-Schickard-Institut für Informatik, Universität Tübingen, Sand 13, 7400 Tübingen, Germany. E-mail: mk@mpi-sb.mpg.de
[3] UMIACS, University of Maryland, College Park, MD 20742, USA. E-mail: raman@umiacs.umd.edu

Abstract. We give algorithms and lower bounds for the problem of routing k-permutations on d-dimensional MIMD meshes with row and column buses We prove a lower bound for routing permutations (the case $k = 1$) on d-dimensional meshes. For $d = 2, 3$ and 4 the lower bound is respectively $0.69 \cdot n$, $0.72 \cdot n$ and $0.76 \cdot n$ steps; the bound increases monotonically with d and is at least $(1 - 1/d) \cdot n$ steps for all $d \geq 5$. Previously, a bisection argument had been used to show that for all $d \geq 1$, $0.66 \cdot n$ steps are required for this problem (i.e., the lower bound did not increase with increasing d). These lower bounds hold for off-line routing as well.
We give a general algorithm that routes k-permutations on d-dimensional meshes in $\min\{(2 - 1/d) \cdot k \cdot n, \max\{4/3 \cdot d \cdot n, k \cdot n/3\}\} + o(d \cdot k \cdot n)$ steps, for all $k, d \geq 1$. This improves considerably on previous results for many values of k and d. In particular, the routing time for permutations is bounded by $2 \cdot n$, for all $1 \leq d < n^{1/3}$, and the routing time is optimal for all $k \geq 4 \cdot d$. More specialized algorithms have better performance for routing on 2-dimensional meshes. A simple algorithm routes 2-permutations in $1.39 \cdot n$ steps, and a more sophisticated one routes permutations in $0.78 \cdot n$ steps. This is the first algorithm that routes permutations on the 2-dimensional mesh in less than n steps. The algorithms are randomized, on-line and achieve the given routing times with high probability.

Keywords: parallel computation, algorithms, packet routing, meshes, buses, lower bounds, randomization, coloring.

1 Introduction

One of the main problems in the simulation of idealized parallel computers by realistic ones is the problem of message routing through the sparse network of links which connect processing units, *PUs*, among each other. The efficiency of a solution to such a problem is usually measured in terms of two parameters, both of which should be as small as possible: the number of steps needed to perform the routing, and the maximal number of packets that may reside in a PU at any given moment during the routing. We consider the case that the PUs are connected by a d-dimensional *mesh* of communication links, which has size $n \times \cdots \times n$. The mesh connects a PU with its (at most) $2d$ neighbors. The links (edges) are bidirectional, one packet of information can go along an edge in both directions in one step. The model of communication we use is the MIMD model. In this model each PU can communicate with all its neighbors in a single step. The communication steps are synchronized. The PUs may store packets in local queues.

The classical routing problem is the problem of transporting one packet from each PU in the mesh to a destination PU, such that every PU finally receives one packet. This problem is called the *permutation* routing problem. In the *k-permutation* routing problem, also known as the k-k routing problem, each processor is the source and destination of k packets. This general problem of routing k-permutations on d-dimensional meshes will be called the (k, d)-routing problem.

A variant of the above described parallel computer is a *mesh with buses*. In this model there is for each 1-dimensional submesh, in addition to the inter-processor links, one *bus*. All PUs of the 1-dimensional submesh are connected to their bus. In one step, a single PU may place a packet on the bus, which may be read by any other PU connected to the bus, without interfering with the use of the inter-processor links. This model has been considered in the early 80's [2, 4] and recently by Leung and Shende [10]. This model is quite practical and we show in this paper that in the context of routing it is far superior to the ordinary mesh: for example, on the d-dimensional mesh with buses for any fixed d, we can route permutations in less than $2 \cdot n$ steps although on the usual mesh this cannot done faster than $d \cdot (n - 1)$ steps.

Many algorithms have been developed for routing permutations on two-dimensional meshes without buses. Rajasekaran and Tsantilas [11] have given a randomized algorithm which needs $2 \cdot n + \mathcal{O}(\log n)$ steps with constant queue size. A deterministic algorithm due to Kunde [7] solves the problem in $2 \cdot n + \mathcal{O}(n/f(n))$ steps with queue size $f(n)$, for any $1 \leq f(n) \leq n$. Finally, Leighton, Makedon and Tollis [9] presented an $2 \cdot n - 2$ algorithm with constant, but large queue size. This matches the "distance" bound of $2 \cdot n - 2$

steps. For the general (k, d)-routing problem on meshes without buses, $\max\{d \cdot (n - 1), k \cdot n/2\}$ is a lower bound on the number of routing steps. This problem has been considered recently. It was solved in [5] with a randomized algorithm in $k \cdot n/4 + o(d \cdot k \cdot n)$ steps for all $k \geq 4 \cdot d$, almost matching the trivial "bisection" bound.

For meshes with buses, the distance bound is no longer a factor, and so both lower bounds and routing times are considerably reduced. Leung and Shende [10] proved a lower bound of $2/3 \cdot n$ steps for the routing of permutations on d-dimensional meshes for any $d \geq 1$. On a 1-dimensional mesh (a linear array) this bound is matched by a simple algorithm. For the 2-dimensional mesh they gave a deterministic algorithm that requires $7/6 \cdot n + \mathcal{O}(\alpha)$ steps, with queue size n/α, for any $1 \leq \alpha \leq n$. They also showed that permutations on d-dimensional meshes can be routed in approximately $7/6 \cdot (d-1) \cdot n$ steps. In this paper we considerably improve the lower bounds and the routing times given by [10]. Furthermore, we give efficient solutions to the general (k, d)-routing problem for meshes with buses.

For the general (k, d)-routing problem a bisection argument gives a lower bound of $k \cdot n/3$ steps. For relatively small values of k, sharper bounds are obtained by considering the distance packets have to go along with the availability of the buses. For proving sharp lower bounds for routing permutations, we consider a class of permutations that lead to heavy use of buses unless packets are "walking" far (i.e., unless packets make many steps using inter-processor communication links). For $k = 1$, the worst of these permutations requires at least $0.69 \cdot n$ steps in two dimensions. Inductively constructing a class of bad permutations for higher dimensions, we prove a lower bound of $0.72 \cdot n$ steps for $d = 3$, $0.76 \cdot n$ steps for $d = 4$, and $(1 - 1/d) \cdot n$ steps for arbitrary d.

We give several randomized routing algorithms. For routing permutations on 2-dimensional meshes we give an algorithm that requires $(\sqrt{141} - 1)/14 \cdot n + o(n) < 0.78 \cdot n$ steps. This is a considerable improvement over the previously known bound and is quite close to the lower bound for this problem. Also this shows that n steps is not a lower bound as was believed for some time. For d-dimensional meshes we give an algorithm whose routing time depends on d only in *low-order* terms. For $d < n^{1/3 - \epsilon}$, for arbitrarily small $\epsilon > 0$, permutations can be routed with less than $2 \cdot n$ steps. This is surprising when contrasted with the distance lower bound of $d \cdot (n - 1)$ steps for d-dimensional meshes without buses.

The mentioned algorithms are also suited for routing k-permutations. On 2-dimensional meshes with buses, a version of the permutation routing algorithm routes 2-permutations in less than $1.39 \cdot n$ steps (for this problem we prove a lower bound of n steps). The algorithm that was used on higher-dimensional meshes can be successfully used for all (k, d). For $k \geq 4 \cdot d$, the performance is optimal: $k \cdot n/3 + o(d \cdot k \cdot n)$ steps.

All algorithms are inspired by the idea from [5] that randomizing the packets is not only useful for bounding the queue sizes, but in many cases also for reducing the number of routing steps. A second crucial idea is to color the packets randomly black or white (generally, with d colors on d-dimensional meshes). The black packets move perpendicularly to the white packets. As the routing phases are uniaxial, this is essential for assuring that the vertical and the horizontal connections and buses are evenly loaded at all times. This idea has been employed successfully before in routing and sorting algorithms [8, 6, 12]. In most algorithms, the decision whether a packet should go by bus or walk is based on the distance this packet (still) has to go. In the fastest algorithm for routing permutations on 2-dimensional meshes, the packets are randomized in an intricate way within $n/2$ steps from the original position. This is necessary for assuring that they are delayed very little before they can take the bus. In this algorithm the phases are *coalesced*, meaning that packets proceed with a next phase as soon as they can.

The remainder of this paper is organized as follows: First, in Section 2, we discuss techniques to allocate the buses and we state the Chernoff bounds. In Section 3 we prove the lower-bound results. In Section 4 we derive important new results on the routing on one-dimensional meshes with buses. These results are used in the analysis of our main algorithms. In Section 5, we present the algorithm for the general (k, d)-routing problem. In the subsequent sections more special algorithms are given for routing on 2-dimensional meshes. The ideas are exposed in Section 6 with a simple algorithm which is easy to analyze. Especially for 2-permutations it performs well. For routing permutations we give better algorithms in Section 7 and Section 8.

2 Preliminaries

2.1 Allocation of the buses

A bus is not allocated by some control unit but taken by a PU which puts a packet onto it. Frequently we will say in this case that the packet "takes the bus." If it does not take the bus, we say that it "goes walking." The algorithm must ensure that each bus is taken by at most one PU at a time. In our algorithms a packet transmitted over the bus has only one destination. This may have practical advantages.

In an off-line routing algorithm the use of the bus may be scheduled ahead of time, thus ensuring that only one PU takes the bus at a time. Avoiding conflicts in an on-line situation requires a little thought. Suppose the PUs in a particular 1-dimensional submesh have a total of P packets they wish to transmit along the associated bus. They can accomplish this by first computing the prefix sum of the number of packets each processor needs to transmit and then using the schedule implied by this prefix sum to transmit all packets in P further steps. The prefix sum can be computed in $2\sqrt{n}$ steps using the bus in such a way that the ith processor ends up knowing the ith prefix sum (the details are left to the reader), and thus the entire task is completed in $P + \mathcal{O}(\sqrt{n})$ steps. This does not suffice for us because of its static character: the parallel prefix only works when it is initially known how many a PU wants to put on the bus. In some of our algorithms packets wanting to use a bus may arrive dynamically at the PUs connected to it.

In dynamic situations, the bus allocation problem can be solved by appointing some r PUs in every 1-dimensional submesh at regular distances as *representatives*, and requiring that only representatives place packets on the bus. For a 1-dimensional mesh, the representatives are the PUs with index $j \cdot \lfloor n/r \rfloor$, $1 \leq j \leq r$. When a packet p residing in P_i has to be transferred over the bus, it walks to $P_{\lceil i \cdot r/n \rceil \cdot \lfloor n/r \rfloor}$, where it is put on the bus. If the packets cannot be placed on the bus immediately, they are put in a queue and taken out in some order. Throughout this paper we assume a FIFO order. The buses are taken by the representatives in some specified order. If M is the maximal number of packets any representative has to transfer, then this scheme requires at most $M \cdot r$ steps to transfer all packets, which may be considerably more than the total number of packets.

We now combine the two techniques above. Suppose that the representatives all have to transmit between m and M packets and that all packets have reached their representative within $m \cdot r$ steps. Then the first m rounds the representatives take the bus every r steps. Subsequently, a parallel prefix is performed by the representatives (which takes r steps in this case), where after the transfer can be completed without loss of steps. As an alternative to the parallel prefix, a representative that was using the bus can use it to send a message to the next representative to notify that it has finished. Either solution requires $T + r$ steps. In In all our algorithms we will apply this hybrid technique and take $r = \sqrt{n}$. This choice of r is motivated by the fact that the use of representatives causes two kinds of losses: the steps required for walking towards and away from the representatives; and the steps required for handling the representatives. With $r = \sqrt{n}$, these losses are of the same order of magnitude. Note that a bus has to be connected to the r representative PUs with this technique, which may have practical advantages. For this reason, we will use representatives in all our algorithms, even where the bus allocation problem could have been solved using a parallel prefix operation.

2.2 Chernoff bounds

Let X be the number of heads in n independent flips of a biased coin, the probability of a head in a single flip being p. Such an X has the binomial distribution $B(n, p)$. The so-called Chernoff bounds provide close approximations to the probabilities in the tail ends of a binomial distribution. We use the form of these bounds given in [1], whereby for all $0 < h < n \cdot p$

$$prob\{X \geq n \cdot p + h\} \leq e^{-h^2/(3 \cdot n \cdot p)}, \tag{1}$$

$$prob\{X \leq n \cdot p - h\} \leq e^{-h^2/(2 \cdot n \cdot p)}. \tag{2}$$

(1) and (2) give bounds on the probability of a deviation of the number of heads from the expected number. The time bounds of our algorithms will all hold "with high probability." We formalize this notion:

Definition 1 *An event a happens with high probability if $prob(a) \geq 1 - n^{-\alpha}$, for some constant $\alpha > 0$.*

Usually we need that a polynomial number of independent binomial random variables, $X_1, \ldots, X_M$, $X_i = B(k \cdot n, p)$, $M \leq n^m$, for some constant m, are all bounded at the same time. (1) and (2) are so strong that the following lemma can be proven easily:

Lemma 1 *There is an $h = \mathcal{O}((p \cdot k \cdot n \cdot \log n)^{1/2})$, such that $p \cdot k \cdot n - h \leq X_i \leq p \cdot k \cdot n + h$, for all $1 \leq i \leq M$ simultaneously, with high probability.*

We will use the Chernoff bounds in this form.

3 Lower bounds

In this section we prove lower bounds for the $(1, d)$ routing problem on meshes with buses. For the general (k, d)-problem it is easy to prove a "bisection" lower bound of $k \cdot n/3$, but we also give results in which both k and d play a role. The following criterion plays a central role in the derivation of lower bounds for meshes with buses:

Lemma 2 *Consider a permutation $\mathcal{P}$ under which a packet p has to travel distance y_i along axis i, $1 \leq i \leq d$, and let δ, j be integers with $1 \leq j \leq d$. If for any set of directions $i_1, \ldots, i_j$, $1 \leq i_i < i_2 < \ldots < i_j \leq d$, $\sum_{l=1}^{j} y_{i_l} \geq \delta$, then $\mathcal{P}$ can be routed with less than δ steps only if p takes the bus at least $d - j + 1$ times.*

Proof: If p were to take a bus less than $d - j + 1$ times, it would have to walk along some j axes. Since the distance it needs to walk along these j axes is at least δ, p would need at least δ steps to reach its destination, and hence $\mathcal{P}$ cannot be routed in less than δ steps. $\qquad\Box$

3.1 The $(1, d)$-routing problem

We analyze a class of permutations that will lead to an improved lower bound for the routing of permutations on d-dimensional meshes, for $d \geq 2$. In Figure 3.1 we give a subdivision of the mesh using a parameter x, $0 \leq x \leq n/2$. The packets starting in A_1, A_2, B_1 or B_2 will be called *corner* packets. The permutation $\mathcal{D}_2(x)$ exchanges as a block the packets from regions with corresponding letters. Note that the corner packets have to travel $n - x$ steps in both directions and all other packets have to go precisely $n - x$ steps in the two directions together.

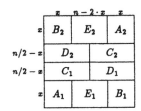

Fig. 1. The subdivision for $\mathcal{D}_2(x)$.

We now inductively construct permutations $\mathcal{D}_d(x)$ for the d-dimensional mesh, for $d \geq 3$. Divide the mesh into n *layers*, where all the processors in a layer have the same coordinate in dimension d. Within layers $1, \ldots, x$, a packet is first mapped within its layer according to $\mathcal{D}_{d-1}(x)$, and subsequently translated $n - x$ units along the positive d-axis (i.e., $n - x$ is added to it's coordinate in dimension d). Similarly, for layers $n - x + 1, \ldots, n$, a packet is first mapped within its layer according to $\mathcal{D}_{d-1}(x)$ and subsequently translated $n - x$ units along the negative d-axis (i.e., $n - x$ is subtracted from it's coordinate in dimension d). A packet in the layers $x + 1, \ldots, n - x$ residing in $P_{a_1, \ldots, a_{d-1}, a_d}$ is routed to the PU with index $((a_1 + n/2) \bmod n, \ldots, (a_{d-1} + n/2) \bmod n, a_d \pm (n/2 - x))$. For $d = 3$ we depict this construction in Figure 3.1. Again, the packets from blocks with corresponding letters are exchanged as a block. Finally, the corner packets are those which originate in processors that are at most $d - 1$ steps along every direction from one of the 2^d corner processors.

It is easy to verify by induction the following properties of $\mathcal{D}_d(x)$: (a) there are $2^d \cdot x^d$ corner packets, each of which travels $n - x$ steps along each dimension; (b) for all remaining packets, the sum of distances travelled along any two distinct dimensions is least $n - x$. This leads to:

Theorem 3 *For all $d \geq 2$, routing $\mathcal{D}_d(x)$ on a d-dimensional mesh with buses requires at least $(1 - \alpha) \cdot n$ steps, where α is the unique solution to the equations $2^d \cdot \alpha^d + d \cdot \alpha - 1 = 0$, $0 \leq \alpha \leq 1/d$.*

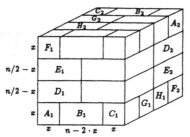

Fig. 2. The subdivision for $\mathcal{D}_3(x)$.

Proof: By Lemma 2, in order to route $\mathcal{D}_d(x)$ within $n - x$ steps, all corner packets must take the bus d times and all other packets must take the bus $d - 1$ times. Thus the number of times the packets take the bus, summed over all packets, must be at least $(d - 1) \cdot n^d + 2^d \cdot x^d$. Since only $d \cdot n^{d-1}$ buses are available, and only one packet can be on the bus during a time step, a lower bound of

$$\min\{n - x, \frac{(d-1) \cdot n^d + 2^d \cdot x^d}{d \cdot n^{d-1}}\}$$

steps is obtained. Equating the two terms, setting $x = \alpha \cdot n$ and simplifying gives that the optimal value of α must satisfy $f(\alpha) = 0$, where $f(\alpha) = 2^d \cdot \alpha^d + d \cdot \alpha - 1$. Since $f(\alpha)$ is negative at $\alpha = 0$, positive at $\alpha = 1/d$ and monotonically increasing in between, it follows that the equation $f(\alpha) = 0$ has exactly one solution in the range $0 \le \alpha \le 1/d$, for all $d \ge 2$. $\qquad \Box$

It is easy to verify that $\alpha = (\sqrt{5} - 1)/4 < 0.31$ for $d = 2$, $\alpha < 0.28$ for $d = 3$ and $\alpha < 0.24$ for $d = 4$. This gives:

Corollary 4 *Routing permutations on 2, 3 and 4-dimensional meshes with buses requires* $0.69 \cdot n$, $0.72 \cdot n$ *and* $0.76 \cdot n$ *steps respectively. For all* $d \ge 2$, $(1 - 1/d) \cdot n$ *is a lower bound for routing permutations on d-dimensional meshes with buses.*

For all $d \ge 2$, the permutation $\mathcal{D}_d(x)$, with x chosen according to Theorem 3, is a worst-case permutation in the sense that no permutation requires the use of more buses for given maximal walking distance. We can show that for every x, $1 \le x \le n/2$, the permutation $\mathcal{D}_2(x)$ can be routed in $(\sqrt{5} - 1)/4 \cdot n$ steps. We conjecture:

Conjecture 1 *Permutations can be routed off-line on 2-dimensional meshes with buses in at most* $(5 - \sqrt{5})/4 \cdot n$ *steps.*

3.2 The (k, d)-routing problem

We give two lower bounds: a general bound depending on k only, and a bound that depends on both k and d, which is better for rather small k.

Lemma 5 *The (k, d)-routing problem requires at least $k \cdot n/3$ steps, for all k and d.*

Proof: Consider the k-permutation k-$\mathcal{S}_d$ on the d-dimensional mesh under which all packets residing in $P_{a_1, a, 2, \ldots, a_d}$ have destination $P_{a_1 + n/2 \bmod n, a_2, \ldots, a_d}$. Under k-$\mathcal{S}_d$ all $k \cdot n^d$ packets have to pass through the central hyperplane perpendicular to axis 1. At most $3 \cdot n^{d-1}$ packets are transmitted every step through this hyperplane by the n^{d-1} bidirectional connections and the n^{d-1} buses. $\qquad \Box$

In analogy to the routing on meshes without buses, we will call this 'the bisection bound'.

For $k \le d$, we can improve on the bound of Lemma 5. We consider the k-permutation k-$\mathcal{E}_d$, under which all packets from $P_{a_1, \ldots, a_d}$ have $P_{(a_1 + n/2) \bmod n, \ldots, (a_d + n/2) \bmod n}$ as destination. Using Lemma 2, we obtain

Lemma 6 *Routing k-$\mathcal{E}_d$ requires at least $n \cdot \min_{\delta \in \mathbb{N}}\{\max\{(d - \delta)/2, k \cdot \delta/d\}\}$.*

This lemma gives that the $(2, 2)$-routing problem requires at least n steps. For a $(k, 2 \cdot k)$-routing problem we obtain a lower bound of $k \cdot n/2$ steps, and for k much smaller than d, the bound lies close to $k \cdot n$ steps.

4 One-dimensional routing

In this section we derive important results concerning the routing on one-dimensional meshes with buses. These result will serve as a basis for the algorithms for routing on higher dimensional meshes in subsequent sections. For the analysis of the time required by the walking packets we use a corollary to Lemma 3.1 from [5]:

Lemma 7 *Let $\mathcal{D}$ be a distribution of packets under which each PU is sending and receiving (an expected number of) k packets. Routing $\mathcal{D}$ with the farthest-first strategy requires the maximum of the maximal distance the packets have to go and the number of packets that pass in either direction through the connection between $P_{n/2-1}$ and $P_{n/2}$.*

4.1 Permutations

For routing permutations, Lemma 5 gives a lower bound of $k \cdot n/3$ steps. It is also easy to check that the permutation under which all packets originating in the leftmost and rightmost $n/(1+2\cdot k)$ PUs are exchanged, requires at least $2 \cdot k/(1+2\cdot k)\cdot n$ steps, when $k \geq 1/2$. For $k \leq 1/2$, $k\text{-}\mathcal{E}_1$, requires at least $k \cdot n$ steps. Hence,

Lemma 8 *The $(k,1)$-routing problem requires at least $\max\{\min\{k, 2\cdot k/(1+2\cdot k)\}, k/3\}\cdot n$ steps, for all $k \geq 0$.*

A basic algorithm for routing on one-dimensional meshes with buses proceeds as follows:

> **Proc** PERMROUTE1;
> Color each packet yellow with probability 1/3, and blue otherwise;
> the yellow packets take the bus, the blue packets walk.

Using Lemma 7 it is easy to show that

Lemma 9 *Routing an k-permutation with PERMROUTE1 requires $\max\{n, k\cdot n/3\}$ steps.*

Thus, PERMROUTE1 is optimal for all $k \geq 3$. For $k \leq 5/2$ the following algorithm is optimal:

> **Proc** PERMROUTE2(k);
> All packets that start in the leftmost or rightmost $n/(1+2\cdot k)$
> PUs take the bus; all other packets walk.

For $k \leq 1/2$, all packets take the bus.

Lemma 10 *Routing a k-permutation with PERMROUTE2 requires $\min\{k, 2\cdot k/(1+2\cdot k)\}\cdot n$ steps, for all $k \leq 5/2$.*

Proof: For $k \leq 1/2$, the lemma is immediate. For $k \geq 1/2$, the number of packets that takes the bus equals the maximal distance a packet has to walk: $2\cdot k/(1+2\cdot k)\cdot n$. For $k \leq 5/2$, this number is larger than the number of packets that pass through the connection between $P_{n/2-1}$ and $P_{n/2}$. □

4.2 Randomizations

With aid of Lemma 1 we derive in this section results on the routing of randomizations on one-dimensional meshes. A *randomization* is a packet distribution under which each PU is sending one packet to a randomly (uniformly and independently) chosen destination PU. We assume that this can be achieved with t steps, for certain t that remains yet to be determined. Packets can walk if they are within distance t from there destination the others take the bus. t is taken so large that the expected number of packets that takes the bus equals t. Notice that the packets walk in 'lock-step mode' without delay.

The algorithm is very simple, but is given for a clear illustration how the representatives can be handled:

> **Proc** RANDROUTE1(t);
> for each packet p going from P_i to P_j do
> if $|j - i| \leq t$ then p walks to P_j
> else p walks to $P_{\lceil i/\sqrt{n}\rceil\cdot\sqrt{n}}$;
> p takes the bus to $P_{\lceil j/\sqrt{n}\rceil\cdot\sqrt{n}}$;
> p walks from $P_{\lceil j/\sqrt{n}\rceil\cdot\sqrt{n}}$ to P_j.

Lemma 11 *On a one-dimensional processor array with bus,* RANDROUTE1 *routes randomizations in* $(3 - \sqrt{5})/2 \cdot n + \mathcal{O}((n \cdot \log n)^{1/2}) < 0.39 \cdot n$ *steps, with high probability.*

Proof: How many packets may have to take the bus? For $i \leq t$, the probability that the packet from P_i has to go further than t steps equals $(n - t - i)/n$; for $t \leq i \leq n - t$ this equals $(n - 2 \cdot t)/n$; and for $i \geq n - t$, $(i - t)/n$. All these probabilities are independent of each other. Over the whole processor array the expected number of packets that takes the bus equals $\sum_{i=1}^{t}(n - t - i)/n + \sum_{i=t+1}^{n-t}(n - 2 \cdot t)/n + \sum_{i=n-t+1}^{n}(i - t)/n \simeq (n - t)^2/n$. By the Chernoff bounds we may assume that the maximal number of packets that will have to use the bus is only slightly larger. Now we can solve the equation $t = (n - t)^2/n$, which gives $t = (3 - \sqrt{5})/2 \cdot n$. □

Queues. Bounding the maximal queue length is not really a problem: the packets that are not welcome yet can be pushed back (hot-potato approach), or the representatives can travel within their $\sqrt{n}$ region. Another much used technique is to smear the packets in small stripes. This means that every row and column is subdivided into small stripes of size say $\log n$. A packet with destination in any of the PUs in such a stripe is moving to the first PU in the stripe that has empty space in its queue. In this way the maximal queue size can be bounded to the expected number of packets $+ \mathcal{O}(1)$. The application of these techniques will be left implicit hereafter.

Low densities. The *density* of packets denotes the (expected) number of packets that resides in all PUs. An k-randomization is a randomization in which each PU is sending (an expected number of) k packets. The time required for routing an k-randomization depends on k. As the packets may be distributed unevenly, also the walking packets may be delayed. However, if the expected number of packets that has to walk in one direction is lower than 1 in each PU, then, with high probability, no packet has to wait more than $\mathcal{O}(\log n)$ steps before it finds a 'hole' in the packet stream. We have the following generalization of Lemma 11:

Lemma 12 *For all* $k \leq 6 - \sqrt{6} \simeq 3.55$, k-*randomizations can be routed with at most* $(1 + 2 \cdot k - \sqrt{1 + 4 \cdot k})/(2 \cdot k) \cdot n + \mathcal{O}((k \cdot n \cdot \log n)^{1/2} + \log n)$ *steps, with high probability.*

Proof: Let m_1 be the expected number of packets that have to use the bus for given maximal walking distance t; let m_2 be the maximal distance packets have to go; and let m_3 be the expected number of packets that goes from left to right through the connection $(n/2 - 1, n/2)$ in the middle. By Lemma 7 and Lemma 1 we know that the total routing time is bounded by $\max\{m_1, m_2, m_3\} + \mathcal{O}((k \cdot n \cdot \log n)^{1/2} + \log n)$. With density k, $m_1 = k \cdot (n - t)^2/n$. Clearly $m_2 = t$. For m_3, it it is easy to show that $m_3 = k \cdot t^2/(2 \cdot n)$ for $t \leq n/2$, and that $m_3 = k \cdot (t - n/4 - t^2)/(2 \cdot n)$ for $t \geq n/2$. For $t \leq (1 - 1/\sqrt{6}) \cdot n$, $m_1 \geq m_3$. Solving $m_1 = m_2$ gives $t = (1 + 2 \cdot k - \sqrt{1 + 4 \cdot k})/(2 \cdot k) \cdot n \leq (1 - 1/\sqrt{6}) \cdot n$, for $k \leq 6 - \sqrt{6}$. □

Corollary 13 *Routing* $1/2$-*randomizations requires at most* $(2 - \sqrt{3}) \cdot n + \mathcal{O}((n \cdot \log n)^{1/2}) < 0.27 \cdot n$ *steps, with high probability. Routing* $1/3$-*randomizations requires at most* $(5 - \sqrt{21})/2 \cdot n + \mathcal{O}((n \cdot \log n)^{1/2}) < 0.21 \cdot n$ *steps if* $k = 1/3$, *with high probability.*

For the analysis of our algorithms we also need

Lemma 14 *Packets reside with density* $k \leq 2$ *in the PUs of a one-dimensional processor array with a bus. They are routed either to a random destination within the half of the array in which they originate or to a random destination in the other half of the array. This routing can be performed with at most* $(\sqrt{1 + 8 \cdot k^2} - 1)/(4 \cdot k) \cdot n + \mathcal{O}((n \cdot \log n)^{1/2})$ *steps, with high probability.*

Proof: Suppose that the total routing time is $t \leq n/2$. A packet walks if it is within distance t from its destination, otherwise it takes the bus. By Lemma 7, all walking packets reach their destination in at most $t + \mathcal{O}(\log n)$ steps. For the analysis of the maximal value of the expected number of packets using the bus m, we consider a PU at position $i \leq n/2$, with a packet p. If p is randomized in the right half, it takes the bus with probability $P = \min\{1, (n - t - i)/(n/2)\}$. If p is randomized in the left half, it takes the bus with probability at most $(n/2 - t)/(n/2) \leq P$. Thus, m is maximal if all packets change sides. In that case $m = 2 \cdot k \cdot (n/2 - t + \sum_{i=n/2-t}^{t}(n - t - i)/(n/2)) = k \cdot n - 2 \cdot k \cdot t^2/n$. Solving $m = t$ gives the result. □

High densities. For k-randomizations, $k \cdot n/6$ is a lower bound on the routing time: the expected number of packets that have to travel from the left half of the processor array to the right half or vice-versa equals $k \cdot n/2$, and there is one two-way connection and one bus that can be used for the transfer of packets between the halves of the processor array. For $k \geq 6$, this bound is matched by PERMROUTE1.

For $k < 6$, no packet should walk more than $k \cdot n/6$ steps, while at most $k \cdot n/6$ packets take the bus. For $k < 6 - \sqrt{6} \simeq 3.55 \cdot n$ this is impossible. We give an algorithm that is optimal for all $k \geq 4$. In its formulation we use the following subdivision of the processor array:

A	B			C	D
$n/6$	$n/6$		Γ	$n/6$	$n/6$

Here Γ denotes the connection between $P_{n/2-1}$ and $P_{n/2}$. The algorithm is a modification of PERMROUTE2:

Proc RANDROUTE2;
All packets going from A to $C \cup D$, all packets going from D to $A \cup B$,
all packets going from B to D and all packets going from C to A take the bus;
all other packets go walking.

Lemma 15 *Routing an k-randomization with* RANDROUTE2 *requires at most* $\max\{2/3 \cdot n, k \cdot n/6\} + \mathcal{O}((k \cdot n \cdot \log n)^{1/2})$ *steps, with high probability.*

Proof: The expected number of packets that take the bus equals $6/36 \cdot k \cdot n$, thus, by the Chernoff bounds, there are at most $k \cdot n/6 + \mathcal{O}((k \cdot n \cdot \log n)^{1/2})$ packets that have to use the bus, with high probability. From [5] we know that in this case (using the farthest-first strategy) the routing time of the walking packets is determined by the maximal distance any packet has to go and the number of packets moving through Γ. Clearly no packet has to walk more than $2/3 \cdot n$ steps. The expected number of packets moving through Γ is $12/36 \cdot k \cdot n$. Another application of Lemma 1 finishes the proof. $\square$

5 A general algorithm

In this section we give an algorithm for the general (k, d)-routing problem. The idea of the algorithm goes back on the algorithm of Valiant and Brebner [13] for routing on hypercubes. In its present form it was applied for routing on meshes in our algorithm of [5]. For permutations, the routing time is almost independent of d and remains below $2 \cdot n$. In view of Theorem 3, this is a quite sharp result.

Initially all packets are randomly (uniformly and independently) given one of the colors $1, \ldots, d$. The algorithm consists of three phases, partitioned into $2 \cdot d - 1$ subphases. It is given only for the packets with color 1, the packets with other colors move perpendicularly to the color-1 packets:

Proc BUSROUTE1(d);
1. **for** $j := 1$ **to** $d - 1$ **do** route the packets to a randomly chosen position on axis j;
2. route the packets to the correct position on axis d;
3. **For** $j := d - 1$ **downto** 1 **do** route the packets to the correct position on axis j.

Let $\alpha = k/d$. Depending on α, the subphases are performed by the various algorithms for routing k-permutations and k-randomizations: the subphases of phase 1 and phase 3 by RANDROUTE1 if $\alpha \leq 6 - \sqrt{6}$, and by RANDROUTE2 otherwise; phase 2 by PERMROUTE1 for $\alpha \leq 5/2$, and by PERMROUTE2 otherwise. As in all algorithms, the number of representatives in any row and column is taken $\sqrt{n}$ and bus-allocation technique 3 is applied (see Section 2.1).

The routing time of BUSROUTE1 is determined by summing the times required by the subphases. We give the results for the most important ranges of α values:

Theorem 16 BUSROUTE1 *is used for routing k-permutations on a d-dimensional mesh with buses. If $\alpha = k/d \leq 1/2$, then the routing time is at most* $(2 \cdot d - 2) \cdot (1 + 2 \cdot \alpha - \sqrt{1 + 4 \cdot \alpha})/(2 \cdot \alpha) \cdot n + \alpha \cdot n + \mathcal{O}(d \cdot (\alpha \cdot n \cdot \log n)^{1/2})$, *with high probability.*
If $\alpha = k/d \geq 4$, then the routing time is at most $k \cdot n/3 + \mathcal{O}(d \cdot (\alpha \cdot n \cdot \log n)^{1/2})$, *with high probability.*

Theorem 16 implies that k-permutations are routed optimally for all $k \geq 4 \cdot d$. For small k, the strength of the result of Theorem 16 is clarified by the following corollaries:

Corollary 17 *Using* BUSROUTE1 *for routing permutations on a 2-dimensional mesh with buses requires* $(9/2 - 2 \cdot \sqrt{3}) \cdot n + \mathcal{O}((n \cdot \log n)^{1/2}) < 1.04 \cdot n$ *steps. On a three-dimensional mesh this requires* $(31/3 - 2 \cdot \sqrt{21}) \cdot n + \mathcal{O}((n \cdot \log n)^{1/2}) < 1.17 \cdot n$ *steps.*

Corollary 18 *Let $d < n^{\beta}$ for some $\beta < 1/3$. Then* BUSROUTE1 *routes permutations on d-dimensional meshes with buses with less than $2 \cdot n$ steps, with high probability.*

Proof: Taylor expansion gives $\sqrt{1 + 4 \cdot \alpha} > 1 + 2 \cdot \alpha - 2 \cdot \alpha^2$. Using this estimate gives a routing time bounded by $(2 - 1/d) \cdot n + \mathcal{O}((d \cdot n \cdot \log n)^{1/2})$. $\square$

6 An alternative algorithm

The algorithm in the previous section is extremely simple and general and it performs good for many instances of the (k, d)-routing problem. However, for $k \leq d$, it is not difficult to give algorithms that perform better. In this section we give an algorithm for the $(2, 2)$-routing problem, that will serve us as a starting point for the fast permutation routing algorithms of the subsequent sections.

Initially all packets are colored black or white as follows: in each PU one packet is selected randomly and colored white; the other packet is colored black. The white packets are routed as follows:

> **Proc BUSROUTE2;**
> Each packet chooses with equal probability any representative in its row as destination;
> 1. each packet walks to the chosen representatives, when it is his turn, a representative puts one of the waiting packets on the bus;
> 2. packets that are within t from their destination go walking, the other packets wait until the bus comes by.

t is determined hereafter. The black packets are routed perpendicularly to the white packets.

Lemma 19 *Phase 1 of BUSROUTE2 takes $n + O(n^{3/4} \cdot \log^{1/2} n)$ steps at most, with high probability.*

Proof: By the coloring the packets walk without delay during phase 1 and all reach the selected representative within n steps. We consider the transfer of the packets over the buses. On the average every representative has to put $n^{1/2}$ packets on the bus, and by Lemma 1 this number is bounded by $n^{1/2} + O(n^{1/4} \cdot \log^{1/2} n)$. A problem is that it may happen that the packets do not reach a representative fast enough and that a bus remains unused. Most difficult is the situation for a representative P on the boundary of the mesh. Every step, P is reached with probability $n^{-1/2}$. Suppose that P waits $O(n^{3/4} \cdot \log^{1/2} n)$ steps before it puts a packet on the bus. Then, after s steps P has received by Lemma 1 at least $s/n^{1/2} - O((s/n^{1/2} \cdot \log n)^{1/2})$ packets and transmitted at most $(s - O((n^{3/4} \cdot \log^{1/2} n))/n^{1/2})$ packets. So, for all $s \leq n$, there still remain some packets in P. Thus, P will not leave the bus unused after the initial waiting. When P does not wait, the transfer of the packets proceeds at least that fast. This implies that after n steps P still has to transfer at most $O(n^{1/4} \cdot \log^{1/2} n)$ packets. $\qquad \square$

At the beginning of phase 2 all packets with destination in a certain row have arrived there and are randomly distributed over the row. This is precisely the reverse of the situation described by Lemma 12. Thus, by Corollary 13, we find that $t = (3 - \sqrt{5})/2 \cdot n$ is optimal. With this t

Theorem 20 *Applying BUSROUTE2 for routing 2-permutations on a 2-dimensional mesh with buses requires $(5 - \sqrt{5})/2 \cdot n + O((n \cdot \log n)^{1/2}) < 1.39 \cdot n$ steps with high probability.*

7 Refinement

For permutations the routing time of BUSROUTE is quite high because phase 1 takes n steps even when the density is low. This is a result of the randomisation within the whole row. It is better to randomize only within the half of the row in which the packets reside. Modifying BUSROUTE2 in this way, gives the first algorithm for routing permutations with less than n steps.

For the white packets the algorithm proceeds as follows (the black packets are routed perpendicularly to the white packets):

> **Proc BUSROUTE3;**
> Every packet chooses with equal probability any representative in the half of the row in which it resides;
> 1. the packets walk to the chosen representatives, every $\sqrt{n}$ steps the representatives put one of the waiting packets on the bus;
> 2. packets that are within t from their destination go walking, the other packets wait until the bus comes by.

BUSROUTE3 can be used for any value of k, but only for small k it performs better than BUSROUTE2. We analyse the algorithm for $k = 1$, with $t = (\sqrt{3} - 1)/2 \cdot n$. First we give an analogue of Lemma 19:

Lemma 21 *Phase 1 of BUSROUTE3 requires $n/2 + O(n^{3/4} \cdot \log^{1/2} n)$ steps, with high probability.*

Proof: We study the problem on a linear array with $n/2$ processors in isolation. We will prove that the leftmost representative P_1 completes in $n/2 + \mathcal{O}(n^{3/4} \cdot \log^{1/2} n)$ steps with high probability, from which it follows that all processors complete with high probability. Since with probability $1/2$ each packet colors itself white and then chooses P_1 with probability $1/n^{1/2}$, the packet originating at processor P_i, $1 \leq i \leq n/2$ is destined for P_1 in with probability $1/(2 \cdot n^{1/2})$. Let p_i be the number of packets destined for P_1 among $P_1, \ldots, P_i$, and let $m = n^{3/4}$. For any fixed $i \geq m$, with high probability, p_i is $i/(2 \cdot n^{1/2}) + \mathcal{O}((i \log n/(2 \cdot n^{1/2}))^{1/2})$. Also, with high probability p_m is $\mathcal{O}(n^{1/4} \log n)$. The packets that contribute to p_m will therefore be transmitted in the first $\mathcal{O}(n^{3/4} \log n)$ stages, and after that, since at most p_i packets will have arrived in the first i stages, and the number of bus cycles available to P_1 is $i/n^{1/2}$, the remaining packets will be transmitted in $n/2$ further steps. $\qquad\square$

After phase 1, all packets with destination in a certain row have arrived there and are randomly distributed over the half of the row in which they originally resided. This is the reverse of the situation described by Lemma 14 with $k = 1/2$. Thus, phase 2 takes $(\sqrt{3} - 1)/2 \cdot n$ steps.

Theorem 22 *Applying* BUSROUTE3 *for routing permutations on a 2-dimensional mesh with buses requires* $\sqrt{3}/2 \cdot n + \mathcal{O}((n \cdot \log n)^{1/2}) < 0.87 \cdot n$ *steps with high probability.*

8 Overlapping the phases

The algorithm of Section 7 is easy to analyze and has good performance. However, a further reduction of the routing time can be achieved when the packets do not wait until the end of phase 1 before they start moving after they have been transferred by the bus: the phases should be coalesced. Packets that may reach their destination by walking, immediately start to do so. The other packets wait for the bus.

It is important now that the packets do not have to wait for the bus that transfers them to the correct row: in BUSROUTE3 the packets that were randomized to the representative in column $n/4$ all reached this representative after $n/4$ steps, but the last of them was transferred only after $n/2$ steps. This is caused by the fact that except for the representatives in column 1 and column $n/2$ the packets enter from two sides. We will correct this by using a non-trivial choice of the regions in which packets are randomized: a packet starting in P_{ij} chooses with equal probability any of the representatives in its row with column index from the set $\mathcal{A}_j$ defined by

$$\mathcal{A}_j = \begin{cases} [0, j] \cup [2 \cdot j, n/2 + j] & \text{if } j \leq n/2 \\ [n/2 - j, n - 2 \cdot j] \cup [n - j, n] & \text{if } j > n/2 \end{cases}$$

In the following picture the $\mathcal{A}_j$ lie within the marked regions:

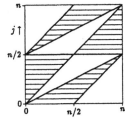

Notice that $\mathcal{A}_j$ consists of exactly $n/2$ elements for all i. Also, the thus chosen representative lies at most $n/2$ steps away form P_{ij}. The $\mathcal{A}_j$ are such that for the first j steps the packets enter a representative from one side (the left for PUs in the left half of the mesh), and then for the remaining $n/2 - j$ steps from the other side. The algorithm is a modification of BUSROUTE3. For the white packets it proceeds as follows:

Proc BUSROUTE4;
For a packet residing in P_{ij}, choose with equal probability a representative P_{ik}, with $k \in \mathcal{A}_j$;
1. the packets walk to the chosen representatives, the representatives put every $\sqrt{n}$ steps one of the waiting packets on the bus;
2. let t_1 be the distance from initial position to the chosen representative; let t_2 be the distance the packet has to cover within the row of its destination; the packet is going to walk towards its destination if $t_1 + t_2 \leq t$; the packets that do not walk to their destination are transferred over the bus.

All phases are coalesced. The packets are put on the bus in FIFO order. But the packets that wait for being transmitted a second time, have lower priority than the packets waiting for the bus for the first time. Walking phase-1 packets move without delay. They are given priority over packets performing phase 2. Among walking phase-2 packets the farthest-first strategy is used.

We aim for a total routing time of t. In order to prove this we must check several things:

1. Packets performing phase 1 have to wait at most $\mathcal{O}(n^{3/4} \cdot \log^{1/2} n)$ steps before they can use the bus,
2. a packet with values t_1 and t_2 as in the algorithm is delayed during its move as phase-2 packet at most $t - t_1 - t_2$ times,
3. during phase 2 in any row there are at most $t - n/2 + \mathcal{O}((n \cdot \log n)^{1/2})$ packets that take the bus.

We prove that this can be realized for $t = (\sqrt{141} - 1) \cdot n/14$. Thus,

Theorem 23 *Applying* BUSROUTE4 *for routing permutations on a 2-dimensional mesh with buses requires* $(\sqrt{141} - 1) \cdot n/14 + \mathcal{O}(n^{3/4} \cdot \log^{1/2} n) < 0.78 \cdot n$ *steps with high probability.*

Proof: See appendix. □

9 Concluding Remarks

We have presented randomized algorithms for routing k-permutations on d-dimensional meshes with buses. We proved new upper and lower bounds that improve all known results. For higher dimensions, the new upper bound depends on d only in low order terms. For the standard case, we presented an algorithm with a runtime smaller than n, which was believed to be a lower bound.

Future research is required for analysing how lower and upper bounds can be brought even closer together. For the higher-dimensional case the time consuming randomization might be left out, reducing the time required for routing permutations to approximately n steps. Also for the 2-dimensional case the routing time might be reduced further. Until now, all packets use the bus during the first phase. Thus, inevitably, this phase requires at least $n/2$ steps. It seems a good idea to let walk certain packets that do not have to go far. **Addendum:** Recently Cheung and Lau [3] have independently proved a lower bound, that is identical to ours, for routing permutations on a two-dimensional mesh with buses. Although their argument is similar, they do not generalise it to d-dimensional meshes for $d > 2$.

References

1. D. Angluin and L. G. Valiant. Fast probabilistic algorithms for Hamiltonian circuits and matchings. *J. Comput. Sys. Sc.*, **18** (1979), pp. 155–193.
2. S. H. Bokhari, Finding maximum on an array processor with a global bus, *IEEE Trans. Comput.*, **33** (1984), pp. 133–139.
3. S. Cheung and F. C. M. Lau, A lower bound for permutation routing on two-dimensional bused meshes, *IPL*, **45** (1993), pp. 225–228.
4. W. M. Gentleman, Some complexity results for matrix computations on parallel processors, *J. ACM*, **25** (1978), pp. 112–115.
5. M. Kaufmann, S. Rajasekaran and J. F. Sibeyn, Matching the bisection bound for routing and sorting on the mesh, in *Proc. 4th ACM SPAA* (1992), pp. 31–40.
6. M. Kaufmann and J. F. Sibeyn, Optimal multi-packet routing on the torus, In *Proc. 3rd Scandinavian Workshop on Algorithm Theory* (1992), LNCS 621, pp. 118–129.
7. M. Kunde, Routing and sorting on mesh connected processor arrays, in *Proc. 3rd Aegean Workshop on Computing* (1988), LNCS 319, pp. 423–433.
8. M. Kunde and T. Tensi, Multi-packet routing on mesh connected processor arrays, in *Proc. 2nd ACM SPAA* (1989), pp. 336–343.
9. F. T. Leighton, F. Makedon, I. G. Tollis, A $2n - 2$ step algorithm for routing in an $n \times n$ array with constant size queues, in *Proc. 2nd ACM SPAA* (1989), pp. 328–335.
10. S. Leung and S. M. Shende, On multi-dimensional packet routing for meshes with buses, in *Proc. 3rd IEEE SPDP* (1991), pp. 834–837. *J. Parl. Dist. Comp.*, to appear.
11. S. Rajasekaran and T. Tsantilas, Optimal routing algorithms for mesh-connected processor arrays, *Algorithmica*, **8** (1992), pp. 21–38.
12. J. F. Sibeyn and M. Kaufmann, k-k sorting on meshes, manuscript, 1992.
13. L. G. Valiant and G. J. Brebner, Universal schemes for parallel communication, in *Proc. 13th ACM STOC* (1981), pp. 263–277.

Proof of Theorem 23

Proof: Point 1 is easy to verify now that we modified the random choices of the representatives: the phase-1 packets move without delay and thus arrive in time. After t steps the expected number of packets that has reached any representative equals $t/\sqrt{n}$. By the Chernoff bounds the actual value will not exceed this number by more than $\mathcal{O}((t/\sqrt{n}\cdot\log n)^{1/2})$. For $t = n/2$ this is largest, $\mathcal{O}((n/\sqrt{n}\cdot\log n)^{1/2})$. As the packets are put on the bus in FIFO order this means that the last arriving packet may be delayed at most $\sqrt{n}\cdot\mathcal{O}((n/\sqrt{n}\cdot\log n)^{1/2})$ steps.

Point 2 does not mean that phase-2 packets are not delayed: packets that were ahead of their 'schedule' may be delayed. E.g., when all packets with destination in the right half of a certain row start in column 1, these packets drop on each other after being transferred by the bus. By the farthest first strategy this results in 'sorting' the packets: the farther a packet has to go, the earlier it moves on.

For the analysis of point 3 we consider the maximal number of packets that may use the bus in a certain row i. The criterion to take the bus for a packet p is that $t_1 + t_2 \geq t$. This is determined by the starting column j_1, the column j_2 of the selected representative and by the destination column j_3. The value of j_1 can be chosen independently, but in every row j_3 must assume each value exactly once. We now determine for each value of j_3 the value of j_1 that maximises the probability that $t_1 + t_2 > t$.

Consider the following subdivision of row i:

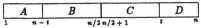

By symmetry we have to consider only the right half of the row i. The packets with destination in the region D use the bus if they are starting in column 1 with probability 1. Consider a packet p with destination in some column j_3 with $n/2 \leq j_3 < t$. It starts in column j_1 and is randomised to some column j_2. We assume that $j_1 \leq n/2 \leq j_3 < t$. The case that $j_1 > n/2$ can be neglected since the probability that it has to take the bus is smaller in that case than in the case $j_1 = n/2$. $t_1 + t_2 \geq t$ iff $j_2 < (j_1 + j_3 - t)/2$ or $j_2 > (j_1 + j_3 + t)/2$. We should find the j_1 which maximises the probability that such a j_2 is chosen. The analysis is complicated by the non-trivial randomization regions. $\#\{j_2 \in [0, j_1] \cup [2 \cdot j_1, n/2 + j_1] | j_2 < (j_1 + j_3 - t)/2 \text{ or } j_2 > (j_1 + j_3 + t)/2\} = \#\{j_2 \in [0, j_1] | j_2 < (j_1 + j_3 - t)/2\} + \#\{j_2 \in [2 \cdot j_1, n/2 + j_1] | j_2 > (j_1 + j_3 + t)/2\} = (j_1 + j_3 - t)/2 + n/2 + j_1 - \max\{2 \cdot j_1, (j_1 + j_3 + t)/2\} = \min\{(-j_1 + j_3 + n - t)/2, j_1 + n/2 - t\}$. This is maximal for $j_1 = (j_3 + t)/3$. Since $j_1 \leq n/2$ this is the starting position with the highest probability for the range $n/2 \leq j_3 \leq 3/2 \cdot n - t$. For $3/2 \cdot n - t < j_3 < t$ we take $j_1 = n/2$. In total the expected number of packets that is going to use the bus during phase 2 (summing only over the packets with destination in C and D to compensate for the coloring) equals $\sum_{j_3=n/2}^{3/2\cdot n-t}(1/2 + j_3/(3\cdot n) - 2\cdot t/(3\cdot n)) + \sum_{j_3=3/2\cdot n-t}^{t}(1 - t/n) + \sum_{t}^{n} 1 \simeq 5/6\cdot(1 - t/n)^2\cdot n + (2\cdot t/n - 3/2)\cdot(n - t) + (n - t) = (2\cdot n + 5\cdot t - 7\cdot t^2/n)/6$. The buses are used $n/2$ steps during phase 1 and the total time the buses are used should equal the walking time t. Thus, solving $n/2 + (2\cdot n + 5\cdot t - 7\cdot t^2/n)/6 = t$, we find $t = (\sqrt{141} - 1)\cdot n/14$. □

On the Distribution of the Transitive Closure in a Random Acyclic Digraph

Klaus Simon, Davide Crippa and Fabian Collenberg*

Institut für Theoretische Informatik
ETH-Zentrum
CH-8092 Zürich

Abstract. In the usual $G_{n,p}$–model of a random acyclic digraph let $\gamma_n^*(1)$ be the size of the reflexive, transitive closure of node 1, a source node; then the distribution of $\gamma_n^*(1)$ is given by

$$\forall\, 1 \leq h \leq n: \quad Pr\big(\gamma_n^*(1) = h\big) = q^{n-h} \prod_{i=1}^{h-1}(1 - q^{n-i}) \ ,$$

where $q = 1 - p$. Our analysis points out some surprising relations between this distribution and known functions of the number theory. In particular we find for the expectation of $\gamma_n^*(1)$:

$$\lim_{n \to \infty} n - E(\gamma_n^*(1)) = L(q)$$

where $L(q) = \sum_{i=1}^{\infty} q^i/(1 - q^i)$ is the so–called Lambert Series, which corresponds to the generating function of the divisor–function. These results allow us to improve the expected running time for the computation of the transitive closure in a random acyclic digraph and in particular we can ameliorate in some cases the analysis of the Goralčíková–Koubek Algorithm.

1 Introduction

Let $G = (V, E)$ be a directed graph, defined by a vertex set $V = \{1, \ldots, n\}$ and an edge set $E \subseteq V \times V - \{(v, v) \mid v \in V\}$. The adjacency list of a node $v \in V$ is given by $\Gamma(v) \overset{\text{def}}{=} \{w \in V \mid (v, w) \in E\}$. Further we define $\Gamma^+(v) \overset{\text{def}}{=} \emptyset$ if $\Gamma(v) \overset{\text{def}}{=} \emptyset$ and $\Gamma^+(v) \overset{\text{def}}{=} \bigcup_{w \in \Gamma(v)} \{w\} \cup \Gamma^+(w)$ otherwise; also we set $\Gamma^*(v) \overset{\text{def}}{=} \{v\} \cup \Gamma^+(v)$. The set $\Gamma^+(v)$ ($\Gamma^*(v)$) is called the (reflexive,) transitive closure of vertex v. With $\gamma(v)$, $\gamma^+(v)$ and $\gamma^*(v)$ we denote $|\Gamma(v)|$, $|\Gamma^+(v)|$ and $|\Gamma^*(v)|$, respectively. The transitive closure $G^+ = (V, E^+)$ of $G = (V, E)$ is given by

$$\forall\, v, w \in V: \quad (v, w) \in E^+ \iff w \in \Gamma^+(v). \tag{1}$$

A subgraph $G_r = (V, E_r)$ of G with the smallest possible number of edges such that $G_r^+ = G^+$ is called the transitive reduction of G. For an acyclic digraph G

* Research supported by the Swiss National Science Foundation, Grant 21–34115.92.

the subgraph G_r is uniquely determined. In particular, the transitive reduction $\Gamma^r(v)$ of the node v is then characterized by

$$\Gamma^r(v) \overset{\text{def}}{=} \{\, w \in \Gamma(v) \mid \text{there is no path of length} \geq 2 \text{ from } v \text{ to } w \,\}. \tag{2}$$

Computing the transitive closure of a digraph is one of the classical topics in computer science. Many algorithms have been proposed for this problem, see [Lee90] for a survey. The best solutions in the worst case are based on the boolean matrix multiplication. But these algorithms are usually hard to implement and their running times contain great constant factors. From the practical point of view simple algorithms with a good average case complexity are more interesting. The first observation in this sense is that our problem on a general digraph can be reduced to a smaller problem on an acyclic digraph in linear time by shrinking strongly connected components[2]. Next for every acyclic digraph $G = (V, E)$ there exists a permutation *ord* of the vertices $V = \{\, 1, \ldots, n \,\}$ such that

$$\forall v, w \in V : \quad (v, w) \in E \;\Rightarrow\; ord(v) < ord(w).$$

The map *ord* is called a *topological ordering* of G and can be computed in linear time. Throughout this paper we will assume that $ord(v) = v$ holds for G, i.e.

$$\forall (i, j) \in E : \quad i < j. \tag{3}$$

In particular this implies that node 1 has indegree 0. Now let $\Gamma(v) = \{\, w_1 < \cdots < w_s \,\}$. Then it is easy to see that (2) is equivalent to

$$\Gamma^r(v) = \{\, w_i \in \Gamma(v) \mid w_i \notin \Gamma^*(w_1) \cup \cdots \cup \Gamma^*(w_{i-1}) \,\}. \tag{4}$$

This statement is the base of Algorithm 1, first described by GORALČÍKOVÁ–KOUBEK [GK79], which can be easily implemented with the worst case running time

$$O(|V| + |E| + |V|\,|E_r|). \tag{5}$$

───────────── **Goralčíková–Koubek** ─────────────

```
(1)   for v ← n downto 1 do
(2)       Γ*(v) ← { v }; Γr(v) ← ∅;
(3)       forall w ∈ Γ(v) (* in increasing order *) do
(4)           if w ∉ Γ*(v)
(5)           then Γ*(v) ← Γ*(v) ∪ Γ*(w);
(6)                Γr(v) ← Γr(v) + { w };
(7)           fi;
(8)       od;
(9)   od;
```

───────────── **Algorithm 1** ─────────────

Despite its simplicity this algorithm runs quite well on average. We will show this for the $G_{n,p}$-model of a random acyclic digraph[3], defined by the vertex set

───────────

[2] see [Lee90] for all statements not proved or cited explicitly in this paper.

[3] An actual introduction to the theory of random graphs is [Pal85].

$V = \{1, \ldots, n\}$ and the set of edges (i, j) with $i < j$, where every edge occurs independently with probability p, $0 < p < 1$.

In this paper we will write $Pr(A)$ for the probability of an event A and $Pr(A \mid B)$ for the conditional probability of A given B. Let X be a random variable, then $E(X)$ will denote the expected value of X and $Var(X)$ its variance. Furthermore, we will use the shortening $q = 1 - p$ in the rest of the paper. The following was proved by SIMON [Sim88].

Theorem 1. *Let $\gamma_n^+(1)$ ($\gamma_n^r(1)$) be the size of $\Gamma^+(1)$ ($\Gamma^r(1)$) in the $G_{n,p}$-model of a random acyclic digraph. Then it holds:[4]*

1. $E(\gamma_n^r(1)) \le \log n + 2$

2. $E(\gamma_n^+(1)) \ge n - \dfrac{|\log p| + 1}{p}$

3. $E(\gamma_n^r(1)) = \dfrac{p}{q}(n - 1 - E(\gamma_n^+(1))) = \dfrac{p}{q}(n - E(\gamma_n^*(1)))$

4. The Algorithm 1 computes the transitive closure of an acyclic digraph in expected time $O(n^2 \log n)$.

In Section 2 we will compute the distribution of $\gamma_n^*(1)$ and give a recursive identity for its expectation and its variance. This distribution is not only of interest in our approach but it usually arises in phenomena such as "contagious diseases", where each occurrence increases the probability of further occurrences, see FELLER [Fel71a]. Many applications in population growth or nuclear research can also be described in this way. In Section 3 the expectation becomes the subject of a deeper analysis; in particular we will show that the asymptotic behaviour of expectation and variance is related the well–known functions of the number theory. In Section 4 we will then focus on the Goralčíková–Koubek Algorithm and improve the bounds for the expected running time.

2 The γ^* Distribution

Let us consider the $G_{n,p}$-model of a random acyclic digraph previously described. In this model the size $\gamma_n^*(1)$ of the reflexive, transitive closure of node 1 is a random variable satisfying the following:

Theorem 2. $\qquad P_{n,h} \overset{\text{def}}{=} Pr(\gamma_n^*(1) = h) = q^{n-h} \prod_{i=1}^{h-1}(1 - q^{n-i})$

Proof. The random variable $\gamma_n^*(1)$ represents a discrete–time, pure–birth process with time $t = n$. Such a process can be described by a sequence of random variables X_t, $t \in \mathbb{N}$, assuming the states $\ell = 1, 2, 3, \ldots$ with probabilities $P_{t,\ell}$ and by a sequence of transition probabilities λ_ℓ, $0 \le \lambda_\ell \le 1$, such that $P_{t,\ell} = 0$ for $t \le 0$ or $\ell \notin \{1, \ldots, t\}$, $P_{1,1} = 1$ and $P_{t,\ell} = (1 - \lambda_\ell) P_{t-1,\ell} + \lambda_{\ell-1} P_{t-1,\ell-1}$ otherwise; this means that the process starts at epoch 1 from state 1, direct

---

[4] Herewith as well as in the rest of this paper "log" stands for the natural logarithm.

transition from state ℓ is only possible to state $\ell + 1$, and this transition has probability λ_ℓ. In SIMON [Sim88] it is shown that $\lambda_\ell = 1 - q^\ell$ holds for $\gamma_n^*(1)$ and therefore the quantities $P_{n,h}$ satisfy the recurrence

$$P_{1,1} = 1 \quad \text{and} \quad P_{n,h} = q^h P_{n-1,h} + (1 - q^{h-1}) P_{n-1,h-1}. \tag{6}$$

We will prove our theorem by induction on n. For $n = 1$ the hypothesis is trivially satisfied; for $n \geq 2$ our induction hypothesis (I.H.) is that Theorem 2 is correct for $n - 1$. Then (6) implies

$$
\begin{aligned}
P_{n,h} &= q^h P_{n-1,h} + (1 - q^{h-1}) P_{n-1,h-1} \\
&\overset{(\text{I.H.})}{=} \underbrace{q^h q^{n-1-h}}_{q^{n-h} q^{h-1}} \prod_{i=1}^{h-1}(1 - q^{n-1-i}) + (1 - q^{h-1}) \underbrace{q^{n-1-(h-1)}}_{q^{n-h}} \prod_{i=1}^{h-2}(1 - q^{n-1-i}) \\
&= q^{n-h} \left(q^{h-1} \prod_{i=1}^{h-1}(1 - q^{n-1-i}) + (1 - q^{h-1}) \prod_{i=1}^{h-2}(1 - q^{n-1-i}) \right) \\
&= q^{n-h} \underbrace{\left(\prod_{i=1}^{h-2}(1 - q^{n-1-i}) \right)}_{\prod_{i=2}^{h-1}(1-q^{n-i})} \left(\underbrace{q^{h-1}(1 - q^{n-1-(h-1)})}_{q^{h-1}-q^{n-1}} + 1 - q^{h-1} \right) \\
&= q^{n-h} \prod_{i=1}^{h-1}(1 - q^{n-i}). \qquad \square
\end{aligned}
$$

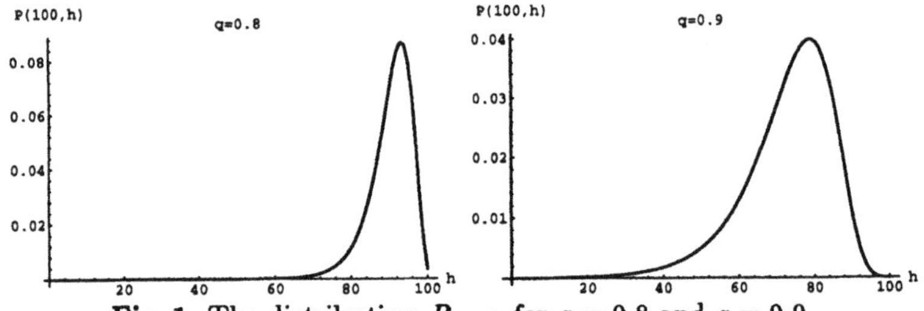

Fig. 1. The distribution $P_{100,h}$ for $q = 0.8$ and $q = 0.9$

Let us now set

$$a_n \overset{\text{def}}{=} E(\gamma_n^*(1)); \tag{7}$$

then, by Theorem 2, we have

$$
\begin{aligned}
a_n &= \sum_{h=1}^{n} h\, q^{n-h} \prod_{i=1}^{h-1}(1 - q^{n-i}) \\
&= q^{n-1} + \sum_{h=2}^{n} h\, q^{n-h} \prod_{i=1}^{h-1}(1 - q^{n-i}) \\
&= q^{n-1} + (1 - q^{n-1}) \sum_{h=2}^{n} h\, q^{n-h} \prod_{i=2}^{h-1}(1 - q^{n-i})
\end{aligned}
$$

$$= q^{n-1} + (1 - q^{n-1}) \sum_{h=1}^{n-1} (h+1)\, q^{n-1-h} \underbrace{\prod_{i=1}^{h-1} (1 - q^{n-1-i})}_{P_{n-1,h}}$$

$$= q^{n-1} + (1 - q^{n-1}) \underbrace{\left(1 + \sum_{h=1}^{n-1} h\, P_{n-1,h}\right)}_{a_{n-1}},$$

that is

$$a_n = 1 + (1 - q^{n-1})\, a_{n-1}. \tag{8}$$

A simple induction now shows:

Lemma 3.
$$a_n = \sum_{h=1}^{n} \prod_{i=1}^{h-1} (1 - q^{n-1}) = n - \sum_{h=1}^{n-1} q^h\, a_h.$$

Let us define further

$$x_n \stackrel{\text{def}}{=} E((\gamma_n^*(1))^2), \tag{9}$$

the second moment of the distribution of $\gamma_n^*(1)$. Then we can prove:

Lemma 4.
$$x_n = 2\, n\, a_n - \sum_{i=1}^{n} (2i-1) \prod_{h=i}^{n-1} (1 - q^h).$$

Proof. Exactly in the same way we derived (8), it is easy to show that x_n satisfies the recursive equation

$$x_n = 2\, a_n - 1 + x_{n-1}(1 - q^{n-1}) \tag{10}$$

with the initial condition $x_1 = 1$. Now we can proceed by induction on n and prove our theorem; for $n = 1$ we have $x_1 = 1 = 2\, a_1 - 1$, and for $n \geq 2$ we can apply (10) and obtain

$$
\begin{aligned}
x_n &= 2\, a_n - 1 + x_{n-1}(1 - q^{n-1}) \\
&\stackrel{\text{I.H.}}{=} 2\, a_n - 1 + (1 - q^{n-1}) \left(2\,(n-1)\, a_{n-1} - \sum_{i=1}^{n-1} (2i-1) \prod_{h=i}^{n-2} (1 - q^h) \right) \\
&= 2\, a_n - 1 + (1 - q^{n-1})\, 2\,(n-1)\, a_{n-1} - \sum_{i=1}^{n-1} (2i-1) \prod_{h=i}^{n-1} (1 - q^h) \\
&\stackrel{(8)}{=} 2\, n\, a_n - (2n-1) - \sum_{i=1}^{n-1} (2i-1) \prod_{h=i}^{n-1} (1 - q^h) \\
&= 2\, n\, a_n - \sum_{i=1}^{n} (2i-1) \prod_{h=i}^{n-1} (1 - q^h). \qquad \square
\end{aligned}
$$

Corollary 5.
$$Var(\gamma_n^*(1)) = a_n\,(2\,n - a_n) - \sum_{i=1}^{n} (2i-1) \prod_{h=i}^{n-1} (1 - q^h).$$

3 The Analysis of the Expression a_n

From the previous discussion it is clear that the expression a_n plays a key role in the distribution of $\gamma_n^*(1)$ and $\gamma_n^r(1)$, and therefore in the estimation of the running time of Algorithm 1. Following the notation of [HW89] we will use $\sigma_m(i)$ for the sum of m-th powers of the divisors of i, i.e.

$$\sigma_m(i) = \sum_{d|i} d^m.$$

In particular then $\sigma_0(i)$ will denote the number of divisors of i. Further we define

$$(K)_n \stackrel{\text{def}}{=} (1 - K)(1 - K q) \ldots (1 - K q^{n-1}) \qquad \text{for } n \geq 1 \qquad (11)$$

and $(K)_0 \stackrel{\text{def}}{=} 1$. Let us now expand a_n for a few values of n:

$$a_2 = 2 - q$$
$$a_3 = 3 - q - 2q^2 + q^3$$
$$a_4 = 4 - q - 2q^2 - 2q^3 + \cdots$$
$$a_5 = 5 - q - 2q^2 - 2q^3 - 3q^4 + \cdots$$
$$a_6 = 6 - q - 2q^2 - 2q^3 - 3q^4 - 2q^5 + \cdots$$

Taking a closer look at the coefficients of $q^i, 1 \leq i \leq n$, of each expression a_n, we notice and can prove the following:

Theorem 6.

$$a_n = n - \sum_{i=1}^{n-1} \sigma_0(i) q^i + R(q^n), \qquad (12)$$

where $R(q^n)$ is a polynomial in q of degree $\geq n$.

Proof. Since $a_i \leq i$, the generating function

$$A(z) \stackrel{\text{def}}{=} \sum_{i=1}^{\infty} a_i z^i \qquad (13)$$

converges absolutely for $|z| < 1$. With (8) we obtain then:

$$A(z) = \sum_{i=1}^{\infty} (1 + a_{i-1} - a_{i-1} q^{i-1}) z^i$$

$$= \sum_{i=1}^{\infty} z^i + z \sum_{i=1}^{\infty} a_i z^i - z \sum_{i=1}^{\infty} a_i (q z)^i$$

$$= \frac{z}{1 - z} + z A(z) - z A(q z)$$

which is equivalent to

$$A(z) = \frac{z}{(1-z)^2} - \frac{z}{1-z} A(q\,z).$$ (14)

Iterative substitution results in

$$A(z) = \sum_{n=1}^{\infty} \frac{(-1)^{n-1} z^n q^{\binom{n}{2}}}{(z)_n (1 - q^{n-1}z)}$$ (15)

so that if we substitute $z = q$ we obtain

$$A(q) = \sum_{n=1}^{\infty} \frac{(-1)^{n-1} q^{\binom{n+1}{2}}}{(q)_n (1 - q^n)}.$$ (16)

UCHIMURA [Uch81] proved that (16) is equivalent to the Lambert series (see KNOPP [Kno51]), which is known to be equivalent to the generating function of the divisor function σ_0, i.e.

$$\sum_{n=1}^{\infty} \frac{(-1)^{n-1} q^{\binom{n+1}{2}}}{(q)_n (1 - q^n)} = \sum_{j=1}^{\infty} \frac{q^j}{1 - q^j} = \sum_{i=1}^{\infty} \sigma_0(i)\,q^i.$$ (17)

Thus, combining Lemma 3, (16) and (17) we obtain

$$a_n = n - \sum_{i=1}^{n-1} q^i\,a_i$$
$$= n - A(q) + R(q^n)$$
$$= n - \sum_{i=1}^{n-1} \sigma_0(i)\,q^i + R(q^n)$$

which completes the proof of Theorem 6. $\qquad\square$

Corollary 7. *For all q, $0 < q < 1$, we have:*

$$\lim_{n\to\infty} (n - E(\gamma_n^*(1))) \equiv \lim_{n\to\infty} n - a_n = \sum_{i=1}^{\infty} \frac{q^i}{1 - q^i} \stackrel{\text{def}}{=} L(q)$$ (18)

$$\lim_{n\to\infty} E(\gamma_n^r(1)) \equiv \lim_{n\to\infty} \frac{p}{q}(n - a_n) = \frac{p}{q} \sum_{i=1}^{\infty} \frac{q^i}{1 - q^i} = \frac{p}{q} L(q).$$ (19)

If in the same way we expand $v_n \stackrel{\text{def}}{=} Var(\gamma_n^*(1))$ for a few values of n, we obtain according to Corollary 5:

$$v_2 = q - q^2$$
$$v_3 = q + 3\,q^2 - 5\,q^3 - \cdots$$
$$v_4 = q + 3\,q^2 + 4\,q^3 - 9\,q^4 - \cdots$$
$$v_5 = q + 3\,q^2 + 4\,q^3 + 7\,q^4 - 19\,q^5 - \cdots$$
$$v_6 = q + 3\,q^2 + 4\,q^3 + 7\,q^4 + 6\,q^5 - 24\,q^6 - \cdots$$

Here again, taking a closer look at the coefficients of $q^i, 1 \leq i \leq n$, of each expression v_n, we notice and can prove the following:

Theorem 8.

$$v_n = \sum_{i=1}^{n-1} \sigma_1(i)\, q^i + R(q^n), \tag{20}$$

where $R(q^n)$ is a polynomial in q of degree $\geq n$.

The proof shows some similarities to the one of Theorem 6, but it is too long to be reported here. It is part of a more general theory which is in development by ANDREWS, CRIPPA AND SIMON [ACS93].

Let now q be fixed and let f be a function satisfying $f(n) = a_n$ for all integers n. Clearly the values $f(n)$ will also satisfy the recursive formula

$$f(0) = 0, \qquad f(n) = n - \sum_{j=0}^{n-1} q^j\, f(j). \tag{21}$$

Applying Euler's summation formula (see e.g. [GKP90]) to the right side of (21) and differentiating we obtain the following differential equation for a function $g(x)$ approximating f:

$$\frac{dg(x)}{dx} = 1 - q^x\, g(x) \tag{22}$$

with initial condition $g(0) = 0$. Integration of (22) results in

$$g(n) = \frac{-Ei(\frac{1}{\log q}) + Ei(\frac{q^n}{\log q})}{\log q \, \exp\left(\frac{q^n}{\log q}\right)} \tag{23}$$

where Ei denotes the Exponential Integral function (see e.g. [AS72]). [5] In Fig. 2 we have plotted the difference $a_n - g(n)$ for $q \in (0,1)$ and $n = 10, \ldots, 60$, and we can see that this function turns out to be a very good approximation of a_n. For $x < 0$ the Exponential Integral function has the following bounds:

$$- e^x \log\left(1 - \frac{1}{x}\right) < Ei(x) < -\frac{1}{2} e^x \log\left(1 - \frac{2}{x}\right) \tag{24}$$

so that for $g(n)$ we find:

$$g(n) < -\frac{\log\left(1 - \frac{\log q}{q^n}\right)}{\log q} + \frac{\log\left(1 - 2\log q\right)}{2\log q} \exp\left(\frac{1 - q^n}{\log q}\right) \tag{25}$$

---

[5] Note that $f(0) = 0$, so that we could also write $f(n) = n - \sum_{j=1}^{n-1} q^j f(j)$. The function $q^j f(j)$ is decreasing in j for $j \geq 1$, therefore we can obtain an **upper bound** h for f by replacing $\sum_{j=1}^{n-1}$ with $\int_1^n$. This will result in a differential equation whose result is $h(n) = (1+q)\, g(n)$. We have preferred to work with g as it gives a better numerical approximation of f (and therefore of a_n), but the reader should pay attention to this difference. In any case, as the later computations will be carried out in the O–notation, the factor $(1+q) \leq 2$ can be neglected.

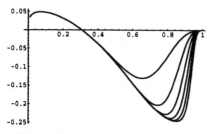
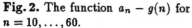

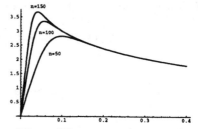

Fig. 2. The function $a_n - g(n)$ for $n = 10, \ldots, 60$.

Fig. 3. The expectation of $\gamma_n^r(1)$.

4 Running Time Analysis

In this section we want to analyse the expected running time of Algorithm 1, taking into consideration the results found in the previous sections. Let us first notice that Algorithm 1 has a complexity which is proportional to the size of the transitive reduction. According to Theorem 1, Part 3, we have

$$E(\gamma_n^r(1)) = \frac{p}{q}\,(n - a_n);$$

by plotting this function for different values of n fixed and q variable, we can see in Fig. 3 that the maximum is taken in the neighborhood of $p = \log n/n$, and therefore we will consider $p = (\log n)^\alpha/n$, $\alpha \le 1$.

Theorem 9. *The expected running time of Algorithm 1 for a $G_{n,p}$-model of a random acyclic digraph is given by*

$$b_n = \frac{p}{q}\sum_{h=1}^{n}\sum_{j=1}^{h-1} q^j\, a_j\, a_{h-j} = \frac{p}{q}\sum_{i=1}^{n} a_i\,[(n-i+1) - a_{n-i+1}]. \qquad (26)$$

Proof. We will first prove the first identity. Let us consider Algorithm 1; following SIMON [Sim88], we can use a bitvector for $\Gamma^*(v)$ in the loop (1)–(9), so that the test "$w \notin \Gamma^*(v)$" takes time $O(1)$. Outside this loop, the set $\Gamma^*(v)$ is kept as a linear list; this implies that the operation "$\Gamma^*(v) \cup \Gamma^*(w)$" has execution time $O(\gamma^*(w))$. If we now consider the $G_{n,p}$-model previously described, we obtain for the expected running time:

$$\begin{aligned}
b_n &= \sum_{h=1}^{n}\sum_{j=h+1}^{n} Pr(j \notin \Gamma^*(h))\, E(\gamma^*(j)) \\
&= \sum_{h=1}^{n}\sum_{j=h+1}^{n} Pr((1, j-h+1) \in E_r)\, E(\gamma_{n-j+1}^*(1)) \\
&= \sum_{h=1}^{n}\sum_{j=2}^{h} Pr((1, j) \in E_r)\, E(\gamma_{h-j+1}^*(1)),
\end{aligned}$$

and following [Sim88], for $j \geq 2$ we have

$$Pr((1,j) \in E_r) = \frac{p}{q}(1 - Pr((1,j) \in E^*)). \tag{27}$$

Let us consider now the probability that $(1,j)$ is an edge in G^+. Since all edges occur independently, for $j \geq 2$ we have

$$Pr((1,j) \in E^+) = \sum_{h=1}^{j-1} \left[\underbrace{Pr((1,j) \in E^+ \mid \gamma_{j-1}^*(1) = h)}_{1-q^h} \underbrace{Pr(\gamma_{j-1}^*(1) = h)}_{P_{j-1,h}} \right]$$

$$= 1 - \sum_{h=1}^{j-1} q^h \, q^{j-1-h} \prod_{i=1}^{h-1}(1 - q^{j-1-i})$$

$$= 1 - q^{j-1} \sum_{h=1}^{j-1} \prod_{i=1}^{h-1}(1 - q^{j-1-i})$$

$$= 1 - q^{j-1} a_{j-1}; \tag{28}$$

accordingly we find

$$b_n = \sum_{h=1}^{n}\sum_{j=2}^{h} \frac{p}{q}(1 - (1 - q^{j-1} a_{j-1}) a_{h-j+1} = \frac{p}{q} \sum_{h=1}^{n}\sum_{j=1}^{h-1} q^j \, a_j \, a_{h-j}. \tag{29}$$

To prove the second identity, let us rewrite the last sum as

$$\sum_{h=1}^{n}\sum_{j=1}^{h-1} q^j \, a_j \, a_{h-j} = q\,a_1\,a_1 +$$
$$q\,a_1\,a_2 + q^2\,a_2\,a_1 +$$
$$q\,a_1\,a_3 + q^2\,a_2\,a_2 + q^3\,a_3\,a_1 +$$
$$\vdots$$
$$q\,a_1\,a_{n-1} + q^2\,a_2\,a_{n-2} + \cdots\cdots + q^{n-1}\,a_{n-1}\,a_1$$

Reading this sum along the diagonals and then applying Lemma 3 we obtain

$$b_n = \frac{p}{q}\left[a_1 \sum_{i=1}^{n-1} q^i\,a_i + a_2 \sum_{i=1}^{n-2} q^i\,a_i + \cdots + a_{n-1}\,q\,a_1 \right]$$

$$= \frac{p}{q}\left[a_1(n - a_n) + a_2(n - 1 - a_{n-1}) + \cdots + a_{n-1}(2 - a_2) \right]$$

$$= \frac{p}{q} \sum_{i=1}^{n} a_i\left[(n - i + 1) - a_{n-i+1}\right]. \quad \Box$$

Using this last characterization of the expected running time we will now prove:

Theorem 10. For $p = (\log n)^\alpha / n$ and $\alpha \leq 1$ the expected running time of Algorithm 1 for a $G_{n,p}$-model of a random acyclic digraph is bounded by $O(n^2)$.

Proof. For $p = (\log n)^{\alpha}/n$ we can approximate $\log q \approx -p$. Further, using $g(n)$ to approximate a_n and the bounds (25) for $g(n)$ and (18) for $n - a_n$ we find:

$$
b_n = \frac{p}{q} \sum_{i=1}^{n} a_i \left(n - i + 1 - a_{n-i+1} \right)
$$

$$
\leq \frac{p}{q} L(q) \cdot \sum_{i=1}^{n} a_i
$$

$$
\leq \frac{p}{q} L(q) \cdot \sum_{i=1}^{n} \underbrace{\frac{-1}{\log q} \log \left(1 - \frac{\log q}{q^i} \right)}_{\leq -\frac{\log q}{q^i}}
$$

$$
\leq \frac{p}{q} L(q) \cdot \sum_{i=1}^{n} e^{-i \cdot \log q}
$$

$$
= O \left(p \, L(q) \frac{e^{-n \cdot \log q}}{-\log q} \right)
$$

$$
= O \left(\frac{\log p}{-p} e^{n \, p} \right)
$$

$$
= O \left(n \frac{(\log n - \alpha \log \log n)}{(\log n)^{\alpha}} e^{(\log n)^{\alpha}} \right)
$$

$$
= O \left(n^2 \right). \quad \square
$$

5 Concluding Remarks

- In this paper we have presented a broad study on the distribution of the transitive closure in a random acyclic digraph, which has led to a better approximation for the expected running time (of the Goralčíková–Koubek Algorithm) for its computation. The distribution $P_{n,h}$ should be of great interest in other domains as well: in particular, for p fixed, Theorem 2 represents an exact solution of the Yule process by discrete time n (see FELLER, [Fel71b], p. 450). In the usual description of this process the state transition's probability $\lambda_h = 1 - q^h$ has been commonly approximated by $\lambda_h \approx h \log q$, and this has led to an (asymptotically wrong) exponential growth rate.
- Theorem 1 Part 1 together with (19) imply that $E(\gamma_n^r(1)) = \Theta(\log n)$ for $p = \log n / n$. This shows that if we use $O(E(\gamma_n^r(1)) n^2)$ as upper bound for the expected running time of the Goralčíková–Koubek Algorithm we cannot show anything better than $O(n^2 \log n)$. On the other side Theorem 10 proved that for $p = \log n / n$ this upper bound can be lowered to $O(n^2)$; considering now that for $p = \log n / n$ the expectation of $\gamma_n^r(1)$ reaches its maximum, it is reasonable to conjecture that the $O(n^2)$ bound should be attained also for $p > \log n / n$.
- Our results also show a lower complexity for the improved algorithm presented by SIMON [Sim88]: $O(n^2 \log \log n)$ for $\log n / n < p < (\log n)^2 / n$ and

$O(n^2)$ otherwise. This follows by the fact that Goralčíková–Koubek is an upper bound for its complexity.

- A final remark is that this entire theory is strongly dominated by the polynomial $a_n = \sum_{h=1}^{n} \prod_{i=1}^{h-1}(1 - q^{n-i})$. The introduction of this term has substantially simplified the notation and further developments concerning a_n will have a great impact on the rest as well. In particular we have improved the results of [Sim88] stated in Theorem 1 concerning $\gamma_n^*(1)$ and $\gamma_n^r(1)$ and we expect the achievement of further results in this direction.

References

[ACS93] G. Andrews, D. Crippa, and K. Simon. Identities Related to the Recursive Equation $a_n(x) = f_n + (1 - x^{n-1}) a_{n-1}(x)$. *Work in progress*, 1993.

[AS72] M. Abramowitz and I. Stegun (Ed.). *Handbook of Mathematical Functions.* Dover Editions, 9 edition, 1972.

[Fel71a] W. Feller. *An Introduction to Probability Theory and Its Applications*, volume 2. John Wiley & Sons, 2. edition, 1971.

[Fel71b] W. Feller. *An Introduction to Probability Theory and Its Applications*, volume 1. John Wiley & Sons, 3. edition, 1971.

[GK79] A. Goralčíková and V. Koubek. A Reduction and Closure Algorithm for Graphs. *Lecture Notes in Computer Science*, 74, 1979.

[GKP90] R.L. Graham, D.E. Knuth, and O. Patashnik. *Concrete Mathematics.* Addison-Wesley, 1990.

[HW89] G. Hardy and E. Wright. *An Introduction to the Theory of Numbers*, chapter 19:Partitions. Oxford Science Publications, 1989.

[Kno51] K. Knopp. *Theory and Applications of Infinite Series.* Blackie and Son, Ltd, London, England, 1951.

[Lee90] J. van Leeuwen, editor. *Handbook of Theoretical Computer Science*, volume A: Algorithms and Complexity. Elsevier, 1990.

[Pal85] E. Palmer. *Graphical Evolution.* Wiley-Interscience, 1985.

[Sim88] K. Simon. *Improved Algorithm for Transitive Closure on Acyclic Digraphs.* *Theoretical Computer Science*, 58, 1988.

[Uch81] K. Uchimura. An Identity for the Divisor Generating Function Arising from Sorting Theory. *Journal of Combinatorial Theory*, 31, 1981.

Complexity of Disjoint Paths Problems in Planar Graphs

Alexander Schrijver

CWI and University, Amsterdam, The Netherlands

Let $G = (V, E)$ be a graph, and let $r_1, s_1, \ldots, r_k, s_k$ be vertices of G. The *disjoint paths problem* is the problem of finding disjoint paths $P_1, \ldots, P_k$, where P_i runs from r_i to s_i $(i = 1, \ldots, k)$.

There are several variants of this problem. "Disjoint" may mean "vertex-disjoint" or "edge-disjoint", and the graph may be directed or undirected. Each of these cases gives an NP-complete problem, as was shown by Knuth (see [2]). On the other hand, it was shown by Robertson and Seymour [7] that for each fixed k the problem is solvable in polynomial time when the graph is undirected. In the directed case, the problem is NP-complete even when fixing $k = 2$ (Fortune, Hopcroft, and Wyllie [1]).

Also when restricting ourselves to planar graphs, the problem for general k is NP-complete, as was shown by Lynch [4] and Kramer and Van Leeuwen [3]. On the other hand, there are some cases where the problem is polynomially solvable for planar graphs. Recently, Wagner and Weihe [11] showed that if all terminals are on the boundary of the infinite face and a certain parity condition is satisfied (the "Okamura-Seymour case"), the edge-disjoint undirected problem can be solved in *linear* time.

We consider some complexity results for problems in planar graphs for fixed k. With B. Reed, N. Robertson, and P.D. Seymour [5] we proved:

> For each fixed k, the vertex-disjoint undirected problem is solvable in linear time.

More generally, for each fixed k there is a linear-time algorithm for the problem of finding vertex-disjoint trees $T_1, \ldots, T_p$ in an undirected planar graph, where T_i covers a given set X_i of vertices $(i = 1, \ldots, p)$, such that $|X_1 \cup \cdots \cup X_p| \leq k$.

Our result extends a result of Suzuki, Akama, and Nishizeki [10] stating that the disjoint trees problem is solvable in linear time for planar graphs for each fixed upper bound k on $|X_1 \cup \cdots \cup X_p|$, when there exist two faces f_1 and f_2 such that each vertex in $X_1 \cup \cdots \cup X_p$ is incident with at least one of f_1 and f_2.

In fact, they showed more strongly that the problem (for nonfixed k) is solvable in time $O(k|V|)$. Indeed, recently Ripphausen, Wagner, and Weihe [6] showed that it is solvable in time $O(|V|)$.

Our proof is based on a lemma of Robertson and Seymour [8] stating that there exists a computable function $g : \mathcal{N} \longrightarrow \mathcal{N}$ with the following property:

> Let $G = (V, E)$ be an undirected plane graph, let $k \in \mathcal{N}$, let $X_1, \ldots, X_p \subseteq V$ such that $|X_1 \cup \cdots \cup X_p| \leq k$ and such that there exist vertex-disjoint trees $T_1, \ldots, T_p$ in G with $X_i \subset T_i$ for $i = 1, \ldots, p$. Moreover, let $v \in V$ be such that each closed curve C in the plane traversing v and intersecting or separating $X_1 \cup \cdots \cup X_p$ has at least $g(k)$ intersections with

G. Then there exist vertex-disjoint trees $T_1', \ldots, T_p'$ in $G - v$ such that $X_i \subseteq T_i'$ for $i = 1, \ldots, p$.

This result makes it possible to remove vertices iteratively until the graph can be decomposed into easier graphs.

For the *directed* case we showed in [9] the following:

For each fixed k there exists a polynomial-time algorithm for the vertex-disjoint directed paths problem in planar graphs.

The proof is based on cohomology over free groups with k generators. Let G be a group and let $D = (V, A)$ be a directed graph. Call two functions $\phi, \psi : A \longrightarrow G$ *cohomologous* if there exists a function $p : V \longrightarrow G$ such that for each arc $a = (u, w)$ of D one has

$$\psi(a) = p(u)^{-1}\phi(a)p(w).$$

Let Γ_k denote the free group generated by $g_1, \ldots, g_k$.
Then:

There exists a polynomial-time algorithm that for given natural number k, directed graph $D = (V, A)$ and function $\phi : A \longrightarrow \Gamma_k$, finds a function $\psi : A \longrightarrow \Gamma_k$ cohomologous to ϕ such that

$$\psi(a) \in \{1, g_1, \ldots, g_k\}$$

for each arc a (provided that such a function ψ exists).

This result is applied to an extension of the dual of the graph for which we want to solve the disjoint paths problem.

It also implies the following:

For each fixed p there exists a polynomial-time algorithm for the vertex-disjoint directed paths problem in directed planar graphs, when all terminals can be covered by the boundaries of p of the faces.

Furthermore, the result can be extended to finding vertex-disjoint rooted directed trees with given roots and covering given sets of vertices, when all roots and given vertices can be covered by the boundaries of a fixed number of faces. We do not know if these problems are solvable in linear time.

Most of the results above can be extended to graphs embedded on any fixed surface.

References

1. S. Fortune, J. Hopcroft and J. Wyllie, The directed subgraph homeomorphism problem, Theoretical Computer Science 10 (1980) 111–121.
2. R.M. Karp, On the computational complexity of combinatorial problems, Networks 5 (1975) 45–68.

3. M.R. Kramer and J. van Leeuwen, The complexity of wire routing and finding the minimum area layouts for arbitrary VLSI circuits, in: "VLSI Theory" (F.P. Preparata, ed.), JAI Press, London, pp. 129–146.

4. J.F. Lynch, The equivalence of theorem proving and the interconnection problem, (ACM) SIGDA Newsletter 5 (1975) 3:31–36.

5. B. Reed, N. Robertson, A. Schrijver and P.D. Seymour, Finding disjoint trees in planar graphs in linear time, in: "Graph Structures" (N. Robertson and P.D. Seymour, eds.), A.M.S. Contemporary Mathematics Series, American Mathematical Society, 1993.

6. H. Ripphausen, D. Wagner, and K. Weihe, The vertex-disjoint Menger problem in planar graphs, preprint, 1992.

7. N. Robertson and P.D. Seymour, Graph minors XIII. The disjoint paths problem, preprint, 1986.

8. N. Robertson and P.D. Seymour, Graph Minors XXII. Irrelevant vertices in linkage problems, preprint, 1992.

9. A. Schrijver, Finding k disjoint paths in a directed planar graph, SIAM Journal on Computing, to appear.

10. H. Suzuki, T. Akama, and T. Nishiseki, An algorithm for finding a forest in a planar graph — case in which a net may have terminals on the two specified face boundaries (in Japanese), Denshi Joho Tsushin Gakkai Ronbunshi 71-A (1988) 1897–1905 (English translation: Electron. Comm. Japan Part III Fund. Electron. Sci. 72 (1989) 10:68–79).

11. D. Wagner and K. Weihe, A linear-time algorithm for edge-disjoint paths in planar graphs, preprint, 1993.

Integer Multicommodity Flows with Reduced Demands

Anand Srivastav[*]   Peter Stangier[**]

Abstract

Given a supply graph $G = (V, E)$, a demand graph $H = (T, D)$, edge capacities $c : E \mapsto \mathbb{N}$ and requests $r : D \mapsto \mathbb{N}$, we consider the problem of finding integer multiflows subject to c, r. Korach and Penn constructed approximate integer multiflows for planar graphs, but no results were known for the general case. Via derandomization we present a polynomial-time approximation algorithm. There are two cases:

a) The main result is: For fractional solvable instances (G, H, c, r) and each $0 < \epsilon \leq \frac{9}{10}$ our algorithm finds in polynomial-time an integer multiflow subject to c, such that for all $d \in D$ the d-th flow value satisfies $f(d) \geq (1 - \epsilon)r(d)$, provided that capacities and requests are not too small, i.e. $c, r = \Omega(\frac{1}{\epsilon^2} \log(|E| + |D|))$. In particular, if $c, r \geq 36\lceil \log 2(|E| + |D| + 1)\rceil$ we have a strongly polynomial-time algorithm and the first $\frac{1}{2}$-factor approximation.

b) If the problem is not fractionally solvable we can reduce it to the case mentioned above decreasing the requests in an optimal way. This can be done by linear programming and the results of a) apply.

The design and analysis of the algorithm require new techniques for randomized rounding as well as for derandomization. One key tool is an *algorithmic* version of the classical Angluin-Valiant inequality (a variant of the well known Chernoff-Hoeffding bound) estimating the tail of *weighted* sums of Bernoulli trials, which was not previously known and might be of independent interest in computational probability theory.

The significance of our rounding algorithm is emphasized by the fact that there is a rich combinatorial theory exhibiting many examples of fractionally solvable problems, but finding approximate integer solutions even for fractionally solvable problems is NP-hard as it is shown in this paper.

1 Introduction

(a) The Problem:
According to [4] the feasibility multicommodity flow problem is stated as follows: Let $G = (V, E)$ be the supply graph and $H = (T, D)$ be the demand graph with $T \subseteq V$. The vertices of H are the terminals and the edges $(q_1, s_1), \ldots, (q_k, s_k)$ of H are called commodities or demand edges. For each demand edge $d = (q_d, s_d) \in D$ let σ_d be an orientation of G forming the directed graph (V, A_d) and let $F(d)$ be an integer (q_d, s_d)-flow in (V, A_d). Then the $|D|$-tuple of flows $F = (F(d))_{d \in D}$ is called a *multicommodity flow*.

Given a capacity function $c : E \mapsto \mathbb{N}$ and a demand (or request) function $r : D \mapsto \mathbb{N}$ the multicommodity flow is said to be subject to c, if for each edge $e \in E$ the sum of the flows through e (in both directions) does not exeed $c(e)$ and is subject to r if for each demand edge $d \in D$ the d-th flow value $f(d)$ is at least $r(d)$. Let henceforth denote (G, H, c, r) an instance of the multicommodity flow problem.

Finding integer multicommodity flows subject to c and r is well-known to be NP-hard [4]. But Korach and Penn [6] found an interesting integer approximate solution of the multicommdity

---

[*]Research Institute of Discrete Mathematics, University of Bonn, Bonn, Germany
[**]Institut für Informatik, Universität zu Köln, Pohligstraße 1, 50969 Köln, Germany

flow problem for planar graphs: If G and $G \cup H$ are planar and if the cut condition holds, then the multicommodity flow problem $(G, H, c, r-1)$, where each request has been reduced by one unit, can be solved in polynomial-time. In fact they proved a stronger result but stated in this form the Korach/Penn result motivates the following interesting approximation problem.

(1.1) Reduced Demand Multiflow Problem

Let (G, H, c, r) be a multicommodity flow problem. Find a rational non-negative function $K :$ $D \mapsto \mathbb{Q}$ with minimum $\sum_{d \in D} K(d)$ such that the reduced problem $(G, H, c, r - K)$ can be integrally solved.

Let K_I denote the minimal (in the sense above) such K and let denote K_R the minimal function, if we allow fractional solvability. In other words, we ask for the best possible approximate integer multiflow.

(b) The Results:

Since (1.1) can be formulated as an integer linear program, its linear programming relaxation can be solved in polynomial-time and gives an efficiently computable fractional "lower" bound K_R on K_I. But we show in section 4 that the decision version of (1.1) is NP-complete, even if (G, H, c, r) is fractionally solvable. In contrast to this we exhibit a large class of non-planar instances of the problem, where surprisingly good approximate integer flows can be constructed in polynomial-time:

Theorem Let (G, H, c, r) be a multicommodity flow problem. Let $0 < \epsilon < \frac{9}{10}$ and suppose that $c(e), r(d) - K_R(d) \geq \frac{6(2-\epsilon)}{\epsilon^2} \lceil \log(2(|E| + |D| + 1)) \rceil$ for all $e \in E$ and $d \in D$. Then an integer multiflow F can be found in polynomial-time such that $f(d) \geq (1 - \epsilon)(r(d) - K_R(d))$ for all demand edges $d \in D$. □

For *fractionally solvable* problems this gives for instances with $c, r \geq \frac{6(2-\epsilon)}{\epsilon^2} \lceil \log(2(|E| + |D| + 1)) \rceil$ the estimate $f(d) \geq (1 - \epsilon)r(d)$ for all demands d and in particular a $\frac{1}{2}$-factor approximation:

Corollary Let (G, H, c, r) be a fractionally solvable multicommodity flow problem. Suppose that $c(e), r(d) \geq 36 \lceil \log(2(|E| + |D| + 1)) \rceil$ for all $e \in E$ and $d \in D$. Then an integer flow F can be found in strongly polynomial-time such that $f(d) \geq \frac{r(d)}{2}$ for all demand edges $d \in D$. □

The dominating part of running time is the time to solve the corresponding linear programming relaxation.

In general, this seems to be the best possible approximation factor. We can construct for each integer $K \in \mathbb{N}$ a non-planar instance for which the fractional problem is solvable, but the integer problem is not solvable even if we decrease each request value by K. In particular, this example shows $K_I(d) \geq \frac{1}{2}(r(d) - 1)$ for all demand edges $d \in D$.

Note that in the RAM-model of computation the theorem above cannot not be proved by the cosh-algorithm of Spencer [1] or the basic pessimistic estimator method of Raghavan [12].

(c) The Methods:

In the design and analysis of our algorithms we wish to use randomized rounding and afterwards derandomization. But here neither randomized rounding nor derandomization are directly applicable: Randomized rounding as proposed by Raghavan and Thompson [11] operates on instances where fractional solutions from the *unit interval* are given. Here the fractional flows can take arbitrary rational values. Representing the fractional flows in terms of directed paths P with associated path values $\lambda(P)$, the direct approach would be to cut off the integer part $\lfloor \lambda(P) \rfloor$ and round the remaining fractional value to 0 or 1. Consequently we must now consider the decreased edge capacities $\tilde{c}(e) = c(e) - \sum_{e \in P} \lfloor \lambda(P) \rfloor$. Unfortunatly, due to the Angluin-Valiant inequality randomized rounding can be analysed only for packing problems, where enough packing space is available. In our problem we must assume $\tilde{c}(e) = \Omega(\log(|E| + |D|))$. So even if $c(e) = \Omega(\log(|E| + |D|))$, it may happen that $\tilde{c}(e)$ is too small and the method fails. An other "solution" would be to split of each P into $2\lfloor \lambda \rfloor + 1$ parallel paths with values from $(0, 1)$, which also reduces the problem to the 0-1 case, but for the prize that we would increase the number of variables in the randomized rounding procedure by the maximal path value. In consequence the complexity of the rounding procedure would depend on the magnitude of numbers appearing in the input and this would not be a polynomial-time procedure anymore. We show that we have to introduce for every $e \in E$ and every $d \in D$ at most $O(\epsilon^{-2}\log(|E| + |D|))$

0 − 1 random variables.

Furthermore the derandomiztion method of pessimistic estimators as introduced by Raghavan [12] for approximating packing integer programs does not give for problems with *rational* weights a polynomial-time algorithm on usual models of computation, for example the RAM model. Unfortunately, the calculation of pessimistic estimators in our algorithm requires exponentiation of rational numbers to the power of rational numbers. There is no obvious way to avoid such numerical problems. We solve this problem extending the derandomization method of conditional probabilities in an interesting way. We introduce the concept of so called *approximate* pessimistic estimators, which are low degree polynomials in polynomial-time computable rational numbers and prove an algorithmic version of the Angluin-Valiant inequality. This enables us to find events concentrated around the mean of *weighted* linear sums of Bernoulli trials in polynomial-time.

(d) Related Work

Raghavan [12] investigated the problem of finding maximal 0-1 multiflows. If $c = \Omega(\log |E|)$ he constructed an integral flow with total sum M satisfying $M \geq \gamma M_{opt}(1 - D)$, where M_{opt} is the integer maximum, $\gamma \geq \frac{1}{2}$ a constant and D a function depending on $|E|$, γ and M_{opt}. If $c = \Omega(\log |E|)$ Raghavan showed that D is constant, hence proved an *implicit* constant factor. We can prove by the methods developed in this paper, especially the algorithmic Angluin-Valiant inequality, a $\frac{1}{2}$-factor approximation of M_{opt}, provided that $c = \Omega(\log |E|)$. This removes the $\gamma D M_{opt}$ term and shows an *explicit* constant factor. In fact, for $0 < \epsilon \leq \frac{9}{10}$ and $c = \Omega(\frac{1}{\epsilon^2} \log |E|)$ we have $M \geq (1 - \epsilon)M_{opt}$ ([15]).

Furthermore the algorithmic version of the *weighted* Angluin-Valiant inequality opens a way to solve *weighted* packing integer programs without the previous restriction to 0-1 cases ([15]).

2 Randomized Flow Generation

For each commodity $d \in D$ and each edge $\{u, v\} \in E$ let us introduce integer variables x_{uv}^d and x_{vu}^d, where x_{uv}^d represents the flow value of the commodity d on edge $\{uv\}$ in direction from u to v and vice versa. The reduced demand multiflow problem (1.1) is equivalent to the following integer linear program:

(2.1) Multicommodity Flow with Reduced Demands as an Integer Linear Program

$$\min \sum_{d \in D} K(d)$$

such that:

$$\sum_{\{v \in V : \{q_d, v\} \in E\}} x_{q_d v}^d - x_{v q_d}^d \geq r(q_d, s_d) - K(q_d, s_d) \qquad (\forall d \in D)$$

$$\sum_{d \in D} x_{uv}^d + x_{vu}^d \leq c(\{u, v\}) \qquad (\forall \{u, v\} \in E)$$

$$\sum_{\{v \in V - \{q_d, s_d\} : \{u, v\} \in E\}} x_{uv}^d = \sum_{\{v \in V - \{q_d, s_d\} : \{u, v\} \in E\}} x_{vu}^d \qquad (\forall d \in D, u \in V - \{q_d, s_d\})$$

Let us denote by (MLP) the fractional relaxation, where the flows x_{uv}^d are rational numbers. The fractional solution K_R together with the coresponding flows can be constructed in polynomial-time with standard LP-algorithms and of course $\sum K_R(d) \leq \sum K_I(d)$ (see [4]).

In the following we use the reformulation of the multicommodity flow problem in terms of *directed paths*. Let $\Gamma = \{P_1, \ldots, P_s\}$ be the set of paths defined as follows: Each path $P \in \Gamma$ is a $(q_d, s_d)-$ path for a commodity $d = (q_s, s_d) \in D$ and can be extended to a circle in $G \cup H$ adding the demand edge (q_d, s_d). Each path $P \in \Gamma$ is associated to a nonnegative integer (in case of fractional flows to a rational number) $\lambda(P)$. The value of the flow for a commodity d is equal to the sum of those $\lambda(P)$ for which P is a (q_d, s_d)-path.

Given edge capacities c and a request function r the multicommodity flow is subject to c, if for each edge $e \in E$ the sum of the $\lambda(P)$ for which P contains e is at most $c(e)$ and it is subject to r, if for each demand edge $d = (q_d, s_d)$ the sum of those $\lambda(P)$, for which P is a $(q_d, s_d)-$path is at least $r(d)$.

Having solved the LP-relaxation of (2.1), we represent the fractional multicommodity flow by directed paths following Malhotra et al. [7]. Raghavan and Thompson [11] used the same idea for randomly *maximizing* multicommodity flow. The idea of the algorithm is very simple: For every commodity d we assign a direction to every edge. Then we try to find a directed path starting at q_d, ending in s_d, such that every edge on the path has a strictly positive weight. We calculate the minimum edge weight on this path. This minimum value is substracted from every edge weight in this path and will be assigned to the path. Edges with zero weight will be deleted and we try to find the next path. After finding at most $|E|$ paths for every commodity the procedure terminates.

In the following let Γ_d be the set of paths representing the commodity d, and let $\Gamma = \bigcup_{d \in D} \Gamma_d$.

(2.2) **Path Generation Algorithm (GENPATH)**

INPUT: A fractional optimal multicommodity flow solving (2.1) and the function $K_R : D \mapsto \mathbb{Q}$.
OUTPUT: For each demand $d \in D$ a set Γ_d, path values $\lambda(P)$ for each path $P \in \Gamma_d$, the set $\Gamma = \bigcup_d \Gamma_d$.

```
begin
for each d in D do
      {* Form a directed graph G_d where E_d is a set of directed edges
         from E as follows:                                              *}
   Let d = (q_d, s_d), set Γ_d := ∅, E_d := ∅.
   while there is v ∈ V with x_{q_d,v} > 0 do
      for each e in E do
         {* assign a direction to e:                                     *}
         let e = {u, v}:
         if x_{uv}^d = x_{vu}^d then next e ∈ E.
         if x_{uv}^d > x_{vu}^d then
            direction(e) = (u, v), x^d(e) = x_{uv}^d - x_{vu}^d.
         if x_{uv}^d < x_{vu}^d then
            direction(e) = (v, u), x^d(e) = x_{vu}^d - x_{uv}^d.
         E_d := E_d ∪ {e}.
      end for
      Discover a directed path P = {q_d, ..., s_d} in G_d
      using depthfirst search discarding loops.
      Set λ(P) = min{x^d(e(j)), 1 ≤ j ≤ p}.
      for 1 ≤ j ≤ p,
         x^d(e_j) := x^d(e_j) - λ(P), Γ_d := Γ_d ∪ {P}.
      for each e in E_d do
         if x^d(e) = 0 then E_d := E_d\{e}
   end while
end for
end
```

It is clear, that the while loop is executed at most $|E|$ times for every demand, as there is always at least one edge which is excluded from E_d. Thus, the algorithm will find a representation of the fractional multicommodity flow with at most $|D||E|$ paths. If we reduce every path value λ by its fractional part $(\lambda - \lfloor \lambda \rfloor)$ we obtain an integer solution where every path value has been reduced by at most 1. So if (G, H, c, r) is fractionally solvable, $(G, H, c, r - |E|)$ trivially has an integer solution.

A simple randomized procedure to turn the fractional path values into integer ones is to flip for each path a biased coin independently deciding whether $\lambda(P)$ should be truncated to $\lfloor \lambda(P) \rfloor$ or rounded up to $\lceil \lambda(P) \rceil$. As mentioned in the introduction such roundings cannot be analysed by the probabilistic methods given so far. An intuitive better idea is to perform a more flexible rounding procedure in which by chance some rounded path values could become much bigger or smaller than $\lceil \lambda(P) \rceil$. One extreme way to do so is to split each path value $\lambda(P)$ into $2\lfloor \lambda(P) \rfloor$ "path segments" of value 0.5 and one segment of value $\lambda(P) - \lfloor \lambda(P) \rfloor$. Then rounding the value of the segments to 0 or 1 randomly with probabilities equal to the segment values the expected total path value is $\lambda(P)$. But this is not a polynomial-time rounding algorithm as the number of trials depends on $r(d)$. Our strategy is to split off each path value into a fixed integer part and a sufficiently big roundable part of size $\Omega(\frac{1}{\epsilon^2}\log(|E|+|D|))$. The following algorithm shows that for each $e \in E$ and $d \in D$ we must introduce at most $O(\frac{1}{\epsilon^2}\log(|E|+|D|))$ 0-1 random variables.

(2.3) Path Splitting Algorithm SPLITPATH(ϵ)

INPUT: The set of directed paths Γ, associated path values $\lambda(P), P \in \Gamma$, and a rational number $0 < \epsilon \leq 1$.

OUTPUT: New path values
$\lambda(P), \lambda_0(P), \ldots, \lambda_{N(P)}(P)$ for each $P \in \Gamma$.

begin
For each $P \in \Gamma$ set
  $\lambda_0(P) := \lambda(P) - \lfloor \lambda(P) \rfloor$,
  $\lambda(P) = \lambda(P) - \lambda_0(P)$ and $N(P) := 0$.
(a) Set $r_\epsilon = \frac{6(2-\epsilon)}{\epsilon^2}\lceil \log(2(|E|+|D|+1)) \rceil$.
for each d in D do
  while $r(d) - K_R(d) - \sum_{P \in \Gamma_d} \lambda(P) < r_\epsilon$ do
    choose $P \in \Gamma_d$ with $\lfloor \lambda(P) \rfloor \geq 1$.
    set $\lambda(P) = \lambda(P) - 1$,
    $\lambda_{N(P)+1}(P) = \lambda_{N(P)+2}(P) = 0.5$,
    $N(P) = N(P) + 2$.
  end while
end for.
(b) Set $c_\epsilon = \frac{6(2-\epsilon)}{\epsilon^2}\lceil \log(2(|E|+|D|+1)) \rceil$.
for each e in E do
  while $c(e) - \sum_{e \in P \in \Gamma} \lambda(P) < c_\epsilon$ do
    choose $d \in D$ and $P \in \Gamma_d$
      with $e \in P$ and $\lceil \lambda(P) \rceil \geq 1$.
    set $\lambda(P) = \lambda(P) - 1$,
    $\lambda_{N(P)+1}(P) = \lambda_{N(P)+2}(P) = 0.5$
    $N(P) = N(P) + 2$.
  end while
end for.
end

As a result of the algorithm SPLITPATH(ϵ) we have a representation of the multicommodity flow with at most $O(|E||D|\epsilon^{-2})$ path values and the original $\lambda(P)$ has been decreased such that for every $e \in E$ and $d \in D$

$$c(e) - \sum_{e \in P \in \Gamma} \lambda(P) \geq c_\epsilon$$

and

$$r(d) - K_R(d) - \sum_{P \in \Gamma_d} \lambda(P) \geq r_\epsilon. \qquad (1)$$

We are ready to perform randomized rounding.

(2.4) Randomized Integer Flow Generation R-FLOW(ϵ)

Let $0 < \epsilon \leq 1$. Let $P \in \Gamma$ and $\lambda(P)$ and $\lambda_i(P)$ $(i = 0, 1, \ldots, N(P))$ generated by SPLITPATH(ϵ). For every path $P \in \Gamma$ and every $i = 0, 1, \ldots, N(P)$ do independently

1. Set
   - $\lambda_i^I(P) = 1$ with probability $(1 - \frac{\epsilon}{2})\lambda_i(P)$.
   - $\lambda_i^I(P) = 0$ with probability $1 - (1 - \frac{\epsilon}{2})\lambda_i(P)$.

2. For each $P \in \Gamma$ set $\lambda^I(P) = \lambda(P) + \sum_{i=0}^{N(P)} \lambda_i^I(P)$.

Algorithm R-FLOW(ϵ) outputs for each path an integer path value. We proceed to the analysis of such an integer multiflow. In Lemma (2.2) we show that the flow is feasible with respect to c and in Lemma (2.3) we prove that it conveys enough commodities. We invoke the Angluin-Valiant inequality in order to estimate deviation of sums of weighted Bernoulli trials from their mean. McDiarmid's proof of the Angluin-Valiant inequality ([8] , proof of corollary 5.1 and 5.2) gives also the following "conditional probability" formulation:

Lemma 2.1 *Let $a_1, \ldots, a_n$ be real numbers with $0 \le a_j \le 1$ for each j and let $\psi_1, \ldots, \psi_n$ be independent $0 - 1$ valued random variables. Let $\tilde{p}_j = E(\psi_j)$, $\tilde{q}_j = 1 - \tilde{p}_j$, $\psi = \sum_{j=1}^n a_j \psi_j, p = \frac{1}{n}E(\psi), q = 1 - p$ and $0 < \beta < 1$. Define $s^+ = \frac{q(1+\beta)}{q - p\beta}$, $s^- = \frac{q + p\beta}{q(1 - \beta)}$ and for $1 \le l \le n$ let $x_1, \ldots, x_l \in \{0, 1\}$. Then we have*

$$(i) Prob(\psi > (1 + \beta)np|\psi_1 = x_1, \ldots, \psi_l = x_l)$$
$$\le e^{-(1+\beta)pn \ln s^+} e^{\sum_{j=1}^l a_j x_j \ln s^+} \prod_{j=l+1}^n [\tilde{p}_j e^{a_j \ln s^+} + 1 - \tilde{p}_j]$$
$$\le e^{-\frac{\beta^2 np}{3}}.$$

$$(ii) Prob(\psi < (1 - \beta)np|\psi_1 = x_1, \ldots, \psi_l = x_l)$$
$$\le e^{-(1-\beta)pn \ln s^-} e^{-\sum_{j=1}^l a_j x_j \ln s^-} \prod_{j=l+1}^n [\tilde{p}_j e^{-a_j \ln s^-} + 1 - \tilde{p}_j]$$
$$\le e^{-\frac{\beta^2 np}{2}}.$$

$\square$

Lemma 2.2 *Let $0 < \epsilon \le 1$. Suppose that $c(e) \ge \frac{6(2-\epsilon)}{\epsilon^2} \lceil \log(2(|E| + |D| + 1)) \rceil$ for all $e \in E$. Then with probability at least $1 - \frac{|E|}{2(|E|+|D|+1)}$ R-FLOW(ϵ) finds for each $P \in \Gamma$ an integer path value $\lambda^I(P)$ such that for all $e \in E$*

$$\sum_{e \in P \in \Gamma} \lambda^I(P) \le c(e).$$

Proof. Since we are rounding only a part of the path values, we have to consider only decreased edge capacities $\tilde{c}(e)$ defined by

$$\tilde{c}(e) := c(e) - \sum_{e \in P \in \Gamma} \lambda(P). \tag{2}$$

Define for every edge $e \in E$ the random variable

$$\chi(e) = \sum_{e \in P \in \Gamma} \sum_{i=1}^{N(P)} \lambda_i^I(P),$$

Then a straight forward calculation shows

$$\mathbb{E}(\chi(e)) \le (1 - \frac{\epsilon}{2})\tilde{c}(e). \tag{3}$$

Taking $\beta := \frac{\epsilon}{2-\epsilon}$ we have $0 < \beta \leq 1$ and with (1)

$$\frac{6(2-\epsilon)}{\epsilon^2}\ln 2(|E| + |D| + 1) \leq \tilde{c}(e). \tag{4}$$

The Angluin-Valiant inequality (Lemma 2.1 (i)), (3) and (4) imply

$$\begin{aligned}
\mathbb{P}(\chi(e) > \tilde{c}(e)) &= \mathbb{P}(\chi(e) > (1+\beta)(1-\frac{\epsilon}{2})\tilde{c}(e)) \\
&\leq \frac{1}{2(|E| + |D| + 1)}.
\end{aligned} \tag{5}$$

This together with (2) implies for all $e \in E$ with probability at least $1 - \frac{|E|}{2(|E|+|D|+1)}$

$$\begin{aligned}
\sum_{e \in P \in \Gamma} \lambda^I(P) &= \sum_{e \in P \in \Gamma} \lambda(P) + \chi(e) \\
&\leq \sum_{e \in P \in \Gamma} \lambda(P) + \tilde{c}(e) = c(e).
\end{aligned} \tag{6}$$

$\square$

In the next lemma we estimate the rounded flows.

Lemma 2.3 *Let* $0 < \epsilon \leq 1$. *Suppose that* $r(d) - K_R(d) \geq \frac{6(2-\epsilon)}{\epsilon^2}\lceil \log(2(|E| + |D| + 1))\rceil$ *for all* $d \in D$. *Then with probability at least* $1 - \frac{|D|}{2(|E|+|D|+1)}$ *R-FLOW(ϵ) finds for each* $P \in \Gamma$ *an integer path value* $\lambda^I(P)$ *such that for all* $d \in D$ *we have*

$$\sum_{P \in \Gamma_d} \lambda^I(P) \geq (1 - \epsilon)(r(d) - K_R(d))$$

Proof. Define the reduced request $\tilde{r}(d)$ by

$$\tilde{r}(d) := r(d) - K_R - \sum_{P \in \Gamma_d} \lambda(P). \tag{7}$$

Define for every edge $e \in D$ the random variable

$$\chi(d) = \sum_{P \in \Gamma_d} \sum_{i=1}^{N(P)} \lambda_i^I(P),$$

Then

$$\mathbb{E}(\chi(d)) = (1 - \frac{\epsilon}{2}) \sum_{P \in \Gamma_d} \sum_{i=1}^{N(P)} \lambda_i(P) = (1 - \frac{\epsilon}{2})\tilde{r}(d). \tag{8}$$

Put $\gamma := \sqrt{\frac{6\ln(|E|+|D|+1)}{(2-\epsilon)\tilde{r}(d)}}$. Then by (1)

$$6\lceil \ln(2(|E| + |D| + 1))\rceil \leq \tilde{r}(d). \tag{9}$$

which implies $0 < \gamma \leq 1$. With the Angluin-Valiant inequality (Lemma 2.1 (ii)), (8), (9) it is not hard to prove

$$\begin{aligned}
\mathbb{P}(\chi(d) &< (1-\epsilon)\tilde{r}(d)) \\
&\leq \mathbb{P}(\chi(d) < (1-\gamma)(1-\frac{\epsilon}{2})\tilde{r}(d)) \leq \frac{1}{2(|E|+|D|+1)}.
\end{aligned} \tag{10}$$

(8) and (10) imply for all $d \in D$ with probability at least $1 - \frac{|D|}{2(|E|+|D|+1)}$

$$\sum_{P \in \Gamma_d} \lambda^I(P) = \sum_{P \in \Gamma_d} \lambda(P) + \chi(d) \geq (1 - \epsilon)(\tilde{r}(d) - K_R(d)).$$

$\square$

Theorem 2.4 *Let (G, H, c, r) be a multicommodity flow problem and let $0 < \epsilon \leq 1$ with $c(e) \geq \frac{6(2-\epsilon)}{\epsilon^2}\lceil\log(2(|E| + |D| + 1))\rceil$ for all $e \in E$ and $r(d) - K_R(d) \geq \frac{6(2-\epsilon)}{\epsilon^2}\lceil\log(2(|E| + |D| + 1))\rceil$ for all $d \in D$. Then with probability at least $1 - \frac{|E|+|D|}{2(|E|+|D|+1)}$ R-FLOW(ϵ) finds an integer multicommodity flow F subject to c such that for all $d \in D$*

$$f(d) \geq (1 - \epsilon)(r(d) - K_I(d)).$$

□

For fractionallly solvable problems we have $f(d) \geq (1 - \epsilon)r(d)$ and for $\epsilon = \frac{1}{2}$:

Corollary 2.5 *Let (G, H, c, r) be a fractionally solvable multicommodity flow problem with $c(e) \geq 36\lceil\log 2(|E| + |D| + 1)\rceil$ for all $e \in E$ and $r(d) \geq 36\lceil\log 2(|E| + |D| + 1)\rceil$ for all $d \in D$. Then with probability at least $1 - \frac{|E|+|D|}{2(|E|+|D|+1)}$ R-FLOW(ϵ) finds an integer multicommodity flow F subject to c such that for all $d \in D$*

$$f(d) \geq \frac{1}{2}r(d).$$

□

3 Algorithmic Angluin-Valiant Inequality and Derandomization

In this section we give a derandomized version of R-FLOW(ϵ). The fundamental inequalities of Hoeffding [5] and Angluin-Valiant[2] gives remarkable tight bounds on the tail of the distribution of the weighted sum of Bernoulli trials. These inequalities prove the existence of certain structures, but does not supply an efficient way of finding such structures, which is the main problem in the theory of derandomization. In his work on approximate packing integer programs Raghavan [12] was able to derive an alogrithmic version of the Angluin-Valiant inequality for *unweighted* sums of Bernoulli trials. The problem in the weighted case remained open, because there the computation of the conditional probabilities under consideration requires the computation of the exponential function (see [12], pg. 138).

In the following we show that this is not necessary and establish an algorithmic version also in the weighted case. This constitutes the essential tool to derandomize R-FLOW(ϵ). We omit the technically difficult proof and refer to [15]

Let $X_1, \ldots, X_n$ be $0-1$ random variables defined through $Prob(X_i = 1) = \tilde{x}_i$ and $Prob(X_i = 0) = 1 - \tilde{x}_i$ for some rational $0 \leq \tilde{x}_i \leq 1$. Let w_{ij} be rational non-negative weights, $1 \leq i \leq m$, $1 \leq j \leq n$, $0 \leq w_{ij} \leq 1$ and denote by ψ_i the random variables

$$\psi_i = \sum_{j=1}^{n} w_{ij} X_j$$

Let $p_i = \frac{\mathbb{E}(\psi_i)}{n}$ and $q_i = 1 - p_i$ and let $0 \leq \beta_i \leq 1$ be a rational number for each $1 \leq i \leq m$. Denote by $E_i^{(+)}$ the event

$$\text{``}\psi_i \leq (1 + \beta_i)\mathbb{E}(\psi_i)\text{''}$$

and by $E_i^{(-)}$ the event

$$\text{``}\psi_i \geq (1 - \beta_i)\mathbb{E}(\psi_i)\text{''}$$

Furthermore let

$$E = E_1 \wedge \ldots \wedge E_m,$$

where E_i is either $E_i^{(+)}$ or $E_i^{(-)}$. For each i, $(i = 1, \ldots, m)$ let δ_i be a rational number $0 < \delta_i \leq 1$ with the property: If E_i is the event "$\psi_i \geq (1 + \beta_i)\mathbb{E}(\psi_i)$" then

$$\exp\left(-\frac{\beta_i^2 \mathbb{E}(\psi_i)}{3}\right) \leq \delta_i$$

and if E_i is the event "$\psi_i \leq (1 - \beta_i)\mathbb{E}(\psi_i)$" then

$$\exp\left(-\frac{\beta_i^2 \mathbb{E}(\psi_i)}{2}\right) \leq \delta_i$$

Hence by the Angluin-Valiant inequality (Lemma 2.1) the event E hold with probability at least $1 - \sum \delta_i$. The basic problem we analyse is to find a $0-1$ vector $x \in \{0,1\}^n$ in *deterministic polynomial-time*, for which the event E holds. This problem can be sovled in the RAM-model of computation by the following theorem. The essential idea of the proof is the use of low degree Taylor-polynomials of elementary functions, like exp, log, $\sqrt{}$ for the construction of a new class of pessimistic estimators. We have

Theorem 3.1 *Let $E = E_1 \wedge \ldots \wedge E_m$ be an event, where E_i denotes either $E_i^{(+)}$ or $E_i^{(-)}$ as defined above. Let $0 < \delta < 1$ be a rational number with $\delta + \sum_{i=1}^n \delta_i < 1$ and suppose that $\beta_i \leq \frac{n-1}{n}$ for all $i = 1, \ldots, m$. Then a vector $x \in \{0,1\}^n$ for which the event E holds can be constructed in $O\left(mn^2 \log \frac{mn}{\delta}\right)$-time.* $\square$

Let $m = |E| + |D|$ and let $L(\epsilon) = \max(L, \frac{1}{\epsilon^4} \log^2 m \log \frac{m}{\epsilon})$, where L is the encoding length of the integer programming formulation of the reduced demand multiflow problem. The deterministic counterpart to Theorem 2.4 then is

Theorem 3.2 *Let (G, H, c, r) be a multicommodity flow problem and let $0 < \epsilon \leq \frac{9}{10}$ with $c(e) \geq \frac{6(2-\epsilon)}{\epsilon^2}\lceil\log(2(|E|+|D|+1))\rceil$ for all $e \in E$ and $r(d) - K_R(d) \geq \frac{6(2-\epsilon)}{\epsilon^2}\lceil\log(2(|E|+|D|+1))\rceil$ for all $d \in D$. Then we can find in time $O\left(L(\epsilon)m^3\right)$ an integer multicommodity flow F subject to c such that for all $d \in D$*

$$f(d) \geq (1 - \epsilon)(r(d) - K_I(d)).$$

Proof. Let n denote the number of Bernoulli trials performed in the randomized rounding procedure. After having executed GENPATH and SPLITPATH(ϵ) n is fixed. Since we introduced for each $e \in E$, $d \in D$ at most $O\left(\frac{\log(|E|+|D|)}{\epsilon^2}\right)$ random variables, we have $n = O(\epsilon^{-2}|E||D|\log(|E| + |D|))$. Let $E = \{e_1, \ldots, e_{|E|}\}$ and $D = \{d_{|E|+1}, \ldots, d_{|E|+|D|}\}$. For $1 \leq i \leq |E|$ let denote E_i the event "$\chi(e_i) > \tilde{c}(e_i)$". For $|E| + 1 \leq i \leq |E| + |D|$ let β_i be a rational number with

$$\sqrt{\frac{3\lceil\log 2(|E| + |D| + 1)\rceil}{(2 - \epsilon)\tilde{r}(d_i)}} \leq \beta_i \leq \sqrt{\frac{6\lceil\log 2(|E| + |D| + 1)\rceil}{(2 - \epsilon)\tilde{r}(d_i)}}. \tag{11}$$

We will later see how to determine such a β_i efficiently. The event E_i, $|E| \leq i \leq |E| + |D|$, then is

$$\text{"}\chi(d_i) \leq (1 - \beta_i)(1 - \frac{\epsilon}{2})\tilde{r}(d_i)\text{"}.$$

Let $E := E_1 \wedge \ldots \wedge E_{|E|+|D|}$. By Theorem 2.4 $Prob(E^c) \leq \frac{1}{2}\frac{|E|+|D|}{|E|+|D|+1}$, so in order to apply Theorem 3.1 we choose $\delta = \frac{1}{2}\frac{|E|+|D|}{|E|+|D|+1}$, m and n as above. Assuming that $m \geq 10$ and using $\epsilon \leq \frac{9}{10}$, (1) and (11) it is not difficult to prove that $\beta_i \leq \frac{n-1}{n}$ for all $i = 1, \ldots, m$. According to Theorem 3.1 we can perform the rounding in time $O(mn^2 \log \frac{mn}{\delta}) = O\left(\frac{m^3}{\epsilon^4}\log^2 m \log \frac{m}{\epsilon}\right)$ and obtain for all $e_i \in E$ and $d_i \in D$

$$\chi(e_i) \leq \tilde{c}(e_i)$$

and

$$\chi(d_i) \geq (1 - \frac{\epsilon}{2})(1 - \beta_i)\tilde{r}(d_i).$$

We add to each fixed integer part of a path value its rounded part. As in the proof of Lemma 2.2 and 2.3 the integer multiflow F is subject to c and $f(d) \geq (1 - \epsilon)(r(d) - K_R(d))$ for all d.

The computation of β_i :

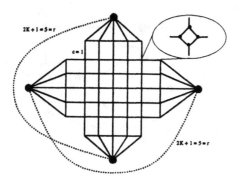

Figure 1: An example with $K = 2$

Choose $\gamma_i = \frac{1}{4}\sqrt{\frac{3\lceil\log 2(|E|+|D|+1)\rceil}{(2-\epsilon)\bar{r}(d_i)}}$. Then iteratively halving the interval $\left[0, \frac{3\lceil\log 2(|E|+|D|+1)\rceil}{(2-\epsilon)\bar{r}(d_i)}\right]$ we can find in $O(\log(\gamma_i^{-1}))$–time a rational β_i such that

$$0 \leq \beta_i - \sqrt{\frac{3\lceil\log 2(|E| + |D| + 1)\rceil}{(2 - \epsilon)\bar{r}(d_i)}} \leq \gamma_i$$

which implies (11). Since $O(\log(\gamma_i^{-1})) = O(\log \frac{m}{\epsilon})$ we are done. $\qquad\square$

Corollary 3.3 *Let* $0 < \epsilon \leq \frac{9}{10}$ *and let* (G, H, c, r) *be a multicommodity flow problem as above but with* $K_R = 0$. *Then we can find in polynomial-time an integer multiflow* F *subject to* c *such that for all* $d \in D$

$$f(d) \geq (1 - \epsilon)r(d).$$

$\qquad\square$

Since the multiflow problem can be fractionally solved in strongly polynomial time by Tardos' algorithm we have:

Corollary 3.4 *Let* (G, H, c, r) *be a fractionally solvable multicommodity flow problem with* $c(e), r(d) \geq 36\lceil\log 2(|E| + |D| + 1)\rceil$ *for all* $e \in E$ *and* $d \in D$. *Then in strongly polynomial time we can find an integer multicommodity flow* F *subject to* c *such that for all* $d \in D$

$$f(d) \geq \frac{1}{2}r(d).$$

$\qquad\square$

4 NP-completeness

Lemma 4.1 *[10] There is no fixed integer* $K \in I\!\!N$ *such that every fractional-solvable multicommodity flow problem posess an integer solution, when the requests are reduced by* K.

Proof. Assume that there is a fixed K, for which every fractionally solvable multicommodity flow problem has an integer solution, when each request is reduced by K. The idea of the proof is visualized by Figure 1:

The figure shows a grid-graph, where each grid-node is blown-up in the described way. Here we have 2 commodities each requesting $r(d) = 2K + 1$. Obviously such many commodities can be delivered using fractional values. But note that for integer values only a request with total

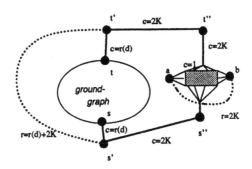

Figure 2: NP-completeness

sum of $2K + 1$ can be conveyed. So the integer multicommodity flow can only be solved, if the requests is reduced by at least $K + 1$. The method can be extended for arbitrary K and arbitrary many pairs of source-sink-pairs by copying the graph in Figure 1. □

Note that in this construction the supply-graph G is planar. This shows that the Korach/Penn result is only valid when the union of the supply graph and the demand graph is planar.

This construction can be used to prove the NP−completness of (1.1), even if the problem is fractionally solvable: We show by a reduction to the original integer multi-flow-problem:

Theorem 4.2 *The demand reduced multiflow problem with fractionally solvable inputs is NP-complete.*

Proof. Let $K \in \mathbb{N}$ be an integer and suppose we are given a fractionally solvable multicommodity-flow problem. For each source-sink-pair (s, t) construct an auxillary graph as shown in Figure 2. The new demand edge now is (s', t') requesting a value of $r(s, t) + 2K$. Connect s'' and t'' to the "grid" with $2K$ edges, all having capacity 1. Introduce a new demand edge (a, b) requesting a value of $2K$. It is clear, that the resulting graph remains ·fractional solvable. It is easy to see that the new graph has an integer solution, where the request is reduced by K if and only if the original problem has a integer solution: To saturate both reduced integer flows form a to b and from s'' to t'' we have to push one flow, say (a, b) through the grid and the other "around" the grid. So there is a flow of $\frac{K}{2}$ from s'' trough a to t'' and the other $\frac{K}{2}$ through b to t''. There is no way to convey any other flow than the original one through the ground graph. □

5 Conclusion and Open Problems

1. We have presented a deterministic approximation algorithm, in particular a $\frac{1}{2}$-factor algorithm finding feasible *integer* multicommodity flows. The running time of our algorithm is dominated by the time needed to solve a LP. The analysis of the algorithm shows that we have to require $c, r = \Omega(\log(|E| + |D|))$. As the example in section 4 shows, there are fractional solvable problems, where the approximation factor is less than $\frac{1}{2}$. The interesting open problem is to give a $\frac{1}{2}$-factor approximation if $c, r = O(\log(|E| + |D|))$ or even $c, r = O(1)$.

2. Note that with similiar methods these results hold also for the integer maximum $0 − 1$ multiflow problem.

3. Better approximation results might be possible in special cases, for example for planar graphs. Here several question arises, for example, whether better Korach/Penn type results can be proved, for planar graphs with stronger cut-conditions.

4. The algorithmic version of the Angluin-Valiant inequality might be usefull to attack other packing-type problems, especially those with weights.

5. The non-intractability of approximation problems from the class MAX-SNP motivates the question, whether or not there is a *polynomial-time* approximation scheme for the integer multiflow problem.

References

[1] N. Alon, J. Spencer, P. Erdös; *The prababilistic method.* John Wiley & Sons, Inc. 1992.

[2] D. Angluin, L.G. Valiant: *Fast probabilistic algorithms for Hamiltonion circuits and matchings.* J. Computer and System Sciences, Vol. 18, (1979), 155–193.

[3] A. Frank: *Packing paths, circuits and cuts - a survey.* In Paths, Flows and VLSI-Layout, B. Korte, L. Lovasz, H.-J. Prömel, A. Schrijver (eds.), Springer Verlag, (1990), pp.47 - 97.

[4] M. Grötschel, L. Lovász, A. Schrijver; *Geometric algorithms and combinatorial optimization.* Springer-Verlag (1988)

[5] W. Hoeffding; *On the distribution of the number of success in independent trials.* Annals of Math. Stat. 27, (1956), 713-721.

[6] E. Korach, M. Penn; *Tight integral duality gap in the chinese postman problem* Technical report, Computer Science Department, Israel Institute of Technology, Haifa, Revised Version, December 1989.

[7] V. M. Malhotra, M. P. Kumar, S. N. Maheshwari; *An $O(|V|^2)$ Algorithm for Finding Maximum Flows in Networks.* Information Processing Letters 7 (1978), 277–278.

[8] C. McDiarmid; *On the Method of Bounded Differences.* Surveys in Combinatorics, 1989. J. Siemons, Ed.: London Math. Soc. Lectures Notes, Series 141, Cambridge University Press, Cambridge, England 1989.

[9] K. Mehlhorn; *Data structures and algorithms 1: Sorting and Searching.* Sringer-Verlag (1984)

[10] F. Pfeiffer; *Personal communication,* Bonn 1991.

[11] P. Raghavan, C. D. Thompson; *Randomized Rounding: A technique for provably good algorithms and algorithmic proofs.* Combinatorica 7 (4), (1987), 365-373.

[12] P. Raghavan; *Probabilistic construction of deterministic algorithms: Approximating packing integer programs.* Jour. of Computer and System Sciences 37, (1988), 130-143.

[13] A. Srivastav, P. Stangier; *The relationship between fractional and integral graph partitioning.* Working Paper; Research Institute of Discrete Mathematics, University of Bonn, (1992).

[14] A. Srivastav, P. Stangier; *Weighted fractional and integral k−matching in hypergraphs.* Working Paper; Research Institute of Discrete Mathematics, University of Bonn (1992).

[15] A. Srivastav; *An Algorithmic Version of the Chernoff-Hoeffding Inequality and New Apllications to Packing Integer Programs.* Working Paper; Research Institute of Discrete Mathematics, University of Bonn (1993).

A Fully Dynamic Data Structure for Reachability in Planar Digraphs*

Sairam Subramanian
Brown University

Abstract. In this paper we investigate the problem of maintaining all-pairs reachability information in a planar digraph G as it undergoes changes. We give a fully dynamic $O(n)$-space data structure to support an arbitrary sequence of operations that consist of adding new edges (or nodes), deleting some existing edge, and querying to find out if a given node v is reachable in G by a directed path from another node u.

We show that using our data structure a reachability query between two nodes u and v can be performed in $O(n^{2/3} \log n)$ time, where n is the number of nodes in G. Additions and deletions of edges and nodes can also be handled within the same time bounds. The time for deletion is worst-case while the time for edge-addition is amortized. This is the first fully dynamic algorithm for the planar reachability problem that uses only sublinear time for both queries and updates.

1 Introduction

Designing dynamic algorithms for graph problems is a challenging area of research, and has attracted much interest motivated by many important applications in network optimization, VLSI layout, and distributed computing. An algorithm for a given graph problem is said to be *dynamic* if it can maintain the solution to the problem as the graph undergoes changes. These changes could be additions or deletions of edges, or a change in the cost of some edge (if applicable). In such a setting an *update* denotes an incremental change to the input, and a *query* is a request for some information about the current solution. We expect the dynamic algorithm to handle both queries and updates quickly i.e. in time that is substantially less than it would take to solve the problem from scratch every time the input changes. We say that the algorithm is *fully dynamic* if it supports both additions and deletions of edges, while it is said to be *semi dynamic* if it supports only one of them. Unfortunately, due to these stringent requirements designing fully dynamic algorithms seems to be considerably harder than designing their sequential counterparts, and very few graph problems have fully dynamic solutions.

In this paper we investigate the problem of maintaining the all-pairs reachability information in planar directed graphs (*digraphs*). The reachability problem on a

* Research supported in part by a National Science Foundation Presidential Young Investigator Award CCR–9047466 with matching funds from IBM, by NSF research grant CCR–9007851, by Army Research Office grant DAAL03–91–G–0035, and by the Office of Naval Research and the Defense Advanced Research Projects Agency under contract N00014–91–J–4052 and ARPA order 8225. Email: ss@cs.brown.edu

directed graph G consists of answering queries of the type "Is there a directed path from u to v in G." The digraph reachability problem is a fundamental research problem and has both theoretical and practical applications. In the complexity-theoretic setting this problem is complete for the class NLOGSPACE, while on the practical side the related problem of computing the transitive closure of a digraph is an important subroutine for answering database queries.

Given a digraph G with n nodes and m edges, a reachability query between two nodes u and v can be easily answered optimally in $O(m)$ time by doing a depth-first search from u (to determine if there is a directed path from u to v). However, in the dynamic realm this problem is much less well-understood. The best previous dynamic results for general digraphs are semi-dynamic data structures with $O(n^2)$ space, constant query time, and $O(n)$ update time. Data structures that support insertions only are given in [2,8,15]. Data structures that are restricted to acyclic digraphs and support only deletions can be found in [2,9].

We can obtain better results if we restrict ourselves to specific classes of graphs. For series-parallel digraphs Italiano, Marchetti-Spaccamela, and Nanni [10] gave an $O(n)$-space data structure that handles both queries and updates in $O(\log n)$ time. For the class of planar st-graphs (acyclic planar digraphs with one source and one sink both of which are on the same face) Tamassia and Preparata [16] give an $O(n)$ space fully dynamic data structure that supports both updates and queries in $O(\log n)$ time. However, their algorithm is embedding specific i.e. only those changes that are consistent with the current embedding of the graph are allowed. Tamassia and Tollis [17] have extended the results of [16] to get a fully dynamic data structure for answering reachability queries in spherical st-graphs and planar digraphs with a single source and a single sink. Here too, the updates are restricted to be embedding specific. To our knowledge there is no previous work that handles general planar digraphs even if updates are restricted to be embedding specific.

The results in [16] and [17] are based on the well known result in dimension theory which (in graph theoretic terms) states that: "Given an n-node planar directed acyclic graph G with one source and one sink, we can find two orderings $<_L$ and $<_R$ of the nodes in G such that a node v is reachable from another node u in G if and only if v occurs after u in both the orderings [1,11]." Unfortunately it seems unlikely that these methods can be extended to cover the entire class of planar digraphs, since Kelly [12] has shown that for general planar digraphs, we need arbitrarily large number of orderings to represent all-pairs reachability information as an intersection of these orderings.

In this paper we give a fully dynamic data structure to maintain the all-pairs reachability information in planar digraphs. Unlike previous algorithms our algorithm handles the entire class of planar digraphs and at the same time allows all updates as long as the graph still remains planar. The time bounds for maintaining the all-pairs reachability are as follows:

Theorem 1. *Given an n-node planar directed graph G, a fully dynamic $O(n)$ space data structure that maintains the all-pairs reachability information in G can be constructed in $O(n \log n)$ time. Any reachability query in G can be answered in $O(n^{2/3} \log n)$ time using this data structure. Updates to the underlying digraph G, like addition* *and deletion of edges and nodes can also be handled within the same time bounds. The time required to modify the data structure for delete operations is worst-case while the time for additions is amortized.*

We do not use the earlier approach based on dimension theory but instead use a clustering technique based on the ideas of Frederickson [4]. The basic idea is to divide the graph G into a number of clusters and use partial solutions for these clusters to derive the solution for the entire graph. This clustering approach was first used in the context of planar graphs by Frederickson [5] for computing single-source shortest paths. Galil and Italiano [6] used this approach in the dynamic setting to derive a fully dynamic data structure for maintaining two and three-vertex connectivity information in planar graphs. Similar ideas were later used by Galil, Italiano, and Sarnak [7] to develop a fully dynamic planarity testing algorithm. In our paper we will use a formulation similar to the one in [7].

To get our results we use the clustering idea along with a novel technique for compactly representing the all-pairs reachability information between a set N of *selected* nodes by a sparse substitute graph that has the following properties:

- Each edge uv in the substitute graph "corresponds" to a directed path π from u to v in G.
- For any pair of selected nodes u, v such that v is reachable from u in G, there exists a directed path from u to v in the substitute graph.

The size of the substitute graph depends upon the distribution of selected nodes in G. In particular, given an n-node directed planar graph G, we have the following bounds on the size of the sparse substitute:

Theorem 2 Face-reachability sparse substitute. *If there are $k = O(\sqrt{n})$ selected nodes, all on the boundaries of a constant number of faces of G, there is a sparse substitute graph having $O(k \log k)$ edges that represents the all-pairs reachability information between the selected nodes. Furthermore, the substitute graph can be constructed in $O(n \log n)$ time.*

Note that the size of the substitute graph is independent of the size of G.

Sparse substitute graphs have been used earlier in [6], [7], and [3] for maintaining various properties in undirected planar graphs. However, ours is the first such substitute for directed planar graphs. Also, in devising our substitute we need to concern ourselves with the distribution of the selected nodes unlike in the previous cases. This implies that the partitioning of G into clusters has to be done more carefully than was the case before. We have used a similar cluster partitioning technique along with a different substitute for constructing a fully dynamic data structure for all-pairs shortest-paths in undirected planar graphs [13].

The remainder of the paper is organized as follows: In section 2 we describe our algorithm for constructing a face-reachability substitute. In section 3 we show how to use our substitute along with the clustering formulation of [7] to develop a fully dynamic data structure for maintaining all-pairs reachability in planar digraphs, and in section 4 we discuss our conclusions and some open problems.

Preliminaries: Throughout the paper, G denotes an n-node planar directed graph. Unless otherwise specified a *path* from u to v in G refers to a directed path from u to v. On the other hand a *face* or a *cycle* refers to a face or a cycle respectively in the underlying undirected graph. A node v is said to be reachable from u in G if there is a path from u to v in G. Given a weight function on the nodes edges and faces of G such that the weights sum up to one, a cycle separator of G is a cycle C in G with $O(\sqrt{n})$ nodes such that the removal of the nodes on C leaves no connected (in the undirected sense) component with weight more than $2/3$. Such a separator is guaranteed to exist if G is a two-connected triangulated graph [14].

2 Constructing a face-reachability sparse substitute

Let G be a planar directed graph, and let N be a set of $k = O(\sqrt{n})$ *selected* nodes in G, distributed over a constant number of faces. In this subsection we show how to efficiently construct a substitute graph that represents the all-pairs reachability information among the selected nodes. The substitute graph we construct has $O(k \log k)$ edges and is called a *face-reachability sparse substitute*. In section 3 we will show how to use the face-reachability substitute along with the clustering technique of Galil and Italiano [3,6,7] to construct a fully dynamic data structure for dynamically maintaining all-pairs reachability information in G. Here we show how to construct a face-reachability substitute when all the selected nodes lie on the boundary of a single face f. The case of multiple faces is similar.

To construct the substitute we follow a simple divide-and-conquer approach. We first divide the selected node set N into two roughly equal halves M and O (see Figure 1). We then find a sparse substitute that represents reachability information from nodes in M to nodes in O and vice-versa. Sparse substitutes for all-pairs reachability among nodes in M (respectively nodes in O) are found recursively. The three substitutes are then unioned together to get the final face-reachability substitute. We now give a detailed description of this procedure:

1. **Finding separating paths:** We use a procedure called **separate** that gives a set Z of one or two paths in G that divides the selected nodes in N into two sets M and O such that neither of them has more than three-fourth of the nodes in N and every directed path from M to O (or from O to M) intersects one of the paths in Z (see Figure 1). Procedure **separate** will be described later.

2. **Sparsifying intersecting paths:** To sparsely represent reachability information between the nodes of M and O we proceed in the following manner: For each path π in Z we perform the following steps:

   (a) We first number the nodes on π in increasing order starting from the source.

   (b) For each node $u \in N$ we then find the lowest numbered node x on π such that there is a directed path from u to x in G. We label the node x to be a *marked node* and call it the *lowest attachment-point* of the node u. To find the lowest attachment-points of all the nodes in N we proceed as follows: We first add a temporary auxiliary node s and construct directed edges from all the nodes on π to s. We then perform an ordered depth-first search from s, exploring the search-tree of a lower numbered node before looking at a higher

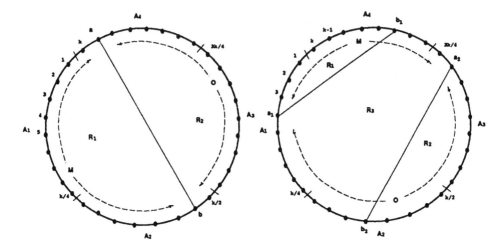

Fig. 1. Using one or two paths to cut the boundary

numbered node, to find all the nodes in N that can reach s. Note that in the depth-first search we only use incoming edges at each node because we are interested in paths that come into π.

(c) As in the previous step, for each node u in N we find the highest point y on π such that there is a path from y to u in G. As before, we label the node y to be marked and call it the *highest attachment-point* of u. This computation can also be done by conducting a depth-first search in a manner analogous to the previous step.

(d) To sparsely represent the paths that cross π we construct a substitute sub_π. The substitute sub_π consists of a *spine* π' that is formed by removing all the unmarked nodes of π and adding directed edges between consecutive marked nodes, directed from the lower numbered to the higher numbered node. Each selected node u is attached to the spine by edges to its lowest and highest attachment-points. We add a directed edge from u to its lowest attachment-point x, and an edge to u from its highest attachment-point y. See Figure 2 for an example.

The sparse substitute sub_1 (called the *crossing substitute*) formed by the union of the sub_π for all the π in Z represents reachability between nodes in M and O.

3. **Dividing G for the recursion:** For each node x in G we determine whether there is a directed path between two nodes u and v in M (respectively in O) that passes through x and does not intersect any separating path in Z. If there is such a path then we place x in V_1 (respectively V_2). We construct G_1 and G_2 by taking node induced subgraphs of V_1 and V_2 respectively. It is easy to see that no node is in both V_1 and V_2 at the same time because that would imply that there is a directed path from some node in M to a node in O that does not intersect any separating path, which is a contradiction.

To find the nodes of G that go into V_1 (respectively V_2) do the following: Combine all the nodes of M (O) into a single auxiliary node s and find the nodes of G that can both reach s and are reachable from s in $G - Z$ (where $G - Z$ is G with the nodes on the separating paths removed).

4. **Recursive computation:** Recursively compute sparse substitutes for all-pairs reachability among the nodes of M in G_1 and the nodes of O in G_2. Union the two recursive substitutes along with the crossing substitute sub_1 to get the face-reachability substitute for all the selected nodes.

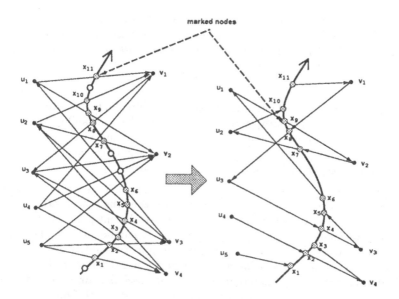

Fig. 2. Sparsifying the paths that cross a separating path

Before analyzing our algorithm for sparsification, we describe procedure **separate:** Consider a division of N into four subsets A_1, A_2, A_3, and A_4 as shown in Figure 1. If there is a path between nodes in A_1 and A_3 or between nodes in A_2 and A_4 then we can use it to divide N into subsets M and O as shown in Figure 1. Note that the nodes of M are topologically separated from those of O by the separator. Thus every path from M to O (and from O to M) crosses the separating path.

When there is no single path that divides N into roughly equal subsets we can use two paths as shown in Figure 1. The end points a_1, b_1 of path π_1 and a_2, b_2 of path π_2 are picked to satisfy the following properties: Node a_1 is in A_1 while node a_2 is in A_3; b_1 and b_2 occur after a_1 and a_2 respectively in the cyclic order around f; and the separation between a_i and b_i, $i = 1, 2$ is the maximum possible (in terms of their placement on the boundary of f) under the previous constraints. In other words there is no path in G between nodes a_1' and b_1' such that $a_1' \in A_1$, b_1' occurs after a_1' in the cyclic ordering, and the separation between a_1' and b_1' is more than that between a_1 and b_1. A similar condition holds for the nodes a_2 and b_2.

As shown in Figure 1 the exterior of face f is topologically separated into three regions R_1, R_2, and R_3 by the two paths π_1 and π_2. Any path between M and O that has one of its end points in R_1 or R_2 is forced to cross one of the separating paths. Also, by the maximality of the separation between a_i and b_i there is no path from M to O that lies entirely in R_3.

It is easy to see that the steps outlined above can be carried out in $O(n + k^2)$ time. Since $k = O(\sqrt{n})$ we have the following lemma:

Lemma 3. *Procedure* separate *constructs a separating set Z that divides N into sets M and O such that neither of them has more than three-fourths of the nodes. Also, every path in P from M to O intersects some path in Z. Moreover, procedure* separate *runs in $O(n)$ time.*

To prove the correctness of our sparsification procedure, all we need to do is prove that the crossing substitute sub_1 correctly represents the reachability information due to paths that go from the nodes in M to those in O and vice-versa. Consider any path ρ from some node u in M (or O) to a node v in O (or M). We now show that there is a corresponding path ρ' in sub_1 from u to v.

By lemma 3 we know that ρ crosses some path in Z. Let us suppose it crosses the path $\pi \in Z$ and the intersection point is p. By construction of sub_π we have edges ux and yv in sub_π where x is the lowest attachment-point of u and y is the highest attachment-point of v. By definition x is numbered no higher than p and y is numbered no lower than p. Furthermore, y is reachable from x in sub_π, since x occurs before y on the spine π'. Therefore, v is reachable from u in sub_π (and therefore in sub_1) by the alternate path ρ' that uses the edges ux and yv and the subpath from x to y in π'. For example in Figure 2 the path from the selected node u_3 to the selected node v_2 is replaced by an alternate path that detours via the marked nodes x_4 and x_9. Note that if v is reachable from u in sub_1 it is also reachable from u in G because every edge in sub_1 corresponds to a path in G.

To bound the size of sub_1 we note that there can be at most $O(k)$ marked nodes since there are at most two paths in Z and each selected node marks no more than two nodes on any path in Z. This also implies that the size of the compressed path π' (for any π in Z) is at most $2k$. Furthermore, a selected node is attached, via directed edges, to at most four nodes in sub_1. Therefore, the total number of nodes and edges in sub_1 is $O(k)$ (see Figure 2). By lemma 3 the number of selected nodes goes down by a constant factor at every stage in the recursion. Therefore, the size of the face-reaching substitute sub is $O(k \log k)$.

To bound the running time we note that the construction of sub_1 involves a call to procedure separate and a constant number of depth-first search operations. The total time needed is therefore the sum of the time needed for the recursive computation and the time needed by procedure separate, which is $O(n)$ by lemma 3. Since the depth of the recursion is proportional to $\log k$ it is easy to see that the total time required by our sparsification procedure is $O(n \log n)$. We thus get the bounds of Theorem 2.

3 A fully dynamic data structure for reachability in planar digraphs

In this section we design a fully dynamic data structure for maintaining the all-pairs reachability in planar digraphs. Our dynamic data structure supports the following operations:

1. is-reachable(u, v): Returns true if the v is reachable from u in G.
2. update($u, v, request$): Results in the addition or deletion of a directed edge between u and v, depending on whether the variable $request$ is set to add or delete.

We use the clustering technique of Galil, Italiano, and Sarnak [6,7] along with the face-reachability sparse substitutes from section 2 to design our data structure. This technique is based upon Frederickson's [4] idea of dividing the graph G into a number of clusters and using the partial solutions for different clusters to derive the solution for the entire graph. Galil and Italiano [6] used this technique to derive a fully dynamic data structure for maintaining two and three-vertex connectivity information in a planar graph. The same technique was later used by Galil, Italiano, and Sarnak [7] to develop a fully dynamic planarity testing algorithm. We will use a formulation similar to the one in [7]. Throughout this section we will assume that all the edge additions are planarity preserving. This can be achieved within the same time bounds by running a fully dynamic planarity testing algorithm [3,7] in the background and allowing only those edge additions that are planarity preserving.

To build our initial data structure we first two-connect and triangulate G by adding dummy nodes and edges. This is done by placing a zero-weight dummy node in each face and connecting it by dummy edges to all the nodes on the boundary of teh face. By applying Miller's algorithm to this graph, we obtain a cycle separator. The set X of non-dummy nodes of the separator need not form a cycle in the original graph. However, it does divide the graph into two pieces such that, in the subgraph G' induced on one piece together with the nodes of X, the nodes of X all lie on the boundary of a single face. These dummy nodes and edges are transient and play no part in the reachability queries. They are only used to partition G into clusters of suitable size.

We now discuss the cluster partition of G that we will use for our data structure.

Definition 4. An $O(k)$ *cluster partition* of an n-node planar graph G is a partitioning of G into k edge-induced subgraphs $G_1, G_2, \cdots, G_j$ (where $j \leq ck$ for some constant c) such that the following properties hold:

1. Each subgraph G_i contains no more than n/k edges.
2. The number of *boundary nodes* in G_i for any i is $O(\sqrt{n/k})$, where a node is a boundary node if it is in more than one subgraph. A node u is a boundary node if and only if edges of more than one cluster are incident on it. Note that u can be a boundary node in G_i only if it is an end point of some edge in G_i.
3. In each subgraph G_i the boundary nodes are present on a constant number of faces (for a suitable constant f). A face of G_i is either a face in G or is a cycle C in G_i, possibly containing some dummy edges, that topologically separates G_i from rest of G.

Lemma 5. *Given an n-node planar graph G an O(k) cluster partition can be obtained in O(n log n) time.*

Proof Sketch: A $O(k)$ cluster partition can be obtained starting with G and repeatedly dividing it using a planar cycle separator [14] until all the pieces have at most n/k edges. We retriangulate subgraphs when faces get too big so that we are guaranteed to have small cycle separators. As mentioned before, these dummy edges are used only to get cycle separators and play no role in the actual data structure. Using techniques due to Frederickson [5] we can also make sure that none of the pieces have too many boundary nodes, and in each G_i the boundary nodes are on a constant number of faces. □

Definition 6. Consider an $O(k)$ cluster partition of G. In such a partition we define the *parent* of an edge uv (denoted by G_{uv}) to be the subgraph G_i that contains it. Similarly, if u is a non-boundary node we define its parent G_u to be the subgraph G_i that contains it. If u is a boundary node we define its parent G_u to be one of the subgraphs G_i that contains it.

The preprocessing step finds an $O(k)$ cluster partition of G, and precomputes a face-reachability substitute $\hat{G}_i$ that represents the boundary-to-boundary reachability information in G_i for $1 \le i \le j$. The substitute graphs $\hat{G}_1, \hat{G}_2, \cdots, \hat{G}_j$ are unioned to form a skeletal graph S.

In our data structure we set k to be $n^{1/3}$. Therefore, in the cluster partition we have $O(n^{1/3})$ clusters each of which has no more than $n^{2/3}$ edges. Furthermore, each cluster has $O(n^{1/3})$ boundary nodes that are distributed on at most f faces. Therefore, by Theorem 2 the size of the skeletal graph $S = f \times O(n^{2/3} \log n) = O(n^{2/3} \log n)$. It also follows from lemma 5 and Theorem 2 that the total amount of time needed for preprocessing is $O(n \log n)$.

The skeletal graph S is a compact representation for the reachability information between the boundary nodes and is used in the query-stage to determine the existence of a path between the query points. To perform the query is-reachable(u, v) we form an auxiliary graph H by unioning the parent regions G_u and G_v of u and v respectively along with the skeletal graph S. We then perform a depth-first search in H from u to see if v is reachable from u in G.

To prove the correctness of our query-procedure, we need to show that v is reachable from u in H if and only if it is reachable from u in G. By construction, if v is reachable from u in H then it is also reachable in G since every edge in H corresponds to some path in G. Suppose v is reachable from u in G; we now show that v is reachable from u in H as well. Let ρ be a path that goes from u to v in G. We show that there exists an alternate path ρ' in H that goes from u to v.

To construct ρ' we mark all the boundary nodes in ρ. This marking divides ρ into a sequence of subpaths such that the first and the last subpaths lie entirely within G_u and G_v respectively and each intermediate subpath is a boundary-to-boundary path that lies entirely within one of the subgraphs G_1 through G_j. By the construction of our face-reachability substitutes any path γ that goes from a boundary node b_1 to another boundary node b_2 and lies entirely inside some region G_i can be replaced by an alternate path γ' from b_1 to b_2 that is in $\hat{G}_i$. Therefore, the path constructed by taking the first and the last piece of ρ along with the alternate paths γ' from each region gives us an alternate path ρ' from u and v that is in H.

To bound the time taken for performing the query we note that H is formed by unioning S and two regions with at most $n^{2/3}$ edges. Therefore, the size of H is $O(n^{2/3} \log n)$. Hence, doing a depth-first search from u in H requires $O(n^{2/3} \log n)$ steps.

We now describe procedure **update**: The basic idea is to recompute the face-reachability substitutes for all the regions that are affected by the update operation. The crux of the proof lies in showing that only a constant number of regions are affected. We discuss the case of an add operation, the delete operation is similar. Consider adding the edge uv: This results in a change in both G_u and G_v since now both u and v become (if they were not already) boundary nodes. Furthermore, the addition could result in changing the distribution of boundary nodes in G_u, G_v, or both. If in any of them more than f faces now contain boundary nodes, we split that region into smaller regions where boundary nodes are distributed over no more than f faces. However, it is clear that doing this would still leave us with at most a constant number of regions. We can therefore, recompute the face-reachability substitutes for all these regions in $O(n^{2/3} \log n)$ time.

A careful look at our **update** algorithm shows that repeated requests for adding edges can result in an excess of boundary nodes, since every time an edge uv is added either u or v or both may become boundary nodes. Therefore, we recompute the cluster partition after every $n^{1/3}$ add operations.

To bound the time taken for each update operation we note that during the entire execution of the algorithm the size of G_i (for any i) is $O(n^{2/3})$. Furthermore, G_i has $O(n^{1/3})$ boundary nodes which are distributed over at most f faces. Thus, the time required to construct the face-reachability substitute for G_i is $O(n^{2/3} \log n)$. This in conjunction with the fact that only a constant number of the subgraphs are affected during any update operation gives us our bound.

To amortize the rebuild operation over *adds* we note that the cluster partitioning and the rebuilding of all the face-boundary substitutes can be done in $O(n \log n)$ time. This coupled with the fact that a rebuild is done once every $n^{1/3}$ *adds* gives us the bounds of Theorem 1.

4 Conclusions and Open Problems

In this paper we have provided the first fully dynamic data structure for maintaining all-pairs reachability information in a planar graph. It would be interesting to see if these techniques can be extended to the problem of maintaining reachability information in bounded genus graphs.

Recently Eppstein, Galil, Italiano, and Spencer [3] gave a fully dynamic data structure for maintaining a minimum spanning tree (among other things) in a planar undirected graph by using sparse substitutes. They were able to use the idea of sparsification to recursively divide the sparse substitutes repeatedly to get smaller and smaller substitutes thus deriving a fully dynamic data structure with $O(\log^2 n)$ time for both queries and updates. For performing such repeated sparsification they used two properties of their substitutes: (1) The substitute graph is planar. and (2) given the sparse substitutes for the subgraphs of G one can construct the sparse substitute for G without looking at G. Unfortunately neither of these properties are obeyed by our substitutes. We conjecture that such substitutes do not exist for planar digraphs. We leave the proof of our conjecture as an open problem.

Acknowledgments

I would like to thank Philip Klein, Bharathi Subramanian, Roberto Tamassia, and Jeff Vitter for useful discussions.

References

[1] G. Birkhoff, "Lattice Theory," *American Mathematical Society Colloquium Publications* 25 (1979).

[2] A.L. Buchsbaum, P.C. Kanellakis, and J.S. Vitter, "A Data Structure for Arc Insertion and Regular Path Finding," *Proc. ACM-SIAM Symp. on Discrete Algorithms* (1990), 22–31.

[3] D. Eppstein, Z. Galil, G.F. Italiano, and T. Spencer, "Separator Based Sparsification for Dynamic Planar Graph Algorithms," *Proc. 25th Annual ACM Symposium on Theory of Computing* (1993).

[4] G.N. Frederickson, "Data Structures for On-Line Updating of Minimum Spanning Trees, with Applications," *SIAM J. Computing* 14 (1985), 781–798.

[5] G.N. Frederickson, "Fast Algorithms for Shortest Paths in Planar Graphs, with Applications," *SIAM Journal on Computing* 16 (1987), 1004–1022.

[6] Z. Galil and G. F. Italiano, "Maintaining biconnected components of dynamic planar graphs," *Proc. 18th Int. Colloquium on Automata, Languages, and Programming.* (1991), 339–350.

[7] Z. Galil, G.F. Italiano, and N. Sarnak, "Fully Dynamic Planarity Testing," *Proc. 24th Annual ACM Symposium on Theory of Computing* (1992), 495–506.

[8] G.F. Italiano, "Amortized Efficiency of a Path Retrieval Data Structure," *Theoretical Computer Science* 48 (1986), 273–281.

[9] G.F. Italiano, "Finding Paths and Deleting Edges in Directed Acyclic Graphs," *Information Processing Letters* 28 (1988), 5–11.

[10] G.F. Italiano, A. Marchetti-Spaccamela, and U. Nanni, "Dynamic Data Structures for Series-Parallel Graphs," *Proc. WADS' 89*, LNCS 382 (1989), 352–372.

[11] T. Kameda, "On the Vector Representation of the Reachability in Planar Directed Graphs," *Information Processing Letters* 3 (1975), 75–77.

[12] D. Kelly, "On the Dimension of Partially Ordered Sets," *Discrete Mathematics* 35 (1981), 135–156.

[13] P. N. Klein and S. Subramanian, "A Fully Dynamic Approximation Scheme for All-Pairs Shortest Paths in Planar Graphs," *Proc. (to appear) 1993 Workshop on Algorithms and Data Structures* (1993).

[14] G. Miller, "Finding Small Simple Cycle Separators for 2-Connected Planar Graphs," *Journal of Computer and System Sciences* 32 (1986), 265–279.

[15] J.A. La Poutré and J. van Leeuwen, "Maintenance of Transitive Closures and Transitive Reductions of Graphs," *Proc. WG '87*, LNCS 314 (1988), 106–120.

[16] R. Tamassia and F.P. Preparata, "Dynamic Maintenance of Planar Digraphs, with Applications," *Algorithmica* 5 (1990), 509–527.

[17] R. Tamassia and I.G. Tollis, "Dynamic Reachability in Planar Digraphs," *Theoretical Computer Science* (1993), (to appear).

A Linear-Time Algorithm for Edge-Disjoint Paths in Planar Graphs*

Dorothea Wagner, Karsten Weihe

Fachbereich Mathematik, Technische Universität Berlin, Straße des 17. Juni 136, D-10623
Berlin, Germany

Abstract. In this paper we discuss the problem of finding edge-disjoint
paths in a planar, undirected graph s.t. each path connects two specified
vertices on the outer face boundary. We will focus on the "classical" case
where an instance must additionally fulfill the so-called *evenness-condition*.
The fastest algorithm for this problem known from the literature requires
$\mathcal{O}(n^{5/3}(\log \log n)^{1/3})$ time, where n denotes the number of vertices. In this
paper now, we introduce a new approach to this problem, which yields an
$\mathcal{O}(n)$ algorithm.

1 Introduction

Throughout this paper, we assume $G = (V, E)$ to be an undirected, connected, planar
graph with a fixed combinatorial embedding, i.e., the adjacency list of any vertex
is sorted according to an embedding in the plane. A *net* $\{s, t\}$ is a pair of vertices
$s, t \in V, s \neq t$, such that both s and t, the *terminals* of net $\{s, t\}$, are incident to the
outer face. An instance of the problem we consider in this paper is a pair (G, N),
where G is a graph as specified above, and $N = \{\{s_1, t_1\}, \ldots, \{s_k, t_k\}\}$ is a set of
nets. Additionally, the so-called *evenness-condition* is fulfilled, i.e., $(V, E + \{s_1, t_1\} + \cdots + \{s_k, t_k\})$ is Eulerian. The problem is to decide whether there are edge-disjoint
paths $p_1, \ldots, p_k$, such that p_i connects s_i with t_i for $i = 1, \ldots, k$, and, if yes, to
determine such a set of paths.

The asymptotically fastest algorithm known so far is due to Kaufmann and Klär
and requires $\mathcal{O}(n^{5/3}(\log \log n)^{1/3})$ time [KK], where $n = |V|$. This algorithm relies
on an $\mathcal{O}(n^2)$-algorithm introduced by Becker and Mehlhorn [BM] and uses decomposition techniques of Frederickson [Fr]. Becker and Mehlhorn also considered a certain
generalization, which arises in VLSI design. There, only the *weak evenness-condition*
must be fulfilled, that is, the vertices that are not incident to the outer face have
even degree. They proposed an $\mathcal{O}(bn + T(n))$ algorithm, where b denotes the number of vertices on the outer face boundary and $T(n)$ the time required to solve any
instance that fulfills the evenness-condition. Clearly, in connection with our result,
this immediately yields an $\mathcal{O}(bn)$ algorithm for that more general problem. The relation to problems from VLSI-design is not surprising at all since any grid graph
without holes fulfills at least the weak evenness-condition. Another generalization
was considered by Hassin [Ha] and by Matsumoto, Nishizeki and Saito [MNS]. Note

* The authors acknowledge the *Deutsche Forschungsgemeinschaft* for supporting this research under grant *Mö 446/1-3*

that the problem of finding edge-disjoint paths between designated pairs of vertices can be seen as an integral multicommodity flow problem with unit capacities. They considered the case where the capacities may be arbitrary positive integer, and they require solutions only to be half-integral, which can be seen as the translation of the eveness-condition into their model.

All these algorithms, as well as many other algorithms tailored to special cases and similar problems (e.g. [Ka], [KM], [NSS]), make extensive use of a theoretical result of Okamura and Seymour concerning the relation of paths and cuts in a even instance [OS]. Obviously, for no *cut* $X \subseteq V$ in a solvable instance does its *density*, i.e. the number of nets having one terminal in X and the other one in $V \setminus X$, exceed its *capacity*, which is the number of edges between X and $V \setminus X$. Okamura and Seymour have shown that this — necessary — *cut condition* is also sufficient for solvability of instances that fulfill the evenness-condition (Theorem of Okamura & Seymour). Moreover, they have shown that one can focus on cuts X s.t. X and $V \setminus X$ are connected in G, and both X and $V \setminus X$ are incident to the outer face boundary. The main action of all algorithms mentioned above is to determine capacities and densities of such cuts *explicitly* and to make decisions basing on this knowledge. In contrast, our algorithm works now as follows. In a preprocessing phase, we determine certain *auxiliary paths*, which help us to determine the solution, i.e. paths $p_1, \ldots, p_k$, correctly. The main part of the algorithm consists in a loop, where in each iteration exactly one of the paths $p_1, \ldots, p_k$ is drawn. We do *not* determine values associated with cuts *explicitly* (not even cuts themselves in fact). For a linear-time realization, we will make use of a technique proposed by Gabow and Tarjan [GT], which will enable our algorithm to perform certain sequences of union-find operations in linear time. Since our algorithm is linear in the worst case, it is optimal. In particular, it improves on some results on special cases [Ka], [KM], and on even instances it is as fast as the algorithm in [NSS], which is in general applicable to weakly even instances with the additional restriction that the underlying graph forms a convex grid.

The algorithm also leads to a new proof of the Theorem of Okamura & Seymour. This proof, as well as some easy proofs which are omitted here, is contained in the full version of the paper [WW].

2 The Algorithm

In the sequel, we will assume that $N \neq \emptyset$, and that $x \in V$ is a fixed terminal, the *start terminal*. We assume that all terminals have degree 1 and all other vertices have even degree. Obviously, a simple modification transforms any instance into a completely equivalent instance that fulfills this assumption. Moreover, for convenience we assume that, according to a reverse clockwise ordering starting with x, s_i precedes t_i for $i = 1, \ldots, k$, and t_i precedes t_{i+1} for $i = 1, \ldots, k-1$. The ith terminal in reverse clockwise ordering, starting with x, is denoted by x_i. If $i < j$, we also write $x_i \prec x_j$.

Before we determine a solution for our input instance (G, N) itself, we will first consider another instance, which is denoted by $(G, N^{()})$. In $N^{()}$, again each s-terminal is paired with a t-terminal, but according to a (unique) "parenthesis structure". That is, consider a $2k$-string with a left parenthesis at the ith position, if x_i is an s-terminal, and a right parenthesis otherwise. Then two terminals are paired if and only if the corresponding parentheses match. Now the following fact is easy to see.

Observation 1 *Let $N' = \{\{s_1', t_1'\}, \ldots, \{s_k', t_k'\}\}$ be a pairing of the s-terminals with the t-terminals such that $s_i' \prec t_i'$ for all i. If $s_i' \prec t_i' \prec s_j' \prec t_j'$ or $s_i' \prec s_j' \prec t_j' \prec t_i'$ for all $i, j = 1, \ldots, k$ with $s_i' \prec s_j'$, then $N' = N^{()}$.*

Lemma 2. *$(G, N^{()})$ is solvable, if (G, N) is.*

2 determines such a solution $(q_1, \ldots, q_k)$ for $(G, N^{()})$. In contrast to the original nets, we denote the nets in $N^{()}$ by $\{s_1^{()}, t_1^{()}\}, \ldots, \{s_k^{()}, t_k^{()}\}$, and we assume w.l.o.g. that $t_i = t_i^{()}$ for $i = 1, \ldots, k$. In this procedure we will use for each path a *right-first search*, i.e., a depth-first search where in each step all possibilities of going forward are searched "from right to left". In principle, we proceed in the same way as the well-known *stack-algorithm* for a similar problem, where *vertex-disjoint* paths are to be drawn [SAN]. Let $v \in V$, and let e be incident to v. In the remainder of this paper, we will say that the *next edge after* e w.r.t. v is the first edge to follow e in the adjacency list of v in reverse clockwise ordering.

Procedure

FOR $i := 1$ TO k DO
    let q_i initially consist of the unique edge incident to $s_i^{()}$;
    $v :=$ the unique vertex adjacent to $s_i^{()}$;
    WHILE v is no terminal DO
        let $\{v, w\}$ be the next free edge after the leading edge of q_i w.r.t. v;
        add $\{v, w\}$ to q_i;
        $v := w$;
    IF $v \neq t_i^{()}$ THEN STOP: return "unsolvable";
return $(q_1, \ldots, q_k)$;

Note that the paths q_i are not cycle-free in general. We need some terminology. Let e_1 and e_2 be two edges of a path p from an s-terminal to a t-terminal. We say that e_2 is an (immediate) successor of e_1, if e_2 is passed (immediately) after e_1, when we walk along p from the s-terminal to the t-terminal. By the Theorem of Jordan, a path p that does not cross itself partitions the set of all vertices and edges not belonging to p into two parts, the *left* side and the *right* side of p. The left side consists of all vertices and edges that appear to the *left*, when we walk along p from the s-terminal to the t-terminal. The right side is defined analogously. The following fact is easy to see. Hence, its straightforward proof is omitted.

Lemma 3. *The auxiliary paths $q_1, \ldots, q_k$ constructed by the procedure neither cross themselves nor each other. In particular, the left and the right sides of all of them are well defined.*

Let p be a path from an s-terminal to a t-terminal. Then X_p denotes the set of all vertices on the right side of p, and $\delta(X_p)$ denotes the set of all edges between a vertex in X_p and a vertex in $V \setminus X_p$ (clearly, a vertex on p in fact).

Lemma 4. *If $(G, N^{()})$ is solvable, the procedure returns a solution for $(G, N^{()})$ and "unsolvable" otherwise.*

For determining the paths $p_1, \ldots, p_k$ for our original instance (G, N), we now construct a *directed* auxiliary graph from $q_1, \ldots, q_k$.

Definition 5. On the paths $q_1, \ldots, q_k$ orient all edges according to the direction in which they are traversed in the procedure. The oriented paths $q_1, \ldots, q_k$ are called the **auxiliary paths**. The **(directed) auxiliary graph** $A(G, N, x)$ of instance (G, N) w.r.t. start terminal x consists of all vertices of G and of all oriented edges on the corresponding auxiliary paths.

An instance (G, N) and its auxiliary graph w.r.t. start terminal $x = s_7$ are shown in Figure 1. As a consequence of the right-first search strategy we have the following corollary.

Corollary 6. *For any auxiliary path q_i the set $\delta(X_{q_i})$ is completely contained in $A(G, N, x)$.*

Let e and e' be two edges incident to non-terminal vertex v. Then the set of all edges appearing after e and before e' in the adjacency list of v in reverse clockwise ordering is denoted by $(e, e')_v$. If e, e', or both are to be included, we instead write $[e, e')_v$, $(e, e']_v$, and $[e, e']_v$, respectively. Let c be a directed, elementary cycle in $A(G, N, x)$. Analogously to paths from s-terminals to t-terminals, the left (right) side of c is the set of all vertices and edges appearing to the left (right), when we walk around c according to its direction. c is called a *right-cycle*, if the interior of c is just the right side of c. Obviously, no auxiliary path contains a right-cycle. Moreover, we can even prove the following, more general fact.

Lemma 7. *The auxiliary graph $A(G, N, x)$ does not contain a right-cycle.*

Proof. Let c be an elementary cycle of $A(G, N, x)$. Consider the stage of the procedure immediately before the first time at all an edge of c is occupied for any auxiliary path. Let v be the current vertex at this stage, and let e_1 be the leading edge of the current auxiliary path. Then e_1 belongs to the exterior of c. As c is elementary, there are only two edges e_2 and e_3, say, of c incident to v. W.l.o.g. assume $e_2 \in (e_1, e_3)_v$. Then e_2 is the immediate successor of e_1, which means that e_2 leaves v in the auxiliary graph. However, if c were a right-cycle in $A(G, N, x)$, e_3 must leave v, and e_2 must enter v.

After constructing the auxiliary graph, we now determine the paths $p_1, \ldots, p_k$ for our original instance (G, N). For each path p_i we use a modified right-first search, starting with s_i. "Modified" means that we use only edges that also belong to $A(G, N, x)$ for going forward, and only according to their orientations in $A(G, N, x)$. For convenience, we will not distinguish between paths in G fulfilling these properties and (directed) paths in $A(G, N, x)$. The adjacency list of v is a cyclic list representing *all* edges incident to v in $A(G, N, x)$, leaving v or entering v.

In Figure 2, the auxiliary graph of Figure 1 and the first path p_1 for the instance of Figure 1 are shown.

The Algorithm

determine $A(G, N, x)$ for an arbitrary start terminal x;
FOR $i := 1$ TO k DO
    let p_i initially consist of the unique edge leaving s_i in $A(G, N, x)$;
    $v :=$ the head of this edge;
    WHILE v is no terminal DO
        let (v, w) be the next free edge leaving v after the leading edge of p_i w.r.t. v;
        add (v, w) to p_i;
        $v := w$;
    IF $v \neq t_i$ THEN STOP: return "unsolvable";
return $(p_1, \ldots, p_k)$;

3 Correctness of the Algorithm

Since for any non-terminal vertex v in $A(G, N, x)$ the number of edges leaving v equals the number of edges entering v, there is always at least one free edge leaving the current vertex left for going forward. This means that the algorithm will actually return either a set of paths or the message "unsolvable". Clearly, since this set of paths is a solution for instance (G, N), we only have to show that the algorithm does not return "unsolvable", if (G, N) is solvable. We will show that, if (G, N) is solvable, the algorithm maintains three invariants through all iterations of the overall FOR-loop. To state these invariants, we denote by G_i, $i = 0, \ldots, k$, the directed graph constructed from G by removing all edges from G that belong to one of the paths $p_1, \ldots, p_i$. In particular, $G_0 = G$. Note that G_i may not be connected for $i > 0$. The first invariant means that algorithm is correct, the second that paths $p_1, \ldots, p_i$ partition G_i into solvable subinstances, and the third means that the procedure has already determined an auxiliary graph for each of these subinstances implicitly.

Invariants for $i = 1, \ldots, k$:

1. Path p_j, $j = 1 \ldots, i$, connects s_j with t_j.
2. All nets $\{s_j, t_j\}$, $j > i$, have both terminals in the same connected component C of G_i, and for each connected component C of G_i the even instance (C, N_C), where $N_C \subseteq N$ is the set of all nets having both terminals in C, is solvable by its own.
3. Let C be a connected component of G_i, and let x_C be the terminal in C s.t. $x_C \prec y$ for any other terminal y in C. (Recall the definition of the relation "$\prec$" for terminals in (G, N) at the beginning of Section 2.) Then the restriction of the auxiliary graph $A(G, N, x)$ to C equals $A(C, N_C, x_C)$, i.e. the auxiliary graph of instance (C, N_C) w.r.t. start terminal x_C.

Observation 8 *Let C be a connected component of G_i. Then the cyclic order of any three terminals on the outer face boundary of C equals their cyclic order on the outer face boundary of G itself.*

This observation means that the order of all terminals is "inherited" from (G, N) to all subinstances (C, N_C). In particular, let C be the connected component of G_i to which $\{s_{i+1}, t_{i+1}\}$ belongs. Then t_{i+1} is the first t-terminal to appear, when we walk along the outer face boundary of C, starting with x_C, because we have $x_C \prec t_{i+1} \prec t_l$ for all t-terminals $t_l \neq t_{i+1}$ in C. Hence, net $\{s_{i+1}, t_{i+1}\}$ is the net for which the algorithm determines a path in the first iteration, when it is applied to instance (C, N_C) with start terminal x_C. In connection with the third invariant, this implies that just path p_{i+1} is the path determined in the first iteration, when the algorithm is applied to (C, N_C) with start terminal x_C. As a result, the three invariants are maintained by the $(i + 1)^{st}$ iteration, if for any instance they are maintained by the first iteration. Consequently, in the following Subsections 3.1, 3.2, and 3.3, we will only show that the three invariants are maintained by the first iteration of the algorithm.

3.1 Proof of Invariant 1 for the First Iteration

Of course, p_1 ends with a t-terminal. We only have to show that this is the correct t-terminal t_1.

Lemma 9. *Path p_1 does not cross itself. In particular, the left side and the right side of p_1 are well defined.*

Lemma 10. *All edges in $\delta(X_{p_1})$ belong to the auxiliary graph and are directed from X_{p_1} into $V \setminus X_{p_1}$.*

Proof. Consider any edge $e \in \delta(X_{p_1})$. Let v be the vertex on p_1 incident to e. There is an edge e_1 entering v and an edge e_2 leaving v that belong to p_1 s.t. e_2 is the immediate successor of e_1 w.r.t. p_1 and $e \in (e_1, e_2)_v$. If e belongs to $A(G, N, x)$ and leaves v, e had been chosen as the immediate successor of e_1 w.r.t. p_1 rather than e_2.

Now assume that e does not belong to $A(G, N, x)$. All edges in $[e_1, e_2)_v$ which belong to $A(G, N, x)$ enter v (e_1 in particular). Hence, we may w.l.o.g. assume that e is the next edge (w.r.t. v) after some edge $e' \in [e_1, e_2)_v$ that belongs to $A(G, N, x)$ and enters v. However, in this case e had been chosen as the immediate successor of e' in the procedure.

Lemma 11. *An auxiliary path that connects an s-terminal in X_{p_1} with a t-terminal in $V \setminus X_{p_1}$ shares exactly one edge with $\delta(X_{p_1})$, and no other auxiliary path shares any edge with $\delta(X_{p_1})$. In particular, no s-terminal in $V \setminus X_{p_1}$ is connected with a t-terminal in X_{p_1}.*

Proof. If one of the former paths shares more than one edge with $\delta(X_{p_1})$ (of course, at least one) or one of the latter paths shares some edge with $\delta(X_{p_1})$, we must switch over from $V \setminus X_{p_1}$ to X_{p_1} at least once via an edge belonging to $\delta(X_{p_1})$, when we walk along this auxiliary path from its s-terminal to its t-terminal. That edge is directed from $V \setminus X_{p_1}$ into X_{p_1}. This contradicts Lemma 10.

Proof of Invariant 1 for the first iteration. Lemmas 10 and 11 together imply that $|\delta(X_{p_1})|$, which is just the capacity of cut X_{p_1}, equals the number of s-terminals in X_{p_1} *minus* the number of t-terminals in X_{p_1}. On the other hand, the number of s-terminals in X_{p_1} s.t. the corresponding t-terminal is in $V \setminus X_{p_1}$ is at least the number of s-terminals in X_{p_1} *minus* the number of t-terminals in X_{p_1}, too. If p_1 does not end with t_1, there is at least one t-terminal in X_{p_1} such that the corresponding s-terminal is in $V \setminus X_{p_1}$, namely t_1. Hence, in this case the density of cut X_{p_1} is at least $|\delta(X_{p_1})| + 1$, which means that instance (G, N) is unsolvable.

3.2 Proof of Invariant 2 for the First Iteration

In this subsection, we show the following theorem, which immediately implies Invariant 2 for the first iteration.

Theorem 12. *If (G, N) is solvable, there is a solution that realizes net $\{s_1, t_1\}$ by path p_1.*

Let P denote the set of all paths from s_1 to t_1 that do not cross themselves and belong to solutions for (G, N). Clearly, $P \neq \emptyset$, if (G, N) is solvable. In this subsection, we assume that for any $p \in P$ there is an edge on p_1 that does not belong to p, because otherwise Theorem 12 is obvious.

First we need some terminology. Let p, q be two paths from s_1 to t_1 that do not cross themselves. We say that p is *more right* than q, if the right side of p is a proper subset of the right side of q. This relation defines a partial order on the set of all paths from s_1 to t_1 that do not cross themselves. In particular, we will consider the suborder induced by P. We will show that the (unique) "rightmost" element of this suborder is just p_1. Clearly, this proves Theorem 12. More precisely, we will show the following two claims.

Claim 1 For any $p \in P$ we have either $p = p_1$, or p_1 is more right than p.

Claim 2 There is a $p \in P$ s.t. p_1 is not more right than p.

Proof of Claim 1. Let μ denote the number of terminals in X_{p_1}. Because of the first invariant, all those terminals are s-terminals. In the last subsection, we have shown that the capacity of X_{p_1} equals the number of s-terminals in X_{p_1} *minus* the number of t-terminals in X_{p_1}, which is just $\mu - 0$, i.e., the density of X_{p_1}. Therefore, $p \in P$ cannot share any edge with $\delta(X_{p_1})$. The claim thus follows.

Now we will show the following two claims, which clearly imply Claim 2. Recall that μ denotes the number of (s-)terminals in X_{p_1}. An element $p \in P$ is called *rightmost*, if there is no element of P more right than p.

Claim 2(a) If p_1 is more right than $p \in P$, then $|\delta(X_p)| > \mu$.

Claim 2(b) If $p \in P$ is rightmost in P, then $|\delta(X_p)| = \mu$.

Proof of Claim 2(a). First note that at least μ edges belong to $\delta(X_p)$ and are directed from X_p to $V \setminus X_p$, namely one edge of any auxiliary path whose s-terminal s_i fulfills $s_1 \prec s_i \prec t_1$. Moreover, recall our assumption that there is an edge of p_1 that does not belong to p. In particular, consider the first such edge in the direction from s_1 to t_1. Since p_1 is more right than p, this edge belongs to $\delta(X_p)$, too, but is directed from $V \setminus X_p$ to X_p. This implies $|\delta(X_p)| > \mu$.

Let S be a solution for (G, N) realizing net $\{s_1, t_1\}$ by a path p that is rightmost in P. Claim 2(b) follows immediately from the following two facts.

Claim 2(b)(i) No two edges of $\delta(X_p)$ are shared by the same path of $S \setminus \{p\}$.

Claim 2(b)(ii) Any edge of $\delta(X_p)$ belongs to a path of $S \setminus \{p\}$ that starts with an s-terminal in X_{p_1}.

Proof of Claim 2(b)(i). Let $e, e' \in \delta(X_p)$ belong to path $p' \in S \setminus \{p\}$, and let v and v' be the vertices on p that are incident to e and e', respectively. W.l.o.g. the subpath of p' with endedges e and e' belongs completely to the right side of p. If $v = v'$, this subpath of p' forms a cycle. Then we remove this cycle from p' and insert it into p. If $v \neq v'$, we exchange the subpaths of p and p' between v and v'. In both cases we obtain a solution where the path for net $\{s_1, t_1\}$ does not cross itself and is more right than p. (W.l.o.g. p' does not cross itself either.)

Proof of Claim 2(b)(ii). Of course, because of Claim 2(b)(i), no edge of $\delta(X_{p_1})$ belongs to an auxiliary path with the s-terminal in $V \setminus X_{p_1}$. So assume that $e \in \delta(X_p)$ does not belong to any of the paths of $S \setminus \{p\}$. As (G, N) fulfills the evenness condition, there is an elementary cycle c that contains e and consists only of edges not belonging to any of the paths of S. If c belongs completely to the right side of p, we simply insert c into p. Else consider a maximal subpath s of c on the right side of p. Then we replace the subpath of p between the endvertices of s with s itself. Again in both cases an element of P is constructed which is more right than p.

3.3 Proof of Invariant 3 for the First Iteration

Now we are going to show that the restriction of $A(G, N, x)$ to any connected component C of G_1 just equals $A(C, N_C, x_C)$. See Figure 3. Remember that $q_1, \ldots, q_k$ denote the (directed) auxiliary paths, in order of construction during the procedure. Path p_1 can be uniquely decomposed into maximal subpaths s.t. each subpath belongs completely to one auxiliary path. For completeness, whenever p_1 crosses an auxiliary path at some vertex, this vertex is also said to be such a maximal subpath of p_1. Let q_λ be the auxiliary path that starts with s_1.

Lemma 13. *Any edge of path p_1 belongs to one of the auxiliary paths $q_1, \ldots, q_\lambda$, and the set of all edges shared by p_1 and q_i, $i = 1, \ldots, \lambda$, is a subpath of both p_1 and q_i. For $i = 1, \ldots, \lambda - 1$, the edges shared by q_{i+1} appear immediately before the edges shared by q_i, if p_1 is traversed from s_1 to t_1.*

In particular, any path $q_i, i = 1, \ldots, \lambda$, decomposes into three subpaths: The subpath from its s-terminal to the first vertex v_i^1 shared by p_1, the subpath shared by p_1, and the subpath from the last vertex v_i^2 shared by p_1 to the t-terminal of q_i. Obviously, we have $v_i^1 = v_{i+1}^2$ for $i = 1, \ldots, \lambda - 1$. Therefore, we may consider the concatenation q_i' of the first subpath of q_i with the third subpath of q_{i+1} at vertex v_i^1. Consider all of the paths $q_1', \ldots, q_{\lambda-1}', q_{\lambda+1}, \ldots, q_k$ which belong to connected component C of G_1. For convenience, we denote these paths by $r_1, \ldots, r_m$.

Lemma 14. *Two terminals in C are connected by one of the paths $r_1, \ldots, r_m$ if and only if they are connected by one of the auxiliary paths in instance (C, N_C) w.r.t. start terminal x_C.*

Proof. Define the relation $\prec_C$ in instance (C, N_C) w.r.t. start terminal x_C analogously to the relation $\prec$ in instance (G, N) w.r.t. start terminal x. Because of Observation 8, we have $x_i \prec_C x_j$ if and only if $x_i \prec x_j$, for all terminals x_i and x_j that belong to C. Therefore, any path r_i starts with an s-terminal s_i' and ends with a t-terminal t_i' s.t. $s_i' \prec_C t_i'$. Moreover, since no two paths r_i and r_j cross, we have either $s_i' \prec_C t_i' \prec_C s_j' \prec_C t_j'$ or $s_i' \prec_C s_j' \prec t_j' \prec t_i'$ for all $i, j = 1, \ldots, m, s_i' \prec t_j'$. Now the claim follows immediately from Observation 1.

Lemma 15. *Let e_1, e_2 be two edges on path r_i incident to vertex v s.t. e_2 is the immediate successor of e_1 w.r.t. r_i. Then any edge $e \in (e_1, e_2)_v$ that is contained in C belongs to one of the paths $r_1, \ldots, r_m$.*

Proof. For any path $q_i, i = \lambda + 1, \ldots, k$, this claim follows immediately from the fact that q_i was drawn using a right-first strategy. The claim is also trivial for $q_i', i = 1, \ldots, \lambda - 1$, if both edges e_1 and e_2 belong to the original path q_i or if both edges belong to q_{i+1}. Thus, assume that e_1 belongs to q_i, and that e_2 belongs to q_{i+1}. Let e_3 be the immediate predecessor of e_2 w.r.t. q_{i+1}. It is easy to see that then we have $e_3 \in (e_2, e_1)_v$. In particular, we have $e \in (e_3, e_2)_v$ for any $e \in (e_1, e_2)_v$. However, in this case e had been chosen as the immediate successor of e_3 w.r.t. the original auxiliary path q_{i+1} rather than e_2. This proves the claim.

The following Lemma 16 immediately implies Invariant 3 for the first iteration.

Lemma 16. *Paths $r_1, \ldots, r_m$ are just the auxiliary paths in instance (C, N_C) w.r.t. start terminal x_C.*

Proof. Let $r_1', \ldots, r_m'$ denote the auxiliary paths in instance (C, N_C) w.r.t. start terminal x_C, in order of construction during the procedure, when it is applied to instance (C, N_C) with start terminal x_C. Because of Lemma 14, we may assume that r_i' connects the same terminals as r_i. Thus we have to show that $r_i = r_i'$ for $i = 1, \ldots, m$. Assume that $r_i \neq r_i'$ for some $i = 1, \ldots, m$ and $r_j = r_j'$ for all $j = 1, \ldots, i - 1$. Then there is a non-terminal vertex v, and there are three different edges e_1, e_2 and e_3 incident to v s.t. r_i' and r_i coincide up to edge e_1, and e_2 resp. e_3 is the immediate successor of e_1 w.r.t. path r_i resp. r_i'. Since r_i' is drawn in right-first manner, we have $e_3 \in (e_1, e_2)_v$. Hence, because of Lemma 15, it suffices to show that in this case e_3 does not belong to any of the paths $r_1, \ldots, r_m$. Because none of the paths $r_{i+1}, \ldots, r_m$ crosses r_i, all those paths belong completely to the left side of r_i, which implies that e_3 does not belong to any of these paths. As $r_j = r'_j$, $j < i$, edge e_3 does not belong to any of the paths $r_1, \ldots, r_{i-1}$ either.

It remains to show that e_3 does not belong to r_i. More precisely, we will show that r_i crosses itself, if e_3 belongs to r_i, which is obviously false. Assume that e_3 belongs to r_i. Either the immediate predecessor or the immediate successor of e_3 w.r.t. r_i is incident to v, too. This edge must belong to $(e_1, e_2)_v$, because otherwise path r_i crosses itself at v. Hence, w.l.o.g. we may assume that e_3 enters v w.r.t. r_i (else take the immediate predecessor instead).

Since r'_i and r_i are incident up to e_1, e_3 cannot be a predecessor of e_1 w.r.t. r_i. In particular, there is a subpath of r_i starting with e_2 and ending with e_3, which forms a cycle c. By removing subcycles from c, this cycle may be reduced to an elementary cycle c' which still hits v and does not cross the subpath of r_i from the corresponding s-terminal up to e_1. Completely analogously to the end of the proof of Lemma 9, this yields a contradiction to Lemma 7.

Corollary 17. *The restriction of $A(G, N, x)$ to any connected component C of G_1 equals $A(C, N_C, x_C)$.*

We conclude this section with the main result.

Theorem 18. *The algorithm is correct.*

4 A Linear-Time Realization

It is not hard to see that the first statement of the algorithm, i.e., constructing the auxiliary graph, can be done in linear time. Since in each iteration of the WHILE-loop one edge is added to a path p_i and no edge is added twice, the total number of iterations of the WHILE-loop is linear. Therefore, realizing the algorithm in linear time amounts to designing data structures for vertices and edges s.t. any simple statement except the first can be done in (amortized) constant time. Recall that we store for each vertex v its adjacency list, which is a cyclic list of *all* edges incident to v, leaving v or entering v.

Obviously, all simple statements except the statement where the next edge after the leading edge of p_i is searched require only constant time. To handle the problem of finding the next edge after a given edge, we need some terminology. At any stage of the algorithm, the *signpost* of $e \in E$ is the next edge after e that leaves the head of e. Conversely, the set of all edges with signpost $e \in E$ is called the *client set* of e. During the algorithm, we will always maintain the client set of each edge. Any edge "knows" the client set it belongs to, and any client set "knows" its signpost. Let $v \in V$ be no terminal. Then the adjacency list of v can be uniquely decomposed into maximal connected components s.t. all edges in such a connected component leave v or all edges enter v. Let $e \in E$ leave v. The client set of e is either empty or just the maximal connected component of edges entering v that appears immediately before e in the adjacency list of v in reverse clockwise ordering. Whenever an edge e is occupied whose client set is non-empty, we have to update this set. If the next edge after e, say e', leaves v, e' must still be a free edge, because e has to be considered before e' for going forward from v. Then the signpost of the client set of e is simply set to e'. In case of e' entering v, there are two different cases. If there is more than one client set consisting of edges entering v, the signpost of the client set of e must from now on be the signpost of the client set to which e' belongs. In other words, we have to unite these two client sets.

On the other hand, if there is only one such client set, there is no more free edge leaving v at all. Since there is also no more free edge entering v in this case, we need not consider v any longer. Therefore, maintaining and using client sets and their signposts amounts to performing for each non-terminal vertex v a sequence of *union-find* operations on the set of all edges entering v. In general, such a sequence cannot be performed in linear time in the worst case [Ta]. However, Gabow and Tarjan have shown that this is possible for a certain special case [GT]. We can show that this special case covers our problem. As a result, we obtain the following theorem.

Theorem 19. *The algorithm can be realized s.t. it requires linear time in the worst case.*

References

[BM] M. Becker and K. Mehlhorn (1986): *Algorithms for Routing in Planar Graphs.* Acta Informatica **23**, 163-176.

[Fr] G.N. Frederickson (1987): *Fast Algorithms for Shortest Paths in Planar Graphs, with Applications.* SIAM J. Comput. **16**, 1004-1022.

[GT] H.N. Gabow and R.E. Tarjan (1985): *A Linear-Time Algorithm for a Special Case of Disjoint Set Union.* J. Comp. System Sciences **30**, 209-221.

[Ha] R. Hassin (1984): *On Multicommodity Flows in Planar Graphs.* Networks **14**, 225-235.

[Ka] M. Kaufmann (1990): *A Linear-Time Algorithm for Routing in a Convex Grid.* IEEE Transact. Computer-Aided Design **9**, 180-184.

[KK] M. Kaufmann and G. Klär (1991): *A Faster Algorithm for Edge-Disjoint Paths in Planar Graphs.* Proc. Int. Symp. on Algorithms (ISA '91), LNCS 557, 336-348.

[KM] M. Kaufmann and K. Mehlhorn (1986): *Generalized Switchbox Routing.* J. Algorithms **7**, 510-531.

[KL] M.R. Kramer and J. van Leeuwen (1984): *The complexity of Wire-Routing and Finding Minimum Area Layouts for Arbitrary VLSI-Circuits.* Advances Comp. Res. **2**, 129-146.

[MNS] K. Matsumuto, T. Nishizeki and N. Saito (1985): *An Efficient Algorithm for Finding Multicommodity Flows in Planar Networks.* SIAM J. Comp. **14**, 289-302.

[NSS] T. Nishizeki, N. Saito and K. Suzuki (1985): *A Linear-Time Routing Algorithm for Convex Grids.* IEEE Transact. Computer-Aided Design **CAD-4**, 68-76.

[SAN] H. Suzuki, T. Akama and T. Nishizeki (1990): *Finding Steiner Forests in Planar Graphs.* First Proc. ACM-SIAM Symp. Discrete Algorithms, 444-453.

[OS] H. Okamura and P.D. Seymour (1981): *Multicommodity Flows in Planar Graphs.* J. Combinatorial Th. **B 31**, 75-81.

[Ta] R.E. Tarjan (1979): *A Class of Algorithms Which Require Non-Linear Time to Maintain Disjoint Sets.* J. Comp. System Sciences **18**, 110-127.

[WW] D. Wagner and K. Weihe (1993): *A Linear-Time Algorithm for Edge-Disjoint Paths in Planar Graphs.* Report No. 344 TU Berlin

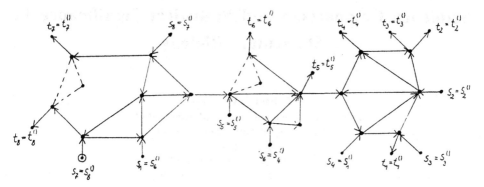

Fig. 1. An instance (G, N) and its auxiliary graph with the start terminal encircled. The edges of the auxiliary graph are solid, and all other edges of G are dotted. The orientations of the edges of the auxiliary graph are indicated by arrows.

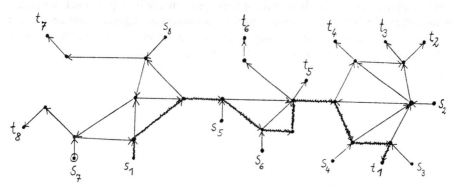

Fig. 2. The first real path p_1 for instance (G, N). Edges that do not belong to the auxiliary graph are omitted.

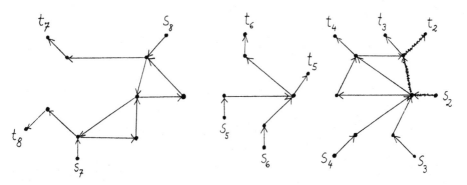

Fig. 3. The restriction of the auxiliary graph of Figure 1 to the "residual" graph G_1 and the second real path p_2 for instance (G, N).

Sequence Comparison and Statistical Significance in Molecular Biology

Michael S. Waterman
Department of Mathematis and Molecular Biology
University of Southern California
1042 W. 36th Place
Los Angeles, CA 90089-113

Abstract. Sequence comparison problems in molecular biology are briefly reviewed. Sequences are optimally aligned allowing for substitutions, insertions, and deletions. Dynamic programming algorithms solving these problems will be presented. So-called local algorithms finding best matching intervals between two sequences are best suited for database searches. A Poisson approximation method for rapid and accurate assignment of statistical significance will be presented.

Surface Reconstruction Between Simple Polygons via Angle Criteria*

EMO WELZL and BARBARA WOLFERS

Institut für Informatik
Freie Universität Berlin
Takustraße 9, 14195 Berlin, Germany
e-mail: emo@tcs.fu-berlin.de, wolfers@tcs.fu-berlin.de

Abstract

We consider the problem of connecting two simple polygons P and Q in parallel planes by a polyhedral surface. The goal is to find an optimality criterion, which naturally satifies the following conditions: (i) if P and Q are convex, then the optimal surface is the convex hull of P and Q (without facets P and Q), and (ii) if P can be obtained from Q by scaling with a center c, then the optimal surface is the portion of the cone defined by P and apex c between the two planes.

We provide a criterion (based on the sequences of angles of the edges of P and Q), which satisfies these conditions, and for which the optimal surface can be efficiently computed. Moreover, we supply a condition, so-called *angle consistency*, which proved very helpful in preventing self intersections (for our and other criteria). The methods have been implemented and gave improved results in a number of examples.

1 Introduction

The reconstruction of a three-dimensional object from its cross sections data is a problem with many applications like clinical medicine (computerized tomography), biomedical research, computer graphics, animation, geology, etc., [Sch].

Here is the set-up we want to consider: P and Q are simple polygons in parallel planes h_P and h_Q, respectively. A *surface between P and Q* is a cyclic sequence of triangles, each triangle is the convex hull of an edge of one of the polygons and a vertex of the other polygon; consecutive triangles share an edge (connecting a vertex from P with a vertex from Q), and the sequence encounters the edges of P in the same counterclockwise order as P, and analogously for Q.

So we ignore the problems arising from the fact that the cross sections of an object may contain several polygons (polygons have to be assigned to each other, and 'branchings' may occur). This can be handled by a preprocessing step by other methods. Moreover, we restrict ourselves by not allowing other vertices in the surface but those in P and Q.

A number of methods have been proposed in the literature. With the exception of the volume based approach, and the work by Boissonnat [B], these methods associate with

---

*Work partially supported by the ESPRIT Basic Research Action Program of the EC under project ALCOM II

every potential connecting surface a parameter (usually a real number), and the surface of choice is one which optimizes (minimizes, maximizes) this parameter. Examples are: (1) surface of minimum area [FKU], [SP], (2) surface where the resulting enclosed solid has maximal volume [K], (3) surface, where the overall edge length is minimal, etc. Other approaches [C], [ChrS], [GD] start the construction at some point and proceed according to local criteria.

It turns out that these methods have drawbacks, which occur already in simple natural examples: probably most striking is the case of two regular n-gons P and Q, where the orthogonal projection of P in h_Q is sufficiently far appart from Q (the optimal surface according to the minimum area criterion is depicted in Figure 1).

Our starting point was to set up general requirements which should be met by a 'good' optimality criterion in a natural way[1]:

Condition C1. If P and Q are convex polygons, then the optimal surface is the convex hull of P and Q (without facets P and Q).

Condition C2. If P can be obtained from Q by scaling with a center c, then the optimal solution is the portion of a cone defined by P with apex c between the two planes h_P and h_Q. Similarly, if P is a translate of Q, then the surface should be a cylindric section.

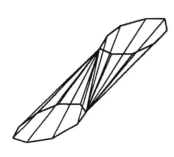

Figure 1: Area-optimal solution for two regular 9-gons.

Surprisingly enough, none of the criteria we found in the literature satisfy these conditions (Figure 1 demonstrates that the minimum area criterion violates both conditions).

Our method starts with the following simple observation. The sequence of triangles from a surface defines a 'merge' of the edges from P and Q, (go through the sequence of triangles and for each one take the edge which is from P or Q, see Figure 2). This sequence yields again a polygon (not necessarily simple!), which has also a geometric interpretation in terms of the surface: If all the edges are halved in length, then we get the polygon obtained by intersecting the surface with the plane half way between h_P and h_Q. For every such merged polygon we add up the absolute values of the 'turning angles' $\delta(e, e')$ between any pair of consecutive edges e and e'. A surface is called optimal if its associated polygon-merge minimizes this sum. The intuition is that we try to keep the surface (or, more precisely, its intersection with planes parallel to h_P) as smooth as possible.

In this way we satisfy conditions C1 and C2, as we will prove in Section 2. It may appear to be more appropriate to consider the sum of squares of $\delta(e, e')$ instead, but, as it turns out, this violates condition C2.

There is the issue of surfaces with self-intersections – definitely an undesired effect – which we have not touched so far. This may very well happen for the optimal surfaces

---

[1] Of course, the requirements can always be fulfilled by adding them as special cases; but that's not what we are aiming for.

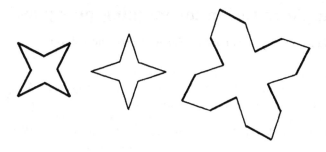

Figure 2: Merge of two polygons.

(also for our criterion). As a matter of fact, Gitlin, O'Rourke and Subramanian [GORS] show that there are instances of polygons which do not allow a connecting surface (in the way we defined it) without self-intersections (one polygon may even be chosen as a triangle). (There is a subtle issue what we call a self-intersection, but we do not elaborate on this; e.g. the surface in Figure 1 has a self-intersection in the sense of [GORS].) Section 3 describes the so-called *angle-consistency condition* for merged polygons. Roughly speaking, this disallows that in the merged sequence between two edges in P there is a sequence of edges in Q which runs into a spiral without 'resolving' it. Experiments show, that the condition prevents self-intersections in many examples, and a violation of the condition *enforces* a self-intersection (i.e. requiring angle-consistency does not exclude any good solutions).

The algorithmic aspects are dealt with in Section 4. We show that the optimal angle-consistent solution with respect to our angle criterion can be computed in time $O((dt)^4 + m + n)$, where m and n are the numbers of edges of P and Q, d is a parameter that indicates to what extent P or Q run into spirals, and t counts the number of edges in P and Q where preceding and succeeding vertex lie on opposite sides of the line through the edge (e.g., for a convex polygon this parameter is 0). In typical instances, d and t are very small compared to the number of edges.

We have implemented our method, and some other methods for the sake of comparison. The angle-consistency condition has been directly motivated by phenomina we observed on results of the implementation in simple natural examples. The figure on the last page shows a heart reconstructed from cross sections[2] with our method.

Clearly, the 'best' surface will always depend on the specific application, and there may even occur applications where our conditions C1 and C2 are not appropriate. Nevertheless, we believe that our method represents an interesting alternative to the existing ones. Moreover, merged polygons raise some mathematically interesting questions. We refer to [GRS] for a paper treating some related aspects. Merged polygons and related methods are used also in 2D-shape blending [SG].

---

[2] We thank J.-D. Boissonnat for supplying the data.

2 An angle criterion for merging polygons.

We first introduce some simple notation for sequences and polygons.

Notation for sequences. Given two sequences $X = (x_0, x_1, \ldots, x_{n-1})$ and $Y = (y_0, y_1, \ldots, y_{m-1})$, we say that X and Y are *cyclically equivalent*, denoted by $X =_{\text{cyc}} Y$, if $n = m$ and there exists an i, $0 \leq i \leq n - 1$, such that $(x_i, x_{i+1}, \ldots, x_{n-1}, x_0, x_1, \ldots, x_{i-1}) = (y_0, y_1, \ldots, y_{m-1})$. We adopt the convention that indices are taken modulo the length of the considered sequence, in particular $x_n = x_0$.

Let $Z = (z_0, z_1, \ldots, z_{n-1})$ be a sequence, and let $I = \{i_1, i_2, \ldots, i_k\}$, $0 \leq i_1 < i_2 < \cdots i_k \leq n - 1$. The *I-restriction*, $Z_{|I}$, of Z is the sequence $(z_{i_1}, z_{i_2}, \ldots, z_{i_k})$.

Z is a *cyclic merge of* sequences X and Y if there is a partition (I, J) of $\{0, 1, \ldots, n-1\}$ such that $X =_{\text{cyc}} Z_{|I}$ and $Y =_{\text{cyc}} Z_{|J}$. Note that I and J are not uniquely determined; in order to be more specific about which elements come from which sequence, we call Z the (I, J)-*indexed cyclic merge of* X and Y.

Polygons. A *polygon* P is a sequence $(p_0, p_1, \ldots, p_{n-1})$ of $n \geq 2$ points in the plane, such that $p_i \neq p_{i+1}$ for all i, $i = 0, 1, \ldots, n - 1$. Two polygons are considered *equivalent* if their defining sequences are cyclically equivalent.

A polygon is *simple* if $n \geq 3$, all points p_i, $i = 0, 1, \ldots, n - 1$, are pairwise distinct, and each open line segment $\overline{p_i p_{i+1}}$, $i = 0, 1, \ldots, n - 1$, is disjoint from all p_j, $j = 0, 1, \ldots, n - 1$, and from all $\overline{p_j p_{j+1}}$, $j = 0, 1, \ldots, n - 1$, $j \neq i$.

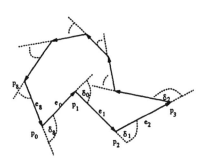

Every polygon $P = (p_0, p_1, \ldots, p_{n-1})$ defines a sequence of *edge vectors* $E_P = (e_0, e_1, \ldots, e_{n-1}) \in (\mathbf{R}^2 - o)^n$ where[3] $e_i = p_{i+1} - p_i$, $o = (0, 0)$ is the zero vector; a sequence of *edge angles* $A_P = (a_0, a_1, \ldots, a_{n-1}) \in (S^1)^n$ where $a_i = \frac{e_i}{\|e_i\|}$; and a sequence of *turning angles* $\Delta_P = (\delta(e_0, e_1), \delta(e_1, e_2), \ldots, \delta(e_{n-1}, e_n = e_0)) \in (-\pi, +\pi]^n$ where $\delta : ((\mathbf{R}^2 - o) \times (\mathbf{R}^2 - o))^n \longrightarrow (-\pi, +\pi]$ and $\delta(e, e')$, is the counterclockwise angle between e and e' in the interval $(-\pi, +\pi]$. $\delta(e_i, e_{i+1})$, for short δ_i, can be seen as the turn of the tangent at point p_{i+1}, where a counterclockwise turn gives a positive value, and a clockwise turn gives a negative value; see Figure 3.

Figure 3: Edge vectors and turning angles, $\delta_0 < 0, \delta_1 > 0$ etc.

Given the edge vector sequence $E_P = (e_0, e_1, \ldots, e_{n-1})$ of a polygon P we write $\delta(P) = \delta(E_P) := \sum_{i=0}^{n-1} \delta_i$ and $\overline{\delta}(E_P) := \sum_{i=0}^{n-1} |\delta_i|$. Note that E_P determines P up to translation, and a sequence $(e_0, e_1, \ldots, e_{n-1})$ in $(\mathbf{R}^2 - o)^n$, $n \geq 2$, is the egde vector sequence of a polygon iff $\sum_{i=0}^{n-1} e_i = o$. We consider the values in Δ_P as real numbers and the arithmetic of these values without equivalence modulo 2π.

Observation 2.1 *(i) For every polygon P, we have $\delta(P) \equiv 0$ modulo 2π.*
(ii) For a simple polygon P, $|\delta| < \pi$ for all turning angles, and $\delta(P) \in \{2\pi, -2\pi\}$.

---

[3] e_i is *not* the edge (segment) connecting p_i and p_{i+1}, it is the vector from p_i to p_{i+1}.

(iii)For a sequence of turning angles $(\delta_0, \ldots, \delta_{n-1})$ of a convex polygon P, $|\delta_i| < \pi$ for all $i = 0, 1, \ldots, n-1$, and $\delta(P) = 2\pi$ and $\delta_i \geq 0$ for all $i = 0, 1, \ldots, n-1$, or $\delta(P) = -2\pi$ and $\delta_i \leq 0$ for all $i = 0, 1, \ldots, n-1$.
(iv) For every polygon P we have $\sum_{i=0}^{n-1} |\delta_i| \geq 2\pi$ with equality iff P is convex.

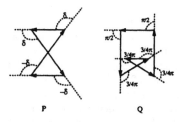

Note that $\delta(P) = 2\pi$ and $\delta(P) = -2\pi$ discriminates wether we run through a simple polygon in counterclockwise or clockwise order, respectively. Without loss of generality, we assume that we run through a simple polygon in counterclockwise order.

Figure 4: $\sum \delta_i = 0$ for P and $\sum \delta_i = 4\pi$ for Q.

Observation 2.2 *For a sequence of edge vectors $(e_i, e_{i+1}, \ldots, e_j)$ with $\sum_{k=i}^{j-1} |\delta_k| < \pi$ we have $\sum_{k=i}^{j-1} |\delta_k| \geq |\delta(e_i, e_j)|$. If, moreover, $\delta_k \geq 0$ for all $k \in i, i+1, \ldots, j-1$ or $\delta_k \leq 0$ for all $k \in i, i+1, \ldots, j-1$, then $\sum_{k=i}^{j-1} |\delta_k| = |\delta(e_i, e_j)|$, (see Figure 5)*

Figure 5: $\sum_{k=i}^{j-1} |\delta_k| = |\delta(e_i, e_j)|$.

L_1-optimal merge. We call $Z = (z_0, z_1, \ldots, z_{n-1})$ an *L_1-optimal* cyclic merge of edge sequences X and Y if Z is a cyclic merge of X and Y and $\overline{\delta}(Z) = \sum_{i=0}^{n-1} |\delta(z_i, z_{i+1})|$ is minimal among all cyclic merges of X and Y.

Lemma 2.3 *For any cyclic merge Z of edge sequences X and Y, we have $\max\{\overline{\delta}(X), \overline{\delta}(Y)\} \leq \overline{\delta}(Z)$.*

Proof. Note that adding an edge into a sequence of edges cannot decrease its $\overline{\delta}$-value (recall Observation 2.2). Since we can obtain Z from X by successively adding the edges from Y, it follows that $\overline{\delta}(X) \leq \overline{\delta}(Z)$. Analogously, we can obtain Z starting from Y which gives $\overline{\delta}(Y) \leq \overline{\delta}(Z)$, and the lemma follows. ▢

With this lemma we are ready to prove the main property of L_1-optimal cyclic merges.

Lemma 2.4 *(i) If X and Y are edge sequences of convex polygons, then every L_1-optimal cyclic merge of X and Y is also convex.*
(ii) If X and Y are sequences of edge vectors of simple polygons and their sequences of edge angles are cyclically equivalent (w.l.o.g. $\frac{x_i}{\|x_i\|} = \frac{y_i}{\|y_i\|}$ for all $i \in 0, 1, \ldots n-1$), then $Z = (x_0, y_0, x_1, y_1, \ldots x_{n-1}, y_{n-1})$ is an L_1-optimal cyclic merge of X and Y. Any L_1-optimal merge can be obtained from Z by successively swapping consecutive edge vectors e and e' with $e = \lambda e', \lambda > 0$.

Before we proceed with the proof, let us remark that (i) implies that condition C1 is satisfied. If X can be obtained from Y by scaling, then Z as described in (ii) corresponds to the cone section as required by condition C2. Since the swappings described do not change the actual surface (only its associated triangulation), this shows that C2 is also fulfilled.

Proof. (i) X comes from a convex polygon, if its angles are cyclically sorted. Two cyclically sorted sequences can be merged to a cyclically sorted sequence, which again describes a convex polygon. Since the $\bar{\delta}$-values of all these sequences are equal 2π, Lemma 2.3 or Observation 2.1 imply the claimed assertion. (ii) Since $\bar{\delta}(Z) = \bar{\delta}(X) = \bar{\delta}(Y)$, the optimality of Z follows immediately from Lemma 2.3. We omit here the proof that any L_1-optimal merge can be obtained from Z by successively swapping consecutive edge vectors e and e' with $e = \lambda e', \lambda > 0$. This fact is somewhat more subtle, as it is perhaps witnessed by the fact that the statement becomes wrong, if we drop the assumption that X and Y come from simple polygons! $\qquad\square$

If the polygons are convex then the L_1-optimal cyclic merge corresponds to the Minkowski sum of the polygons. No such correspondence exists as soon as the polygons are not convex (e.g. the L_1-optimal merge is in general not unique).

L_2-**optimal merge.** A cyclic merge $Z = (z_0, z_1, \ldots, z_{m+n-1})$ of $X = (x_0, x_1, \ldots x_{n-1})$ and $Y = (y_0, y_1, \ldots y_{m-1})$ is L_2-*optimal* if $\sum_{i=0}^{m+n-1} \delta(z_i, z_{i+1})^2$ is minimal.

Lemma 2.5 *If X and Y are sequences of convex polygons then an L_2-optimal cyclic merge of X and Y is also convex.*

The proof of Lemma 2.5 is not as simple as the one for Lemma 2.4, because adding an edge vector to a sequence may actually decrease the sum of squares of turning angles. We omit the proof here, since L_2-optimal solutions may violate condition C2. To this end consider the example of two stars in Figure 6. Let the acute angle in the polygons be

Figure 6: L_1-optimal merge and L_2-optimal merge of two cyclically equivalent polygons

$\epsilon, 0 < \epsilon < \pi/2$. Then the L_2-value of the L_1-optimal solution is $4((\pi - \epsilon)^2 + (\pi/2 - \epsilon)^2)$. The alternative merge (given by a program as the L_2-optimal merge for $\epsilon \approx \pi/5$) has an L_2-value of $4(2(\pi/2 - \epsilon)^2 + (\pi/2 + \epsilon)^2 + \epsilon^2)$. As ϵ approaches 0, the first value converges to $5\pi^2$, while the second one converges to $3\pi^2$. So for some ϵ small enough ($\epsilon < (\sqrt{2} - 1)\pi/2$), the solution suggested by condition C2 will not be L_2-optimal.

3 Angle consistency

Let us right go back to the example in Figure 6. The solution suggested as L_2-optimal obviously leads to a surface with self-intersection, since even the merged polygon Z is not simple; even without looking at the picture, we could compute $\delta(Z) = -4\pi$, a value which contradicts the simplicity of the underlying polygon (no matter what the lengths of the edge vectors are). In this section we will suggest a criterion which eliminates such obviously bad solutions.

Before we start with the key definition, we want to point out that a cyclic merge Z of two edge vector sequences X and Y does not necessarily determine the surface. However, the surface is determined if we give Z as an indexed merge[4], when it is clear which vector in Z comes from X and which one comes from Y.

Let $Z = (z_0, z_1, \ldots, z_{n-1})$ be an (I, J)-indexed merge of two edge vector sequences X and Y. Let $i < j$ be two consecutive indices in I. We define $\delta_{i,j}^{(X)} := \delta(z_i, z_j)$, and $\delta_{i,j}^{(Z)} := \sum_{k=i}^{j-1} \delta(z_k, z_{k+1})$; analogously, we define $\delta_{i,j}^{(Y)}$ for consecutive indices in J. Moreover, we agree on the obvious cyclic extension for indices $i > j$, where i is the largest index in I and j is the smallest index in I (and similar for J).

We say that Z is *angle consistent*, if $\delta_{i,j}^{(X)} = \delta_{i,j}^{(Z)}$ and $\delta_{i,j}^{(Y)} = \delta_{i,j}^{(Z)}$ for all pairs of cyclically consecutive indices in I and J, respectively.

Note that if $0 \leq i_0 \leq i_1 \leq \cdots i_{k-1} \leq n - 1$ are indices in I, and X comes from a simple polygon, then $2\pi = \delta(X) = \sum_{l=0}^{k-2} \delta(z_{i_l}, z_{i_{l+1}}) + \delta(z_{i_{k-1}}, z_{i_0})$. This sum equals $\delta(Z)$, if Z is an angle consistent merge; and hence $\delta(Z) = 2\pi$. However, it may very well be that $\delta(Z) = 2\pi$ (it may even be simple), but it is not angle consistent, see Figure 7.

Although, a cyclic merge which is not angle consistent may be simple, the resulting surface will always contain self-intersections (as will be shown below). So it is justified to exclude such merges for our surfaces. This will eliminate also self-intersections for our L_1-angle criterion; see Figure 8 for an example where an L_1-optimal merge violates angle consistency, and go back to Figure 2 for the L_1-optimal angle consistent merge.

Figure 7: A cyclic merge which is simple but not angle consistent.

Theorem 3.1 *An indexed cyclic merge of edge vector sequences of two simple polygons which is not angle consistent leads to a surface with self-intersections.*

Proof. Let us assume that P lies in the xy-plane, and Q lies in a parallel plane at height 1 (i.e. in the plane $z = 1$). We have argued before in the introduction, that the cyclic merge Z defined by a surface is the intersection of the surface with the plane at height $1/2$, scaled with a factor 2. Let us be more specific, saying that $Z = (z_0, z_1, \ldots, z_{n-1})$ is the (I, J)-indexed cyclic merge of the edge vector sequences of P and Q. If we consider now the intersection of the surface with a plane at height λ, $0 \leq \lambda \leq 1$, then this can be obtained from Z by multiplying all edge vectors from P (with index in I) by $1 - \lambda$, and the edge vectors from Q (with index in J) by λ. This gives a family of polygons with

---

[4]Recall definition in the beginning of Section 2.

Figure 8: L_1-optimal cyclic merge which violates angle consistency.

edge vector sequences Z_λ. The surface is free of self-intersections, if all polygons Z_λ are simple.

For the remaining proof let us multiply the length of the edges in Z_λ, $0 < \lambda \le 1$ by $1/\lambda$ to obtain edge vector sequences Z'_λ where the edge vectors from Q have constant length, and the edge vectors from P are multiplied by $(1 - \lambda)/\lambda$.

Consider now a violation of angle consistency, i.e. a pair $i < j$ of consecutive indices in I where $\delta_{i,j}^{(X)} \ne \delta_{i,j}^{(Z)}$ (other cases of violation are symmetric). Hence, the sequences Z'_λ contain as a subsequence $S_\mu = (\mu z_i, z_{i+1}, \ldots, z_{j-1}, \mu z_j)$ with $\mu := (1 - \lambda)/\lambda \longrightarrow \infty$ as $\lambda \longrightarrow 0$.

If $0 < \delta_{i,j}^{(X)} < \pi$, let $v_\mu = -(\mu z_i + z_{i+1} + \cdots + z_{j-1} + \mu z_j)$, i.e. $(\mu z_i, z_{i+1}, \ldots, z_{j-1}, \mu z_j, v_\mu)$ is the edge vector sequence of a polygon, unless $v_\mu = o$; if $v_\mu = o$, then this immediately reveals a self-intersection. Let $\mu \longrightarrow \infty$, v_μ only intersects μz_i and μz_j in the polygon. If the polygon is simple then $\delta_{i,j}^{(X)} = \delta_{i,j}^{(Z)}$ which is a contradiction to the assumption. Else there is a part of Z_λ which is not simple which yields a self-intersection. In a similar way the case $\delta_{i,j}^{(X)} = 0$ can be handled (let $v_\mu^1 + v_\mu^2 = -(\mu z_i + z_{i+1} + \cdots + z_{j-1} + \mu z_j)$ and $\delta(v_\mu^1, v_\mu^2) = \epsilon \longrightarrow 0$.) $\qquad \square$

If $X = (x_0, x_1, \ldots, x_{n-1})$ and $Y = (y_0, y_1, \ldots, y_{m-1})$ are edge vector sequences of simple polygons then there always exists a cyclic merge $Z = (x_0, \ldots, x_i, y_0, y_1, \ldots, y_{m-1}, x_{i+1}, \ldots, x_{n-1})$ of X and Y which is angle consistent. For example take the leftmost vertex of X (or the uppermost of these if there are more than one) as the i-th vertex and the rightmost (the lowermost of those) of Y as the $(m-1)$-th then $\delta(Z) = 2\pi$ and it directly follows from the construction that angle consistency is fulfilled.

4 Algorithm

If one polygon is convex it is easy to find an L_1-optimal cyclic merge.

Lemma 4.1 *An angle consistent L_1-optimal cyclic merge Z of an edge sequence X of a convex polygon with n vertices and an edge sequence Y of a simple polygon with m vertices can be constructed in $O(n + m)$ time.*

Proof. The angles of X are cyclically sorted. Edges of X can be successively inserted into the edge sequence of Y without increasing its $\bar{\delta}$-value because $\delta(Y) = 2\pi$. If this is done in a greedy way (insert as soon as possible), angle consistency is guaranteed. $\square$

Otherwise the problem can be formulated as a shortest path problem in a directed graph.
    Description of the algorithm:
Every possible triangle in a connecting surface (defined by an edge of one polygon and a vertex of the other) is represented by a node in the graph. The node set of the graph has cardinality $2 \cdot m \cdot n$. A node is labeled $(i, j, 0)$ if the triangle is defined as the convex hull of the edge between the $(i-1)$-th and i-th vertex of polygon P and the j-th vertex of polygon Q; $(i, j, 1)$ is defined analogously by the i-th vertex of P and the $(j-1)$-th and j-th of Q. Arcs in the graph connect nodes of consecutive triangles which share an edge. The graph is a torus graph. Indegree and outdegree of a vertex are 2. Arc weights are assigned according to the absolute value of the turning angle between the polygon edges of the two consecutive triangles. Fixing a starting triangle (w.l.o.g. $(0, j, \cdot)$), we are looking for a cycle of minimum weight passing node $(0, j, \cdot)$ containing $n + m$ triangles. A global optimal solution is the minimum among all minimum weight cycles in the torus graph passing $(0, 0, 0), (0, 0, 1), (0, 1, 0), (0, 1, 1), (0, 2, 0), \ldots, (0, m-1, 0)$ or $(0, m-1, 1)$, respectively. For a fixed starting triangle, w.l.o.g. $(0, 0, 0)$, we regard a subgraph of the torus graph which is a directed acyclic graph with $2 \cdot (n+1) \cdot (m+1)$ nodes $(i, j, 0)$ and $(i, j, 1)$; $0 \leq i \leq n$, $0 \leq j \leq m$, where $(n, m, \cdot)$ is a copy of $(0, 0, \cdot)$. A minimum weight cycle in the torus graph passing $(0, 0, 0)$ corresponds to a shortest path from $(0, 0, 0)$ to $(n, m, 0)$ in the subgraph. A shortest path can be computed in $O(n \cdot m)$ time since the subgraph is a directed acyclic graph of this size. But we have to compute a shortest path for each of the $2 \cdot m$ starting triangles $(0, 0, 0), (0, 0, 1), (0, 1, 0), \ldots, (0, m-1, 1)$. So the overall running time to compute the value of an L_1-optimal cyclic merge is $O(n \cdot m^2)$ and space requirements are $O(n \cdot m)$ (the size of the union of the subgraphs is $2 \cdot (n+1)(2m)$). The L_1-optimal merge itself can be obtained by backtracking through the graph.

Theorem 4.2 *An L_1-optimal merge of two polygons with n and m vertices can be computed in $O(n \cdot m^2)$ time.*

The algorithm can be used to compute other angle dependent optimal merges like the L_2-optimal merge.
Remark. This solution is based on two papers, one of the first papers written on contour triangulation [K], it employs a smaller directed graph to compute a maximal volume contour triangulation; Fuchs, Kedem and Uselton [FKU] refined the modeling of the graph to accelerate the algorithm. They gave a faster algorithm with running time $O(n \cdot m \cdot \log m)$ but they need graph planarity and our subgraphs are not planar.

The L_1-optimal merge produced by the algorithm may not fulfill angle consistency. To guarantee that the solution is angle consistent we have to extend the algorithm.
    Suppose starting vertex $(0, 0, 0)$ is fixed. (We proceed analogously for all $2m$ starting vertices.) Guaranteeing angle consistency, the algorithm successively computes shortest paths to all vertices of the graph. Reaching a vertex we test if the path represents an angle consistent part of a solution. For example if the vertex corresponds to a triangle with a polygon edge from edge vector z_j with j in I and $i < j$ consecutive indices in I, we test if $\delta_{i,j}^{(X)} = \delta_{i,j}^{(Z)}$. To do this test in constant time per vertex we compute two entries

δ^X and δ^Y at every vertex. Reaching a vertex corresponding z_j, δ^X denotes $\delta^{(Z)}_{k,j}$ with k is the largest index in I smaller than j, and δ^Y denotes $\delta^{(Z)}_{l,j}$ with l the largest vertex smaller j in J. δ^X and δ^Y are computed and updated in constant time per vertex. If j is from I then $\delta^{(Z)}_{i,j}$ is given by δ^X. If j is from J and $i < j$ preceding j in J then we have to test if $\delta^{(Y)}_{i,j} = \delta^{(Z)}_{i,j}$ and $\delta^{(Z)}_{i,j}$ is given by δ^Y.

For every vertex $(i, j, 0)$ $[(i, j, 1)]$ in the graph the shortest angle consistent path from $(0, 0, 0)$ passing $(i, j, 0)$ $[(i, j, 1)]$ to $(i + 1, j, 0)$ and $(i, j + 1, 1)$ is computed. This means that we check what will happen if the next edge vector from X or from Y is taken. It is easy to compute these angle consistent paths for $i = 0$ or $j = 0$ (assuming $\delta^{(Y)}_{k,l} = \delta^{(Z)}_{k,l}$ with $z_k = y_0$ and $z_l = y_1$). Now we compute the paths to $(i, j, .)$ vertex by vertex in rows, what means before j is increased all values for $(i, j, .)$ for all $0 < i \leq n$ are computed. At every vertex the shortest angle consistent paths are computed as the shortest paths in the algorithm above; only if the shortest angle consistent paths passing $(i, j, 0)$ $[(i, j, 1)]$ to $(i, j + 1, 1)$ $[(i + 1, j, 0)]$ are computed angle consistency may be violated. Suppose angle consistency is violated at $(i, j, 0)$, i.e. passing $(i, j, 0)$ taking the arc to $(i, j + 1, 1)$. Only the pair of edge vectors $y_j (= z_k)$ and $y_{j+1} (= z_l)$ violates angle consistency, $\delta^{(Y)}_{k,l} \neq \delta^{(Z)}_{k,l}$. The shortest angle consistent path we are looking for contains a shortest angle consistent subpath passing $(i', j, 1)$ taking the arc to $(i' + 1, j, 0)$ which we already computed for all $i' < i$. For each i' compute the length of the path passing $(i', j, 1)$, $(i' + 1, j, 0)$, ..., $(i, j, 0)$, $(i, j + 1, 1)$ and test if $\delta^{(Y)}_{k',l} = \delta^{(Z)}_{k',l}$ where Z is the merge corresponding to the path and $x_{i'} = z_{k'}$. All together for all i' this can be done in $O(n)$ time and also finding the shortest angle consistent among these $O(n)$ paths takes the same time. With this algorithm we find the shortest angle consistent path, we guaranteed angle consistency for all pairs of consecutive indices in I and J but k, l with $z_k = y_0$ and $z_l = y_1$ (see above). But $\delta^{(Y)}_{k,l} = \delta^{(Z)}_{k,l}$, because $\delta(X) = \delta(Z) = 2\pi = \sum \delta^{(Z)}_{i,j}$ with summation over all pairs i, j of consecutive indices in J.

At each vertex we spend at most $O(n)$ time. The resulting running time for a fixed starting vertex is $O(n^2 \cdot m)$ time, the overall running time to compute an L_1-optimal angle consistent merge ,i.e. L_1-optimal among the angle consistent, is $O(n^2 \cdot m^2)$ time.

For many polygons it is possible to compute an L_1-optimal angle consistent merge in less time. We exploit the degree of convexity of a polygon in a similar way to Lemma 4.1. Given an edge vector sequence $(e_0, e_1, \ldots, e_{n-1})$, an edge vector e_i is called *turning-edge vector* if $\delta_{i-1} \cdot \delta_i \leq 0$. In a polygon, the vertices preceding and succeeding an edge corresponding to a turning-edge vector do not lie on the same side of the line along the edge. Given an edge vector sequence $X = (x_0, x_1, \ldots, x_{n-1})$, we define

$$d(X) = \max_{i,j} \left\{ \sum_{k=i}^{j-1} |\delta_k| \right\} \text{ with } \begin{array}{l} \delta_k > 0 \text{ for all } k \in i, i+1, \ldots, j-1 \text{ or} \\ \delta_k < 0 \text{ for all } k \in i, i+1, \ldots, j-1 \end{array}$$

The *distortion* d_X of X is defined as $d_X := \lfloor d(X)/\pi \rfloor$. (This is a notion related e.g. to the winding number in [GRS].) The number of turning-edge vectors describes the degree of "convexity" of an edge vector sequence and the distortion describes how "spiral" it is.

Theorem 4.3 *If t is the number of turning-edge vectors of the edge vector sequences of two simple polygons with n and m points and d is the maximum of their distortion then an L_1-optimal angle consistent cyclic merge can be constructed in $O((dt)^4 + n + m)$ time.*

Proof. $X = (x_0, x_1, \ldots, x_{n-1})$ and $Y = (y_0, y_1, \ldots, y_{m-1})$ are the edge vector sequences of two simple polygons. If X or Y is an edge vector sequence of a convex polygon then Lemma 4.1 proves the statement of this theorem.

X decomposes into maximal convex chains, i.e. subsequences $(x_i, x_{i+1}, \ldots, x_j)$ with $\delta_k > 0$ for all $k \in \{i, i+1, \ldots, j-1\}$ with $\delta_{i-1} < 0$ and $\delta_{j+1} < 0$, or $\delta_k < 0$ for all $k \in \{i, i+1, \ldots, j-1\}$ with $\delta_{i-1} > 0$ and $\delta_{j+1} > 0$. (Y analogously). Notice that the number of maximal convex chains in X and Y is t.

We will proceed as follows: First X and Y are reduced to at most dt edge vectors. Then a partial solution for the reduced problem is computed with the algorithm above in $O((dt)^4)$ time. In the second step the removed edge vectors are merged into the partial solution in $O(n+m)$ time and we get an L_1-optimal angle consistent cyclic merge of X and Y.

Reduction of X to X' and Y to Y': X' contains all turning-edge vectors of X, these are the first and last edge vectors of the maximal convex chains, together with some additional edge vectors which witness the spirals of the polygon. Suppose x_i and x_j are the first and last edge vector of a maximal convex chain (consecutive turning-edge vectors) and $\sum_{k=i}^{j-1} \delta(x_k, x_{k+1}) > \pi$. Beginning with $x_i = x_{i_0}$ (walking in direction x_j) we take from X the next possible edge vector x_{i_1} with $\sum_{k=i_0}^{i_1-1} \delta(x_k, x_{k+1}) < \pi$ and $\sum_{k=i_0}^{i_1} \delta(x_k, x_{k+1}) > \pi$. If $\sum_{k=i_1}^{j-1} \delta(x_k, x_{k+1}) > \pi$ then beginning with x_{i_1}, we take the last possible edge vector x_{i_2} with $\sum_{k=i_1}^{i_2-1} \delta(x_k, x_{k+1}) < \pi$ etc. until we have taken x_{i_l} with $\sum_{k=i_l}^{j-1} \delta_x(x_k, x_{k+1}) < \pi$. ($Y'$ analogously). In X' and Y' we have added at most d edge vectors per turning-edge vector. With the algorithm an L_1-optimal angle consistent cyclic merge Z' of X' and Y' is computed. Assume Z is the L_1-optimal angle consistent cyclic merge of X and Y. $\delta(Z') \leq \delta(Z)$ because $X' \subseteq X$ and $Y' \subseteq Y$. More precisely: Let Z'' be generated from Z by deleting the edge vectors lying in X but not in X' and those lying in Y but not in Y'. $\delta(Z'') \leq \delta(Z)$ (Observation 2.2) and $\delta(Z') \leq \delta(Z'')$ because Z' is the L_1-optimal angle consistent cyclic merge of X' and Y' and Z'' is an angle consistent cyclic merge of the same edge vectors.

Merging step: The edges removed from X, $X - X'$, consist of sorted sequences which are merged into Z' in a way described in Lemma 4.1 such that the ordering of the edge vectors relative to X is preserved. For the resulting cyclic merge Z'_X it holds: $\delta(Z'_X) = \delta(Z')$. The edge vectors of $Y - Y'$ are merged into Z'_X in the same way. $\delta(Z'_{XY}) = \delta(Z') \leq \delta(Z)$ and also $\delta(Z) \leq \delta(Z'_{XY})$ because of the optimality of Z. It follows $\delta(Z) = \delta(Z'_{XY})$. $\boxdot$

References

[B]   J. D. Boissonnat. Shape reconstruction from planar cross-sections. *Computer Vision, Graphics and Image Processing 44*, 1988, 1-29

[ChrS]   H. N. Christiansen, T. W. Sederberg. Conversion of complex contour line definitions into polygonal element mosaics. *Computer Graphics 13*, 1978, 187-192

[C]   P. N. Cook et al. Three-dimensional reconstruction from cross-sections for medical applications. *Proceedings, 14th Hawaii Int. Conf. on System Sci.*, 1981, 358-389

[FKU]   H. Fuchs, Z. M. Kedem, S. P. Uselton. Optimal surface reconstruction from planar contours. *Communications of the ACM 20*, 1977, 693-702

[GD]   S. Ganapathy, T. G. Dennehy. A new general triangulation method for planar contours. *ACM Trans. Computer Graphics 16*, 1982, 69-75

[GORS]   C. Gitlin, J. O'Rourke, V. Subramanian. On Reconstructing Polyhedra from Parallel Slices. *Technical Report 025, Smith College, Dept. Computer Science*, 1993

[GRS] L. Guibas, L. Ramshaw, J. Stolfi. A kinetic framework for computational geometry. *Proc. 24th IEEE Foundations of Computer Science*, 1983, 100-111

[K] E. Keppel. Approximating complex surfaces by triangulation of contour lines, *IBM Journal of Research and Development 19*, 1975, 2-11

[Sch] L. L. Schumaker. Reconstructing 3D objects from cross-sections, *in: W. Dahmen et. al. (eds.), COMPUTATION OF CURVES AND SURFACES*, 1990, 275-309

[SG] T. W. Sederberg, E. Greenwood. A physically based approach to 2-D shape blending. *SIGGRAPH '92 Conference Proceedings, Computer Graphics, Vol. 26, No. 2*, 1992, 25-34

[SP] K. R. Sloan, J. Painter. Pessimal guesses may be optimal: A counterintuitive search result. *IEEE Transactions on Pattern Analysis and Machine Intelligence 10*, 1988, 949-955

[WA] Y. F. Wang, J. K. Aggarwal. Surface reconstruction and representation of 3D scenes. *Pattern Recognition 19*, 1986, 197-207

A Linear Algorithm for Edge-Coloring

Partial k-Trees

Xiao Zhou, Shin-ichi Nakano and Takao Nishizeki

Department of System Information Sciences

Graduate School of Information Sciences

Tohoku University, Sendai 980, Japan

E-mail: (zhou|nakano|nishi)@ecei.tohoku.ac.jp

Abstract. Many combinatorial problems can be efficiently solved for partial k-trees. The edge-coloring problem is one of a few combinatorial problems for which no linear-time algorithm has been obtained for partial k-trees. The best known algorithm solves the problem for partial k-trees G in time $O(n\Delta^{2^{2(k+1)}})$ where n is the number of vertices and Δ is the maximum degree of G. This paper gives a linear algorithm which optimally edge-colors a given partial k-tree for fixed k.

1 Introduction

This paper deals with the edge-coloring problem which asks to color, using a minimum number of colors, all edges of a given graph so that no two adjacent edges are colored with the same color. The *chromatic index* $\chi'(G)$ of a graph G is the minimum number of colors used by an edge-coloring of G. This problem arises in many applications, including various scheduling and partitioning problems [FW]. Since the edge-coloring problem is NP-complete [Hol], it seems unlikely that there exists a polynomial-time algorithm to solve the problem for general graphs. On the other hand, it is known that many combinatorial problems can be solved very efficiently, say in linear time, for series-parallel graphs or partial k-trees [ACPD, AL, BPT, C, TNS]. Such a class of problems has been characterized in terms of "forbidden graphs" or "extended monadic logic of second order" [ACPD, AL, BPT, C, TNS]. The edge-coloring problem does not belong to such a class, and is indeed one of the "edge-covering problems" which, as mentioned in [BPT], do not appear to be solved efficiently for partial k-trees. However the following three partial results have been known. First, Terada and Nishizeki have given an $O(n^2)$ algorithm for series-parallel simple graphs G, i.e., partial 2-trees [TN]. In the paper n denotes the number of vertices in G. Second, Zhou, Suzuki and Nishizeki have given a linear-time algorithm for series-parallel multigraphs [ZSN]. Third, Bodlaender has given a polynomial-time algorithm for partial k-trees G [B] but the complexity $O(n\Delta(G)^{2^{2(k+1)}})$ is very high, where the maximum degree $\Delta(G)$ of G is not always a constant although k is assumed to be a constant.

In this paper we give a linear algorithm for partial k-trees, which determines the chromatic index $\chi'(G)$ of a given partial k-tree G and actually finds an edge-coloring of G using $\chi'(G)$ colors. Note that k is assumed to be a constant. Our algorithm greatly improves the complexity: for example, for partial 3-trees, Bodlaender's algorithm requires time $O(n^{257})$, but ours requires time $O(n)$. Our idea is twofold: first, we prove that $\chi'(G) = \Delta(G)$ holds for every partial k-tree G of large maximum degree, say $\Delta(G) \geq 5k$; and second, we show that such a graph G can be decomposed into several subgraphs $G_1, G_2, \cdots, G_s$ of small maximum degrees such that $\Delta(G) = \sum_{i=1}^{s} \Delta(G_i)$ and $\chi'(G_i) = \Delta(G_i) < 5k$ for each i, and hence an optimal edge-coloring of G can be obtained simply by extending those of $G_1, G_2, \cdots, G_s$ which can be found in linear time by Bodlaender's algorithm.

2 Terminology and Definitions

In this section we give some definitions. Let $G = (V, E)$ denote a graph with vertex set V and edge set E. We often denote by $V(G)$ and $E(G)$ the vertex set and the edge set of G, respectively. The paper deals with *simple* graphs without multiple edges or self-loops. An edge joining vertices u and v is denoted by (u, v). The class of k-*trees* is defined recursively as follows:

(a) A complete graph with k vertices is a k-tree.

(b) If $G = (V, E)$ is a k-tree and k vertices $v_1, v_2, \cdots, v_k$ induce a complete subgraph of G, then $H = (V \cup \{w\}, E \cup \{(v_i, w)|1 \leq i \leq k\})$ is a k-tree where w is a new vertex not contained in G.

(c) All k-trees can be formed with rules (a) and (b).

A graph is a *partial k-tree* if and only if it is a subgraph of a k-tree. Thus partial k-trees are simple graphs. In this paper we assume that k is a constant.

The *degree* of vertex $v \in V(G)$ is denoted by $d(v, G)$ or simply by $d(v)$. The *maximum degree* of G is denoted by $\Delta(G)$ or simply by Δ. For a vertex $v \in V(G)$, denote by $n_\Delta(v)$ the number of vertices which are adjacent to v and have degree $\Delta(G)$. The graph obtained from G by deleting all edges in $E' \subseteq E(G)$ is denoted by $G - E'$. Similarly the graph obtained from G by deleting all vertices in $V' \subseteq V(G)$ is denoted by $G - V'$.

3 Determining the Chromatic Index

By the classical Vizing's theorem, $\chi'(G) = \Delta(G)$ or $\Delta(G) + 1$ for any simple graph G [FW]. In this section we first show that $\chi'(G) = \Delta(G)$ holds for any partial k-tree G with $\Delta(G) \geq 2k$, and then show that the chromatic index $\chi'(G)$ can be determined in linear time for any partial k-tree G.

Hoover [Hoo] has claimed that $\chi'(G) = \Delta(G)$ holds for any partial k-tree G with $\Delta(G) \geq 4k$, but his proof contains a flaw. His "proof" is based on "Theorem 4.5" in [Hoo]: if the chromatic index of a general graph G is $\Delta(G) + 1$ then

$$|E| \geq \frac{n\Delta(G)}{4}.$$

However this "Theorem" is incorrect as seen from the following counterexample. Let G be a graph obtained from K_7, a complete graph of seven vertices, by inserting many vertices, say seventy vertices, in an arbitrary edge e of K_7. Then $\Delta(G) = 6$, $n = 77$ and $|E| = 91$. Clearly $\chi'(G) = \Delta(G) + 1 = 7$ since $7 \leq \chi'(K_7 - e) \leq \chi'(G)$. However

$$|E| < \frac{n\Delta(G)}{4},$$

contrary to the "Theorem." This flaw looks to stem from an incorrect interpretation of a result on "critical graphs," Theorem 13.6 in [FW].

We prove a claim slightly stronger than his: $\chi'(G) = \Delta(G)$ holds for any partial k-tree G with $\Delta(G) \geq 2k$. An edge (u, v) of G is *eliminable* [TN, NC] if the following equations hold:

$$d(u) + n_\Delta(v) \leq \Delta \text{ if } d(u) < \Delta; \text{ and}$$

$$n_\Delta(v) = 1 \text{ if } d(u) = \Delta.$$

The following lemma is an expression of a classical result on "critical graphs," called "Vizing's adjacency lemma" (see, for example, [FW], [TN] or [NC]).

Lemma 3.1   *If (u,v) is an eliminable edge of a simple graph G and $\chi'(G - (u,v)) \leq \Delta(G)$, then $\chi'(G) = \Delta(G)$.*

For partial k-trees we have the following lemma.

Lemma 3.2   *If a partial k-tree $G = (V,E)$ has maximum degree $\Delta(G) \geq 2k$, then G has an eliminable edge.*

Proof.   Let $S_1 = \{v \in V(G) |\ d(v,G) \leq k\}$ and $S_2 = \{v \in V(G - S_1) |\ d(v,G - S_1) \leq k\}$. Then $S_1, S_2 \neq \phi$ since $\Delta(G) \geq 2k$. Furthermore there exists an edge joining vertices $u \in S_1$ and $v \in S_2$, because $k + 1 \leq d(v,G)$ and $d(v, G - S_1) \leq k$. Every vertex $w \in S_1$ has degree $d(w,G) \leq k < \Delta(G)$, and $d(v, G - S_1) \leq k$. Therefore $d(u) \leq k < \Delta$, $n_\Delta(v) \leq k$, and hence $d(u) + n_\Delta(v) \leq 2k \leq \Delta$. Thus edge (u,v) is eliminable. $Q.\mathcal{E}.\mathcal{D}.$

Using Lemmas 3.1 and 3.2, we have the following theorem.

Theorem 3.3   *If a partial k-tree G has maximum degree $\Delta(G) \geq 2k$, then $\chi'(G) = \Delta(G)$.*

Proof.   By Lemma 3.2 G has an eliminable edge e_1. Since $G - \{e_1\}$ is also a partial k-tree, $G - \{e_1\}$ has an eliminable edge e_2 if $\Delta(G - \{e_1\}) \geq 2k$. Thus there exists a sequence of edges $e_1, e_2, \cdots, e_m$ such that

   (a) $\Delta(G') = \Delta(G) - 1$ where $G' = G - \{e_1, e_2, \cdots, e_m\}$; and

   (b) e_i, $1 \leq i \leq m$, is eliminable in $G - \{e_1, e_2, \cdots, e_{i-1}\}$.

By the classical Vizing's theorem [FW], $\chi'(G') \leq \Delta(G') + 1 = \Delta(G)$. Therefore, applying Lemma 3.1 repeatedly, we have $\chi'(G) = \Delta(G)$. $Q.\mathcal{E}.\mathcal{D}.$

Since $\Delta(G)$ can be computed in linear time, the chromatic index of a partial k-tree G with $\Delta(G) \geq 2k$ can be determined in linear time. On the other hand Bodlaender [B] has given an algorithm which determines $\chi'(G)$ of a partial k-tree G and obtains an edge-coloring of G with $\chi'(G)$ colors total in time $O(n\Delta^{2^{2(k+1)}})$. Clearly his algorithm runs in linear time if $\Delta(G) < 2k$. Note that k is a constant. Thus we have the following theorem.

Theorem 3.4   *The chromatic index of a partial k-tree can be determined in linear time if k is a constant.*

4 Obtaining an Edge-Coloring

In Section 3 we have shown that the chromatic index $\chi'(G)$ of a given partial k-tree G can be determined in linear time. In this section we give a linear algorithm which actually obtains an edge-coloring of G with $\chi'(G)$ colors. Using Bodlaender's algorithm [B], one can obtain an edge-coloring of G with $\chi'(G)$ colors in linear time if $\Delta(G)$ is a constant. Therefore it suffices to give a linear algorithm only for the case $\Delta(G) \geq 5k$.

The proofs in the previous section do not yield a linear algorithm for the case $\Delta(G) \geq 5k$, as follows. Lemma 3.3 implies that a partial k-tree G with $\Delta(G) \geq 5k$ necessarily has an eliminable edge. If (u,v) is an eliminable edge in a graph G and an edge-coloring of $G - (u,v)$ with $\Delta(G)$ colors is known, then, using a standard technique of "shifting a fan sequence," one can obtain an edge-coloring of G with $\chi'(G) = \Delta(G)$ ($> 2k$) colors in time $O(|E|)$ [NC, TN]. By Lemma 3.3 there exists an edge-sequence $e_1, e_2, \cdots, e_m$ such that $\Delta(G - \{e_1, e_2, \cdots, e_m\}) = 5k$ and e_i is an eliminable edge in $G - \{e_1, e_2, \cdots, e_{i-1}\}$ for every i, $1 \leq i \leq m$. Using Bodlaender's algorithm, one can obtain an edge-coloring of $G' = G - \{e_1, e_2, \cdots, e_m\}$ with $\chi'(G') = 5k$ ($> 2k$) colors in time $O(n)$. Add edges $e_m, e_{m-1}, \cdots, e_2, e_1$ to G' in this order, and modify the edge-coloring of G' to an edge-coloring of G with $\Delta(G)$ colors by repeatedly using the technique of "shifting a fan sequence." Such a repetition of recoloring would require time $O(n^2)$.

Our idea is to decompose G into several subgraphs when $\Delta(G)$ is large, say $\Delta(G) \geq 5k$, as in the following lemma.

Lemma 4.1 *If a partial k-tree $G = (V, E)$ has maximum degree $\Delta(G) \geq 5k$, then E can be partitioned into subsets $E_1, E_2, \cdots, E_s$ such that the subgraphs G_j, $1 \leq j \leq s$, of G induced by E_j satisfy*

(a) $\Delta(G_j) = 2k$ *for each* j, $1 \leq j \leq s - 1$, *and*

(b) $3k \leq \Delta(G_s) = \Delta(G) - 2k(s-1) < 5k$.

Furthermore such a partition of E can be found in time $O(n)$.

Such a partition $E_1, E_2, \cdots, E_s$ of E is said Δ-*bounded*. Since $2k \leq \Delta(G_j) < 5k$ for each j, $1 \leq j \leq s$, by Theorem 3.4 $\chi'(G_j) = \Delta(G_j)$. Using Bodlaender's algorithm, one can obtain an edge-coloring of G_j with $\Delta(G_j)$ colors in time $O(|E_j|)$. Since $\Delta(G) = \sum_{j=1}^{s} \Delta(G_j)$, one can immediately extend these edge-colorings of $G_1, G_2, \cdots, G_s$ to an edge-coloring of G with $\Delta(G)$ colors in linear time.

In order to prove Lemma 4.1, we need the following two lemmas. Let $S_1, S_2, \cdots, S_l$ be a partition of $V(G)$. For each $v \in S_i$, $1 \le i \le l$, let

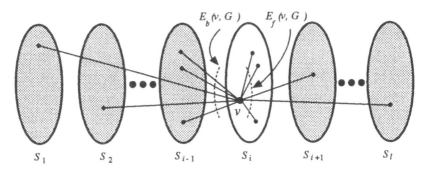

Figure 1. Illustration for notations.

$$E_b(v, G) = \{(v, w) \in E | \ w \in S_j \text{ and } j < i\},$$

$$E_f(v, G) = \{(v, w) \in E | \ w \in S_j \text{ and } j \ge i\},$$

$$d_b(v, G) = |E_b(v, G)|, \text{ and}$$

$$d_f(v, G) = |E_f(v, G)|.$$

Thus $d(v, G) = d_b(v, G) + d_f(v, G)$. (See Figure 1.) The partition $S_1, S_2, \cdots, S_l$ of V is d_f-bounded if $d_f(v, G) \le k$ for every vertex $v \in V$.

Lemma 4.2 *Every partial k-tree G has a d_f-bounded partition.*

Proof. Since G is a partial k-tree, G has a vertex of degree at most k. Let S_1 be the set of all such vertices, and delete all vertices in S_1 from G. Since the resulting graph G_1 is also a partial k-tree, G_1 has a vertex of degree at most k. Let S_2 be the set of all such vertices, and delete all vertices in S_2 from G_1. By repeating the same operation above, one can obtain a d_f-bounded partition S_1, $S_2, \cdots, S_l$ of V. $\mathcal{Q.E.D.}$

Lemma 4.3 *Let $G = (V, E)$ be a partial k-tree, and let $S_1, S_2, \cdots, S_l$ be a d_f-bounded partition of V. Let $I = \{i_1, i_2, \cdots, i_{l'}\}$, $1 \le i_1 < i_2 < \cdots < i_{l'} \le l$, and let S_i', $i \in I$, be a nonempty subset of S_i such that $d_b(v, G) \ge 2k$ for every vertex $v \in S_i'$. Then G has a subgraph G' such that $\Delta(G') = 2k$ and $V_\Delta(G') = \bigcup_{i \in I} S_i'$, where $V_\Delta(G') = \{v \in V(G') | \ d(v) = \Delta(G')\}$. Furthermore G' can be found in* *time $O(|E(G')|)$ if $E_b(v, G)$ for all vertices $v \in V$ are known.*

Proof. A required subgraph G' can be constructed as follows.

```
1  Procedure Subgraph;
2    begin
3      let G' = (⋃ᵢ∈ᵢ S'ᵢ, φ);
4      for j := l' downto 1 do
5        for each vertex v ∈ S'ᵢⱼ do
6          add to G' any 2k − d(v, G') edges in Eᵦ(v, G)
7    end.
```

Whenever line 6 is going to be executed, $d(v, G') \leq k$ for a current graph G' since $d_f(v, G) \leq k$. Therefore $k \leq 2k - d(v, G') \leq 2k$. Furthermore $d_b(v, G) \geq 2k$, and none of edges in $E_b(v, G)$ has not been added to G' so far. Thus one can always add to G' $2k - d(v, G')$ $(\geq k)$ edges in $E_b(v, G)$ which have not been added to G' so far.

Clearly $d(v, G') = 2k$ holds for the final graph G' if $v \in \bigcup_{i \in I} S'_i$. On the other hand $d(v, G') \leq d_f(v, G) \leq k$ holds if $v \in V(G') - \bigcup_{i \in I} S'_i$. Thus $\Delta(G') = 2k$ and $V_\Delta(G') = \bigcup_{i \in I} S'_i$. Given lists containing $E_b(v, G)$ for all vertices $v \in V$, one can easily execute the procedure above in time $O(|E(G')|)$.

$$Q.\mathcal{E}.\mathcal{D}.$$

We are now ready to prove Lemma 4.1.

Proof of Lemma 4.1 The following algorithm finds a required decomposition $G_1, G_2, \cdots, G_s$ of G.

```
1  Procedure Subgraphs;
2    begin
3      Δ := Δ(G);
4      find a d_f-bounded partition S₁, S₂, ···, Sₗ of V(G);
                                                    { Lemma 4.2 }
5      for each i, 1 ≤ i ≤ l, do S'ᵢ := {v ∈ Sᵢ| d(v, G) ≥ 3k};
                            { dᵦ(v, G) ≥ 2k for every vertex v ∈ S'ᵢ, 1 ≤ i ≤ l }
6      I := {i| 1 ≤ i ≤ l and S'ᵢ ≠ φ};
7      s := ⌊(Δ−k)/2k⌋;                    { 3k ≤ Δ − 2k(s − 1) < 5k }
8      for j := 1 to s − 1 do
9        begin                             { Δ(G) = Δ − 2k(j − 1) ≥ 5k }
```

```
10        find a subgraph $G_j$ of $G$ such that $\Delta(G_j) = 2k$
              and $V_\Delta(G_j) = \bigcup_{i \in I} S'_i$;              {Lemma 4.3 }
11        $G := G - E(G_j)$;              { $\Delta(G)$ decreases exactly by $2k$ }
12        $S'_i := \{v \in S_i \mid d(v, G) \geq 3k\}$ for all $i$, $i \in I$;       { update $S'_i$ }
13        $I := \{i \in I \mid S'_i \neq \phi\}$              { update $I$ }
14      end;
15      $G_s := G$;
16      return $G_1, G_2, \cdots, G_s$
17    end.
```

Whenever line 10 is executed for a current graph G, $d_f(v, G) \leq k$ holds for every vertex $v \in V$, and $d_b(v, G) \geq 2k$ holds for every $v \in S'_i$, $i \in I$. Therefore by Lemma 4.3 G has a subgraph G_j such that $\Delta(G_j) = 2k$ and $V_\Delta(G_j) = \bigcup_{i \in I} S'_i$. Since $\Delta(G) \geq 5k$, $\Delta(G)$ decreases exactly by $2k$ whenever line 11 is executed. Thus we have $3k \leq \Delta(G_s) = \Delta - 2k(s-1) < 5k$. Hence the algorithm above correctly finds subgraphs $G_1, G_2, \cdots, G_s$.

We now analyze the time complexity. Lines 4 and 5 can be done in time $O(|E|)$. By Lemma 4.3 line 10 can be done in time $O(|E(G_j)|)$ for every j. Therefore the **for** loop at lines 8–14 can be done total in time $O(\sum_{j=1}^{s-1} |E(G_j)|) \leq O(|E|)$. Since $|E| \leq kn$, the algorithm above runs in time $O(n)$. $Q.\mathcal{E}.\mathcal{D}.$

This paper concludes the following theorem.

Theorem 4.4 *The edge-coloring problem can be solved in linear time for partial k-trees if k is a constant.*

5 Conclusion

In this paper we gave an efficient algorithm for the edge-coloring problem on partial k-trees. The algorithm runs in linear time for fixed k and in $O((\min\{5k, \Delta\})^{2^{2(k+1)}} n)$ time for general k.

Our algorithm solves a single particular problem, that is, the edge-coloring problem. However the methods which we developed in this paper appear to be useful for many other problems, especially for the "edge-partition problem with respect to property π" which asks to partition the edge set of a given graph into a minimum number of subsets so that the subgraph induced by each subset satisfies the property π. For the edge-coloring problem, π is indeed a matching.

Consider for example a property π: the degree of each vertex v is not greater than $f(v)$, where $f(v)$ is a positive integer assigned to v. Clearly the edge-partition problem with respect to such a property π is the same as the so-called f-coloring problem [NNS, HK]. Our algorithm can be generalized to solve the f-coloring problem on partial k-trees in linear time.

Another direction of generalization is to parallelize the sequential algorithm of this paper. Indeed we have recently obtained an optimal parallel algorithm for edge-coloring partial k-trees [ZNN]. It is the first NC parallel algorithm, and runs in $O(\log n)$ time using $O(n/\log n)$ processors for a partial k-tree G given by its decomposition tree. It is known that a decomposition tree of G can be found in $O(\log^3 n)$ time using $O(n)$ processors [BK].

Acknowledgment

We would like to thank Dr. Hitoshi Suzuki for various discussions. This research is partly supported by Grant in Aid for Scientific Research of the Ministry of Education, Science, and Culture of Japan under a grant number: General Research (C) 05650339.

References

[ACPD] S. Arnborg, B. Courcelle, A. Proskurowski, and D. Seese, *An algebraic theory of graph reduction*, Tech. Rept. 91-36, Laboratoire Bordelais de Recherche en Informatique, Bordeaux, 1991.

[AL] S. Arnborg and J. Lagergren, *Easy problems for tree-decomposable graphs*, Journal of Algorithms, 12, 2, pp.308-340, 1991.

[B] H. L. Bodlaender, *Polynomial algorithms for graph isomorphism and chromatic index on partial k-trees*, Journal of Algorithms, 11, 4, pp.631-643, 1990.

[BK] H.L. Bodlaender and T. Kloks, *Better algorithms for the pathwidth and treewidth of graphs*, Proceedings of 18'th International Colloquium on Automata, Languages and Programming, Springer Verlag, Lecture Notes in Computer Science, 510, pp.544-555, Berlin, 1991.

[BPT] R.B. Borie, R.G. Parker and C.A. Tovey, *Automatic generation of linear-time algorithms from predicate calculus descriptions of problems on recursively constructed graph families*, Algorithmica, 7, pp.555-581, 1992.

[C] .B. Courcelle, *The monadic second-order logic of graphs I: Recognizable sets of finite graphs*, Information and Computation, 85, pp.12-75, 1990.

[FW] S. Fiorini and R.J. Wilson, *Edge-Coloring of Graphs*, Pitman, London, 1977.

[HK] S. L. Hakimi and O. Kariv, *On a generalization of edge-coloring in graphs*, Journal of Graph Theory, 10, pp.139-154, 1986.

[Hol] I.J. Holyer, *The NP-completeness of edge-coloring*, SIAM J. on Computing, 10, pp.718-720, 1981.

[Hoo] M. Hoover, *Complexity, structure, and algorithms for edge-partition problems*, Technical Report No. CS90-16, Dept. of Computer Science, The University of New Mexico, 1990.

[NC] T. Nishizeki and N. Chiba, *Planar Graphs: Theory and Algorithms*, North-Holland, Amsterdam, 1988.

[NNS] S. Nakano, T. Nishizeki and N. Saito, *On the f-coloring multigraphs*, IEEE Transactions on Circuits and Systems, Vol. 35, No. 3, pp. 345-353, 1988.

[TN] O. Terada and T. Nishizeki, *Approximate algorithms for the edge-coloring of graphs*, Trans. Inst. of Electronics and Communication Eng. of Japan, J65-D, 11, pp. 1382-1389, 1982.

[TNS] K. Takamizawa, T. Nishizeki, and N. Saito, *Linear-time computability of combinatorial problems on series-parallel graphs*, J. of ACM, 29, 3, pp. 623-641, 1982.

[ZSN] X. Zhou, H. Suzuki and T. Nishizeki, *Sequential and parallel algorithms for edge-coloring series-parallel multigraphs*, Proc. of Third IPCO, pp. 129-145, 1993.

[ZNN] X. Zhou, S. Nakano and T. Nishizeki, *A parallel algorithm for edge-coloring partial k-trees*, Tech. Rept. TRECIS 93001, Dept. of Inf. Eng., Tohoku Univ., 1993.

Author Index

Lecture Notes in Computer Science

For information about Vols. 1–655
please contact your bookseller or Springer-Verlag